AF335084

Principles of Machine Learning

Wenmin Wang

Principles of Machine Learning

The Three Perspectives

 Springer

Wenmin Wang
School of Computer Science
and Engineering
Macau University of Science
and Technology
Macao, China

ISBN 978-981-97-5332-1   ISBN 978-981-97-5333-8   (eBook)
https://doi.org/10.1007/978-981-97-5333-8

The original submitted manuscript has been translated into English. The translation was done using artificial intelligence. A subsequent revision was performed by the author(s) to further refine the work and to ensure that the translation is appropriate concerning content and scientific correctness. It may, however, read stylistically different from a conventional translation.

This Springer imprint is published by the registered company Springer Nature Singapore Pte Ltd.
The registered company address is: 152 Beach Road, #21-01/04 Gateway East, Singapore 189721, Singapore

If disposing of this product, please recycle the paper.

Preface

This book, *Principles of Machine Learning*, is written after my another one, *Principles of Artificial Intelligence*. Both books use the term principles.

"Principles," usually refer to formal axioms, theoretical postulates, and fundamental doctrines. They are derived from many observations and practices, through abstraction, induction, and generalization.

Therefore, to discuss the principles of machine learning, it is necessary to collate, induce, abstract, and systematize for the discipline, to provide certain guidance for the study and practice of machine learning.

Learning, in English, refers to the process of acquiring knowledge or skill; while the Chinese word "learning" comes from *The Analects of Confucius*, Confucius said:

> Learning and practicing often, isn't it a pleasure?

This means that in addition to "learning," there is also the meaning of "practicing."

Humans have the ability to learn. Machines can also be given the ability to learn.

In fact, before the term "artificial intelligence" was officially used at the historic workshop "Dartmouth Summer Research Project on Artificial Intelligence" in 1956, Alan Turing published the famous paper "Computing Machinery and Intelligence" in 1950, where he also discussed "learning machines."

The paper titled "Some Studies in Machine Learning Using the Game of Checkers" published by Arthur Samuel in 1959 can be regarded as the pioneering work of "machine learning," and the term has become popular since then.

There are several branches in the field of artificial intelligence named after "machine," such as machine translation, machine vision, and machine reasoning.

Machine learning is a branch of artificial intelligence, and deep learning is a branch of machine learning. This is because deep learning is one of the ways to implement learning functions using deep networks—mainly artificial neural networks.

The reason I wrote this book *Principles of Machine Learning* mainly comes from the following considerations:

First, there are many books that discuss the algorithms, methods, and practices of machine learning, but there are few monographs that systematically discuss machine learning from its principles.

Second, the source code of some typical algorithms of machine learning is easy to find, especially since some open-source libraries or software platforms—such as PyTorch and TensorFlow—have implemented these algorithm functions. However, their theoretical foundations, basic paradigms, functional categories, research methodologies, etc. still need to be further sorted out.

This book discusses the principles of machine learning around three perspectives based on a systematic explanation of those perspectives. It is mainly divided into 4 parts with 15 chapters. The organization, the titles of each part and chapter, and their relationship in this book are shown in the following flow diagram.

Part I Perspectives

1 Introduction	2 On Perspectives

Part II Frameworks

3 Probabilistic Framework	4 Statistical Framework
5 Connectionist Framework	6 Symbolic Framework
7 Behavioral Framework	

Part III Paradigms

8 Supervised Learning Paradigm	9 Unsupervised Learning Paradigm
10 Reinforcement Learning Paradigm	11 Other Learning Quasi-Paradigms

Part IV Tasks

12 Classification Task	13 Regression Task
14 Clustering Task	15 Dimensionality Reduction Task

The points of each part and chapter are as follows.

Part I   Perspectives , includes two chapters, namely introduction and on perspectives. The former gives an overview of machine learning, and the latter discusses the three perspectives of studying machine learning which are learning frameworks, learning paradigms, and learning tasks. These three perspectives are the themes of the last three parts.

Part II   Frameworks, is at the theoretical perspective. It is divided into five chapters, mainly discussing the probabilistic, statistical, connectionist, symbolic, and behavioral frameworks of machine learning.

Part III Paradigms, is at the methodological perspective. It is divided into four chapters, the first three chapters focus on the three paradigms of supervised learning, unsupervised learning, and reinforcement learning, and then one more chapter is used to introduce several quasi-paradigms that emerged in machine learning.

Part IV Tasks, is at the practical perspective. It is the common tasks abstracted from application problems of machine learning: divided into four chapters, explaining the typical tasks of classification, regression, clustering, and dimensionality reduction of machine learning.

Although I have been engaged in the research and teaching of artificial intelligence and machine learning for many years, and have referred to many books, papers, blogs, Wikipedia, etc., there will inevitably be omissions and errors in the book, some views may be also biased and generalized. Please criticize and correct any inappropriate places. Borrowing the famous saying

doubts should be analyzed together,

from Tao Yuanming (365–427 CE), who is one of the best-known poets in the Eastern Jin dynasty.

If this book can become a stepping stone for readers to learn and research machine learning, I would be quite relieved.

Macao, China   Wenmin Wang
March 2024

Acknowledgments

I have to start by thanking Dr. Celine Chang, editorial director of computer science at Springer Nature. My original manuscript of this book was written in Chinese, and it was precisely because of her guidance and assistance that the English edition of the book was made available to readers.

I am also grateful to everyone on the book's publishing team at Springer Nature; moreover, the English draft of my book was translated by their auto-translator so that I could revise it based on the draft, which greatly accelerated the translation progress.

About This Book

This book proposes three perspectives, namely learning frameworks, learning paradigms, and learning tasks, for studying machine learning. The learning frameworks are the theoretical perspectives, the learning paradigms are the methodological perspectives, and the learning tasks are the practical perspectives.

The book is divided into 4 parts, comprising 15 chapters. Part I, "Perspectives," consists of two chapters: introduction and perspectives. Part II, "Frameworks," contains five chapters that discuss probabilistic, statistical, connectionist, symbolic, and behavioral frameworks. Part III, "Paradigms," includes four chapters that explain the three paradigms of supervised learning, unsupervised learning, and reinforcement learning, as well as several quasi-paradigms that emerged in machine learning. Part IV, "Tasks," covers four chapters that examine the most commonly used tasks of classification, regression, clustering, and dimensionality reduction.

This book attempts to provide a systematic and comprehensive interpretation of machine learning, making it suitable as a textbook for senior undergraduates or graduate students in artificial intelligence, machine learning, data science, computer science, and related fields, as well as a reference book for scientific researchers and technical professionals.

Contents

About the Author

Wenmin Wang is a professor and program director in the School of Computer Science and Engineering within the Faculty of Innovation Engineering at Macau University of Science and Technology (MUST), China, from 2019. Previous to the MUST, he held the position of professor and executive vice dean/dean with the School of Electronic and Computer Engineering at Peking University (PKU).

In PKU, he taught a course on Principles of Artificial Intelligence to graduate students. And in MUST, he has been teaching the two core courses Machine Learning and Principles of Artificial Intelligence to graduate students.

This book was started to be written after his Chinese edition of *Principles of Artificial Intelligence* was published by Higher Education Press (China) in August 2019. In recognition of his accomplishments in the online open course "Principles of Artificial Intelligence," he was honored with the "National Excellent Online Open Course" award by the Chinese Ministry of Education in 2018. Additionally, he was bestowed with the "Teaching Excellence Award" by PKU, in 2017.

His journey into the field of artificial intelligence during his doctoral studies culminated in his PhD thesis entitled *A Member System Model Supporting AI Problem Solving*. Then he received a PhD degree in computer science from Harbin Institute of Technology (HIT), China, in March 1989.

Notation

x	A scalar (integer or real)
$\boldsymbol{x}$	A vector
$\mathcal{X}$	The input space
$\mathcal{Y}$	The output space
S	A set
H	The hypothesis set
$\boldsymbol{\theta}$	A set of parameters
$\mathbb{N}$	The set of natural numbers
$\mathbb{N}_+$	The set of non-negative natural numbers
$\mathbb{Z}$	The set of integers
$\mathbb{Z}_+$	The set of non-negative integers
$\mathbb{R}$	The set of real numbers
$\mathbb{R}_+$	The set of non-negative real numbers
$\mathbb{R}^m$	The set of m-dimensional real-valued vectors
$f : \mathcal{X} \to \mathcal{Y}$	The function f maps from input space $\mathcal{X}$ to output space $\mathcal{Y}$
$\mathcal{L}(\cdot)$	A loss function
$\mathbb{I}(\cdot)$	The indicator function
$[a,b]$	The real interval including a and b
(a, b)	The real interval excluding a and b
$\{0, 1\}$	The set containing 0 and 1
$\{x_1, x_2, \cdots, x_n\}$	The set containing $x_1, x_2, \cdots, x_n$
X	A random variable
$P(X)$	A probability mass function of the discrete random variable X
$p(x)$	A probability density function of the continuous random variable x
$\mathbb{E}[X]$	The expectation of the random variable X
$x \sim P(\cdot)$	Random variable x has distribution P
$\mathrm{Var}(X)$	The variance of the random variable X
$\mathrm{Cov}(X, Y)$	The covariance of the random variable X
$\mathrm{Corr}(X, Y)$	The correlation of the random variable X and variable Y
$P(Y|X)$	The conditional probability of Y given X

$P(X, Y)$	The joint probability of X and Y
$D_{KL}(P \parallel Q)$	Kullback-Leibler (KL) divergence from P to Q
μ	The mean
σ^2	The variance
$\mathcal{N}(\mu, \sigma^2)$	The normal distribution, also called Gaussian distribution
$\|\mathbf{x}\|$ or $\|\mathbf{x}\|_2$	The L_2 norm of $\mathbf{x}$
$\|\mathbf{x}\|_p$	The L_p norm of $\mathbf{x}$
$\langle \cdot, \cdot \rangle$	The inner product
$\kappa(\cdot, \cdot)$	The kernel function
$\mathbf{x} \odot \mathbf{y}$	The element-wise product
$\mathbf{x} \otimes \mathbf{y}$	The convolution
$\mathbf{X}$	A matrix
$\mathbf{x}^\mathsf{T}$ or X^T	Transpose of the vector $\mathbf{x}$ or the matrix $\mathbf{X}$
$\mathbf{I}$	The identity matrix
$\mathbb{P}[\cdot]$	A transition matrix

Acronyms

A3C	Asynchronous Advantage Actor-Critic
ABC	Artificial Bee Colony
ACO	Ant Colony Optimization
AD	Automatic Differentiation
AdaBoost	Adaptive Boosting
AE	Autoencoder (Auto-Encoder)
AGNES	AGglomerative NESting
AI	Artificial Intelligence
AlexNet	A convolutional neural network, by Alex Krizhevsky, et al.
AM	Autoregressive Model
AMI	Adjusted Mutual Information
ANN	Artificial Neural Network
AR	Association Rule
ARES	Atomic Rotationally Equivariant Scorer
ARI	Adjusted Rand Index
BDN	Bayesian Decision Network
BDT	Bayesian Decision Theory
BERT	Bidirectional Encoder Representations from Transformers
BM	Boltzmann Machine
BN	Bayesian Network
BNN	Bayesian Neural Network
BP	Backpropagation (Back-Propagation)
BSVM	Bayesian Support Vector Machine
CapsNet	Capsule Network
CASP	Critical Assessment of Structure Prediction
CDF	Cumulative Distribution Function
CE	Cross-Entropy
CHI	Calinski-Harabasz index
CLIQUE	Clustering in Ques
CLS	Concept Learning System
CLT	Central Limit Theorem

CNN	Convolutional Neural Network
CR	Conditioned Response
CRF	Conditional Random Field
CS	Conditioned Stimulus
CV	Computer Vision
DAE	Denoising Autoencoder
DAM	Deep Autoregressive Model
DBI	Davies-Bouldin Index
DBM	Deep Boltzmann Machine
DBN	Deep Belief Network
DBSCAN	Density-Based Spatial Clustering of Applications with Noise
DDPG	Deep Deterministic Policy Gradient
DenseNet	Densely Connected Convolutional Networks
DIANA	DIvisive ANAlysis
DL	Deep Learning
DNN	Deep Neural Network
DP	Dynamic Programming
DQC	Dynamic Quantum Clustering
DQN	Deep Q-Network
DRL	Deep Reinforcement Learning
DT	Decision Tree
ECM	Expectation Conditional Maximization
ELM	Extreme Learning Machine
EM	Expectation Maximization
ENIAC	Electronic Numerical Integrator and Computer
ERM	Empirical Risk Minimization
F1 Score	harmonic mean of the precision and recall
FC	Fully-Connected
FDA	Fisher Discriminant Analysis
FMI	Fowlkes-Mallows Index
FNN	Feedforward Neural Network
FNR	False Negative Rate
FPR	False Positive Rate
FSL	Few-Shot Learning
GA	Genetic Algorithm
GAN	Generative Adversarial Network
GD	Gradient Descent
GMM	Gaussian Mixture Model
GNN	Graph Neural Network
GoogLeNet	A convolutional neural network based on inception architecture
GPT	Generative Pre-trained Transformer
GRU	Gated Recurrent Unit
HAI	Human-centered Artificial Intelligence
HMM	Hidden Markov Model
HNN	Hopfield Neural Network

i.i.d.	independent and identically distributed
ICA	Independent Component Analysis
ICM	Independent Causal Mechanisms
ID3	Iterative Dichotomiser 3
ILP	Inductive Logic Programming
ILSVRC	ImageNet Large Scale Visual Recognition Challenge
ImageNet	A large-scale image dataset
IoU	Intersection over Union
Isomap	Isometric Mapping
KDD	Knowledge Discovery in Databases
KDE	Kernel Density Estimation
KL Divergence	Kullback Leibler Divergence
kNN	k-Nearest Neighbours
LDA	Linear Discriminant Analysis
LE	Laplacian Eigenmaps
LLE	Locally Linear Embedding
LLM	Large Language Model
LLN	Law of Large Numbers
LP	Logic Programming
LSTM	Long Short-Term Memory
LTSA	Local Tangent Space Alignment
LTU	Linear Threshold Unit
LVQ	Learning Vector Quantization
MAE	Mean Absolute Error
mAP	Mean Average Precision
MAP	Maximum A Posteriori
MAPE	Mean Absolute Percentage Error
MBE	Mean Bias Error
MC	Monte Carlo
MCC	Multi-Class Classification
MCMC	Markov Chain Monte Carlo
MCTS	Monte Carlo Tree Search
MDP	Markov Decision Process
MDS	Multi-Dimensional Scaling
MEU	Maximum Expected Utility
MI	Mutual Information
MICA	Multilinear Independent Component Analysis
ML	Machine Learning
MLC	Multi-Label Classification
MLDA	Multilinear Linear Discriminant Analysis
MLE	Maximum Likelihood Estimation
MLP	Multi-Layer Perceptron
MPCA	Multilinear Principal Component Analysis
MSE	Mean Squared Error
NEC	Neural Episodic Control

NF	Normalizing Flow
NFL	No Free Lunch
NLP	Natural Language Processing
NMI	Normalized Mutual Information
NN	Neural Network
OCC	One-Class Classification
OCR	Optical Character Recognition
OPTICS	Ordering Points To Identify the Clustering Structure
OSL	One-Shot Learning
P-R	Precision-Recall (Curve)
PAC	Probabilistic Approximately Correct
PCA	Principal Component Analysis
PCoA	Principal Coordinates Analysis
PLP	Probabilistic Logic Programming
POMDP	Partially Observable Markov Decision Process
PSM	Parameterized Statistical Model
PSO	Particle Swarm Optimization
QC	Quantum Clustering
RAE	Relative Absolute Error
RBM	Restricted Boltzmann Machine
ReLU	Rectified Linear Unit
ResNet	Residual Network
RF	Random Forest
RI	Rand Index
RL	Reinforcement Learning
RMSE	Root Mean Square Error
RNA	Ribonucleic Acid
RNN	Recurrent Neural Network
ROC Curve	Received Operating Characteristic Curve
ROI	Region of Interest
RSE	Relative Squared Error
RSS	Residual Sum of Squares
SAE	Sparse Autoencoder
SARSA	State-Action-Reward-State-Action
SC	Silhouette Coefficient
SFLA	Shuffled Frog-Leaping Algorithm
SGD	Stochastic Gradient Descent
SL	Supervised Learning
SLP	Single-Layer Perceptron
SM	Statistical Model
SMAE	Smooth Mean Absolute Error (also known as Huber Loss)
SMS	Sparse Mechanism Shift
SNE	Stochastic Neighbor Embedding
SPD	Symmetric Positive Definite
SRM	Structural Risk Minimization

SSL	Self-Supervised Learning
STING	Statistical Information Grid
SVC	Support Vector Clustering
SVM	Support Vector Machine
SVR	Support Vector Regression
t-SNE	t-Distributed Stochastic Neighbor Embedding
TCC	Two-Class Classification
TD	Temporal Difference
TLU	Threshold Logic Unit
TNR	True Negative Rate
TPR	True Positive Rate
U-Net	A convolutional neural network, for image segmentation
UAE	Undercomplete Autoencoder
UAV	Unmanned Aerial Vehicle
UL	Unsupervised Learning
UMAP	Uniform Manifold Approximation and Projection
UML	Unified Modeling Languages
UNREAL	Unsupervised Reinforcement and Auxiliary Learning
US	Unconditioned Stimulus
VAE	Variational Autoencoder
VC Dimension	Vapnik-Chervonenkis Dimension
VGGNet	A convolutional neural network, by Visual Geometry Group
WPE	Weighted Prediction Error
ZSL	Zero-Shot Learning

Part I
Perspectives

The concept of perspectives encompasses mental viewpoints that are closely tied to the angles of observation. Viewing an object from a single angle, we can only perceive one side of it. However, examining an object from multiple angles allows us to see its whole entirety. Different perspectives yield varying forms of the same object and lead to distinct conclusions about the same phenomenon.

> Looking a ridge from the front and a peak from the side,
> Far or near and high or low there are no alike.
> Don't know the true color of Mount Lu,
> Because we are just in this mountain sight.

This is a seven-character quatrain "Written on the Wall at West Forest Temple" by Su Shi (also known as Su Dongpo) of the Northern Song Dynasty, a poem written after touring Mount Lu in May of the seventh year of Yuanfeng (1084 AD).

According to records, there are 171 named peaks in Mount Lu since ancient times, scattered among them are 26 ridges, 20 ravines, 16 caves, and 22 peculiar rocks. In addition, there are 22 waterfalls, 18 streams, and 14 lakes and ponds. Mount Lu is so diverse and colorful that the artistic conception of the poet Su Shi is endless and profound upon careful appreciation. This poem by Su Shi, inspired by objects, uses objects to convey meaning and expresses philosophical ideas in a simple and unadorned manner.

The poet points out that to observe things, one should be objective, comprehensive, and multifaceted; otherwise it is difficult to draw correct conclusions.

Machine learning is a branch of artificial intelligence. Since its inception in 1959, it has developed rapidly over more than 60 years, just like Mount Lu, which has formed over hundreds of millions of years, with a myriad of forms. To explore the grandeur, strangeness, danger, and beauty of machine learning, one must not only gaze at its peaks, ridges, ravines, caves, rocks, and waters but also understand the process and context of its formation, as well as the internal relationships and differences between them. This requires climbing each peak and ridge, entering each ravine and cave, looking at each waterfall from afar, observing each stream up close, and playing in each lake and pond. That is to say, machine learning is like a majestic Mount Lu, which needs to be observed and analyzed from multiple perspectives, directions, and levels.

This is the thesis of this part, with "perspectives" being its main argument.

This part is divided into two chapters, the first chapter is "Introduction", and the second chapter is "On Perspectives."

Chapter 1 provides a definition of machine learning, discusses the relationship between machine learning and artificial intelligence and other related disciplines, narrates the history of machine learning from the five main development trajectories of machine learning, and introduces some application examples of machine learning.

Chapter 2 starts from the basic disciplines of machine learning, abstracts machine learning on this basis, discusses its functions and roles, and identifies the difficulties in studying machine learning. We then discuss why machine learning is classified into three perspectives, and provide a detailed discussion on the division of these three perspectives, the relationship between the perspectives, the content of each perspective, and the views of related researchers. This is also the origin of the title of Chap. 2, "On Perspectives."

Chapter 1
Introduction

Abstract This chapter leads us into the field of machine learning. We first retrospect the origins of machine learning, introduce some typical definitions, and demystify the distinction between machine learning and human learning. Next, we survey the related fields such as artificial intelligence, deep learning, statistical learning, pattern recognition, and data mining. We then turn to review the historical trajectory of machine learning, spotlighting the pivotal development paths: neural networks, decision trees, boosting methods, support vector machines, and reinforcement learning. After that, we delve into the practical realm by showcasing representative applications like computer vision, speech recognition, natural language processing, data mining, computer games, autonomous technologies, medicine, biology, and so on. At the end of this chapter, we finally invoke "The Too-Small World" for a succinct discussion on machine learning.

1.1 From "Queen" to "King"

In recent years, machine learning has gradually become a hot research field in artificial intelligence.

In June 2015, a researcher named Tomasz Malisiewicz, after attending the 2015 *IEEE Conference on Computer Vision and Pattern Recognition* (CVPR), wrote a blog titled "Deep down the rabbit hole: CVPR 2015 and beyond."[1]

CVPR is one of the top three international conferences in the field of computer vision, the other two being the *IEEE International Conference on Computer Vision* (ICCV) and the *European Conference on Computer Vision* (ECCV).

Tomasz Malisiewicz has been attending CVPR almost every year since 2004, but the 2015 CVPR left a deeper impression on him. He wrote on his blog:

> "Machine learning used to be the Queen.
> Machine learning is now the King."

---

[1] https://www.computervisionblog.com/2015/06/deep-down-rabbit-hole-cvpr-2015-and.html

Let's first understand machine learning and see how did it change from the "Queen" to the "King."

First of all, we may have such questions: about machine learning, what is its origin, what is its definition, and what is the difference between machine learning and human learning?

This section will try to answer those questions.

1.1.1   Origins of Machine Learning

In 1946, ENIAC (Electronic Numerical Integrator and Computer), the first electronic general-purpose digital computer, was officially released. The programmability, stored program, and logical control features of this machine inspired some scientists to start exploring the possibility of building machine intelligence.

In 1950, Alan Turing, the father of computer science, published a groundbreaking paper titled "Computing Machinery and Intelligence" in *Mind*, a psychological and philosophical journal (Turing 1950). In this paper, he not only proposed a method later known as the "Turing test" to provide a satisfactory operational definition of machine intelligence but also proposed the topic of "learning machines." Therefore, Alan Turing is also known as the father of artificial intelligence.

At the *Western Joint Computer Conference* held in Los Angeles, California, in March 1955, a "Session on Learning Machines" (Ware 1955) was set up. The topics covered included research on the learning ability of machine systems, the complexity of chess, and pattern recognition. The session chair emphasized that the label of learning machines "is not the machine itself but the program that guides the machine toward its goal, which learns."[2]

In the summer of 1956, a group of researchers gathered at Dartmouth College in the United States to attend a workshop titled *Dartmouth Summer Research Project on Artificial Intelligence*, the famous Dartmouth meeting. The use of the name artificial intelligence (AI) as the theme of the workshop marked the official birth of artificial intelligence as a research field.

In 1959, American computer scientist Arthur Samuel published a paper in the *IBM Journal of Research and Development* titled "Some Studies in Machine Learning Using the Game of Checkers" (Samuel 1959). Because of his paper, the term machine learning (ML) became popular; the year 1959 can also be thought the opening year of machine learning. At that time, Arthur Samuel was working at the IBM Poughkeepsie Laboratory in the United States, and in 1966, he moved to Stanford University. In 1987, Arthur Samuel was awarded the "Computer Pioneer Award" by the IEEE Computer Society.

Arthur Samuel was also one of the main participants in the 1956 Dartmouth meeting, organized by John McCarthy from the Dartmouth College, Marvin Minsky

---

[2] https://dl.acm.org/doi/10.1145/1455292.1455308

from the Harvard University, Nathaniel Rochester from IBM Corporation, and Claude Shannon from Bell Telephone Laboratories, along with other notable figures such as Ray Solomonoff, Oliver Selfridge, Trenchard More, Allen Newell, and Herbert A. Simon.

Arthur Samuel passed away in 1990. John McCarthy, one of the founders of artificial intelligence, and Edward Feigenbaum, the father of expert systems, coauthored a commemorative article titled "Arthur Samuel: Pioneer in Machine Learning" (McCarthy and Feigenbaum 1990).

1.1.2   Definitions of Machine Learning

Since the inception of machine learning in 1959, several researchers have proposed various definitions. Here are a few representative descriptions.

In 1959, Arthur Samuel defined machine learning as "the field of study that gives computers the ability to learn without being explicitly programmed" (Samuel 1959).

In 1983, Herbert Simon, one of the founders of artificial intelligence, described machine learning as "a process by which a system improves its performance" (Simon 1983).

In 1997, Thomas Mitchell, a professor at Carnegie Mellon University, provided a more formal definition in his book *Machine Learning*: "A computer program is said to learn from experience E with respect to some class of tasks T and performance measure P, if its performance at tasks in T, as measured by P, improves with experience E" (Mitchell 1997).

Mitchell's definition included the three elements: task (T), which is the clearly defined task that machine learning needs to accomplish; performance (P), the performance measure that machine learning needs to improve; and experience (E), the experience accumulated through training and other methods. Many machine learning problems can be reduced to these three elements. For example, for the problem of handwriting recognition, task T is to recognize handwritten text images, performance P is the accuracy of handwriting recognition, and experience E is obtained by training the classifier with given handwritten text and its classification labels.

In 2004, Ethem Alpaydin, a professor in the Department of Computer Engineering at Özyegin University, wrote a textbook called *Introduction to Machine Learning* (Alpaydin 2009). He defined machine learning as "programming computers to optimize a performance criterion using example data or past experience."

In 2012, the MIT Press published a book titled *Foundations of Machine Learning*. The book has three authors: Mehryar Mohri, Afshin Rostamizadeh, and Ameet Talwalkar from New York University (Mohri et al. 2012). They defined machine learning as "computational methods that use experience to improve performance or make accurate predictions."

Peter Flach, a professor in the Department of Computer Science at the University of Bristol, in his 2012 book *Machine Learning: The Art and Science of Algorithms*

Table 1.1 Comparison of machine learning definitions

Who	When	Definition
Samuel	1959	The field of study that gives computers the ability to learn without being explicitly programmed
Simon	1983	Learning is any process by which a system improves performance from experience
Mitchell	1997	A computer program is said to learn from experience E with respect to some class of tasks T and performance measure P, if its performance at tasks in T, as measured by P, improves with experience E
Ethem Alpaydin	2004	Programming computers to optimize a performance criterion using example data or past experience
Mohri et al.	2012	The computational methods using experience to improve performance or to make accurate predictions
Flach	2012	Machine learning is about using the right features to build the right models that achieve the right tasks
This book	Present	A branch of AI that builds models that can learn from data or the environment to make predictions or decisions

that Make Sense of Data (Flach 2012), described the relationship between the features, tasks, and models of machine learning. The definition of machine learning given in the book is: "Machine learning is about using the right features to build the right models that achieve the right tasks."

Based on the above definitions, I myself in this book also propose a definition of machine learning, which is given below.

Definition 1.1 (Machine Learning)   Machine learning is a branch of artificial intelligence that studies and builds models that can learn from data or the environment to make predictions about unknown data or make decisions based on environmental information.

All of the above definitions of machine learning are summarized in Table 1.1.

It is worth emphasizing that my definition in this book also considers the relationship between machine learning and artificial intelligence, as well as the factor of learning from the environment.

This book is developed around my above definition, which will be further discussed in the next section.

1.1.3   Machine Learning vs. Human Learning

First, let's compare the similarity and difference between machine learning and human learning.

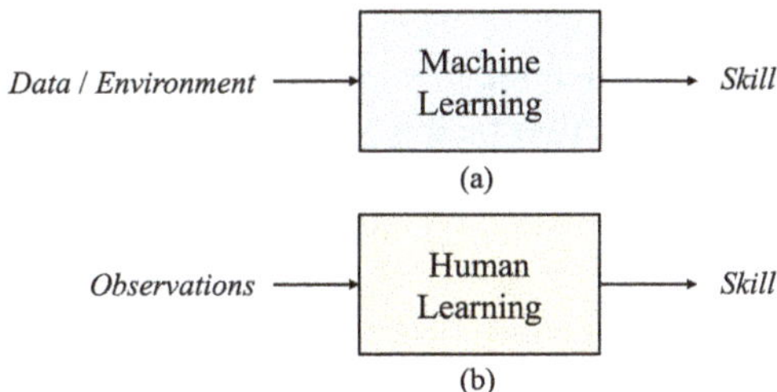

Fig. 1.1 Analogy between machine learning and human learning

As we know from Definition 1.1, the input of machine learning is data or environmental information, and the output is the skill of prediction or decision-making, as shown in Fig. 1.1a.

Humans have a high learning ability, as shown in Fig. 1.1b, which is characterized by using observation information as input and obtaining skills after brain processing.

Obviously, the difference between machine learning and human learning is that machine learning requires ready-made data or information feedback from the environment, while humans can use their senses to observe the external world. In addition, the skills of machine learning mainly manifest in prediction or decision-making, that is, to improve its prediction accuracy or make corresponding decisions through learning, while the range covered by human skills is much wider.

Many researchers have conducted in-depth research on human learning. Edward L. Thorndike, American psychologist and the father of educational psychology, compiled a series of lectures he gave at Cornell University into a book, published in 1931, titled *Human Learning*.

Thorndike's research on human learning originated from his animal psychology experiments. It is worth noting that his research on animal operant conditioning and other aspects not only made important contributions to the theory of behavior decision-making but also became the soil for the birth of connectionism.[3] This will be introduced in Chap. 7 of this book.

1.2   Related Fields of Machine Learning

Since its inception in 1959, machine learning has developed for over sixty years. Just from the books, papers, and other materials I have consulted, various methods and algorithms in machine learning have not yet been collated and systematized. Therefore, for beginners, it is difficult to judge the relationships between machine learning and statistical learning, between machine learning and deep learning,

---

[3] https://en.wikipedia.org/wiki/Connectionism

between machine learning and artificial intelligence, and so on. In especial, some people even think that machine learning is nothing but modern statistics.

This section mainly discusses the relationships between machine learning and artificial intelligence, deep learning, statistical learning, pattern recognition, data mining, and other related fields.

1.2.1   Artificial Intelligence

Definition 1.1 describes machine learning as a branch of artificial intelligence, which is the mainstream view in the AI and machine learning community. Further interpretation about this will be given as below.

First, let's review and analyze the definitions of artificial intelligence and machine learning.

Artificial intelligence is the study of how to build an intelligent agent that can perceive the external environment and take corresponding actions to achieve a certain goal (Wang 2019), while machine learning is how to build a model that can learn from data or the environment and use it to make predictions or decisions.

From the definition of artificial intelligence, we can abstract out the three elements of the intelligent agent:

- The percepts $\mathcal{P}^*$ through sensors
- The actions $\mathcal{A}$ with actuators
- The transition function f that maps percepts to actions

Their relationship can be represented as $f : \mathcal{P}^* \rightarrow \mathcal{A}$.

Furthermore, the following relationship can be derived from the definition of machine learning: machine learning requires information from data or the environment, and the percepts $\mathcal{P}^*$ of the intelligent agent can provide data or perceive environmental information for machine learning; machine learning can learn to make predictions or decisions, which are equivalent to enhancing the performance of transition function f of the intelligent agent, enabling it to produce the correct actions $\mathcal{A}$.

Through the above analysis, we can draw a relationship diagram between artificial intelligence and machine learning, as shown in Fig. 1.2. This diagram is also my interpretation of the academic view that "machine learning is a branch of artificial intelligence."

With the continuous development of machine learning, its research methods and application scopes are constantly expanding, especially its ability to learn from data or the environment to make predictions on unknown data or make decisions based on environmental information, which has established a relationship with many research fields.

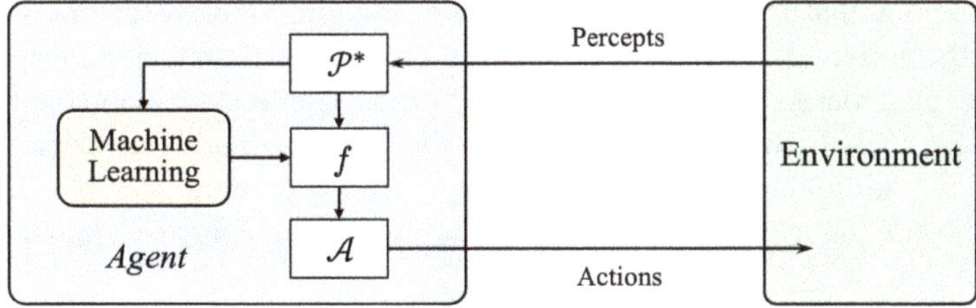

Fig. 1.2 Relationship between machine learning and artificial intelligence

Fig. 1.3 Relationship
between deep learning,
machine learning, and
artificial intelligence

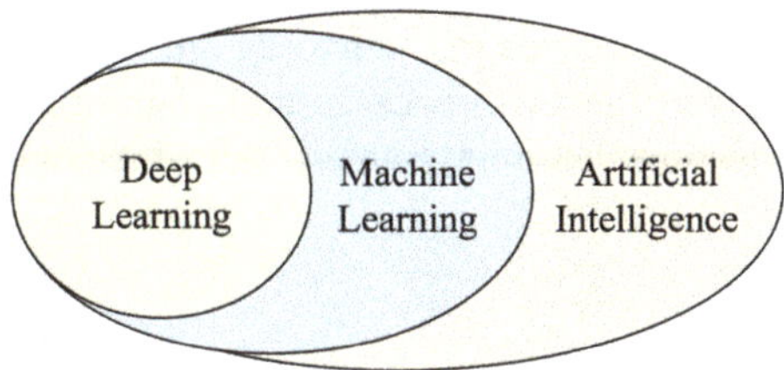

1.2.2 Deep Learning

Deep learning (DL) is a member of the machine learning family. In other words, deep learning is a subfield of machine learning. The relationship between deep learning, machine learning, and artificial intelligence is depicted in Fig. 1.3.

This book has been conducted a little to investigate the term "deep learning."

In 1986, Rina Dechter from the Cognitive Systems Laboratory at the University of California, Los Angeles, proposed deep learning as a learning scheme in a paper titled "Learning While Searching in Constraint-Satisfaction Problems (Dechter 1986)." Later, Igor Aizenberg et al. linked deep learning with artificial neurons in their book *Multi-Valued and Universal Binary Neurons: Theory, Learning and Applications* published in 2000 (Aizenberg et al. 2000).

Most modern deep learning models are built on artificial neural networks (ANNs), which are inspired by the information processing structure in the brain, and each ANN is a type of network consisting of multiple layers of neurons. In addition to the input and output layers, there are several hidden layers in the artificial neural network. An artificial neural network with many hidden layers is called a deep neural network (DNN). The machine learning method using deep neural networks is called deep learning.

The three giants of deep learning: Yann LeCun, Yoshua Bengio, and Geoffrey Hinton published a review article titled "Deep Learning" in the journal *Nature* in 2015, in which they wrote: "Deep learning allows computational models that are composed of multiple processing layers to learn representations of data with multiple levels of abstraction" (LeCun et al. 2015). This sentence depicts the essence and function of deep learning.

This implies that another meaning of deep learning is representation learning, which is the automatic extraction of features needed for classification or prediction from raw data based on deep neural networks. The features obtained through learning are called learned features, while the features that need to be manually extracted in traditional machine learning are called handcrafted features. Therefore, the role of automatic feature learning is self-evident compared to manual feature extraction.

There are various types of deep neural networks, such as convolutional neural networks (CNNs), recurrent neural networks (RNNs), deep belief networks (DBNs), deep Boltzmann machines (DBMs), and autoencoders (AEs).

Artificial neural networks are used by connectionism as an approach to study human mental processes and cognition, and the connectionist approach is also a machine learning framework, which will be discussed, respectively, in Chaps. 2 and 5 in detail.

1.2.3 Statistical Learning

Statistical learning is a framework of machine learning (Vapnik 1999). Its theoretical foundation is constantly being improved, forming a relatively complete statistical learning theory (Vapnik 1998; James et al. 2013).

To corroborate the above point, I have searched for the term "statistical learning" on *Wikipedia* multiple times in recent years, and it has always been redirected to "machine learning." This was still the case when I confirmed it again in March 2024.

I also confirmed the term "statistical learning theory" on *Wikipedia* multiple times, and the first sentence is still "Statistical learning theory is a framework for machine learning drawing from the fields of statistics and functional analysis."

It should be pointed out that statistical learning is one of theoretical frameworks in machine learning, although it is very important, which will be discussed, respectively, in Chaps. 2 and 4 in detail.

1.2.4 Pattern Recognition

Pattern recognition focuses on how to recognize patterns and regularities in input data. It usually uses feature vectors to represent patterns and processes them through algorithms. Pattern recognition has many real-world applications, including speech recognition, speaker identification, optical character recognition (OCR), handwriting recognition, fingerprint analysis, and face detection.

Traditional pattern recognition originates from statistics and engineering. With the continuous development of machine learning and the rapid increase in data volume, pattern recognition has gradually turned to processing patterns in data based on machine learning. Therefore, pattern recognition can be seen as "data patterns + machine learning."

The flagship journal of the IEEE in the field of artificial intelligence, namely, the *IEEE Transactions on Pattern Analysis and Machine Intelligence* (TPAMI), covers the content of pattern analysis and machine learning.

Pattern recognition also includes the formalization of patterns, interpretability, and visualization processing, which has been widely used in data analysis, signal processing, image analysis, information retrieval, bioinformatics, data compression, computer graphics, etc.

Because processing patterns in images and videos occupy a large part, computer vision has a certain association with pattern recognition. One of the top international academic conferences in computer vision, the *IEEE/CVF Conference on Computer Vision and Pattern Recognition* (CVPR), includes these two fields in its name.

With the continuous development of pattern recognition, it has extended to biometric recognition and developed into biometrics technology. It mainly includes two categories, namely, physiological characteristics recognition and behavioral characteristics recognition. Among them, physiological characteristics include fingerprints, palm prints, irises, and faces, while behavioral characteristics include voiceprints, handwriting, gait, and body movements.

1.2.5   Data Mining

Data mining is the process of extracting and discovering meaningful patterns and useful information in large datasets.

It is worth noting that data mining is not mining data itself, although the term data mining has become a popular professional term.

Data mining belongs to the cross-research field of database systems and machine learning. Therefore, data mining can be thought equivalent to "database system + machine learning."

Data mining uses many machine learning methods, but the goals are different. From the data perspective, data mining focuses on extracting and discovering meaningful patterns and useful information in data, but in contrast, machine learning focuses on extracting certain features from data and using those features for making predictions.

Data mining is often associated with knowledge discovery, known as knowledge discovery and data mining. This means that it is not only to mine the meaningful patterns in the data but also to discover the useful knowledge in large datasets.

1.2.6 Others

In addition to the above research fields, there are also some other fields related to machine learning, such as computer vision, natural language processing, computer speech, robotics, bioinformatics, and smart medical diagnosis. The relevant contents will be introduced in "1.4 Applications of Machine Learning."

1.3 History of Machine Learning

Machine learning has developed in the process of constantly exploring how to build effective machine learning models, and it has also developed in the situation where the demand for machine learning in many fields is constantly increasing, so many kinds of algorithms for machine learning have been produced.

Through refinement and in-depth exploration, it can be found that machine learning mainly has five major development trajectories; those are neural networks, decision trees, boosting algorithms, support vector machines, and reinforcement learning. The following are divided into five subsections for separate narration.

1.3.1 Neural Networks

Neural networks refer to artificial neural networks (ANNs) usually. A neural network can be said to be a computing system built inspired by the biological neural network in the brain, and it is also an implementation approach to machine learning.

Here is a timeline of the development of neural networks, as shown in Fig. 1.4, and then a detailed introduction.

Neural networks can be traced back to a classic paper "A Logical Calculus of the Ideas Immanent in Nervous Activity" published in *The Bulletin of Mathematical Biophysics* in 1943 by Warren S. McCulloch and Walter Pitts (McCulloch and Pitts 1943). The paper attempted to make clear how the brain generates highly complex patterns by using many basic cells being connected. These basic brain cells are

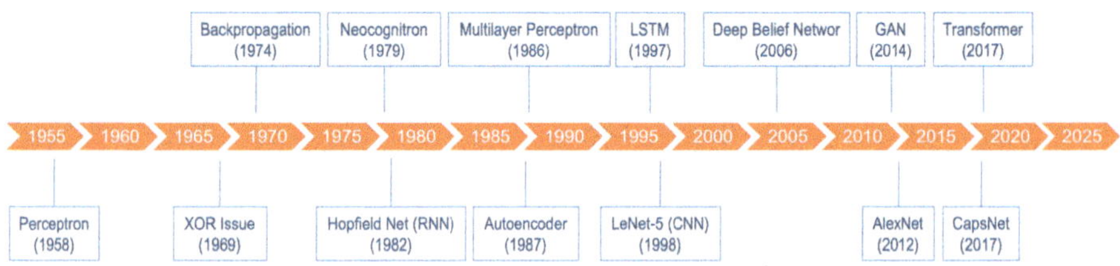

Fig. 1.4 Main development history of neural networks

called neurons, and the paper presents a highly simplified model of a neuron, the threshold logic unit (TLU), also known as the linear threshold unit (LTU) (Anthony 2001). This was the first artificial neuron proposed in history, also known as the McCulloch-Pitts neuron.

In 1949, Canadian psychologist Donald O. Hebb proposed a neuropsychological learning theory in his book *The Organization of Behaviour: A Neuropsychological Theory* (Hebb 1949). His theory is known as Hebbian theory, and the models established following Hebbian theory are known as Hebbian learning. Donald Hebb is referred to as the father of neuropsychology and neural networks.

In 1954, Belmont G. Farley and W. A. Clark published a paper titled "Simulation of Self-Organizing Systems by Digital Computer," which was the first to use a computer to simulate a neuron network (Farley and Clark 1954).

In 1958, Frank Rosenblatt invented the perceptron at the Aeronautical Laboratory of Cornell University in the United States (Rosenbaltt 1957, 1958). The perceptron is a pattern recognition algorithm, a binary classifier with only one output layer; thereby, it is called the single-layer perceptron.

In 1960, Bernard Widrow and Marcian E. Hoff invented the Delta Rule (Widrow and Hoff 1988), a gradient descent learning rule used for updating the weight of neuron inputs in a single-layer perceptron, i.e., the training of a single-layer perceptron.

However, in 1969, Marvin Minsky and Seymour A. Papert published the controversial book *Perceptrons: An Introduction to Computational Geometry* (Minsky and Papert 1969). They pointed out that single-layer perceptrons can only be used for linearly separable functions but cannot for solving a wide variety of interesting problems such as XOR functions. Their assertion, like a bucket of ice water, cooled down the research on neural networks.

In 1970, Seppo Linnainmaa, in his master's thesis, first fully described the automatic differentiation (AD) method (Linnainmaa 1970) for discrete connected networks of nested differentiable functions; his work was the prototype of the famous algorithm, backpropagation (BP) algorithm. However, it did not attract attention in the environment at that time.

In 1974, Paul J. Werbos, in his Ph.D. thesis, proposed the possibility of using Seppo Linnainmaa's backpropagation algorithm for artificial neural networks (Werbos 1974). It was not until 1982 that Werbos' automatic differentiation method was applied to multilayer artificial neural networks, which accelerated the development of multilayer neural networks.

In 1979, inspired by the structure of animal neural visual systems, Kunihiko Fukushima et al. from the NHK Broadcasting Technology Research Institute in Japan published the neocognitron, a multilayer network composed of neuron-like cells, used for visual pattern recognition (Fukushima and Miyake 1982; Fukushima 1988).

In 1982, John Hopfield proposed a content-addressable memory system with binary threshold units (Hopfield 1982), later known as Hopfield network. Unlike perceptrons, the neurons in this network can also perform back-coupling, so it is considered a type of recurrent neural network (RNN). Hopfield's paper became the foundational literature for recurrent neural networks.

In 1985, Geoffrey Hinton et al. invented the Boltzmann machine, another symmetrically connected network with neuron-like units that can randomly decide whether to turn on or off (Hinton and Sejnowski 1983; Ackley et al. 1985).

In the 1980s, some neural network researchers published research results using the backpropagation algorithm to train multilayer perceptron (MLP) (Rumelhart et al. 1985). David E. Rumelhart, Geoffrey Hinton, and Ronald Williams pointed out in their paper "Learning Representations by Back-propagating Errors" published in the journal *Nature* in 1986 that the backpropagation algorithm method repeatedly adjusts the weights of connections in the network to minimize the difference between the actual output vector and the expected output vector of the network and can make the data in the hidden layer of the neural network form useful learning representations (Rumelhart et al. 1986).

In 1987, Yann LeCun proposed the concept of autoencoders (AE) in his doctoral thesis (LeCun 1987). It is a type of unsupervised learning neural network, consisting of an encoder and a decoder. The encoder converts the input data into an effective representation, and the decoder then transforms it back into the original data format.

In 1997, Sepp Hochreiter and Jürgen Schmidhuber proposed long short-term memory (LSTM) (Hochreiter and Schmidhuber 1997). An LSTM is a unit in a recurrent neural network (RNN), and an RNN composed of LSTM units is often referred to as an LSTM network.

In 1998, Yann LeCun et al. published LeNet-5, a pioneering seven-layer convolutional neural network (CNN), used for checking handwritten numbers on checks (LeCun et al. 1998).

In 2006, Geoffrey Hinton et al. published the deep belief network (DBN) (Hinton et al. 2006). In the same year, Geoffrey Hinton and Ruslan Salakhutdinov, then a doctoral student, published the paper "Reducing the Dimensionality of Data with Neural Networks" in *Science* magazine (Hinton and Salakhutdinov 2006), in which autoencoder is used for dimensionality reduction. These two papers can be seen as a prelude to deep learning.

In 2011, Jeff Dean, Greg Corrado, and Andrew Ng formed the Google Brain team to build a large-scale deep learning system consisting of 16,000 computers capable of recognizing high-level concepts,[4] such as cats, from unlabeled images. This achievement was reported in *The New York Times* on June 25, 2012.

In 2012, Alex Krizhevsky, Geoffrey Hinton, and others designed a convolutional neural network called AlexNet that achieved excellent results in the *ImageNet Large Scale Visual Recognition Challenge* (ILSVRC). AlexNet was presented in their paper "ImageNet Classification with Deep Convolutional Neural Networks" (Krizhevsky et al. 2012), at the *Conference on Neural Information Processing Systems* (NIPS), a leading international conference on machine learning and neural

---

[4] https://www.geeksforgeeks.org/what-is-google-brain/

networks. The great success of AlexNet has attracted the attention of researchers and industries, which has set off a boom in deep learning around the world.

In 2014, Ian Goodfellow et al. proposed the generative adversarial network (GAN), a novel unsupervised deep learning algorithm. This is achieved by two neural networks playing zero-sum games with each other (Goodfellow et al. 2014). The innovation of GAN has set off a new upsurge in deep learning.

In May 2015, Yann LeCun, Yoshua Bengio, and Geoffrey Hinton published a review article titled "Deep Learning" in the journal *Nature*. It states that "ConvNets have their roots in the neocognitron, the architecture of which was somewhat similar but did not have an end-to-end supervised-learning algorithm such as backpropagation" (LeCun et al. 2015).

In October 2015, Olaf Ronneberger et al. published a paper titled "U-Net: Convolutional Networks for Biomedical Image Segmentation" that is a modified fully CNN architecture and is used for fast and precise biomedical image segmentation (Ronneberger et al. 2015).

In June 2016, Kaiming He et al. at Microsoft Research Asia published a paper titled "Deep Residual Learning for Image Recognition" at the *Conference on Computer Vision and Pattern Recognition* (CVPR). This paper proposes a residual network with shortcut connections between layers, referred to as ResNet, which solves the problem that deep neural networks become more complex and difficult to train with the increase in the number of layers (He et al. 2016).

In June 2017, Ashish Vaswani et al. at Google Brain published the Transformer, a deep learning architecture that relies on the parallel multi-head attention mechanism (Vaswani et al. 2017). This model is mainly used in natural language processing and has also been used in the field of computer vision. The introduction of Transformer has led to the development of large language models (LLMs), such as BERT (Bidirectional Encoder Representations from Transformers) and GPT (Generative Pre-trained Transformers).

In August 2017, Jeffrey Hinton proposed the capsule network (CapsNet). This network is based on the theory of cortical mini-columns in neuroscience, where each capsule is composed of a group of neurons, thus capable of representing complex information such as the position, direction, and color of objects. The dynamic routing in the capsule network can overcome the limitations of pooling in convolutional neural networks (Sabour et al. 2017). And in 2019, Hinton's team proposed stacked capsule autoencoders (Kosiorek et al. 2019).

In February 2019, OpenAI, a US-based AI research and deployment organization, released a large-scale unsupervised language model called GPT-2. This model, which was trained on a dataset of eight million web pages, demonstrated the ability to generate coherent and contextually relevant sentences by predicting the next word in a given piece of text. The release of GPT-2 sparked a debate about the ethical implications of AI technology, as the model's ability to generate realistic and persuasive text could potentially be used for malicious purposes.

In July 2019, Facebook AI Research (FAIR) released a model called RoBERTa (Liu et al. 2019), which is a variant of BERT that was trained with more data and for a longer period of time. RoBERTa demonstrated superior performance on several benchmark tasks, further highlighting the potential of large-scale language models in natural language processing tasks.

In June 2020, OpenAI released GPT-3, a language model with 175 billion parameters, which is more than 100 times larger than its predecessor, GPT-2. GPT-3 demonstrated the ability to generate even more realistic and contextually relevant text, and it also showed the ability to perform tasks that it was not explicitly trained to do, such as translation and arithmetic, simply by providing it with a few examples, a technique known as "few-shot learning."

In June 2021, the so-called three giants of deep learning, also the winners of the 2018 Turing Award, Yoshua Bengio, Yann LeCun, and Geoffrey Hinton, published a Turing lecture titled "Deep Learning for AI" in *Communications of the ACM*, in which they review the basic concepts and origins of deep learning and discuss the breakthrough progress of deep learning, the existing limitations of deep learning, and its direction of improvement (Bengio et al. 2021).

In November 2022, OpenAI launched ChatGPT based on GPT-3.5, which is credited with starting the boom of large language models (LLMs).

The development of artificial neural networks advances at a rapid pace, with new models and techniques being developed all the time. These advancements are driving progress in a wide range of fields, from computer vision and natural language processing to healthcare, autonomous vehicles, etc.

1.3.2 Decision Trees

Decision trees are based on the multistage decision-making method that breaks down complex decision problems into a series of simple decisions. It is composed of a treelike structure of all decision options and subsequent chance events related to specific options.

A decision tree is also a decision support tool that uses a treelike graph to make decisions about possible outcomes, including chance event outcomes, resource costs, and utilities. It is a way to present algorithms through conditional judgment control mechanisms. Decision trees can also be seen as a generative model of inductive rules from empirical data, often used in operations research, especially decision analysis, to help determine the strategy most likely to achieve the goal. It is also a popular tool in machine learning.

The main development of decision trees is shown in Fig. 1.5.

The idea of decision trees can be traced back to the concept learning system (CLS) framework proposed by Earl B. Hunt et al. in 1962. The core of this framework is to construct a decision tree that minimizes the cost of classifying objects (Hunt 1962; Quinlan 1986).

In January 1969, Patrick Winston of MIT published a paper titled "A Heuristic Program That Constructs Decision Trees" (Winston 1969). Stanford University also proposed the idea of searching for probabilistic decision trees in a technical report in 1969.

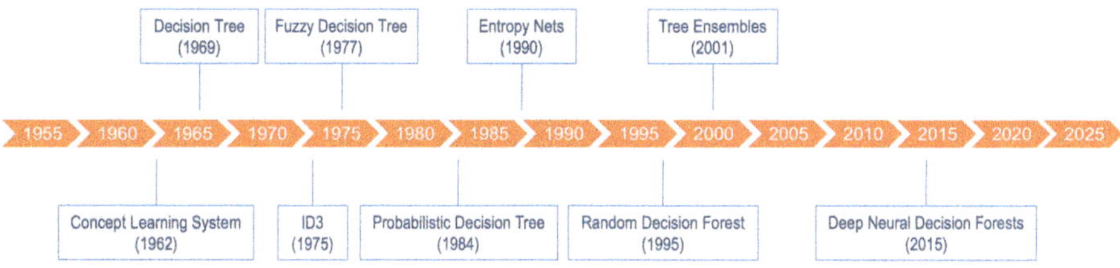

Fig. 1.5 Main development history of decision trees

In 1973, William S. Meisel of the University of Southern California and Demetrios A. Michalopoulos of California State University, Fullerton, proposed a partitioning algorithm that uses a decision tree for pattern classification and optimization (Meisel and Michalopoulos 1973).

In 1975, John R. Quinlan published the ID3 (Iterative Dichotomiser 3), which is a classification algorithm for generating decision trees from fixed samples in a dataset (Quinlan 1986).

In 1977, Robin L. Chang and Theodosios Pavlidis published a paper titled "Fuzzy Decision Tree Algorithms" (Chang and Pavlidis 1977), introducing fuzzy decision functions into decision trees and proposing a branch-bound-backtrack algorithm.

In 1981, Galigekere R. Dattatreya and V. S. Sarma discussed the "Bayesian and Decision Tree Approaches for Pattern Recognition Including Feature Measurement Costs" in their paper (Dattatreya and Sarma 1981).

In 1984, Richard Casey of the IBM Research Institute in the United States and George Nagy of the University of Nebraska published a paper titled "Decision Tree Design Using a Probabilistic Model" (Casey and Nagy 1984), which was used for optical character recognition.

In 1989, Paul E. Utgoff published a paper titled "Incremental Induction of Decision Trees" (Utgoff 1989) in the journal *Machine Learning*, proposing an incremental induction decision tree algorithm equivalent to the non-incremental ID3 algorithm.

In 1990, Ishwar K. Sethi studied the mapping problem of tree to network and published entropy networks, which converts decision trees into neural networks (Sethi 1990). An entropy network is a type of network with very few connections.

In 1995, the random decision forest, also known as random forest, was developed and derived based on decision trees. The random forest is an ensemble model of multiple decision trees, trained on different parts of the same training set, with the aim of reducing variance.

The study of random forests began with Tin Kam Ho, a researcher at AT&T Bell Laboratories in the United States. She published a paper on decision forests at the *International Conference on Document Analysis and Recognition* (ICDAR) in 1995. To address the issue of the limited complexity of decision trees, she constructed multiple trees that could be randomly expanded in feature space based on random modeling to improve the accuracy of the classifier (Ho 1995). She also published a paper titled "The Random Subspace Method for Constructing Decision Forests" in the *IEEE Transactions on Pattern Analysis and Machine Intelligence* (TPAMI) in 1998 (Ho 1998). Since 2014, Tin Kam Ho[5] has been at the IBM Thomas J. Watson Research Center at Yorktown Heights, New York.

In 2001, Leo Breiman of the University of California, Berkeley, published a paper titled "Random Forest" in the journal *Machine Learning* (Breiman 2001). The paper discussed that random forests belong to the ensemble of multiple trees and added extensions to random forests, including the introduction of bagging and the concept of random features, to build a decision tree collection with controlled variance.

In 2015, Peter Kontschieder of Microsoft Research, Madalina Fiterau of Carnegie Mellon University, and other researchers published a paper titled "Deep Neural Decision Forests" at the *IEEE International Conference on Computer Vision* (ICCV), proposing a classification method that integrates deep convolutional neural networks and decision trees (Kontschieder et al. 2015).

1.3.3 Boosting Methods

In machine learning, the basic idea of the boosting methods is to integrate several weak learners into a strong learner. This is because one has found that it is often much easier to find multiple low-precision empirical rules than a high-precision empirical rule.

The main development trajectory of the boosting methods is shown in Fig. 1.6.

The boosting method originated from an extension of the theory of learnability—strong learnability and weak learnability.

In 1988, Michael Kearns gave a formal definition of strong learnability and weak learnability in a technical report titled "Thoughts on Hypothesis Boosting" and proposed the so-called hypothesis boosting problem (Kearns 1998).

In 1990, Robert E. Schapire published "The Strength of Weak Learnability" which affirmatively answered and detailedly proved the hypothesis boosting problem and proposed a method to boost weak learning algorithms into any strong

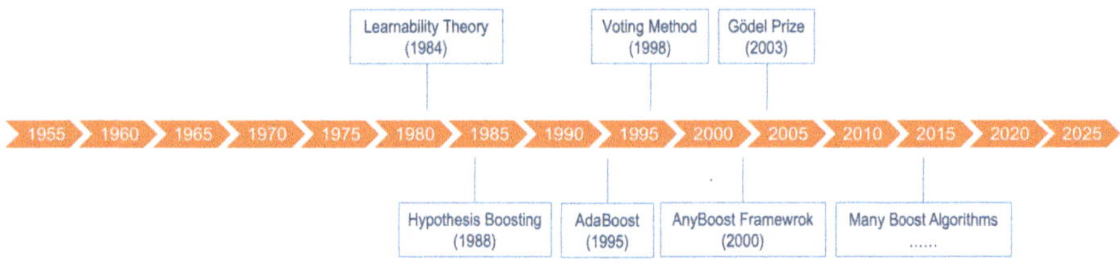

Fig. 1.6 Main development history of the boosting method

---

[5] https://researcher.watson.ibm.com/researcher/view.php?person=us-tho

learner (Schapire 1990). His work has advanced the development of the boosting method.

In 1995, Adaptive Boosting (AdaBoost) algorithm was proposed by Yoav Freund and Robert Schapire (Freund and Schapire 1995). The AdaBoost can be used with many other types of learning algorithms, combining the outputs of other weak classifiers into a weighted sum, boosting it into the final output of a strong classifier. The adaptability of AdaBoost is reflected in the appropriate combination of weak classifiers used to improve classification performance. Their outstanding work on AdaBoost won the *Gödel Prize* in 2003.

The *Gödel Prize* is an annual outstanding paper award in the field of theoretical computer science, established in 1993 by the *European Association for Theoretical Computer Science* (EATCS) and the *ACM Special Interest Group on Algorithms and Computational Theory* (ACM SIGACT). This prize is named in honor of Kurt Gödel (1906–1978), an Austrian mathematician, logician, and philosopher, in recognition of his outstanding contributions to modern logic and so on.

In 1998, Robert Schapire et al. referred to the boosting method as the voting method in their published paper (Schapire et al. 1998).

In 2000, Llew Mason et al. proposed a general boosting method framework called AnyBoost and proved that many successful voting methods, including AdaBoost, are special cases of AnyBoost (Mason et al. 1999).

Afterward, many new or improved boosting methods emerged, such as LPBoost, TotalBoost, SoftBoost, ERLPBoost, BrownBoost, XGBoost, MadaBoost, and LogitBoost.

The continuous development and improvement of boosting methods have given rise to ensemble learning.

1.3.4   Support Vector Machines

Support vector machines (SVMs) are a milestone in the machine learning community. Its development history is shown in Fig. 1.7.

The concept of SVM originated from the Vapnik-Chervonenkis theory (Vapnik 1999) proposed from the late 1960s to the 1990s, also known as VC theory. This

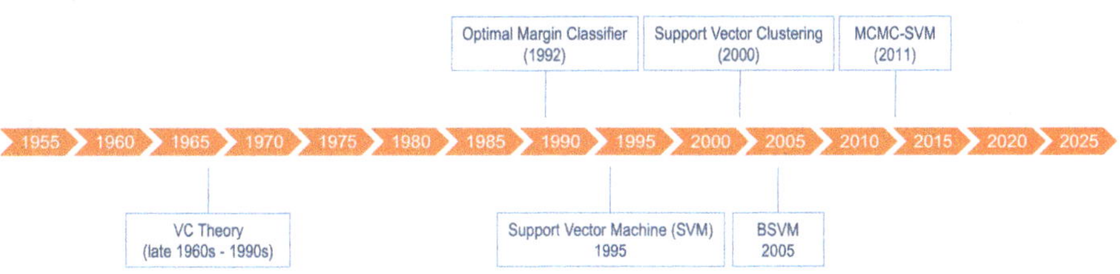

Fig. 1.7 Development history of support vector machines

theory attempts to explain the learning process from a statistical perspective and has become part of statistical learning theory.

In 1992, Bernhard E. Boser from the University of California, Berkeley, and Isabelle M. Guyon and Vladimir N. Vapnik from Bell Labs published a paper titled "A Training Algorithm for Optimal Margin Classifiers" at the *Fifth Annual Workshop on Computational Learning Theory*. They proposed a method for maximizing the decision boundary of classification, which became the precursor of SVM (Boser et al. 1992).

In 1995, a paper titled "Support-Vector Networks" by Corinna Cortes from Bell Labs and Vladimir Vapnik was published in the journal *Machine Learning* (Cortes and Vapnik 1995).

SVM belongs to supervised learning, and its initial design was for binary linear classification. Given a set of training data, each sample is labeled as belonging to one of two categories, and then, the SVM algorithm is trained to become a non-probabilistic binary linear classifier. This method maps vector points in space, finds a boundary line for classification in two-dimensional space, and forms a separation band as wide as possible centered on this boundary line. The vectors on the two parallel lines on both sides of this separation band support these two lines.

The solid theoretical foundation and excellent experimental results of SVM have attracted the attention of the machine learning academic community. The flexibility of the kernel function has made the functions of SVM rich and colorful, deriving different SVM algorithms and making SVM support both linear and nonlinear classification. For more than ten years after 1995, SVM became one of the hotspots of machine learning, supporting half of the field of machine learning.

In 1996, the year after the publication of SVM, Bell Labs applied SVM to regression tasks; a paper titled "Support Vector Regression Machines" (Drucker et al. 1996) was published at the *Conference on Neural Information Processing Systems* (NIPS).

As mentioned earlier, SVM was initially designed as a binary classifier. How to use SVM for multiclass classification, that is, multiclass SVM, also became a research topic.

In 1999, Erin J. Bredensteiner and Kristin P. Bennett proposed a method using linear programming approaches to implement multicategory classification by support vector machines (Bredensteiner and Bennett 1999). Reference (Hsu and Lin 2002) also compared several types of multi-classification support vector machines.

In 2000, Asa Ben-Hur et al. proposed a kernel method for data clustering, using support vectors to describe the data, called support vector clustering (SVC) algorithm (Ben-Hur et al. 2000).

By the beginning of the 2000s, those conventional support vector machines belong to non-probabilistic models.

In 2005, three researchers proposed the Bayesian support vector machine (BSVM) based on the Bayesian theorem in probability theory, used for classifying tumors based on gene expression data (Mallick et al. 2005).

In 2011, two researchers proposed a latent variable representation of regularized SVMs that enables expectation-maximization (EM), expectation-conditional maximization (ECM), or Markov chain Monte Carlo (MCMC) algorithms (Polson and Scott 2011).

1.3.5 Reinforcement Learning

Reinforcement learning (RL) originally comes from behaviorist psychology, studying how intelligent agents improve their behavior according to the state and rewards of the environment. Due to the universality of the mechanism of reinforcement learning, it has also become the research topic of some other disciplines, such as game theory, control theory, operations research, and information theory.

The main development progress of reinforcement learning is shown in Fig. 1.8.

Computational reinforcement learning can be traced back to the research of Marvin L. Minsky, who wrote his doctoral thesis in 1954, titled "Theory of Neural-Analog Reinforcement Systems and Its Application to the Brain-Model Problem" (Minsky 1954).

In 1965, the *IEEE Transactions on Automatic Control* published a paper titled "A Heuristic Approach to Reinforcement Learning Control Systems," which can be used to control nonlinear and nonstationary factory equipment, authored by M. D. Waltz and King-sun Fu (Waltz and Fu 1965).

In 1970, Jerry M. Mendel and Robert W. McLaren published a paper titled "Reinforcement Learning Control and Pattern Recognition Systems" in *Mathematics in Science and Engineering* (Mendel and Mclaren 1970).

Richard S. Sutton, a distinguished research scientist at Google DeepMind and a professor at the University of Alberta, Canada, is considered the founder of modern computational reinforcement learning. His doctoral thesis completed in 1984, titled *Temporal Credit Assignment in Reinforcement Learning*, proposed the Actor-Critic architecture and temporal credit assignment in reinforcement learning (Sutton 1984). Due to his contributions to the field of reinforcement learning and artificial intelligence, the *International Neural Network Society* awarded Richard Sutton the *President Award* in 2003.

Richard Sutton's PhD thesis advisor was professor Andrew G. Barto of the University of Massachusetts Amherst. By this account, Professor Barto should also be considered a godfather-like figure in reinforcement learning. A quick online

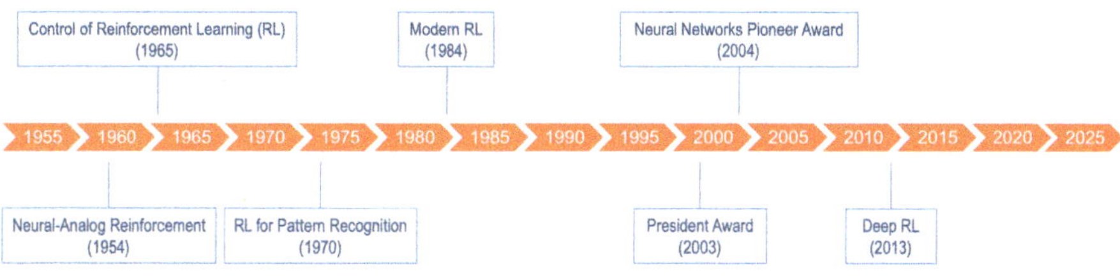

Fig. 1.8 Development history of reinforcement learning

search confirms this: due to Andrew Barto's contributions to the fundamental research field of reinforcement learning, he was awarded the *Neural Networks Pioneer Award* for 2004 by the *IEEE Neural Networks Society*.

The book *Reinforcement Learning: An Introduction* coauthored by Richard Sutton and Andrew Barto in 1998 is considered a classic textbook in reinforcement learning. The second edition of the book was released in October 2018 (Sutton and Barto 2018).

In the field of machine learning, reinforcement learning is to learn and improve behavior through a trial-and-error process of dynamic interaction with the environment (Kaelbling et al. 1996). This is different from supervised learning, which requires a large amount of labeled training data, and also different from unsupervised learning, which directly analyzes a large amount of input data.

It's worth pointing out that, in the field of artificial intelligence, the intelligent agent is usually defined as an intelligent rational entity that interacts with the environment and, in the field of machine learning, the paradigm that interacts with the environment is such that reinforcement learning. The intelligent agent in reinforcement learning not only perceives its state but also receives rewards from the environment.

By the way, the meaning of "reward" in the English dictionary is "a recompense for worthy acts or retribution for wrongdoing,"[6] and in reinforcement learning, the value of "reward" can be positive or negative.

In reinforcement learning, the intelligent agent decides the next action based on the current state and rewards. This is a continuous iterative process, learning from the environment through this iterative process. An important part of reinforcement learning is the construction of action policies, which can be obtained through extensive error-and-trial procedure, and the policies about action choices are made based on state and rewards from the environment during actual operation.

In 2013, DeepMind company in the United Kingdom first combined the *Q-learning algorithm* in reinforcement learning with the convolutional neural network (CNN) in deep learning to build an intelligent video game system using the Atari video game console released in the end of 1977, which outperformed existing methods (Mnih et al. 2013).

In 2014, DeepMind published an improved reinforcement learning algorithm (Silver et al. 2014). And in this year, DeepMind was acquired by Google.

A year later, in February 2015, Google DeepMind published a paper in the journal *Nature*, building an intelligent video game system through deep reinforcement learning (DRL) that can reach human control levels (Mnih et al. 2015).

Since the second half of 2015, Google DeepMind proposed several versions of AlphaGo that have won against top professional Go players in the world, namely,

---

[6] https://www.powerthesaurus.org/reward/definitions

AlphaGo Fan, AlphaGo Lee, AlphaGo Master, AlphaGo Zero, and AlphaZero that can play three types of board games (Go, chess, and Japanese shogi). Those great successes are by combining deep reinforcement learning with Monte Carlo tree search (Silver et al. 2016, 2017a).

1.4 Applications of Machine Learning

This section categorizes and organizes the landmark events of machine learning applications in recent years by their fields and in chronological order. Some of these applications involve not only machine learning but also other artificial intelligence technologies.

1.4.1 Computer Vision

Computer vision is the research field of how to make computers process images or videos at a higher level. It includes recognition, understanding, and generation of visual information such as images or videos.

Specifically, image or video recognition involves identifying objects, actions, and locations; image or video understanding involves converting them into descriptions of their content; and image or video generation involves creating new images or videos based on text and/or other information.

According to the analysis of researchers at Harvard Business School, the proportions of information that humans need to process daily are taste 1%, touch 1.5%, smell 3.5%, hearing 11%, and vision 83%. Therefore, the importance of computer vision is self-evident.

The tasks of computer vision include image restoration, event detection, video tracking, object segmentation, object recognition, pose estimation, motion estimation, scene modeling, scene reconstruction, 3D reconstruction, image understanding, and visual generation.

The launch of the ImageNet dataset and its challenge promoted the development of computer vision.

ImageNet is a large-scale visual dataset, officially launched at the *Conference on Computer Vision and Pattern Recognition* (CVPR) in 2009. The dataset has over 14 million manually annotated images and over a million images with object bounding boxes, covering over 20,000 categories. Since 2010, the *ImageNet Large Scale Visual Recognition Challenge* (ILSVRC) has been held annually.

In the 2012 ILSVRC challenge, the team led by Geoffrey Hinton from the University of Toronto used a deep convolutional neural network to classify 1.2M high-resolution images from ILSVRC-2010 into 1,000 classes, achieving a top-5 error rate of 15.3% and winning the ILSVRC 2012 championship, far ahead of the second place's 26.2% (Krizhevsky et al. 2012). This deep convolutional neural network is AlexNet, named after the first author Alex Krizhevsky, and has become an important image classification deep network model.

In 2015, the deep residual networks (ResNet) proposed by Kaiming He et al. from Microsoft Research Asia achieved a good result of 3.57% error rate in the aforementioned ImageNet image classification competition, which is lower than the error rate of human classification of these images (5.1%).

The ILSVRC ended in 2017. After that, the WebVision challenge took place, with its dataset composed of noisy web visual data.

1.4.2   Computer Speech

Computer speech mainly includes speech recognition and speech synthesis.

1.4.2.1   Speech Recognition

Speech recognition is to develop methodologies and technologies that enable the recognition and translation of spoken language into text. Its iconic achievements are as follows.

In 2011, Apple introduced the intelligent personal assistant Siri, which has now become a part of various operating systems of the Apple company. Siri is a voice-controlled intelligent personal assistant. It uses a natural language user interface to answer questions, make suggestions, and perform actions. It supports more than 20 languages, including English and Chinese.

In April 2014, Microsoft launched Cortana, a virtual assistant that runs on Windows, iOS, Android, and other systems.

In June 2014, Microsoft also launched Xiaoice, a natural language chatbot.

Researchers from Microsoft Research and the University of Toronto have made a breakthrough, using deep neural network technology that simulates human brain behavior, to make speech recognizers more discerning, reducing the word error rate of speech recognition by 30%.

A year later, on September 8, 2015, Baidu launched an intelligent personal assistant "Duer" at the 2015 *Baidu World Conference*, which can provide users with secretarial services.

1.4.2.2   Speech Synthesis

Speech synthesis is the artificial production of human speech, mostly used for translating text information into audio information.

In 2012, Microsoft's chief research officer Rick Rashid demonstrated a real-time English-to-Chinese universal translator. He spoke a paragraph in English, and the translator not only played the translated Chinese but also maintained his own voice and tone. This is the role of speech synthesis.

1.4.3   Natural Language Processing

Natural language processing (NLP) is another field where artificial intelligence and machine learning have made breakthrough progress.

1.4.3.1   Machine Translation

In recent years, machine translation has made breakthrough progress. In March 2018, a machine translation system jointly developed by Microsoft Research Asia and Redmond claimed to have reached the level of human professional translators on the Chinese-English test set of NewsTest 2017.

1.4.3.2   Large Language Models

A large language model (LLM) is such a type of language model that is notable for its ability to achieve general-purpose language understanding and generation.

OpenAI, an artificial intelligence research laboratory based in San Francisco, USA, launched the Generative Pre-trained Transformer 3 (GPT-3) in June 2020. It is the third generation in the GPT-n series, an autoregressive language model based on Transformer, which can generate humanlike text, i.e., generate natural language text similar to human creation based on user input prompts or requests or questions posed in sentence form. The full version of GPT-3 has 175 billion parameters.

Before the release of GPT-3, the largest language model was Microsoft's Turing NLG, which was launched in February 2020, with a capacity of 17 billion parameters, only 10% of the capacity of GPT-3.

The quality of the text generated by GPT-3 is so high that it is difficult to distinguish it from text written by humans. Some philosophers have called GPT-3 one of the most interesting and important AI systems in history.

In November 2022, OpenAI launched ChatGPT based on GPT-3.5. And in March 2023, OpenAI launched GPT-4, a multimodal large language model that can accept images as well as text as its input.

Other LLMs include Google's Gemini, Meta's LLaMa, Baidu's Ernie, Huawei's Pangu, Anthropic's Claude, and Amazon's Olympus.

1.4.4   Data Mining

Data mining is the process of extracting and discovering unknown patterns from large datasets, and it is a data processing method that combines machine learning and database systems.

Specific tasks of data mining include obtaining new data record groups through cluster analysis, discovering abnormal records through anomaly detection, and finding association rules through dependency relationships.

Data preprocessing is often required before data mining, and result verification is needed after data mining.

The most well-known branch of data mining is knowledge discovery, also known as knowledge discovery in databases (KDD).

Currently, data mining and knowledge discovery are often put together, collectively referred to as data mining and knowledge discovery.

1.4.5   Computer Games

There are many types of computer games; here, we only introduce board games, card games, tile-based games, and video games.

1.4.5.1   Board Games

Most board games are two-player games with relatively simple game rules and strategies. However, due to their vast game space, the gameplay itself becomes very complex. If we were to rank representative board games from easy to difficult, they would be checkers, chess, and Go.

Here, we only introduce the progress of computer Go, the most difficult game in board games.

In December 2015, Google DeepMind's AlphaGo, the best computer Go at that time, defeated the European Go champion, Fan Hui, with a score of 5 wins in 5 games. This was the first time a computer Go using artificial intelligence defeated a human professional Go player. However, this news was not announced until January 27, 2016, to coincide with the publication of the paper on AlphaGo (Silver et al. 2016) in the journal *Nature*.

A few months later, from March 8 to 15, 2016, the world-famous Go AI versus human player was held in Seoul, South Korea, where AlphaGo played against Korean nine-dan professional Go player Lee Sedol, winning the match with 4 wins in 5 games.

According to the website Go Ratings[7] in July 2016, AlphaGo was ranked first in the world Go rankings at the time, surpassing Ke Jie. The reason was that Ke Jie performed poorly in the "Cross-Strait Go Championship" and his points dropped, being overtaken by AlphaGo.

---

[7] https://www.goratings.org/en/

But, while I checked this Go Ratings again in 2018, circumstances changed with the passage of time, and this picture was no longer visible. The reason is said to be that AlphaGo has withdrawn from this world Go ratings.

On December 29, 2016, a mysterious player registered under the username Magister, on the Tygem server in South Korea, and then changed its account name to "Master" on December 30. It moved to the famous Chinese FoxGo Server[8] on January 1, 2017. As of January 5, 2017, it played 60 online Go games with the world's top Go players, winning 60 consecutive victories. Its opponents included many of the world's top Go players, such as Ke Jie, Park Jeong-hwan, Yuta Iyama, Gu Li, and Nie Weiping. It was later revealed that Master was an upgraded version of AlphaGo, later known as AlphaGo Master. This allowed human Go players to experience the different playing styles of the upgraded version of AlphaGo. However, this round of competition was a fast-paced Go game, compared to the full-length Go game; due to the short response time, human players are prone to make mistakes in fast-paced games.

From May 23 to 27, 2017, AlphaGo played a match at the Future of Go Summit held in Wuzhen of Zhejiang province, China, with Ke Jie, the world's top-ranked professional Go player at the time. The match was a foregone conclusion, with AlphaGo defeating Ke Jie 3-0. After the match, Google DeepMind announced that this Go summit was the last event AlphaGo participated in and AlphaGo would retire after that. Therefore, it can be said that this was the ultimate showdown between computer Go and humans.

On October 19, 2017, in a research paper (Silver et al. 2017a) published in the journal *Nature*, Google's DeepMind introduced a new version of the computer Go system called AlphaGo Zero, where Zero means learning from scratch. With only knowledge of the rules of Go and without any human Go data or human guidance, it defeated all previous versions of AlphaGo after several days of self-learning training using its Monte Carlo tree search (MCTS) and deep reinforcement learning methods. That is, after 3 days of training, it defeated AlphaGo Lee, after 21 days of training, it defeated AlphaGo Master, and after 40 days of training, it defeated all previous versions, including AlphaGo Master.

On December 5, 2017, Google's DeepMind company submitted a paper on AlphaZero (Silver et al. 2017b) to *arXiv*, an open-access repository of electronic preprints and postprints. AlphaZero's algorithm is similar to AlphaGo Zero, but it not only plays Go but also chess and Japanese shogi.

Japanese shogi, also known as Japanese chess, is a popular board game in Japan, said to have evolved from Chinese chess during the Tang dynasty.

AlphaZero is also a type that can start from scratch, using the general reinforcement learning algorithm proposed in AlphaGo Zero, to self-play reinforcement learning, and can surpass the human level in various board games.

AlphaZero's record is that it defeated the strongest chess program Stockfish after only 4 hours of training, defeated the strongest Shogi program Elmo after 2 hours of

---

[8] https://www.foxwq.com/

Table 1.2 AlphaGo all versions and AlphaZero

Version	When	Match
AlphaGo Fan	Oct 2015	5:0 Fan Hui
AlphaGo Lee	Mar 2016	4:1 against Lee Sedol
AlphaGo Master	Dec 2016–Jan 2017	60:0 professional players
AlphaGo Ke	May 2017	3:0 Ke Jie
AlphaGo Zero	Oct 2017	100:0 AlphaGo Lee (trained 3 days)
		89:11 AlphaGo Master (trained 21 days)
		Exceeded all old versions (trained 40 days)
AlphaZero	Dec 2017	Master Chess, Shogi, and Go:
		Defeated strongest Chess bot Stockfish (trained 4 hours)
		Defeated strongest Shogi bot Elmo (trained 2 hours)
		Defeated strongest Go bot AlphaGo Zero (trained 34 hours)

training, defeated AlphaGo Lee after 8 hours of training, and defeated the strongest
Go program AlphaGo Zero after 34 hours of training.

AlphaGo all versions and AlphaZero are depicted in Table 1.2.

On December 7, 2018, one year after the birth of AlphaZero, it was published in
the *Science* as a cover paper, titled "A General Reinforcement Learning Algorithm
that Masters Chess, Shogi, and Go through Self-Play" (Silver et al. 2018).

1.4.5.2   Card Games

Card games, including Texas Hold'em and Bridge, are also other important topics
studied by the community of artificial intelligence and machine learning. Among
them, Texas Hold'em has made great progress, and here are three representative AI
Texas Hold'em systems.

In December 2016, in a heads-up no-limit Texas Hold'em match, the artificial
intelligence poker DeepStack developed by the team of Professor Michael Bowling
of the University of Alberta, Canada, defeated 11 professional poker players. This
was the first time that a poker bot defeated human players in the card game with
imperfect information.

Subsequently, the paper "DeepStack: Expert-level Artificial Intelligence in
Heads-Up No-Limit Poker" with Professor Michael Bowling as the corresponding
author was published in the *Science* magazine in March 2017 (Moravčík et al. 2017).

From January 11 to 31, 2017, the poker bot from Carnegie Mellon University
participated in a heads-up no-limit Texas Hold'em match. After more than 20 days
of competition, it defeated four top human Texas Hold'em players. This poker bot
was named Libratus, developed by Professor Tuomas Sandholm and Noam Brown,
then a doctoral student, at Carnegie Mellon University. After winning, Professor
Sandholm said that this was the first time that artificial intelligence had defeated the
best human players in heads-up no-limit Texas Hold'em.

From April 6 to 10, 2017, Libratus, translated as Lengpudashi (cold poker master) in Chinese, competed against the China Dragon team in Hainan, China, and won a prize of two million yuan with a total score of 792,327.

The paper "Safe and Nested Subgame Solving for Imperfect-Information Games" by Dr. Brown and Professor Sandholm of Carnegie Mellon University (Brown and Sandholm 2017a) won the Best Paper Award at the *Conference on Neural Information Processing Systems* (NIPS) held in Long Beach, USA, in December 2017. In addition, their paper "Superhuman AI for Heads-Up No-Limit Poker: Libratus Beats Top Professionals" was also published in the *Science* magazine in December 2017 (Brown and Sandholm 2017b).

It is worth noting tha: Go, chess, and shogi belong to board games, where the pieces are placed on the board and can be seen by the two players, which are perfect information games, while Texas Hold'em belongs to card games, where some cards cannot be seen by the opponent, which is an imperfect information game. Therefore, it can be said that the significance of the victories of those poker bots, DeepStack and Libratus, is no less than that of the computer Go, AlphaGo.

Subsequently, Carnegie Mellon University collaborated with Facebook to create an AI poker called Pluribus, which defeated five top human players in the complex game of six-player no-limit Texas Hold'em. Pluribus, from one-on-one no-limit to one-to-many no-limit, seems to just involve more people, but there is a fundamental difference. What's even more shocking is that this version of AI poker was trained from scratch in just 8 days, with a training cost of only $150.

On July 11, 2019, both *Nature* and *Science* magazines published a news and a paper on this. That is, *Nature* published a news article titled "No limit: AI poker bot is first to beat professionals at multiplayer game" (Heaven 2019), while *Science* published a research paper titled "Superhuman AI for Multiplayer Poker," authored by Dr. Brown and Professor Sandholm, the inventors of Pluribus (Brown and Sandholm 2019).

1.4.5.3 Tile-Based Games

The most difficult game in tile-based games is Mahjong, which, like poker, is a game of imperfect information and multiple players.

At the *World Artificial Intelligence Conference* held in Shanghai on August 29, 2019, Harry Shum (Shen Xiangyang), then Executive Vice President of Microsoft, officially announced that Microsoft Research Asia has developed the most powerful Mahjong AI in history—Suphx, which became the first Mahjong AI to be promoted to the tenth level on the internationally renowned professional Mahjong platform *Tenhou*. This was the best achievement in the field of Mahjong at the time, surpassing 99.9% of human players.

On March 30, 2020, Microsoft Research Asia submitted a paper titled "Suphx: Mastering Mahjong with Deep Reinforcement Learning" (Li et al. 2020) to the electronic preprint repository *arXiv*, revealing some technical details of Suphx.

1.4.5.4   Video Games

Video games based on artificial intelligence and machine learning can be traced back to Google DeepMind's AI video game system built on deep reinforcement learning, which can reach the level of human control of video game consoles. This achievement was published in the journal *Nature* (Mnih et al. 2015).

Their development was carried out on an Atari video game console produced in the late 1970s to early 1980s. The developers tested 49 video games on this AI video game system, and the result was that about 59.18% reached or exceeded the human level (Mnih et al. 2013).

OpenAI Five is the AI system that plays the five-on-five video game Dota 2, an upgraded version of OpenAI's victory over top professional player Dendi in a one-on-one Dota 2 match in 2017. Dota 2 is a multiplayer online competitive video game, and the five-on-five team battle is a type of Dota 2 event, with two teams of five players each. On April 13, 2019, OpenAI Five Finals defeated the previous TI8 International champion team "OG" with a score of 2:0.

It's worth noting that Dota 2's five-on-five team battle is not only a game of imperfect information and multiple players but also a complex real-time strategy game, which is of extraordinary significance.

In January 2019, AlphaStar, a computer game program developed by DeepMind, defeated two human professional players in the real-time strategy game StarCraft II, one of whom was a top player ranked in the top ten in the world.

In December 2020, DeepMind announced MuZero and published a paper in the journal *Nature* titled "Mastering Atari, Go, Chess and Shogi by Planning with a Learned Model" (Schrittwieser et al. 2020). In addition to the three board games, MuZero also "stir-fried" the video game Atari.

1.4.6   Autonomous Technologies

Autonomous technologies are those that can function and execute tasks without being controlled by a human. They are designed to operate independently and adapt to changing environments and scenarios. Some of the most common autonomous technologies include self-driving vehicles, drones, and robots.

1.4.6.1   Self-Driving Vehicles

Self-driving vehicles are also known as driverless cars.

As early as 2005, Stanley, an autonomous vehicle developed by Stanford University, won the 2005 *DARPA Grand Challenge* for autonomous vehicles.

DARPA, the *Defense Advanced Research Projects Agency*, is an administrative agency under the US Department of Defense, established in 1958, responsible for developing high tech for military use.

The head of Stanford University's autonomous vehicle development team was Sebastian Thrun, then director of Stanford's Artificial Intelligence Lab. He joined Google in 2009 and became the head of its autonomous vehicle project. At the end of 2016, he and Anthony Levandowski founded Waymo, an autonomous vehicle company, which became a subsidiary of Alphabet. As is well known, Alphabet is the parent company of Google and several subsidiaries, established after Google's reorganization in October 2015.

In addition, car companies such as BMW, Nissan, Ford, General Motors, Tesla, and Mercedes-Benz have also conducted research on autonomous vehicles.

Baidu's autonomous vehicle project started in 2013, and domestic companies such as Alibaba and Tencent have also entered the field of autonomous vehicles.

1.4.6.2 Drones

Drones are another example of autonomous technology. They are unmanned aerial vehicles that can be remotely controlled or programmed to fly autonomously. Drones have a wide range of applications, including aerial photography, surveying, search and rescue operations, and package delivery.

The combination of the technologies of drones and vehicles has led to the emergence of flying cars and flying car drones, which are futuristic modes of transportation that are being developed to help create a faster, cheaper, cleaner, safer, and more integrated transportation system.

Flying cars are essentially road-legal cars that can also fly. They are designed to be driven on the road like a regular car and can take off and land vertically like a helicopter.

On the other hand, flying car drones, also known as passenger drones, are unmanned aerial vehicles (UAVs) that can carry passengers. They are designed to transport people in city centers to avoid traffic congestion.

1.4.6.3 Robots

Robots are also a form of autonomous technology. They can be programmed to perform a wide range of tasks, from manufacturing and assembly line work to cleaning and maintenance.

On June 27, 2019, the *Nature* magazine on its cover page under the title "Flying Solo" introduced the lightest autonomous flying robot in history, weighing only 259 milligrams, which can achieve sustained flight with solar power alone. The robot, a four-winged bee robot named RoboBee X-Wing, was developed by the Harvard University Microrobotics Lab led by Professor Robert J. Wood (Jafferis et al. 2019).

1.4.7   Machine Learning in Medicine

Medicine is one of the important application fields of machine learning.

1.4.7.1   Medical Diagnosis

In 2016, *The Journal of the American Medical Association* published an article on the diagnosis of diabetic retinopathy using deep learning (Gulshan et al. 2016).

In 2017, the journal *Nature* published a paper on the classification of skin cancer using deep neural networks (Esteva et al. 2017).

Additionally, after the journal *Nature* added the *Nature Biomedical Engineering*, one of Nature Portfolio journals, in 2017, it published several papers related to medical artificial intelligence, for example, rapid intraoperative histology of brain tumors (Orringer et al. 2017), an artificial intelligence platform for managing congenital cataracts (Long et al. 2017), as well as a deep learning method for predicting cardiovascular disease risk based on retinal fundus photos (Poplin et al. 2018).

In March 2018, nearly 150 scientists from over 100 laboratories in the United States, Germany, Italy, and other countries collaborated to publish an article in *Nature*, announcing their developed AI system. Based on the methylation data of tumor tissue DNA, it can accurately distinguish nearly 100 different central nervous system tumors. In addition, this AI system can also discover new classifications not found in clinical guidelines (Wong and Yip 2018).

In June 2020, a paper titled "Clinically Applicable AI System for Accurate Diagnosis, Quantitative Measurements and Prognosis of COVID-19 Pneumonia Using Computed Tomography" was published in the journal *Cell* (Zhang et al. 2020). The authors are 45 scientists from Macau, mainland China, and Hong Kong, with the first and corresponding author being Professor Zhang Kang from Macau University of Science and Technology.

1.4.7.2   Drug Synthesis

On March 29, 2018, *Nature* published online a research paper by Professor Mark P. Waller's team from Shanghai University, "Planning Chemical Syntheses with Deep Neural Networks and Symbolic AI," which uses deep learning and symbolic artificial intelligence for retrosynthetic analysis to design drugs (Segler et al. 2018).

Retrosynthetic analysis is the reverse process of small molecule forward reaction prediction. The American organic chemist, Elias Corey, first proposed the concept of retrosynthetic analysis in the 1960s, and he won the *Nobel Prize in Chemistry* for this in 1990.

Waller's team proposes three neural networks combined with Monte Carlo tree search (3N-MCTS) to perform chemical synthesis planning. Their research results are a breakthrough in the current field of chemical synthesis, especially for compound synthesis and drug synthesis.

1.4.8 Machine Learning in Biology

Machine learning has been making strides in solving biological problems.

1.4.8.1 Protein Structure Determination

Google's DeepMind company has made significant breakthroughs in solving a 50-year problem in the field of biology. In the 14th Critical Assessment of Structure Prediction (CASP) in November 2020, the company's AlphaFold 2 protein structure prediction achieved the best score of 92.4 in the *Global Distance Test* (GDT), surpassing the level of complex laboratory instruments, and is said to have achieved unparalleled accuracy.

On December 30, 2020, *Nature* titled its report "It will change everything: DeepMind's AI makes gigantic leap in solving protein structures,"[9] providing a detailed account of the news. The subtitle reads, "Google's deep-learning program for determining the 3D shapes of proteins stands to transform biology, say scientists."

For the past 50 years, the "protein folding problem" has been a major challenge in the field of biology. The cost of determining protein structures with complex laboratory equipment is high, often taking several years to depict the shape of a protein, and success is not guaranteed. However, AlphaFold 2 only takes a few days. The information contained in protein structures carries life information, and the breakthrough in this problem will strongly promote the development of life sciences and greatly accelerate the discovery of new drugs for difficult diseases.

1.4.8.2 Ribonucleic Acid Structure Determination

For half a century, determining the three-dimensional structure of biological molecules has puzzled scientists and has been one of the major challenges in biology. Because ribonucleic acid (RNA) presents as a complex folded three-dimensional structure, it is difficult to determine through experiments or calculations. The known RNA structures are less than 1% of the known protein structures.

On August 27, 2021, the journal *Science* published a cover paper titled "Geometric Deep Learning of RNA Structure" (Townshend et al. 2021). The paper has seven authors, with the main contributors being Raphael Townshend, then a

---

[9] https://www.nature.com/articles/d41586-020-03348-4

doctoral student, at Stanford University, and Stephan Eismann, under the guidance of Associate Professor Ron Dror.[10]

They developed a neural network called the atomic rotationally equivariant scorer (ARES) for predicting RNA structures. With only limited data for training, it can accurately predict the three-dimensional shapes of drug targets and other important biological molecules. Moreover, because it only uses atomic coordinates as input to the deep neural network and does not include RNA-specific information, its method is applicable to various problems in structural biology, chemistry, materials science, and other fields. The study achieved the best results in the community-wide blind RNA structure prediction challenges.

1.4.9 Others

There are also a lot of other applications of machine learning, such as sentiment analysis, fraud detection, spam filtering, recommender systems, time-series forecasting, and cybersecurity.

1.5 The Too-Small World

It is worth noting that in recent years, including machine learning, artificial intelligence has entered its heyday. This is because both the number of international journal or conference papers and the number of participants in international conferences are growing exponentially.

Since 2017, the Institute for Human-Centered Artificial Intelligence (HAI) at Stanford University has been publishing the *AI Index Report*[11] annually, which tracks, collates, distills, and visualizes data relating to artificial intelligence, including journal articles, conference papers, and repositories.

However, some researchers of machine learning and artificial intelligence in their heyday have also begun to think calmly. On December 16, 2019, Sergii Shelpuk, CEO and cofounder of a startup called DeepTrait that is engaged in biotechnology and artificial intelligence, published an article on the *Towards AI* website titled "The Too-small World of Artificial Intelligence," with the subtitle "Overcrowded and overlooked parts of the AI world" (Shelpuk 2019).

Machine learning and artificial intelligence are sciences, and science cannot tolerate bubbles; machine learning and artificial intelligence are also technologies, and only those that can be applied to real life and bring economic or social benefits can be called technologies.

---

[10] https://news.stanford.edu/2021/08/26/ai-algorithm-solves-structural-biology-challenges/

[11] https://aiindex.stanford.edu/report/

Further Reading

1. David Silver, Aja Huang, Chris Maddison, et al. "Mastering the Game of Go with Deep Neural Networks and Tree Search." *Nature*, 529(7587), 2016.
   [Notes] This is a paper published by Google DeepMind, written after defeating the European Go champion Fan Hui in December 2015. AlphaGo combines deep learning, reinforcement learning, and Monte Carlo tree search, allowing computer Go to defeat human professional Go players for the first time.
2. David Silver, Julian Schrittwieser, Karen Simonyan, et al. "Mastering the Game of Go without Human Knowledge." *Nature*, 550(7676), 2017.
   [Notes] This paper is published by Google DeepMind. It introduces AlphaGo Zero, an algorithm entirely based on reinforcement learning, which, apart from the rules of the game of Go, requires no human data, human guidance, or other domain knowledge. AlphaGo Zero defeated all previous versions of AlphaGo that had defeated human Go champions.
3. Noam Brown and Tuomas Sandholm, "Superhuman AI for Heads-Up No-Limit Poker: Libratus Beats Top Professionals." *Science*, Dec. 2017.
   [Notes] This paper was published by Dr. Noam Brown and Professor Tuomas Sandholm of Carnegie Mellon University. No-Limit Texas Hold'em is a popular form of poker, which, unlike Go, is a game of imperfect information and is more challenging in some ways. The paper introduces several game-solving methods in Libratus: blueprint strategy, nested safe subgame, and self-improvement.
4. Noam Brown and Tuomas Sandholm. "Superhuman AI for Multiplayer Poker." *Science*, Aug. 2019.
   [Notes] This paper is the second one published by Dr. Noam Brown and Professor Tuomas Sandholm. The paper describes how Carnegie Mellon University and companies like Facebook jointly created a poker bot called Pluribus, which defeated five top human players in the complex game of six-player no-limit Texas Hold'em. This version of AI poker, without human knowledge, started from scratch and was trained for only 8 days, with a training cost of only 150 dollars.
5. Julian Schrittwieser, Ioannis Antonoglou, Thomas Hubert, Karen Simonyan, Laurent Sifre, Simon Schmitt, Arthur Guez, et al. "Mastering Atari, Go, Chess and Shogi by Planning with a Learned Model." *Nature*, 588(7839), 2020.
   [Notes] Google DeepMind announced MuZero in December 2020 and published the paper. As can be inferred from the title of the paper, MuZero encompasses not only three board games but also Atari video games.
6. Jumper, John, Richard Evans, Alexander Pritzel, Tim Green, Michael Figurnov, Olaf Ronneberger, Kathryn Tunyasuvunakool, et al. "Highly Accurate Protein Structure Prediction with AlphaFold." *Nature*, 596(7873), 2021.
   [Notes] This is a paper published by Google DeepMind in August 2021. AlphaFold solved the protein structure prediction problem that has plagued the biological community for 50 years.

References

Ackley, D. H., G. E. Hinton, and T. J. Sejnowski. (1985). A learning algorithm for boltzmann machines. *Cognitive Science* 9(1): 147–169.

Aizenberg, I., N. N. Aizenberg, and J. P. L. Vandewalle. (2000). *Multi-valued and universal binary neurons: Theory, learning and applications.* Berlin: Springer Science & Business Media.

Alpaydin, E. (2009). *Introduction to machine learning* (2nd Edn.). Cambridge: MIT Press.

Anthony, M. (2001). *Discrete mathematics of neural networks: selected topics.* Society for Industrial and Applied Mathematics.

Bengio, Y., Y. Lecun, and G. Hinton. (2021). Deep learning for AI. *Communications of the ACM* 64(7): 58–65.

Ben-Hur, A., D. Horn, H. T. Siegelmann, and V. Vapnik. (2000). A support vector clustering method. In *International Conference on Pattern Recognition (ICPR).*

Boser, B. E., I. M. Guyon, and V. N. Vapnik. (1992). A training algorithm for optimal margin classifiers. In *Proceedings of the fifth annual workshop on Computational learning theory.*

Bredensteiner, E. J., and K. P. Bennett. (1999). Multicategory classification by support vector machines. *Computational Optimization and Applications* 1(12): 53–79.

Breiman, L. (2001). Random forests. *Machine Learning* 45(1): 5–32.

Brown, N., and T. Sandholm. (2017a). Safe and nested subgame solving for imperfect-information games. In *Conference on Neural Information Processing Systems (NIPS).*

Brown, N., and T. Sandholm. (2017b). Superhuman AI for heads-up no-limit poker: Libratus beats top professionals. *Science* 359(6374): 418–424.

Brown, N., and T. Sandholm. (2019). Superhuman AI for multiplayer poker. *Science* 365(6456): 885–890.

Casey, R., and G. Nagy. (1984). *Decision tree design using a probabilistic model.* Piscataway: IEEE Press.

Chang, R. L. P., and T. Pavlidis. (1977). Fuzzy decision tree algorithms. *IEEE Transactions on Systems Man & Cybernetics* 7(1): 28–35.

Cortes, C., and V. Vapnik. (1995). Support vector networks. *Machine Learning* 20(3): 273–297.

Dattatreya, G., and V. Sarma. (1981). Bayesian and decision tree approaches for pattern recognition including feature measurement costs. *IEEE Transactions on Pattern Analysis and Machine Intelligence* 3(3): 293–298.

Dechter, R. (1986). Learning while searching in constraint-satisfaction problems. In *Fifth AAAI National Conference on Artificial Intelligence (AAAI-86).*

Drucker, H., C. J. Burges, L. Kaufman, A. Smola, and V. Vapnik. (1996). Support vector regression machines. In *Conference on Neural Information Processing Systems (NIPS).*

Esteva, A., B. Kuprel, R. A. Novoa, J. Ko, S. M. Swetter, H. M. Blau, and S. Thrun. (2017). Corrigendum: dermatologist-level classification of skin cancer with deep neural networks. *Nature* 542(7639): 115.

Farley, B. G., and W. A. Clark. (1954). Simulation of self-organizing systems by digital computer. *IEEE Transactions on Information Theory* 4(4): 76–84.

Flach, P. (2012). Machine learning: The art and science of algorithms that make sense of data. Cambridge: Cambridge University Press.

Freund, Y., and R. E. Schapire. (1995). A decision-theoretic generalization of on-line learning and an application to boosting. In *Computational Learning Theory: Second European Conference, EuroCOLT '95.*

Fukushima, K. (1988). Neocognitron: A hierarchical neural network capable of visual pattern recognition. *Neural Networks* 1(2): 119–130.

Fukushima, K., and S. Miyake. (1982). Neocognitron: A new algorithm for pattern recognition tolerant of deformations and shifts in position. *Pattern Recognition* 15(6): 455–469.

Goodfellow, I. J., J. Pouget-Abadie, M. Mirza, B. Xu, D. Warde-Farley, S. Ozair, A. Courville, and Y. Bengio. (2014). Generative adversarial nets. *Conference on Neural Information Processing Systems (NIPS)* 3: 2672–2680.

Gulshan, V., L. Peng, M. Coram, M. C. Stumpe, D. Wu, A. Narayanaswamy, S. Venugopalan, K. Widner, T. Madams, and J. Cuadros. (2016). Development and validation of a deep learning algorithm for detection of diabetic retinopathy in retinal fundus photographs. *The Journal of the American Medical Association (JAMA)* 316(22): 2402–2410.

He, K., X. Zhang, S. Ren, and J. Sun. (2016). Deep residual learning for image recognition. In *Proceedings of the IEEE Conference on Computer Vision and Pattern Recognition (CVPR)*.

Heaven, D. (2019). No limit: AI poker bot is first to beat professionals at multiplayer game. *Nature* 571: 307–308.

Hebb, D. O. (1949). The organization of behaviour. *Journal of Applied Behavior Analysis* 25(3): 575–577.

Hinton, G., S. Osindero, and Y.-W. Teh. (2006). A fast learning algorithm for deep belief nets. *Neural Computation* 18(7): 1527–1554.

Hinton, G. E., and R. R. Salakhutdinov. (2006). Reducing the dimensionality of data with neural networks. *Science* 313(5786): 504–507.

Hinton, G. E., and T. J. Sejnowski. (1983). *Optimal Perceptual Inference*. Citeseer.

Ho, T. K. (1995). Random decision forests. In *International Conference on Document Analysis and Recognition*.

Ho, T. K. (1998). The random subspace method for constructing decision forests. *IEEE Transactions on Pattern Analysis and Machine Intelligence (TPAMI)* 20(8): 832–844.

Hochreiter, S., and J. Schmidhuber. (1997). Long short-term memory. *Neural Computation* 9(8): 1735–1780.

Hopfield, J. J. (1982). Neural networks and physical systems with emergent collective computational abilities. In *Proceedings of the National Academy of Sciences*. 79(8): 2554–2558.

Hsu, C.-W., and C.-J. Lin. (2002). A simple decomposition method for support vector machines. *Machine Learning* 46: 291–314.

Hunt, E. B. (1962). *Concept learning: An information processing problem*. New York: Wiley.

Jafferis, N. T., E. F. Helbling, M. Karpelson, and R. J. Wood. (2019). Untethered flight of an insect-sized flapping-wing microscale aerial vehicle. *Nature* 570(7762): 491–495.

James, G., D. Witten, T. Hastie, and R. Tibshirani. (2013). An introduction to statistical learning: With applications in R. New York: Springer.

Kaelbling, L. P., M. L. Littman, and A. W. Moore. (1996). Reinforcement learning: A survey. *Journal of Artificial Intelligence Research* 4(1): 237–285.

Kearns, M. (1988). Thoughts on hypothesis boosting. Technical report, Machine Learning Class Project.

Kontschieder, P., M. Fiterau, A. Criminisi, and S. R. Bulo. (2015). Deep neural decision forests. In *IEEE International Conference on Computer Vision (ICCV)*.

Kosiorek, A., S. Sabour, Y. W. Teh, and G. E. Hinton. (2019). Stacked capsule autoencoders. In *Conference on Neural Information Processing Systems (NeurIPS)*.

Krizhevsky, A., I. Sutskever, and G. E. Hinton. (2012). Imagenet classification with deep convolutional neural networks. In *Conference on Neural Information Processing Systems (NIPS)*.

LeCun, Y. (1987). Modeles Connexionnistes de L'apprentissage (Connectionist Learning Models).

LeCun, Y., L. Bottou, Y. Bengio, and P. Haffner. (1998). Gradient-based learning applied to document recognition. *Proceedings of the IEEE* 86(11): 2278–2324.

LeCun, Y., Y. Bengio, and G. Hinton. (2015). Deep learning. *Nature* 521(7553): 436.

Li, J., S. Koyamada, Q. Ye, G. Liu, C. Wang, R. Yang, L. Zhao, T. Qin, T.-Y. Liu, and H.-W. Hon. (2020). Suphx: Mastering mahjong with deep reinforcement learning. arXiv preprint arXiv:2003.13590.

Linnainmaa, S. (1970). The representation of the cumulative rounding error of an algorithm as a taylor expansion of the local rounding errors (in Finnish). Master, University of Helsinki.

Liu, Y., M. Ott, N. Goyal, J. Du, M. Joshi, D. Chen, O. Levy, M. Lewis, L. Zettlemoyer, and V. Stoyanov. (2019). RoBERTa: A robustly optimized BERT pretraining approach. arXiv preprint arXiv:1907.11692.

Long, E., H. Lin, Z. Liu, X. Wu, L. Wang, J. Jiang, Y. An, Z. Lin, X. Li, and J. Chen. (2017). An artificial intelligence platform for the multihospital collaborative management of congenital cataracts. *Nature Biomedical Engineering* 1(2): 0024.

Mallick, B. K., D. Ghosh, and M. Ghosh. (2005). Bayesian classification of tumours by using gene expression data. *Journal of the Royal Statistical Society* 67(2): 219–234.

Mason, L., J. Baxter, P. Bartlett, and M. Frean. (1999). Boosting algorithms as gradient descent. In *Conference on Neural Information Processing Systems (NIPS)*.

McCarthy, J., and E. Feigenbaum. (1990). Arthur Samuel: Pioneer in machine learning. *AI Magazine* 11(3): 10–11.

McCulloch, W. S., and W. Pitts. (1943). A logical calculus of the ideas immanent in nervous activity. *The Bulletin of Mathematical Biophysics* 5(4): 115–133.

Meisel, W. S., and D. A. Michalopoulos. (1973). A partitioning algorithm with application in pattern classification and the optimization of decision trees. *IEEE Transactions on Computers* C-22(1): 93–103.

Mendel, J. M., and R. W. Mclaren. (1970). Reinforcement learning control and pattern recognition systems. *Mathematics in Science and Engineering* 66: 287–318.

Minsky, M. L. (1954). *Theory of neural-analog reinforcement systems and its application to the brain model problem*. Princeton: Princeton University.

Minsky, M., and S. Papert. (1969). *Perceptrons: An introduction to computational geometry*. Cambridge: The MIT Press.

Mitchell, T. M. (1997). *Machine learning*. New York: McGraw-Hill Science.

Mnih, V., K. Kavukcuoglu, D. Silver, A. Graves, I. Antonoglou, D. Wierstra, and M. Riedmiller. (2013). Playing atari with deep reinforcement learning. arXiv preprint arXiv:1312.5602.

Mnih, V., K. Kavukcuoglu, D. Silver, A. A. Rusu, J. Veness, M. G. Bellemare, A. Graves, M. Riedmiller, A. K. Fidjeland, and G. Ostrovski. (2015). Human-level control through deep reinforcement learning. *Nature* 518(7540): 529.

Mohri, M., A. Rostamizadeh, and T. Ameet. (2012). *Foundations of machine learning*. Cambridge: MIT Press.

Moravčík, M., M. Schmid, N. Burch, V. Lisý, D. Morrill, N. Bard, T. Davis, K. Waugh, M. Johanson, and M. Bowling. (2017). Deepstack: Expert-level artificial intelligence in heads-up no-limit poker. *Science* 356(6337): 508.

Orringer, D. A., B. Pandian, Y. S. Niknafs, T. C. Hollon, J. Boyle, S. Lewis, M. Garrard, S. L. Herveyjumper, H. J. L. Garton, and C. O. Maher. (2017). Rapid intraoperative histology of unprocessed surgical specimens via fibre-laser-based stimulated Raman scattering microscopy. *Nature Biomedical Engineering* 1(2): 0027.

Polson, N. G., and S. L. Scott. (2011). Data augmentation for support vector machines. *Bayesian Analysis* 6(1): 1–23.

Poplin, R., A. V. Varadarajan, K. Blumer, Y. Liu, M. V. Mcconnell, G. S. Corrado, L. Peng, and D. R. Webster. (2018). Prediction of cardiovascular risk factors from retinal fundus photographs via deep learning. *Nature Biomedical Engineering* 2(2): 158–64.

Quinlan, J. R. (1986). Induction on decision tree. *Machine Learning* 1(1): 81–106.

Ronneberger, O., P. Fischer, and T. Brox. (2015). U-Net: Convolutional networks for biomedical image segmentation. In *International Conference on Medical Image Computing and Computer-Assisted Intervention, 2015*. Berlin: Springer.

Rosenbaltt, F. (1957). The perceptron: A perciving and recognizing automation. Cornell Aeronautical Laboratory, Ithaca, Tech. Rep.

Rosenblatt, F. (1958). The perceptron: A probabilistic model for information storage and organization in the brain. *Psychological Review* 65(6): 386–408.

Rumelhart, D. E., G. E. Hinton, and R. J. Williams. (1985). *Learning internal representations by error propagation*. California: Univ San Diego La Jolla Inst for Cognitive Science.

Rumelhart, D. E., G. E. Hinton, and R. J. Williams. (1986). Learning representations by back-propagating errors. *Nature* 323(6088): 533–536.

Sabour, S., N. Frosst, and G. E. Hinton. (2017). Dynamic routing between capsules. In *Conference on Neural Information Processing Systems (NIPS)*.

Samuel, A. (1959). Some studies in machine learning using the game of checkers. *IBM Journal of Research and Development* 3(3): 210–229.

Schapire, R. E. (1990). The strength of weak learnability. *Symposium on Foundations of Computer Science*.

Schapire, R. E., Y. Freund, P. Bartlett, and W. S. Lee. (1998). Boosting the margin: A new explanation for the effectiveness of voting methods. *The Annals of Statistics* 26(5): 1651–1686.

Schrittwieser, J., I. Antonoglou, T. Hubert, K. Simonyan, L. Sifre, S. Schmitt, A. Guez, E. Lockhart, D. Hassabis, and T. Graepel. (2020). Mastering atari, Go, chess and Shogi by planning with a learned model. *Nature* 588(7839): 604–609.

Segler, M. H., M. Preuss, and M. P. Waller. (2018). Planning chemical syntheses with deep neural networks and symbolic AI. *Nature* 555(7698): 604–610.

Sethi, I. K. (1990). Entropy nets: From decision trees to neural networks. *Proceedings of the IEEE* 78(10): 1605–1613.

Shelpuk, S. (2019). The too-small world of artificial intelligence: Overcrowded and overlooked parts of the AI world.

Silver, D., G. Lever, N. Heess, T. Degris, D. Wierstra and M. Riedmiller. (2014). Deterministic policy gradient algorithms. In *International Conference on Machine Learning (ICML)*.

Silver, D., A. Huang, C. J. Maddison, A. Guez, L. Sifre, d. D. G. Van, J. Schrittwieser, I. Antonoglou, V. Panneershelvam, and M. Lanctot. (2016). Mastering the game of Go with deep neural networks and tree search. *Nature* 529(7587): 484–489.

Silver, D., T. Hubert, J. Schrittwieser, I. Antonoglou, M. Lai, A. Guez, M. Lanctot, L. Sifre, D. Kumaran, and T. Graepel. (2017a). Mastering Chess and Shogi by Self-Play with a General Reinforcement Learning Algorithm. arXiv Preprint arXiv:1712.01815.

Silver, D., J. Schrittwieser, K. Simonyan, I. Antonoglou, A. Huang, A. Guez, T. Hubert, L. Baker, M. Lai, A. Bolton, Y. Chen, T. Lillicrap, F. Hui, L. Sifre, G. van den Driessche, T. Graepel, and D. Hassabis. (2017b) Mastering the game of Go without human knowledge. *Nature* 550(7676): 354.

Silver, D., T. Hubert, J. Schrittwieser, I. Antonoglou, M. Lai, A. Guez, and M. Lanctot. (2018). A general reinforcement learning algorithm that masters chess, shogi, and Go through self-play. *Science* 362(6419): 1140–1144.

Simon, H. A. (1983). Why should machines learn? In R. Michalski, J. Carbonell, & T. Mitchell (Eds.), *Machine Learning*: 25–37. Berlin: Springer.

Sutton, R. S. (1984). *Temporal credit assignment in reinforcement learning*, University of Massachusetts Amherst.

Sutton, R. S., and A. G. Barto. (2018). *Reinforcement learning: An introduction*. Cambridge: The MIT Press.

Townshend, R. J., S. Eismann, A. M. Watkins, R. Rangan, M. Karelina, R. Das, and R. O. Dror. (2021). Geometric deep learning of RNA structure. *Science*, 373(6558): 1047–1051.

Turing, A. M. (1950). Computing machinery and intelligence. *Mind* 59(236): 433–460.

Utgoff, P. E. (1989). Incremental induction of decision trees. *Machine Learning* 4: 161–186.

Vapnik, V. N. (1998). *Statistical learning theory*. New York: Wiley.

Vapnik, V. N. (1999). *The nature of statistical learning theory* (2nd Edition). New York: Springer.

Vaswani, A., N. Shazeer, N. Parmar, J. Uszkoreit, L. Jones, A. N. Gomez, L. Kaiser, and I. Polosukhin. (2017). Attention is all you need. In *Conference on Neural Information Processing Systems (NIPS)*.

Waltz, M. D., and K.-s. Fu. (1965). A heuristic approach to reinforcement learning control systems. *IEEE Transactions on Automatic Control* 10(4): 390–398.

Wang, W. (2019). *Principles of artificial intelligence (Chinese edition)*. China: Higher Education Press.

Ware, W. H. (1955). Introduction to session on learning machines. In *Western Joint Computer Conference*.

Werbos, P. (1974). Beyond Regression: New Tools for Prediction and Analysis in The Behavioral Sciences. PhD thesis, Committee on Applied Mathematics, Harvard University, Cambridge, MA.

Widrow, B., and M. E. Hoff. (1988). *Adaptive switching circuits*. Cambridge: MIT Press.

Winston, P. (1969). *A heuristic program that constructs decision trees*. AI Memo No.173. Cambridge: Artificial Intelligence Lab Publications, MIT.

Wong, D., and S. Yip. (2018). Machine learning classifies cancer. *Nature* 555(7697): 446–447.

Zhang, K., X. Liu, J. Shen, Z. Li, Y. Sang, X. Wu, Y. Zha, W. Liang, C. Wang, and K. Wang. (2020). Clinically applicable AI system for accurate diagnosis, quantitative measurements, and prognosis of Covid-19 pneumonia using computed tomography. *Cell* 181(6): 1423–1433.

Chapter 2
On Perspectives

Abstract This chapter comes to show the three perspectives on machine learning, i.e., the subtitle of this book. We start from the discussion between learning and perspectives. Next, we review the foundations of machine learning such as mathematics, logic, information theory, decision theory, and neuroscience. Discussed on the difficulties in studying machine learning, we propose the viewpoints on machine learning. We then focus on the three perspectives: learning frameworks, learning paradigms, and learning tasks, which corresponds to six learning frameworks, three main machine learning paradigms, and seven representative machine learning tasks, respectively. Finally, we give the logical and hierarchical illustrations between the three perspectives.

2.1 Learning and Perspectives

Learning and perspectives may seem like two unrelated behaviors, but what we want to discuss are the perspectives on studying machine learning, so they have a close relationship with each other.

This section first discusses the general concepts of learning, introduces learning theories in education and machine learning, and then discusses the perspectives on studying machine learning.

2.1.1 Learning

Learning refers to the process of acquiring knowledge or skills.

The term "learning" in Chinese includes two concepts, "learning" and "practicing," which are related but distinct: "learning" is the process of acquiring knowledge or skills, while "practicing" is the process of consolidating knowledge or improving skills.

Research on learning can be divided into two types: narrow and broad. Narrow learning refers to the process of acquiring knowledge or skills through means such

W. Wang, *Principles of Machine Learning*,
https://doi.org/10.1007/978-981-97-5333-8_2

as textbooks, while broad learning also includes the process of gaining experience or skills in life.

Learning styles are a series of theories aimed at explaining individual learning differences. These theories believe that all people can be classified according to their learning styles, although different theories define and classify learning styles differently. A common concept is that different people have different learning styles.

The fields to study the nature and process of learning cover educational psychology, neuropsychology, experimental psychology, and pedagogy. Learning is also a key constituent part of artificial intelligence, known as machine learning.

Looking back at history, the term "learning machines" was used in the famous paper "Computing Machinery and Intelligence" by Turing in 1950. After about a decade, the paper published by Arthur Samuel in 1959 entitled "Some Studies in Machine Learning Using the Game of Checkers" (Samuel 1959) made machine learning (ML) a popular term, which can also be thought the opening year of machine learning.

2.1.2 Learning Theory

German-American psychologist Kurt Lewin (1890–1947) once said, "There is nothing more practical than a good theory."

A theory is a universal principle derived from the abstraction and induction of things, which has an important guiding and promoting effect on practice and is often the cornerstone of innovation and creation.

Learning theory is a topic studied in both education and machine learning.

The learning theory in education mainly studies the basic principles of human learning. It includes four aspects:

(1) Based on behavior analysis, it studies how to absorb, process, and retain knowledge in the learning process.
(2) Based on cognition, it studies the impact of cognition, emotion, and environment on learning.
(3) Based on constructivism, it studies how learners actively participate in knowledge construction.
(4) Based on transfer learning, it studies how to learn new knowledge based on existing experience.

The learning theory in machine learning is to demonstrate the basic principles of learning from the perspective of computational process models, where the machine refers to the agent composed of computer software and hardware.

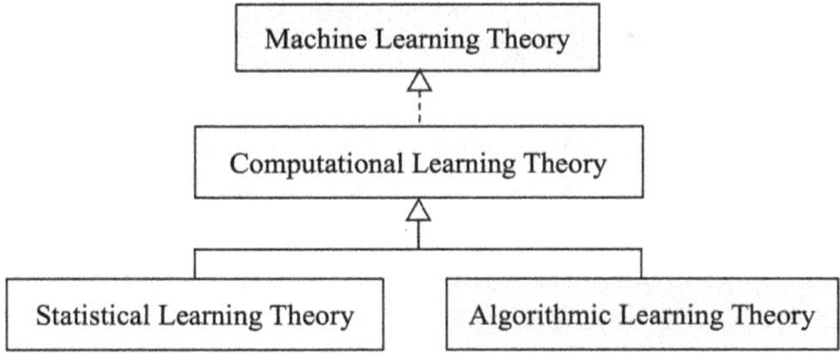

Fig. 2.1 Logical relationship diagram of machine learning theories

2.1.3 Machine Learning Theories

The theories of machine learning mainly include three types, namely, computational learning theory,[1] statistical learning theory,[2] and algorithmic learning theory.[3]

The differences among the three machine learning theories mentioned above are as follows:

- Computational learning theory:
  a formalized learning theory based on probability theory
- Statistical learning theory:
  a machine learning framework based on statistics
- Algorithmic learning theory:
  a mathematical framework for analyzing machine learning problems and algorithms

Just as computational theory is considered the theoretical foundation of computer science, computational learning theory can be seen as the theoretical foundation of machine learning, while statistical learning theory and algorithmic learning theory are considered the branches of computational learning theory. Because computational learning theory uses probability theory as a tool and takes supervised learning as an example, it theoretically explains why machines can learn based on existing experience. The logical relationship between the three is shown in Fig. 2.1.

This book will discuss computational learning theory in Chap. 3, while statistical learning theory will be covered in Chap. 4.

In August 1988, the first *Workshop on Computational Learning Theory* (COLT) was held in Cambridge, Massachusetts, USA. This annual academic conference was later renamed the *Conference on Computational Learning Theory* (COLT) and the *Conference on Learning Theory* (COLT) after that, but the acronym of the conference name has always remained the same. The changes in the conference

---

[1] https://en.wikipedia.org/wiki/Computational_learning_theory

[2] https://en.wikipedia.org/wiki/Statistical_learning_theory

[3] https://en.wikipedia.org/wiki/Algorithmic_learning_theory

name reflect the expansion of its scale and the shift of its theme from "computational learning theory" to "learning theory."

Based on computational learning theory, several representative theoretical frameworks of machine learning have been formed. In addition to statistical learning, machine learning has also drawn nutrients from other disciplines such as probability theory, logic, information theory, neuroscience, and behavioral science. Looking at the massive and various types of algorithms that have emerged in the more than sixty years since the birth of machine learning, it is necessary to have systematic collation, organization, and explanation for machine learning from multiple perspectives and in all directions.

2.1.4 The Perspectives

As mentioned earlier, perspectives have a close relationship with the angle and dimension of looking at a problem. One angle can only see one side, and one dimension can only see one line. Only from multiple angles and dimensions can we see the full picture of things.

For machine learning, studying it from different perspectives can bring it closer to its essence.

It is not uncommon for books to explain machine learning from a certain perspective. In 2011, Stephen Marsland published a textbook titled *Machine Learning: An Algorithm Perspective* (Marsland 2011). In 2012, Kevin P. Murphy published a textbook called *Machine Learning: A Probabilistic Perspective* (Murphy 2012). And, in 2015, Sergios Theodoridis published a textbook named as *Machine Learning: A Bayesian and Optimization Perspective* (Theodoridis 2015).

This book attempts to expound machine learning from three perspectives, namely, theoretical, methodological, and practical perspectives. After introducing the basic knowledge of machine learning in the next section, this chapter will elaborate on its arguments.

2.2 Foundations of Machine Learning

Machine learning is an interdisciplinary field; its foundations include mathematics, logic, information theory, decision theory, and neuroscience. On top of these foundational disciplines, several theoretical frameworks of machine learning have been formed. Based on these frameworks, various algorithms of machine learning and even deep learning have been constructed and applied in many fields.

Below, these foundational disciplines of machine learning are discussed separately.

2.2.1 Mathematics

The mathematics used in machine learning mainly includes probability theory, statistics, linear algebra, calculus, geometry, and functional analysis.

2.2.1.1 Probability Theory

Probability theory is a mathematical theory about random phenomena and uncertain problems, used to study the inherent laws in these phenomena and problems and to analyze the possibility of various outcomes using mathematical methods.

The term "probability" can be traced back to the research on gambling by the Italian polymath Gerolamo Cardano (1501–1576). He referred to the possible outcomes of a game event as probabilities and explained the basic concept of probability through dice gambling. In 1564, he wrote a book in Italian titled *Liber de ludo aleae* (Book on Games of Chance); it is regrettable that his book was not published until 1663, nearly a hundred years after his death. His book is considered the first monograph on probability theory.

In 1654, two French mathematicians, Blaise Pascal (1623–1662) and Pierre de Fermat (1607–1665), studied the "problem of points"[4] in gambling through correspondence. The so-called problem of points, also known as "division of the stakes," is how to distribute the profits when the game ends prematurely for two players with the same stake. They are considered the founders of probability theory because Cardano's book had not been published at that time.

In 1657, Dutch mathematician Christiaan Huygens (1629–1695) published his monograph *De Ratiociniis in Ludo Aleae* (On Reasoning in Games of Chance), which is the first mathematical work on games of chance. The book expanded the concept of expected value and added rules on how to calculate expectations in more complex situations than the original problem and is considered the first successful attempt to lay the foundation of probability theory.

Swiss mathematician Jacob Bernoulli (1654–1705) spent more than twenty years studying the law of large numbers (LLN), proving its validity under the condition of binary random variables, which is a special form of the law of large numbers. He published it in his monograph *Ars Conjtandi* (The Art of Conjecturing) in 1713, naming it the golden theorem, which later generations called Bernoulli's theorem.

In 1733, French mathematician Abraham de Moivre (1667–1754) first used the normal distribution to estimate the distribution of the number of heads in multiple coin tosses in his paper. His work laid the foundation for the definition and proof of the central limit theorem.

British mathematician Thomas Bayes (1701–1761), while studying how to calculate the probability distribution of binomial distribution parameters, found

---

[4] https://en.wikipedia.org/wiki/Problem_of_points

that the posterior probability of an event can be calculated based on the prior and likelihood related to the event and wrote it into a manuscript. His friend, British mathematician Richard Price (1723–1791), organized the manuscript after Bayes' death and published Bayes's major work in 1763 as "An Essay towards Solving a Problem in the Doctrine of Chances" in the *Philosophical Transactions*, where it contains Bayes' theorem.

The theoretical foundation of the *Bayesian school* of probability theory is Bayes' theorem.

French mathematician Pierre-Simon Laplace (1749–1827) published his French-written monograph *Theorie Analytique des Probabilites* (Analytic Theory of Probability) in 1812, which systematically summarized the work of his predecessors, extended de Moivre's normal distribution for estimating binomial distribution, and introduced statistical analysis methods into probability theory, pushing probability theory to a new stage of development.

In 1837, French mathematician Siméon Denis Poisson (1781–1840) refined the law of large numbers, thereby promoting its widespread use. Subsequently, several mathematicians also contributed to the law of large numbers, including Russian mathematicians Pafnuty Chebyshev (1821–1894) and Andrey Markov (1856–1922), French mathematician Émile Borel (1871–1956), Italian mathematician Francesco Paolo Cantelli (1875–1966), and Soviet mathematicians Aleksandr Khinchin (1894–1959) and Andrey Kolmogorov (1903–1987). Among them, Khinchin's proof is known as the weak law of large numbers (weak LLN), while Kolmogorov's proof is known as the strong law of large numbers (strong LLN). The law of large numbers is the first theorem in probability theory, which states that the probability of a random event is approximately equal to its expected value.

In 1901, Russian mathematician Aleksandr Lyapunov (1857–1918) defined the central limit theorem (CLT) in general terms and proved it mathematically.

The law of large numbers and the central limit theorem became the theoretical foundation of the *frequentist school* of probability theory.

With the continuous development and improvement of probability theory, a complete mathematical system has been formed, which is widely used in statistics, science, engineering, economics, game theory, computer science, artificial intelligence, and machine learning.

2.2.1.2   Statistics

Statistics is a discipline about the collection, organization, representation, analysis, interpretation, and prediction based on data. Mathematical statistics is the application of probability theory in statistics.

In 1589, Italian scholar Girolamo Ghilini (1589–1668) proposed the term statistics in an unpublished paper.

In 1662, British demographers John Graunt (1620–1674) and William Petty (1623–1687) proposed early census and statistical methods. By the eighteenth century, the term statistics referred to the systematic collection of population and economic data.

The mathematical foundation of statistics was developed from the research of some mathematicians on games of chance, including Gerolamo Cardano, Blaise Pascal, Pierre de Fermat, Christiaan Huygens, and Jacob Bernoulli, who have been introduced in the previous subsection.

From the late nineteenth to the early twentieth century, statistics went through the following three stages of development:

First stage	It was at the turn of the eighteenth and nineteenth centuries, represented by British polymath Francis Galton (1822–1911) and British mathematician Karl Pearson (1857–1936), who turned statistics into a rigorous mathematics. Galton's contributions include the concepts of standard deviation, correlation, and regression analysis, while Pearson proposed product-moment correlation, used for fitting sample distribution, Pearson distribution, and so on.
Second stage	It was from the 1910s to the 1920s, initiated by British statistician William S. Gosset (1876–1937) and reached its peak with the insights of British polymath Ronald Fisher (1890–1962). Fisher almost built the foundation of modern statistics; he first used the term variance in statistics, published *Statistical Methods for Research Workers* in 1925 and *The Design of Experiments* in 1930. He also created the statistical term null hypothesis.
Third stage	It was the 1930s, starting with the collaboration of British statistician Egon Pearson (1895–1980) and Polish mathematician Jerzy Neyman (1894–1981). They improved and expanded the early work of statistics and proposed the concepts of type II error, power of a test, and confidence intervals. They also proposed the famous Neyman-Pearson lemma in 1933, which became the theoretical basis for hypothesis testing.

With the birth and development of probability theory, statistics has a solid theoretical foundation. Therefore, some terms in statistics are consistent with probability theory, such as variables in statistics are called random variables in probability theory. In addition, statistics has some field-specific terms: the data to be statistically analyzed is called observation or observed data, the collection of all data is called population, and the part drawn from the population is called sample.

The continuous development of statistics has formed two main paths, namely, descriptive statistics and inferential statistics. Descriptive statistics targets the samples and performs the calculation, description, and analysis through numerical calculations or visualization methods. Inferential statistics starts from the sample, infers the potential probability distribution characteristics of the sample, and then obtains the properties of the population.

If the terms population and sample are used to describe the relationship between probability and statistics, probability theory infers the probability of the sample from the population, while statistics inducts the properties of the population from the sample.

Because statistics is particularly suitable for making accurate inferences from organized data and making decisions based on statistical methods when facing uncertainty, its application field is very wide, and statistical learning has also become one of the theoretical frameworks of machine learning.

2.2.1.3   Linear Algebra

Linear algebra is a branch of mathematics that studies systems of linear equations and their linear mappings. It involves the operations of scalars, vectors, matrices, and even tensors. Unlike *discrete mathematics*, a branch of mathematics that is concerned with "discrete" mathematical structures, linear algebra is a *"continuous" mathematics*.

Linear algebra can be used to model many natural phenomena and problems and effectively operate on the model. Therefore, it is used in many scientific and engineering fields and naturally becomes one of the foundational subjects of mathematics in machine learning.

Some machine learning algorithms, such as principal component analysis (PCA), can be derived using basic linear algebra knowledge.

2.2.1.4   Calculus

Calculus is the study of differential calculus and integral calculus and related concepts and applications.

Differential calculus mainly studies the derivative of functions, providing a set of theories about the instantaneous rate of change and slope of curves, which can be used to calculate the rate of change of physical quantities such as speed and acceleration.

Integral calculus mainly studies the combination of minimal data in a given interval, providing a universal method for defining and calculating displacement, area, and volume.

For example, differential calculus is used in gradient descent calculations in machine learning, and integral calculus is used when calculating the expected value of continuous random variables.

2.2.1.5   Geometry

Geometry is a branch of mathematics that studies shapes, sizes, relative positions of figures, and spatial properties. Its basic concepts include points, lines, surfaces, angles, distances, and manifolds.

There are many branches in geometry, including Euclidean geometry, non-Euclidean geometry, differential geometry, algebraic geometry, and topology. Among them, non-Euclidean geometry is further divided into elliptic geometry, hyperbolic geometry, and so on.

This book considers geometry as one of the mathematical foundations of machine learning due to two main reasons:

- Some representation for data requires geometry, such as representing data with manifolds or point clouds.
- Some processing for data requires geometry, such as using hyperplane for data classification, using Euclidean distance for data clustering, and using manifold for dimensionality reduction.

Since 2016, there has been an emerging field called geometric deep learning, which focuses on developing deep learning models for structured data with geometric structures (Masci et al. 2016; Bronstein et al. 2017). These geometric structures include graphs, point clouds, grids, geodesics, and manifolds in Euclidean and non-Euclidean geometry.

Geometric deep learning provides a general framework from the perspective of geometric symmetry, representability, and invariance and is a general term for a large class of deep learning methods based on geometry. Based on first principles, geometric deep learning has derived several geometric neural network architectures, such as convolutional neural networks (CNNs), recurrent neural networks (RNNs), graph neural networks (GNNs), and Transformers.

2.2.1.6 Functional Analysis

Functional analysis is a branch of mathematical analysis that studies the transformation of functions and their algebraic and topological properties.

One example of functional analysis particularly relevant to machine learning is the study of kernel methods. In kernel support vector machines, the theory of reproducing kernel Hilbert space in functional analysis plays a significant role.

2.2.2 Logic

Logic is the systematic study of the form of reasoning used to draw correct conclusions. That is, it studies the relationship that leads to the acceptance of another proposition (conclusion) based on a series of propositions (premises). Broadly speaking, logic is the analysis and evaluation of arguments.

There are mainly two types of logic, namely, informal logic and formal logic.

The concept of formal logic can be traced back to the famous syllogism of the ancient Greek philosopher Aristotle, also known as syllogistic logic.

In formal logic, the logical system represented by symbols is called symbolic logic, which began with George Boole (1815–1864). In 1847, he introduced propositional logic and proposed Boolean algebra based on it. Propositional logic is also known as zeroth-order logic.

Subsequently, Augustus De Morgan (1806–1871) introduced De Morgan's laws, also known as De Morgan's rules, into propositional logic and Boolean algebra.

In the years following 1870, Charles S. Peirce (1839–1914) developed Boole's logic system, adding relations and quantifiers.

In 1879, Gottlob Frege (1848–1925) proposed first-order logic, which is an extension of propositional logic, adding objects and relations.

Alfred Tarski (1902–1983) proposed the causal theory of reference, which reveals how to associate objects in logic with objects in the real world.

In 1929, Kurt Gödel (1906–1978) proved the completeness theorem in his doctoral thesis, thereby establishing the correspondence between syntax and semantics in first-order logic.

Formal logic continues to develop and enrich, and some researchers study formal logic systems from a mathematical perspective, namely, mathematical logic. This includes modal logic, fuzzy logic, and description logic.

The manifestation of symbolic logic is symbols, which led to the birth of symbolism. Symbolism views the objective world as a combination of symbols, representing information employing symbols and their relationships, and completes human cognition through the storage, extraction, reasoning, and transformation of symbols. Specific algorithms are used to process these symbols to solve problems and derive new knowledge.

With the continuous development of machine learning, logic will undoubtedly play a more important role in machine learning.

Currently, propositional logic, first-order predicate logic, and fuzzy logic are commonly used in machine learning.

2.2.3 Information Theory

Information theory is a research field for the quantification, storage, and communication of digital information and is also an interdisciplinary field of probability theory, statistics, computer science, and information engineering.

In the 1920s, Harry Nyquist (1889–1976) and Ralph Hartley (1888–1970) studied the ability of communication systems to transmit information at Bell Labs, proposed the quantification of information, and are considered the precursors of information theory.

In the early 1940s, when Claude Shannon came to Bell Labs after his doctorate, Nyquist and Hartley were the academic leaders of Bell Labs. In 1948, Claude Shannon published a paper titled "A Mathematical Theory of Communication" in the *Bell Labs Technical Journal*, which pioneered information theory and is known as the father of information theory.

Entropy, the average amount of information, is a key indicator in information theory, quantifying the uncertain information involved in the values of random variables or the results of random processes. Many concepts in information theory, such as joint entropy, conditional entropy, cross entropy, mutual information, and Kullback-Leibler divergence, have played important roles in machine learning and its application fields.

2.2.4 Decision Theory

In psychology, decision-making is considered a cognitive process, a process of judgment and selection based on the values and beliefs of the decision-maker. The result is a judgment among several possibilities, choosing a belief or a course of action. The decision-making process can be rational or irrational.

Decision-making is an important skill for the subject to continuously learn and improve behavior in the process of interaction with the external environment. The concept of decision theory has been widely applied in behavioral science, psychology, and neuroscience experiments and their computational models.

Some researchers have conducted in-depth research on decision-making from the following three views:

(1) Psychological views:
    examining the decision-making process in the context of a series of needs and values that the subject possesses or seeks.
(2) Cognitive views:
    the decision-making process is considered a continuous process of interaction with the environment.
(3) Normative views:
    the analysis of decision-maker involves the logic or communicative rationality of decision-making and the invariant choices it leads to.

A major part of decision-making involves analyzing a finite set of alternatives described according to evaluation criteria. This task is to rank these alternatives according to their attractiveness to the decision-maker when considering all criteria simultaneously. In addition, it is also to find the best alternative or determine the relative overall priority of each alternative when considering all criteria simultaneously.

The environment of the decision-maker often plays an important role in the decision-making process, and the complexity of the environment is an important factor affecting the cognitive function of the decision-maker.

Decision theory is a theory studying the agent's choice. Decision theory is a formal framework for choosing the optimal or near-optimal behavior in an uncertain environment.

The foundational disciplines of decision theory are probability theory and utility theory, used to obtain the probability distribution of possible results of an action in any given state and the rationality function of the results. By defining the utility function of the results, the behavior of an intelligent agent has a higher expected utility. The task of the decision-maker is simple, that is, to choose the decision plan with the maximum expected utility (MEU).

The two-dimensional representation of multistage decisions in decision theory has become the theoretical basis for machine learning methods such as decision trees and decision forests.

Decision theory also provides a theoretical basis for behaviorism. Behaviorism simulates human intelligent activities and behavioral characteristics in the control process, that is, self-optimization, self-adaptation, and self-learning. Behaviorism is also a model of the interaction between an intelligent agent and its environment, which has had a significant impact on decision-theoretic planning and reinforcement learning. Reinforcement learning is a machine learning approach to sequential decision based on decision theory.

In the field of machine learning, the intelligent agent decides the next action based on the state of each step and the feedback from the environment each time, this process is called sequential decision, and its theoretical basis is decision theory. Sequential decision can be seen as procedural decision-making or step-by-step decision-making.

2.2.5 *Neuroscience*

Neuroscience is the science that studies the nervous system (brain, spinal cord, and peripheral nerves) and its functions. It combines physiology, anatomy, molecular biology, developmental biology, cytology, psychology, and mathematical modeling to understand the characteristics of neurons and neural circuits.

The neuron is the basic unit of the brain's nervous system. A neuron is a nerve cell with dendrites, axons, and synapses, and the cell itself is composed of a soma (cell body) and a nucleus. Each neuron's dendrites can receive stimuli and transmit this signal into the cell body, the axon transmits the signal from the cell body to the synapse, and then, the synapse transmits the signal to another neuron or other tissues.

Some researchers start by studying the nervous system to study how the brain processes information. They organically combine behavior, cognition, and brain mechanisms to try to comprehensively reveal the information processing process and its neural mechanisms of humans and animals in perceiving objects, forming representations, using language, remembering information, and reasoning in decision-making.

Brain science is one of the main research contents of neuroscience. The brain is very superior in rational decision-making, although it is not perfect. The brain has decision-making functions, and prediction and inference are part of decision-making.

Connectionism is an approach to the study of human cognition with neural network theory at its core, advocates establishing its computational model by simulating the structure and function of biological neural networks. Artificial neural networks (ANNs) can be seen as an artificial manifestation of the brain, used to simulate the human learning process. Connectionism has many manifestations in practice, but the most common is the use of artificial neural networks.

The perceptron is the earliest artificial neural network; it only has one layer of neurons, so it is also called a single-layer perceptron. Later, the number of layers of neurons increased, forming a multilayer perceptron (MLP), that is, in addition to the input and output layers, there are several intermediate hidden layers. Artificial neural networks with multiple hidden layers are called deep neural networks (DNNs).

Deep learning is a subfield of machine learning based on deep neural networks.

2.3   Difficulties in Studying Machine Learning

Let's temporarily put aside the large number of obscure mathematical formulas in machine learning and discuss where are the difficulties in studying machine learning.

Since its inception in 1959, machine learning has undergone more than sixty years of development, with too many machine learning algorithms to count or even estimate how many machine learning algorithms there are. Moreover, numerous machine learning algorithms emerge every year, every month, and even every day.

After we studied several machine learning algorithms, we may find that we still haven't gotten out of the maze of machine learning, because we may feel that machine learning is just a few isolated islands of algorithms, and we are still unclear about the internal relationship of these algorithms, unable to grasp the overall picture of machine learning. It is like what the famous poem said:

Don't know the true color of Mount Lu,
Because we are just in this mountain sight.

It can be considered that there are several common difficulties in the process of studying machine learning.

2.3.1   Different Problems

Machine learning is to learn from data or the environment. As for data, there are different modalities, such as text, image, audio, or video. The purpose of learning from data, that is, the output result, depends on the modalities of data and the task to be solved. As for learning from the environment, it is necessary to know the current state and the feedback rewards from the environment and how to decide the next action. These different modalities of data, tasks, or environments, as well as different output results, present a diversity of problems.

Therefore, we need to collate and systematize these problems to understand how to solve them with machine learning, in short, "what to do."

The so-called collating and systematizing are to find out the common problems in machine learning and ignore unnecessary details. For example, classification in machine learning can be text classification, image classification, or speech classification. By collating and systematizing, we can remove the details related to the domain and focus on how to solve the substantive problem of classification.

2.3.2   *Different Methods*

The so-called different methods refer to different machine learning algorithms, some of which can use a large amount of labeled data to train the machine learning model in order to make predictions on the actual unknown data, some do not have labeled data but only need to focus on discovering similarities in the data, and some can decide the next action based on the current state and feedback reward from the environment, and so on.

Therefore, it is necessary to analyze what methods these machine learning algorithms belong to, in other words, "what type."

2.3.3   *Different Theories*

Even for the same problem and the same method, there are many machine learning algorithms to consider or choose from. It is necessary to analyze the characteristics and internal structures of these algorithms, which is called "how to do."

The previous section of this chapter introduced the foundational disciplines of machine learning, including various mathematics, logic, information theory, decision theory, and neuroscience. With these foundational disciplines as the framework, many machine learning algorithms and models have been formed. Therefore, they should be collated and systematized.

2.4   Viewpoints on Machine Learning

This section will propose the viewpoints of three perspectives for studying machine learning, from the three key points of what to do, what type, and how to do it.

The chain of thought of the three perspectives on machine learning is shown in Fig. 2.2.

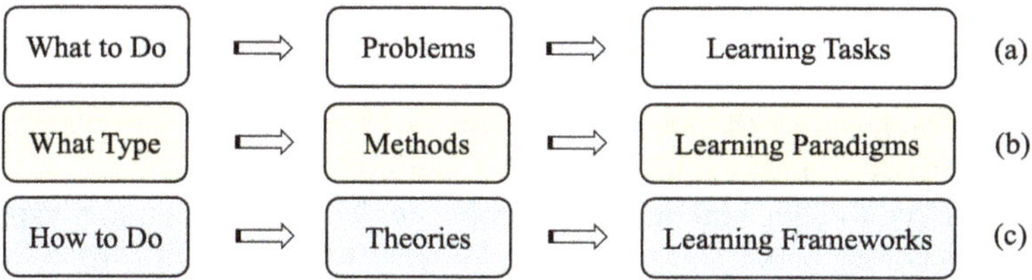

Fig. 2.2 Viewpoints of the three perspectives

2.4.1 *What to Do*

The first thing to consider in machine learning is "what to do," involving what "problems" to solve with machine learning, what kind of processing to do on the input data, what the output form is, and what action needs to be taken on the current state and reward. From this, we can derive the task that machine learning needs to complete, referred to as "learning tasks," as shown in Fig. 2.2a.

2.4.2 *What Type*

The next thing to consider is "what type" a machine learning algorithm belongs to, as shown in Fig. 2.2b. It is related to the "methods" taken by machine learning algorithms and also needs to consider which machine learning model to use, how to improve the ability of prediction, how to improve the accuracy of analysis, and how to interact with the environment. From this, we can derive the paradigms that machine learning needs to adopt, referred to as "learning paradigms."

2.4.3 *How to Do*

After determining the tasks and paradigms of machine learning, it also needs to consider "how to do" for the machine learning algorithm, as shown in Fig. 2.2c. That is, how to use suitable theories to design machine learning algorithms, from which the theoretical framework that machine learning algorithms depend on is derived, is referred to as "learning frameworks."

2.5 Perspectives on Machine Learning

Through the analysis of the previous section, we have derived the viewpoints of three perspectives of machine learning, as shown in Fig. 2.3.

That is to say, the machine learning can be studied from the three perspectives, i.e., the frameworks, paradigms, and tasks. For ease of memory, it is represented as a triple, namely, $ML = \langle \mathcal{F}, \mathcal{P}, \mathcal{T} \rangle$, where $\mathcal{F}$ represents the collection of learning frameworks, $\mathcal{P}$ represents the collection of learning paradigms, and $\mathcal{T}$ represents the collection of learning tasks.

The brief descriptions of the learning frameworks, learning paradigms, and learning tasks are shown in Table 2.1.

The following three subsections will elaborate on these three perspectives.

2.5.1 Learning Frameworks

Definition 2.1 (Learning Frameworks) A learning framework refers to a theoretical framework for the design of machine learning algorithms. It belongs to the theory level of machine learning rather than the data level or task level, and it is also different from the software framework used to build machine learning algorithms.

See Table 2.2; this book abstracts representative machine learning frameworks into six categories, namely, probabilistic framework, statistical framework, geometric framework, connectionist framework, symbolic framework, and behavioral framework. Each framework contains several sub-frameworks, each of which corresponds to several algorithms.

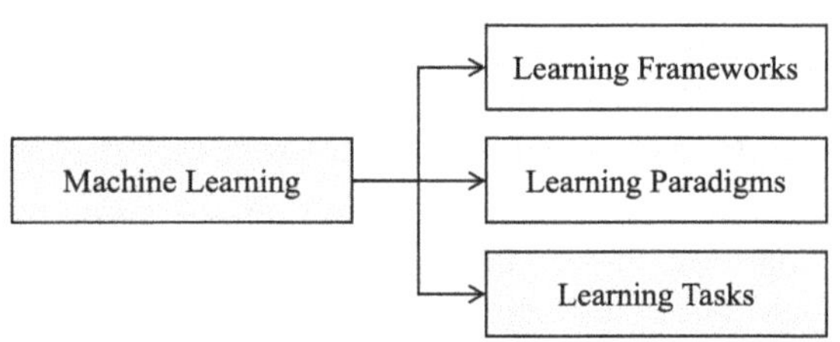

Fig. 2.3 The three perspectives on machine learning

Table 2.1 Meanings of the three perspectives

Perspective	Brief description
Learning frameworks	Represent the theoretical frameworks for the design of machine learning algorithms
Learning paradigms	Represent the patterns or styles taken by machine learning algorithms
Learning tasks	Indicating the basic problems that can be solved with machine learning

Table 2.2 Representative frameworks for machine learning

Framework	Brief description	Sub-framework
Probabilistic	Using probability to represent the conditional correlation of random variables	Generative
		Discriminative
Statistical	Using statistics for constructing machine learning algorithms	Parametric
		Nonparametric
Geometric	Using Euclidean or non-Euclidean geometry for data or algorithms	Hyperplane
		Manifold
		Point cloud
Connectionist	Based on the connectionist model of artificial neural networks	Shallow/deep
		Forward/backward
Symbolic	Based on the physical symbolic system	Logic based
		Rule based
		Causality based
Behavioral	Based on the behavioral decision theory	Multistep
		Sequential

(1) The probabilistic framework:

It is based on probability to represent the conditional correlations between random variables. Its core is probabilistic learning theory, especially the computational learning theory. It mainly includes two sub-frameworks, namely, generative models and discriminative models. The typical algorithm of generative models is Bayesian network, and the typical algorithm of discriminative models is conditional random field.

(2) The statistical framework:

It is based on statistics to construct machine learning algorithms. It can be further divided into two sub-frameworks: parametric models and nonparametric models. The typical algorithm of parametric models is linear regression, and the typical algorithm of nonparametric models is k-nearest neighbors.

(3) The geometric framework:

It is based on Euclidean geometry or non-Euclidean geometry as the representation for datasets or algorithms of machine learning. It can be further divided into hyperplane, manifold, and point cloud. The typical algorithm based on hyperplane is support vector machine (SVM), the typical algorithm based on the manifold is Isomap, and the application field based on point cloud is 3D modeling and so on.

(4) The connectionist framework:

It uses artificial neural networks as the connectionist model to mimic the brain's nervous system. The framework can be further divided into two main aspects, namely, shallow networks vs. deep networks, as well as feedforward networks vs. backward networks. The typical shallow network is perceptron, the typical feedforward deep network is convolutional neural network (CNN), and the typical backward network is recurrent neural network (RNN).

(5) The symbolic framework:

It uses the physical symbolic system such as logic, rules, or other symbolic processing methods to construct learning algorithms, so it mainly includes the three sub-frameworks: logic-based models, rule-based models, and causality-based models. The typical models are inductive logic, association rules, and causality learning, respectively.

(6) The behavioral framework:

It is based on behavioral decision theory that mainly includes three sub-frameworks: single-stage decision, multistage decision, and sequential decision. The typical algorithm of multistage decision is decision tree, and the typical algorithm of sequential decision is Q-learning.

More details about the learning frameworks will be discussed in Part II of this book.

2.5.2 *Learning Paradigms*

At the time of writing this book, upon investigation, the term "learning paradigms" can be found in similar statements in learning theory which is one of the theories of education. The learning theory in education describes how students receive, process, and retain knowledge during the learning process. The relevant description of learning paradigms in the learning theory is as follows:

Learning theories are usually divided into several paradigms, which represent different perspectives on the learning process. Theories within the same paradigm share the same basic point of view[5]

Similarly, machine learning is also divided into several paradigms. This book gives its definition as follows.

Definition 2.2 (Learning Paradigms) A learning paradigm in machine learning refers to the pattern or style taken by learning algorithms. It is a division of the paradigm of machine learning, independent of specific applications. The main basis to distinguish a learning paradigm is how to learn from data or how to interact with the environment.

In practice, the design of a machine learning algorithm may consider different paradigms, such as whether labeled data is needed or whether there is a dynamic interaction process between the intelligent agent and the environment.

---

[5] https://www.learning-theories.org/doku.php?id=learning_paradigms

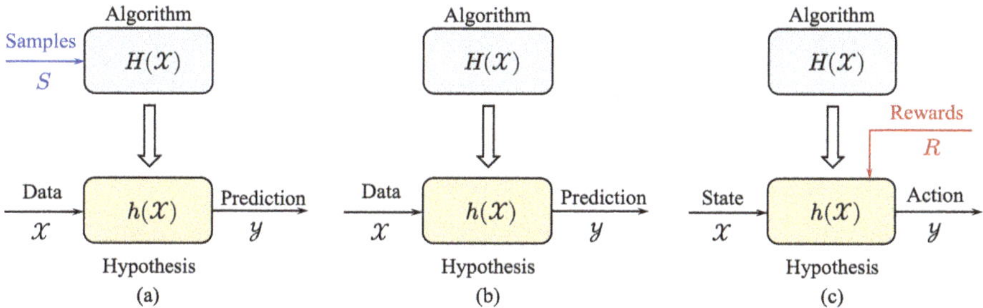

Fig. 2.4 Comprehensive schematic diagram of typical machine learning paradigms

The importance of studying learning paradigms lies in:

- Firstly, understanding various machine learning algorithms from the perspective of paradigms, thereby deepening the understanding of machine learning at the methodological level
- Secondly, analyzing the machine learning tasks that need to be solved from the perspective of learning paradigms, in order to better design machine learning algorithms and achieve the best machine learning results

The comprehensive schematic diagram of typical machine learning paradigms is shown in Fig. 2.4. The so-called comprehensive refers to the representation of the working principles of three typical machine learning paradigms, namely, supervised learning, unsupervised learning, and reinforcement learning, in a graphical way, intending to facilitate comprehensive comparative analysis.

In Fig. 2.4, $\mathcal{X}$ represents the input space, $\mathcal{Y}$ represents the output space, S represents the training samples, R represents the rewards, $H(\mathcal{X})$ represents the learning algorithm, also known as the hypothesis set, and $h(\mathcal{X})$ is one of the hypotheses in it, that is, $h \in H$.

Below, we analyze the working principles of the three machine learning paradigms shown in Fig. 2.4 from a macro viewpoint.

These three paradigms all require the design of a machine learning algorithm that satisfies the paradigm, that is, the hypothesis set $H : \mathcal{X} \to \mathcal{Y}$. They all aim to obtain an optimal hypothesis $h \in H$, and this hypothesis h is the learner actually used in machine learning, also known as the model of machine learning.

The difference between Fig. 2.4a, b lies in whether there are labeled training samples S in the process of forming the hypothesis, which is the fundamental difference between the supervised learning paradigm and the unsupervised learning paradigm. Figure 2.4c includes environmental feedback rewards R, which belong to the reinforcement learning paradigm.

Based on the above working principles, typical machine learning paradigms can be divided into three types, namely, supervised learning, unsupervised learning, and reinforcement learning. The introduction and representative algorithms of these three paradigms are shown in Table 2.3.

Table 2.3 Representative paradigms for machine learning

Paradigm	Brief description
Supervised	The algorithm H is trained with manually labeled samples S, and the obtained hypothesis h is then used to predict all unknown data
Unsupervised	The algorithm H is trained only with unlabeled data, and the obtained h is then used to process all unknown data
Reinforcement	The algorithm H is optimized to obtain h, and then the next action is determined based on the current state and the reward R of the environment

These three typical machine learning paradigms are also the mainstream views in the machine learning academic community. The purpose of the above analysis in this book is to systematically discuss the individuality and commonality between those three paradigms.

(1) Supervised learning:
It is to train the algorithm with a set of labeled data and then use the trained hypothesis to make predictions on unknown data. The labeled training data is a set of input-output pairs used to train the algorithm, so the labeled data is also called training examples. Supervised learning can be seen as a "teacher-guided learning." One of the typical algorithms is support vector machine (SVM).
(2) Unsupervised learning:
It is to receive unlabeled data for analysis or training and then make predictions on all unknown data. Unsupervised learning is a learning paradigm similar to "self-learning." One of the typical algorithms is k-means.
(3) Reinforcement learning:
It is that the intelligent agent interacts with the external environment and each action receives reward feedback from the environment and then decides the next action based on the current state and reward value. It can be seen as a "online learning." One of the typical algorithms is Q-learning.

The learning paradigms will be detailed in Part III of this book.

2.5.3 Learning Tasks

Definition 2.3 (Learning Tasks) A learning task refers to a basic problem that can be solved by machine learning. It is a common problem abstracted from domain problems, and the algorithms of machine learning with the same mechanism can be used to complete this task.

In the application field of artificial intelligence and machine learning, such as computer vision, pattern recognition, natural language processing, and computer speech, there are many basic and common tasks that need to be solved. For example, identifying whether an email is spam is a binary classification task, and the output is the result of email classification; recognizing handwritten Arabic numerals, i.e., the ten numbers from 0 to 9, is a multi-classification task that divides handwritten Arabic numerals into ten categories; processing various data such as the brand, model, displacement, and mileage of used cars can yield the price curve of used cars (such tasks are called regression); in the process of data analysis, similar or close data are divided into several groups (such tasks are called clustering); data in high-dimensional space is reduced to low-dimensional space without significant changes in its main features (such tasks are called dimensionality reduction).

Through the induction of various basic problems that machine learning can solve, it can be concluded that there are seven main learning tasks in machine learning:

(1) Classification:
    It is based on the labeled samples (known categories) to train the machine learning algorithm, used to identify which class the unknown input data belongs to; its output space is a set of discrete classes. The representative algorithm is the support vector machine (SVM).
(2) Regression:
    It is based on the labeled sample to estimate the regression function of its variable relationship and then used to predict the given input data; its output space is a set of real continuous numbers. The representative algorithm is ridge regression.
(3) Clustering:
    It is based on certain criteria to divide the input data into several clusters, these clusters are unknown in advance, but the data within the same cluster have some similarity. The representative algorithm is k-means.
(4) Ranking:
    It transforms the order of input data and replaces their values with their rankings; its output space is a set with relative order. The representative algorithm is PageRank.
(5) Dimensionality reduction:
    It maps data from high-dimensional space to low-dimensional space while retaining the basic features of the original data. The representative algorithm is principal component analysis (PCA).
(6) Association:
    It is to discover meaningful relationships between data in the input dataset. The representative method is association rule learning.
(7) Decision-making:
    It includes multistage decision and sequential decision. The former divides the decision into several stages, and the representative method is the decision tree; the latter is in the process of interacting with the environment, deciding the next action based on the current state and rewards of the environment, and its representative algorithm is Q-learning.

For ease of comparison, the above seven representative learning tasks are summarized in Table 2.4.

Regarding learning tasks, they will be explained in Part IV of this book in detail.

Table 2.4 Representative tasks for machine learning

Tasks	Brief description
Classification	Dividing input data into known categories
Regression	Outputting continuous rather than discrete value
Clustering	Dividing input into groups unknown in advance
Ranking	Transforming order of input data to their rankings
Dimension reduction	Map high-dimensional data to low-dimensional data
Association	Discovering meaningful relationships between data
Decision	Including multistep and sequential decisions

2.6   Relations Between the Perspectives

Through in-depth analysis of the three perspectives and their related frameworks, paradigms, and tasks, we can find that they contain logical relationships and hierarchical relationships, where the logical relationships can be further divided into the logical relationship of the three perspectives and the logical relationship between paradigms and tasks.

2.6.1   Logical Relationships

Here, unified modeling languages (UML) are used to depict the relationship between machine learning perspectives, mainly using its aggregation, inheritance, and other graphic symbols (Rumbaugh et al. 2004).

2.6.1.1   Logical Relationship of the Perspectives

Figure 2.5 is the logical relationship diagram of the three perspectives.

From this figure, machine learning can be studied from the perspectives of learning framework, learning paradigm, and learning task. Among them:

- The learning frameworks refer to the theoretical framework used for machine learning algorithm design, which has six sub-frameworks: probabilistic framework, statistical framework, geometric framework, connectionist framework, symbolic framework, and behavioral framework.
- The learning paradigms refer to the patterns or styles of machine learning algorithms. There are three academically recognized learning paradigms: supervised learning, unsupervised learning, and reinforcement learning. In addition, some other learning quasi-paradigms have emerged in recent years, which will be detailed in Chap. 11.
- The learning tasks are common problems abstracted from domain problems, which can be solved by machine learning algorithms with the same mechanism.

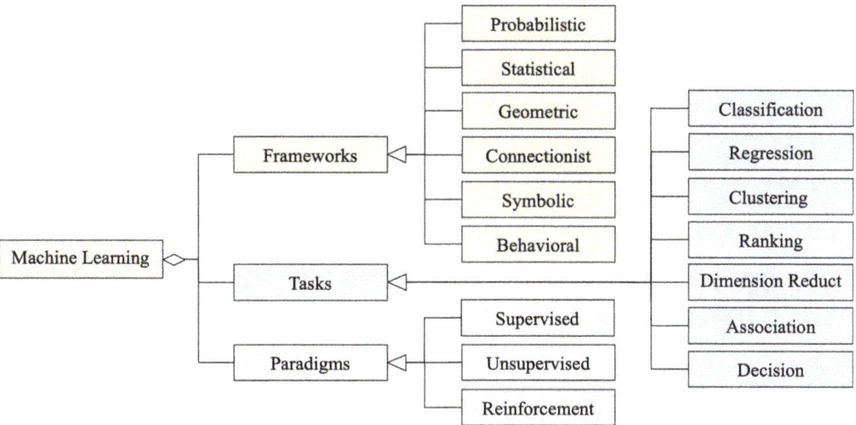

Fig. 2.5 Logical relationship of the three perspectives

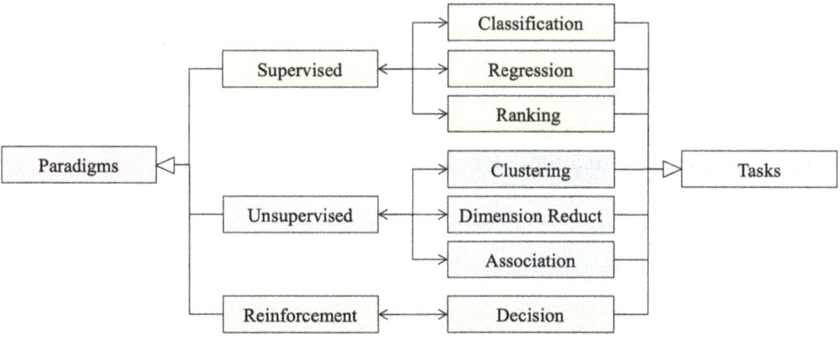

Fig. 2.6 Logical relationship between learning paradigms and tasks

The representative machine learning tasks mainly include seven types: classification, regression, clustering, ranking, dimensionality reduction, association, and decision.

2.6.1.2   Logical Relationship Between Paradigms and Tasks

There is a corresponding logical relationship between the typical learning paradigms and the learning tasks.

The logical relationship diagram between typical learning paradigms and learning tasks is illustrated in Fig. 2.6, represented using unified modeling language (UML). Its purpose is to provide a systematic understanding of the relationship between paradigms and tasks in machine learning.

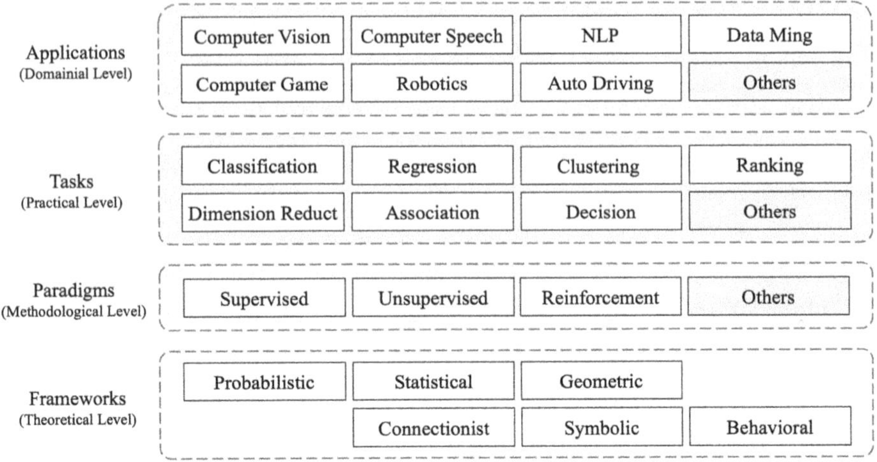

Fig. 2.7 Hierarchical relationship of the three perspectives and applications

From Fig. 2.6, it can be seen that:

- The machine learning tasks corresponding to the supervised learning paradigm mainly include classification, regression, and ranking.
- The machine learning tasks corresponding to the unsupervised learning paradigm mainly include clustering, dimensionality reduction, and association.
- The machine learning task corresponding to the reinforcement learning paradigm is sequential decision.

2.6.2 Hierarchical Relationship

Figure 2.7 is a hierarchical relationship diagram of the three perspectives of machine learning and their applications.

Taking a look at Fig. 2.7 from bottom to top, we can see four levels as below:

Level 1:   Frameworks—it is an abstraction at the theoretical level of machine learning, so it is the theoretical basis of machine learning.

Level 2:   Paradigms—it is an abstraction at the methodological level of machine learning, built on top of the learning framework.

Level 3:   Tasks—it is an abstraction at the practical level of machine learning, built on top of the learning paradigm.

Level 4:   Applications—it is an abstraction at the domain level, built on top of the learning task.

Here, only some major application fields of machine learning are listed, namely, computer vision, computer speech, natural language processing (NLP), data mining, computer games, robotics, and autonomous driving. Some other applications that are not listed fall under the category of "Others."

Further Reading

1. Michael I. Jordan, and Tom M. Mitchell. "Machine Learning: Trends, Perspectives, and Prospects." *Science*, 349(6245), 2015.
   [Notes] The authors of the paper published in the journal *Science* are Michael Jordan and Tom Mitchell. The former is a professor at the University of California, Berkeley, and one of the pioneers of machine learning, while the latter is a professor at Carnegie Mellon University, the former head of the Machine Learning Department and the author of the first textbook *Machine Learning*. The paper introduces the core algorithms of machine learning, the latest developments, emerging trends, opportunities, and challenges.
2. Trevor Hastie, Robert Tibshirani, Jerome H. Friedman, and Jerome H. Friedman. *The Elements of Statistical Learning: Data Mining, Inference, and Prediction (2nd Edition)*. Springer, 2009.
   [Notes] This book introduces various algorithms of supervised learning and unsupervised learning based on the statistical learning framework. The first edition of the book was published in 2001.
3. Kevin P. Murphy. *Machine learning: A probabilistic perspective*. MIT Press, 2012.
   [Notes] The book believes that the best way to enable machines to learn from data is to use the tool of probability theory, which has been the mainstream method adopted by statistics and engineering for centuries. The author also published two books in 2022 and 2023, titled *Probabilistic Machine Learning: An Introduction* and *Probabilistic Machine Learning: Advanced Topics*, respectively.
4. Stephen Marsland. *Machine Learning: An Algorithmic Perspective (2nd Edition)*. Chapman and Hall/CRC, 2014.
   [Notes] This book was first published in 2011, and the second edition was in 2014. The author focuses on explaining machine learning methods from an algorithmic perspective.
5. Richard S. Sutton, and Andrew G. Barto. *Reinforcement Learning: An Introduction (2nd Edition)*. MIT Press, 2020.
   [Notes] This book is considered a classic textbook on reinforcement learning. The book was first published in 1998, and the second edition was in 2020.
6. Ian Goodfellow, Yoshua Bengio, and Aaron Courville. *Deep Learning*. MIT Press, 2016.
   [Notes] The textbook provides a detailed introduction to the foundational knowledge of applied mathematics and machine learning, delves into various deep neural networks and their usage, and extensively discusses research topics related to machine learning.

References

Bronstein, M.M., J. Bruna, Y. LeCun, A. Szlam, and P. Vandergheynst. (2017). Geometric deep learning: going beyond euclidean data. *IEEE Signal Processing Magazine* 34(4): 18–42.

Marsland, S. (2011). *Machine learning: An algorithmic perspective.* Boca Raton: Chapman and Hall/CRC.

Masci, J., E. Rodolá, D. Boscaini, M. M. Bronstein, and H. Li. (2016) Geometric deep learning. *SIGGRAPH ASIA 2016 Course Notes.*

Murphy, K. P. (2012). *Machine learning: A probabilistic perspective.* Cambridge: The MIT Press.

Rumbaugh, J., I. Jacobson, and G. Booch. (2004). *The unified modeling language reference manual.* Addison-Wesley Professional.

Samuel, A. (1959). Some studies in machine learning using the game of checkers. *IBM Journal of Research and Development* 3(3): 210–229.

Theodoridis, S. (2015). *Machine learning: A bayesian and optimization perspective.* Cambridge: Academic press.

Part II
Frameworks

A framework generally refers to an essential supporting structure such as a building, bridge, or vehicle and has also been extended to a fundamental supporting architecture such as a theory, technology, or system, so that other things can build on top of a framework.

This part discusses the frameworks of machine learning, specifically the theoretical frameworks for machine learning algorithms or models. The frameworks of machine learning are the theoretical perspective, but not methodological or problematical perspective, and are different from the software frameworks that are used to build machine learning algorithms.

Based on the above standpoint, this book has sorted out six representative machine learning frameworks, namely: probabilistic framework, statistical framework, geometric framework, connectionist framework, symbolic framework, and behaviorist framework.

Among these six machine learning frameworks, the first three, i.e., the probabilistic framework, statistical framework, and geometric framework, belong to the foundations of mathematics, and the last three, i.e., the connectionist framework, symbolic framework, and behaviorist framework, originate from the three schools of artificial intelligence.

Under the same theoretical framework of machine learning, there are often several sub-frameworks, and each sub-framework corresponds to several algorithms. Some one of sub-frameworks is also called model in this book, its reason is to follow the term commonly used in academia, and the model is a mathematical model but not a machine learning model.

Moreover, there are some natural relations between some frameworks. For instance, probability theory is the theoretical basis to statistics, and statistics is the application to probability theory. In other viewpoints, probability theory is one of theoretical mathematics, and statistics is an applied mathematics. For this reason, this part places the chapter probabilistic framework before the statistical framework.

The connectionist framework, symbolic framework, and behaviorist framework originate from the three schools of artificial intelligence, namely: connectionism, symbolism, and behaviorism. Upon deep thinking and analysis, we may find that these three schools also have some interesting relationships: connectionism employs the neural network in the brain as its computing and learning model, symbolism represents the mental thinking process as a symbolic system for learning and inference, and behaviorism focuses on the behavior learning and decision-making in the interaction process between the intelligent agent and the environment. Therefore, it can be considered that these three schools have an interesting hierarchical relationship: connectionism belongs to the brain level, symbolism belongs to the mental level, and behaviorism belongs to the agent level.

It should be pointed out that some machine learning algorithms or models are not based solely on one of the above six types of frameworks but are a combination of several frameworks. For example, deep reinforcement learning is built on deep neural networks and reinforcement learning, the former is based on the connectionist framework, and the latter is based on the behaviorist framework.

This part is divided into five chapters, discussing the probabilistic framework, statistical framework, connectionist framework, symbolic framework, and behaviorist framework, respectively. The geometric framework will be touched upon in Chap. 15.

Chapter 3
Probabilistic Framework

Abstract The probabilistic framework comes in handy when we need to quantify the uncertainty or stochastic processes in the observed data or the environment of machine learning. We in this chapter first take an overview on probability theory and probability learning theory. Reviewed the basic concepts and theorems of probability theory, we explain the theoretical foundation of machine learning, i.e., computational learning theory, that adopted the basic ideas of probability theory. Then we, respectively, introduce Bayesian models, Markov models, probabilistic models, probabilistic graphical models, and Monte Carlo methods, as they are all the important components in the probabilistic framework and the theoretical models for many machine learning algorithms.

3.1 Overview

The theoretical foundation of the probability framework is probability theory, the core of which is probability learning theory, especially computational learning theory based on probabilistic thinking.

3.1.1 Probability Theory

Probability theory is the branch of mathematics about uncertain problems and random phenomena, used to study the inherent laws in the uncertainty and stochasticity and to analyze the possibility of various outcomes using mathematical methods.

Two main schools have formed in the development of probability theory: the frequentist school and the Bayesian school. The former bases probability on the frequency of the same event repeated many times, such as flipping a coin millions of times to determine whether the chances of the coin landing on heads or tails are equal. The latter is based on prior knowledge and the likelihood of an event, aiming to infer the result closest to the truth.

     69
W. Wang, *Principles of Machine Learning*,
https://doi.org/10.1007/978-981-97-5333-8_3

The frequentist school has two of the most important theoretical foundations. One is the law of large numbers (LLN), which is a theorem describing the probabilistic properties of a large number of identical experimental results. The other is the central limit theorem (CLT), which characterizes the tendency of random variables to approach a normal distribution, meaning that probabilistic and statistical methods effective for normal distributions can be applied to many other types of distribution problems.

The Bayesian school has one of most important theoretical foundations, i.e., Bayes' theorem, also known as Bayes' law or Bayes' rule, which calculates the posterior probability of an event based on prior knowledge related to the event, i.e., likelihood probability and prior probability.

In probability theory, discrete and continuous random variables, probability distributions, stochastic processes, etc. can be used for the mathematical abstraction of uncertainty and stochastic processes.

Machine learning is learning from data or the environment. Probability theory is used as a mathematical framework to solve uncertainty and stochastic problems in machine learning.

3.1.2 Probability Learning Theory

Many researchers are dedicated to the research of machine learning methods based on probability theory.

Kevin B. Korb and Ann E. Nicholson published *Bayesian Artificial Intelligence* in 2004, and the second edition of the book was released in 2011 (Korb and Nicholson 2011). The book devotes four chapters to learning problems based on Baye' theorem.

In 2011, Alexander Clark and Shalom Lappin published a book with an interesting title: *Linguistic Nativism and the Poverty of the Stimulus* (Clark and Lappin 2011). This book focuses on the problem of language acquisition and discusses the learning mechanisms, computational linguistics, and learning theories. Chapter 5 of the book is titled "Probabilistic Learning Theory for Language Acquisition." It is worth pointing out that they adopted the term probabilistic learning theory in the chapter title.

Kevin P. Murphy published a book in 2012 while being a professor at the University of British Columbia, titled *Machine Learning: A Probabilistic Perspective* (Murphy 2012). The book discusses the best ways to learn from data from a probabilistic perspective.

Zoubin Ghahramani, a professor at the University of Cambridge and chief scientist at Uber Technology Company in the United States, published a paper in the journal *Nature* in 2015 titled "Probabilistic Machine Learning and Artificial Intelligence" (Ghahramani 2015), in which it is pointed out that the probabilistic framework describes how to represent and manipulate the uncertainty about models and predictions and plays a central role in machine learning, cognitive science, and artificial intelligence.

In 2016, Ian Goodfellow, Yoshua Bengio, and Aaron Courville published a famous book titled *Deep Learning* (Goodfellow et al. 2016). There is a chapter in the book titled "Probability and Information Theory," where most contents of the chapter are probability theory. In explaining why probability is used, the book states:

> This is because machine learning must always deal with uncertain quantities, and sometimes may also need to deal with stochastic (non-deterministic) quantities. Uncertainty and stochasticity can arise from many sources.

Some researchers studied the problems of machine learning based on the idea of probability theory and proposed some learning theories, for instance, probably approximately correct (PAC) learning in computational learning theory and PAC-Bayesian learning (Shawe-Taylor and Williamson 1997; McAllester 1998, 1999; McAllester and Akinbiyi 2013; Guedj 2019). Other researchers combined Occam's razor with machine learning and proposed Bayesian Occam's razor for probabilistic learning, which is used to describe how to deal with a larger range of data under the framework of Bayesian learning (Murray and Ghahramani 2005; Blanchard et al. 2018).

In addition, some researchers also combine probability with computer programming and proposed probabilistic programming[1] for machine learning, which is a programming paradigm in which probabilistic models are specified and inference for these models is performed automatically (Gordon et al. 2014; Ghahramani 2015). Those languages based on the probabilistic programming paradigm are called probabilistic programming languages.

3.2   Basics of Probability Theory

Probability theory is the branch of mathematics about uncertain and stochastic processes, studying the regularities implied in these processes and analyzing the possibility of various results appearing with mathematical methods.

To facilitate the discussion of the probability framework, this section first reviews the basics of probability theory, namely, probability space, random variables, discrete and continuous distribution, the probability relationship between random variables, independence, expectation and variance, and convergence including the law of large numbers and the central limit theorem.

Readers familiar with the above concepts can skip the content of this section.

---

[1] https://en.wikipedia.org/wiki/Probabilistic_programming

3.2.1 Probability Space

In probability theory, a probability space is a mathematical structure that models uncertainties, randomly occurring states, or "experiments." The probability space is established on the basis of specific states or experiments. Whenever the same case occurs, the set of its results may be the same, and the probability is also the same.

Definition 3.1 (Probability Space) A probability space is defined as a 3-tuple $\langle \Omega, \mathcal{F}, P \rangle$, where Ω denotes the sample space, $\mathcal{F}$ denotes the event space, and P is the probability function.

The three elements of the probability space are specified as follows:

(1) Ω is a non-empty sample space, which is a set of all possible outcomes in an experiment.
(2) $\mathcal{F}$ is a measurable event space, which is modeled as σ algebra of Ω, denoted as $\mathcal{F} \subseteq 2^{\Omega}$, is the collection of all subsets of Ω, which has the following properties:

- The empty set $\emptyset \in \mathcal{F}$, and the sample space $\Omega \in \mathcal{F}$.
- If $A \in \mathcal{F}$, then $\Omega \backslash A \in \mathcal{F}$.
- If $A_i \in \mathcal{F}$ $(i = 1, 2, \ldots, n)$, then $\left(\bigcup_{i=1}^{n} A_i \right) \in \mathcal{F}$, and $\left(\bigcap_{i=1}^{n} A_i \right) \in \mathcal{F}$.

(3) P is a probability function on $\langle \Omega, \mathcal{F} \rangle$, $P : \mathcal{F} \to [0, 1]$. $\forall A, B \in \mathcal{F}$, the following axioms exist:

- $P(A) \geq 0$.
- $P(\Omega) = 1$.
- If $A \cap B = \emptyset$, then $P(A \cup B) = P(A) + P(B)$.

The above three axioms act as the theoretical foundation for probability theory, which implicates the three important properties: $P(A) \geq 0$ means nonnegativity, $P(\Omega) = 1$ means normalization, and $P(A \cup B) = P(A) + P(B)$ means additivity.

On the basis of the above three axioms, the following results can also be obtained:

- $P(\emptyset) = 0$.
- $P(A \cup B) = P(A) + P(B) - P(A \cap B)$.

Exercise 3.1 (Dice and Probability Space) Using the example of rolling a dice to catch on the concept of a probability space.

A dice has six faces, each with one to six dots, representing the numbers 1 to 6. Its sample space is $\Omega = \{1, 2, 3, 4, 5, 6\}$. If the events we want to know are odd and even numbers, then the event space is $\mathcal{F} = \{\emptyset, \{1, 3, 5\}, \{2, 4, 6\}, \Omega\}$.

3.2.2 Random Variable

In probability theory, random variables play an important role. It is worth noting that a random variable is not a variable in the traditional sense but a function defined on a probability space, so the random variables are usually represented by capital letters.

Definition 3.2 (Random Variable) A random variable X on the probability space is a measurable function, $X : \Omega \to \Sigma$, where Σ be a measurable space. That is, it is a mapping from the sample space to the measurable space.

In many cases, the measurable space Σ is a real-valued space, that is, $\Sigma = \mathbb{R}$.

If the random variable X takes the value a, and $a \in \Sigma$, it is denoted as $X = a$. Please note that the lowercase letter a is not a variable but a value of random variable X.

The range of random variable X is represented as $\mathrm{Val}\,(X)$, which refers to the range of values that this random variable can take. $a \in \mathrm{Val}\,(X)$ indicates that a is a value in the range of X. The value of random variable depends on the result of random phenomenon.

Exercise 3.2 (Dice and Random Variables) Using the examples of rolling a dice to catch on the meaning of random variables. Consider the following two situations of dice and their random variables:

- Let the range of X be the number of points on the dice, that is, $\mathrm{Val}\,(X) = \{1, 2, 3, 4, 5, 6\}$. Therefore, if the number on the dice after rolling is 3, it is represented as $X = 3$.
- Let the range of X be the parity of the dice, where the odd number is 1 and the even number is 0, that is, $\mathrm{Val}\,(X) = \{1, 0\}$. Therefore, if the number on the dice after rolling is odd, it is represented as $X = 1$; otherwise, $X = 0$.

By the way, the random variable in the second example of the above exercise is a binary random variable, also known as an indicator variable, which is often used in machine learning.

For the probability framework of machine learning, this book will mainly discuss the probability of random variables. Although some concepts of probability can be defined without random variables, random variables facilitate the characterization of uncertainty and provide a consistent description of the probability framework.

For the random variable X taking the value a, its probability is expressed as $P\,(X = a)$. And for X taking a certain value in the measurable set $S \subseteq \Sigma$, its probability is denoted as

$$P\,(X \in S) = P\,(\{\omega \in \Omega\} \mid X\,(\omega) \in S). \tag{3.1}$$

Definition 3.3 (Independent and Identically Distributed) A set of random variables is called independent and identically distributed (i.i.d.), only if each random variable within it has the same probability distribution and the random variables are mutually independent.

3.2.3 Discrete and Continuous

The probability of an event is used to describe a random phenomenon, which is called a probability distribution. In other words, a probability distribution is the probability of a random variable taking a certain value.

Random variables are divided into two types: if the value of X is countable, it is called a discrete random variable; else if the value of X is uncountable and infinite, it is called a continuous random variable.

Depending on the type of random variable, there are two types of probability functions.

3.2.3.1 Discrete Distribution

Definition 3.4 (Discrete Probability Distribution) If X is a discrete random variable, then its distribution is discrete, also known as a discrete probability distribution.

For discrete random variables, we can simply enumerate the probability of random variable taking each possible value. This probability function is called *probability mass function*. That is, let X be a discrete random variable, the function

$$P_X(x) = P(X = x) \tag{3.2}$$

is the probability mass function, where $x \in \text{Val}(X)$ and $\sum_x P_X(x) = 1$.

The probability mass function can be extended to conditional probability distributions and joint probability distributions.

3.2.3.2 Continuous Distribution

Definition 3.5 (Continuous Probability Distribution) If X is a continuous random variable, then its distribution is expressed as a continuous distribution, also known as a continuous probability distribution.

The *probability density function* is used to assign a probability to the range of a continuous random variable. The probability density function is a nonnegative integrable function, expressed as

$$\int_{\text{Val}(X)} p(x)\, dx = 1. \tag{3.3}$$

The probability density function of the random variable X in the interval (a, b) is denoted as

$$P(a \leq X \leq b) = \int_a^b p(x)\,dx. \tag{3.4}$$

Similar to the probability mass function, the probability density function can also be extended to conditional probability distributions and joint probability distributions.

3.2.4 Probability Relationship

The so-called probability relationship refers to the relationship between random variables, including conditional probability, joint probability, and marginal probability.

3.2.4.1 Conditional Probability

Conditional probability is a measure of the probability of one random variable, given the value of another random variable.

Let $\forall X, Y \in \mathcal{F}$, and $P(Y) \neq 0$, then

$$P(X|Y) = \frac{P(X \cap Y)}{P(Y)}. \tag{3.5}$$

$P(X|Y)$ means the conditional probability of X given the value of Y. For instance, $P(X|Y = y)$ means the conditional probability of X given $Y = y$.

The concept of conditional distribution can be extended to measure the probability of one random variable, given multiple other random variables. For example, given $Y = y$ and $Z = z$, the conditional probability of $X = x$ is denoted as

$$P(X|Y, Z) = \frac{P(X \cap Y \cap Z)}{P(Y \cap Z)}. \tag{3.6}$$

Conditional probability is one of the important methods in probability theory for inference or learning under uncertainty.

3.2.4.2 Joint Probability

Joint probability refers to the probability of two or more random variables.

The joint probability of two random variables can be expressed as $P(X, Y)$ and $P(X, Y) = P(X \cap Y)$.

From Eq. 3.5, it can be known that the following joint distribution equation holds

$$P(X, Y) = P(X|Y) P(Y).\tag{3.7}$$

According to the commutative law of probability theory, Eq. 3.7 can also be written as

$$P(X, Y) = P(Y|X) P(X).\tag{3.8}$$

Equations 3.7 and 3.8 are also known as the product rules.

Extending the above joint distribution to n random variables, the following chain rule can be formed:

$$P(X_1, \ldots, X_n) = P(X_1) P(X_2|X_1) P(X_3|X_2, X_1) \cdots P(X_n|X_{n-1}, \ldots, X_1)$$

$$= P(X_1) \prod_{i=2}^{n} P(X_i|X_{i-1}, \ldots, X_1).\tag{3.9}$$

The chain rule is often used for the joint probability of multiple random variables, especially be useful when there is a property of conditional independence between some variables.

3.2.4.3　Marginal Probability

Marginal probability refers to the probability distribution of the random variable itself. Given a joint distribution, for random variables X and Y, we can calculate either the marginal distribution of X or Y. To find the marginal distribution of one random variable, all the distributions of another random variable need to be summed. This can be formally represented as

$$P(X) = \sum_{y \in \mathrm{Val}(Y)} P(X, Y = y).\tag{3.10}$$

Equation 3.10 is also known as the sum rule.

The marginal distribution is called so because of this viewpoint: if we add up all the items in a row or column of a joint distribution and place the result on the last side, it is the probability of the random variable taking that value, and the last side is called the margin.

Of course, this viewpoint only holds when the joint distribution includes two random variables.

3.2.5 *Independence*

There are mainly two types of independence between random variables, which will be described separately as below.

3.2.5.1 Unconditional Independence

For two random variables X and Y, if their joint probability distribution can be expressed as the product of their probabilities, that is,

$$P(X, Y) = P(X) P(Y), \tag{3.11}$$

then the X and Y can be called unconditionally independent random variables, denoted as $X \perp Y$. Therefore, we have

$$X \perp Y \Leftrightarrow P(X, Y) = P(X) P(Y). \tag{3.12}$$

After comparing Eq. 3.12 with Eq. 3.11, it can be found that the so-called X and Y are unconditionally independent, which means that when the random variable Y changes, the distribution of the random variable X does not change.

3.2.5.2 Conditional Independence

In addition to the above unconditional independence, many random variables often influence each other, and their influence usually comes with certain conditions.

Given the random variable Z, it is said that X and Y are conditionally independent random variables, if and only if

$$X \perp Y | Z \Leftrightarrow P(X, Y | Z) = P(X|Z) P(Y|Z). \tag{3.13}$$

3.2.6 *Expectation*

Expectation, also known as expected value, is one of the most common operations performed on random variables. It is the weighted average of all possible values of a random variable. The expectation of the random variable X is denoted as $\mathbb{E}[X]$.

The expectation of a discrete random variable is the probability-weighted average of all possible values, that is,

$$\mathbb{E}[X] = \sum_{a \in X} a \cdot P(X = a). \tag{3.14}$$

This is because the sum of all probabilities equals one, that is, $\sum_{a \in X} P(X = a) = 1$.

Exercise 3.3 (Dice and Expectation) Take the dice as an example to explain the meaning of expectation. The range of the dice is $\mathrm{Val}(X) = \{1, 2, 3, 4, 5, 6\}$; its expectation is as follows:

$$\mathbb{E}[X] = \sum_{a=1}^{6} a \cdot P(X = a) = 3.5.$$

Where $P(\cdot)$ is a probability mass function.

For a continuous random variable, its expectation is the Lebesgue integral as follows:

$$\mathbb{E}[X] = \int_{a \in X} x \cdot p(x)\, dx. \tag{3.15}$$

Where $p(\cdot)$ is a probability density function.

The expectation of the random variable X can also be denoted as μ.

3.2.7 Variance and Standard Deviation

If X is a random variable, its variance and standard deviation are as follows.

3.2.7.1 Variance

The variance of a random variable X is a measure of the range of the variable's distribution, defined as

$$\mathrm{var}(X) = \mathbb{E}\left[(X - \mathbb{E}[X])^2\right]. \tag{3.16}$$

From the above definition, it can be seen that the variance is the expectation of the square of the difference between the random variable X and its expectation. It provides the degree of deviation between the random variable and its expectation.

3.2.7.2 Standard Deviation

The standard deviation of a random variable X is usually denoted by σ; its relationship with variance is $\sigma = \sqrt{\mathrm{var}(X)}$.

Due to the above relationship, in addition to being denoted by var (X), variance can also be denoted by the symbol σ^2.

3.2.8 Covariance and Correlation

If X and Y are two random variables, their covariance and correlation are as follows.

3.2.8.1 Covariance

Covariance is used to measure the joint error between two random variables X and Y, defined as

$$
\begin{aligned}
\mathrm{cov}\,(X, Y) &= \mathbb{E}\,[(X - \mathbb{E}\,[X])\,(Y - \mathbb{E}\,[Y])] \\
&= \mathbb{E}[XY - X\mathbb{E}\,[Y] - \mathbb{E}\,[X]\,Y + \mathbb{E}\,[X]\,\mathbb{E}\,[Y]] \\
&= \mathbb{E}\,[XY] - \mathbb{E}\,[X]\,\mathbb{E}\,[Y] - \mathbb{E}\,[X]\,\mathbb{E}\,[Y] + \mathbb{E}\,[X]\,\mathbb{E}\,[Y] \\
&= \mathbb{E}\,[XY] - \mathbb{E}\,[X]\,\mathbb{E}\,[Y]\,.
\end{aligned}
\tag{3.17}
$$

3.2.8.2 Correlation

Correlation is used to measure the linear relationship between two random variables X and Y, defined as follows:

$$
\mathrm{corr}\,(X, Y) = \frac{\mathrm{cov}\,(X, Y)}{\sqrt{\mathrm{var}\,(X)\,\mathrm{var}\,(Y)}}\,.
\tag{3.18}
$$

Correlation is a normalized quantity, and the result is always between -1 and 1. The random variables X and Y can be represented by their expectations as μ_X and μ_Y, respectively, and their standard deviations as σ_X and σ_Y, respectively. Similarly, the expectation of XY is represented as μ_{XY}, the covariance $\mathrm{cov}\,(X, Y)$ is represented as σ_{XY}, and correlation $\mathrm{corr}\,(X, Y)$ is denoted as ρ_{XY}. Therefore, we have

$$
\sigma_{XY} = \mu_{XY} - \mu_X \mu_Y,
$$

$$
\rho_{XY} = \frac{\sigma_{XY}}{\sigma_X \sigma_Y}\,.
$$

3.2.9　Convergence

In probability theory, convergence refers to the convergence of random variables, which is a very important probabilistic property. The two main theorems of convergence are the *law of large numbers* and the *central limit theorem*.

This subsection only introduces their core concepts and omits the proof processes.

3.2.9.1　Law of Large Numbers

The law of large numbers (LLN) is a theorem that describes the probabilistic properties presented by a large number of identical experimental results, and it is also one of the most important theoretical foundations of probability theory and modern statistics.

The weak law of large numbers and the strong law of large numbers are introduced below.

Suppose there is a set of independent and identically distributed (i.i.d.) random variables $X_1, X_2, \ldots, X_n$, with an expected value of $\mathbb{E}[X_1] = \mathbb{E}[X_2] = \cdots = \mathbb{E}[X_n] = \mu$, the average is denoted as $\overline{X}_n = \frac{1}{n}\sum_{k=1}^{n} X_k$, then this set of random variables presents the following two laws, namely, the weak law of large numbers and the strong law of large numbers.

Weak Law of Large Numbers

The weak law of large numbers (weak LLN) states that the mean of the random variables converges in probability (P) to its expected value, that is,

$$\bar{X}_n \xrightarrow{P} \mu \quad \text{when} \quad n \to \infty. \tag{3.19}$$

In other words, for any positive number, $\varepsilon > 0$, then we have

$$\lim_{n \to \infty} P\left(\left|\bar{X}_n - \mu\right| > \varepsilon\right) = 0. \tag{3.20}$$

The above Eq. 3.20 indicates that for any given nonzero margin, no matter how small, as long as there are enough samples, the average within a certain range is likely to approach the expected value.

Strong Law of Large Numbers

The strong law of large numbers (strong LLN) states that the mean of random variables almost surely (a.s.) converges to its expected value, that is,

$$\bar{X}_n \xrightarrow{\text{a.s.}} \mu \quad \text{when} \quad n \to \infty. \tag{3.21}$$

It can also be written as

$$P \left(\lim_{n \to \infty} \bar{X}_n = \mu \right) = 1. \tag{3.22}$$

This means that when the number of trials n approaches infinity, the probability that the mean of the random variable converges to its expected value equals 1.

The importance of the law of large numbers lies in that it appears as a theorem after strict proof and a large number of experimental verifications, rather than a hypothesis in probability theory. Since the law of large numbers links the probability derived from theory with the frequency actually occurring in the real world, it is considered a theoretical pillar of probability theory and modern statistics and has had a very wide impact.

3.2.9.2 Central Limit Theorem

The central limit theorem (CLT) states that, as the number of independent random variables increases, the distribution of all random variables tends toward a normal distribution, even if these random variables themselves are not normally distributed.

In general, the premise of the normal distribution is that the random variables must satisfy the condition of independent and identically distributed (i.i.d.). For nonidentical distributions or nonindependent observations, if they meet certain conditions, their averages will also converge to a normal distribution.

The central limit theorem implies that methods using normal distribution can also be applied to many other types of distribution problems.

There are several versions of the central limit theorem, including classical CLT, Lyapunov CLT, Lindeberg-Lévy CLT, and multidimensional CLT.

Below is the definition of the Lindeberg-Levy central limit theorem.

Suppose there is a set of independent and identically distributed (i.i.d.) random variables, $X_1, X_2, \ldots, X_n$, the average is denoted as $\overline{X}_n = \frac{1}{n} \sum_{k=1}^{n} X_k$, the expected value is denoted as $\mathbb{E}[X_k] = \mu$, the variance is denoted as $\text{var}(X_k) = \sigma^2 < \infty$.

Lindeberg-Levy central limit theorem states that, when $n \to \infty$, the random variable converges in distribution to a normal distribution, that is,

$$\sqrt{n} \left(\bar{X}_n - \mu \right) \xrightarrow{a} \mathcal{N} \left(0, \sigma^2 \right). \tag{3.23}$$

In the case where $\sigma > 0$, convergence in distribution means that for every real number z, the cumulative distribution function (CDF) of $\sqrt{n} \left(\overline{X}_n - \mu \right)$ converges pointwise to the cumulative distribution function of normal distribution $\mathcal{N}(0, \sigma^2)$, that is,

$$\lim_{n \to \infty} P \left(\sqrt{n} \left(\bar{X}_n - \mu \right) \leq z \right) = \lim_{n \to \infty} P \left(\frac{\sqrt{n} \left(\bar{X}_n - \mu \right)}{\sigma} \leq \frac{z}{\sigma} \right) = \Phi \left(\frac{z}{\sigma} \right). \tag{3.24}$$

Where $\Phi(z)$ is the cumulative distribution function of the standard normal distribution for z. As for the following expression, the convergence of z is uniform:

$$\lim_{n \to \infty} \sup_{z \in \mathbb{R}} \left| P\left(\sqrt{n}\left(\bar{X}_n - \mu\right) \leq z\right) - \Phi\left(\frac{z}{\sigma}\right) \right| = 0. \tag{3.25}$$

Where $\sup(\cdot)$ represents the least upper bound.

The central limit theorem is another important law in probability theory and also serves as another theoretical foundation for probability theory and mathematical statistics.

3.3 Computational Learning Theory

Computational learning theory is a mathematical analysis theory about learnability, mainly used for the design and analysis of machine learning algorithms. Its mathematical analysis process adopts the basic ideas of probability theory.

Early machine learning research was mainly aimed at supervised learning. That is, some pre-labeled samples are given to train the machine learning model, which is then used to predict unknown data. Computational learning theory is also the case.

Therefore, it can be said that computational learning theory is a theory about the learnability of supervised learning. According to this theory, if a learning algorithm can complete its learning task in polynomial time, it is considered feasible. The issues mainly involved in computational learning theory are the performance boundaries of learning, the complexity and feasibility of learning, the number of training samples required, the presentation of training samples, the complexity of the hypothesis space, the approximation accuracy of the target function, the possibility of successful learning, and so on.

Computational learning theory originated from learnability theory and later evolved into probably approximately correct (PAC) learning, as well as Occam learning.

An important idea in computational learning theory is the introduction of concepts from computational theory into machine learning. Therefore, it can be said that computational learning theory is to machine learning, as computational theory is to computer science.

In the book *An Introduction to Computational Learning Theory* written by Michael J. Kearns and Umesh V. Vazirani, the VC (Vapnik-Chervonenkis) dimension is included as part of the content of computational learning theory (Kearns and Vazirani 1994). However, Vladimir Vapnik, one of the founders of VC theory and its VC dimension, discussed VC theory and VC dimension in his books *Statistical Learning Theory* (Vapnik 1998) and *The Nature of Statistical Learning Theory* (Vapnik 1999). Therefore, the VC dimension will be introduced in the statistical learning theory of Chap. 4 of this book.

This section mainly introduces the theory of learnability, probably approximately correct (PAC) learning, and Occam learning.

3.3.1 Learnability Theory

There is a rich theory of computability in computer science. However, although machine learning officially debuted in 1959, there was no corresponding theory to explain why machines could learn before 1984.

The learnability theory is the theoretical basis of computational learning theory that concerns the questions such as why is machine learnable, what information is required to support learning, and what computation is required for learning to be possible.

3.3.1.1 Origin of the Theory

Leslie G. Valiant, a British-American computer scientist born in Hungary and a professor at Harvard University and the University of Edinburgh, published a paper in 1984 titled "A Theory of the Learnable" at the *ACM Symposium on Theory of Computing* (Valiant 1984a) and proposed a methodology of studying learning from a computational viewpoint. After that, he revised this paper and published it under the same title in the journal *Communications of the ACM* in the same year (Valiant 1984b). His related research work pioneered computational learning theory.

The main contribution in the above paper is that it proved one can design learning machines that have all three of the following properties:

(1) The machines can provably learn whole classes of concepts. And these classes can be characterized.
(2) The classes of concepts are appropriate and nontrivial for general-purpose knowledge.
(3) The computational process by which the machines deduce the desired programs requires a polynomial number of steps.

It is worth emphasizing that the probably approximately correct (PAC) learning was summarized later based on the core ideas of the above paper. Some books or papers directly quote Valiant's 1984 paper when introducing PAC learning, but in fact, the term "probably approximately correct" or "PAC learning" did not appear in the above paper.

By the way, Leslie G. Valiant won the *Turing Award* in 2010; one of the reasons for his award was his outstanding contribution to computational learning theory.

The inspiration for the theory of learnability mainly comes from the theory of computability, because Valiant was originally one of the important contributors to the theory of computer science. The theory of computability is a branch of computational theory that studies which problems are computable within polynomial time.

Similarly, the theory of learnability is a branch of learning theory that studies which problems are learnable within polynomial time.

Humans can learn new concepts, unlike traditional computers that need to be explicitly programmed, so machine learning should also learn without the need for explicit programming. Valiant wrote in the paper:

> This paper is concerned with precise computational models of the learning phenomenon. We shall restrict ourselves to skills that consist of recognizing whether a concept (or predicate) is true or not for given data. We shall say that a concept Q has been learned if a program for recognizing it has been deduced (i.e., by some method other than the acquisition from the outside of the explicit program).

In summary, the theory of learnability is a precise computational model of concept learning.

Valiant gave an example in the paper, narrowing the problem to learning how to recognize its given concept from a given dataset. He proposed that a learning machine can be composed of a learning protocol and a deduction procedure, the former specifies how to obtain information from the outside, and the latter is the algorithm for concept learning. Therefore, he proposed that the study of learning methods is to define a reasonable learning protocol and explore the use of this protocol to deduce the program for learning this concept class within polynomial time.

Next, we will introduce the learning protocol, learnability, combination boundaries, and expressions in the theory of learnability.

3.3.1.2 Learning Protocol

Without loss of generality, only the learning protocol of Boolean functions is considered here, the values of its variables are $\{0, 1\}$, and the case where the value of the variable is undetermined is not considered.

Let there be n Boolean variables, $x_1, x_2, \ldots, x_n$, each variable can take the value 1 or 0, representing the related proposition as true or false, respectively. No assumptions are made about the independence of the variables. Please note that this is why the computational learning model is later referred to as "distribution-free."

A Boolean vector x is an instance formed by assigning n Boolean variables, respectively, to $\{0, 1\}$. For example, $x = (x_1 = 1, x_2 = 0, \ldots, x_n = 1)$, that is, $x = (1, 0, \ldots, 1)$.

The set of Boolean vectors is denoted as X, and then, the size of this Boolean vector set X is $|X| = \{0, 1\}^n = 2^n$.

Let the Boolean function F be the mapping from the Boolean vector set X to $\{0, 1\}$, represented as $F : X \to \{0, 1\}$. For a Boolean vector $x \in X$, there exists $F(x) = 1$; if and only if for any Boolean vector $w \in X$ also satisfies $F(w) = 1$, then we call F a concept.

Given a concept F, let P denote any probability distribution of the Boolean vector set X. If $x \in X$, such that $F(x) = 1$, then $P(x) \geq 0$, and $\sum_{x \in X} P(x) = 1$.

The learning protocol proposed by Valiant is a way to specify how to obtain information from the outside. For a new concept F, this learning protocol provides two types of information and completes it through the following two routines:

EXAMPLES() This routine has no input. Its output is a Boolean vector x, such that $F(x) = 1$. For each Boolean vector x, the probability of getting output x from any single call to this routine is $P(x)$.
In short, this routine generates random samples.

ORACLE(x) The input is a Boolean vector x. If $F(x) = 1$, then the output is 1; otherwise, it is 0.
Obviously, this routine is used to check whether the input sample is a positive sample of this concept.

The above two routines can be linked with the teacher-student model: each time the student calls the EXAMPLES() routine, it will randomly generate a sample according to the unknown distribution of the problem domain, so as to create the required sample set. And ORACLE(x) plays the role of the teacher, telling the student whether these samples are positive samples of a certain concept.

3.3.1.3 Learnability

Consider an input as a Boolean vector $x = (x_1, x_2, \ldots, x_n)$, the class C of programs exists, and there are $f, g \in C$.

For the given learning protocol, the class C of programs is learnable, if and only if there exists an algorithm A, which has the following two properties:

- First, the algorithm runs in polynomial time and includes adjustable parameters h and various quantified parameters.
- Second, for the distribution $P(x)$, it satisfies: i) for all vector x, $g(x) = 1$ implies $f(x) = 1$; and ii) the sum of $P(x)$ overall x, such that $f(x) = 1$ but $g(x) \neq 1$ is at most h^{-1}.

3.3.1.4 Combinatorial Bound

Define a function $L(h, S)$, where h is any real number greater than 1 ($1 < h \in \mathbb{R}$), and S is any positive integer ($S \in \mathbb{Z}_+$). Let $L(h, S)$ be the smallest integer, such that in $L(h, S)$ independent Bernoulli trials each with a probability of success is not less than h^{-1}, and the probability of having fewer than S successes is less than h^{-1}.

Valiant's paper provides the following example to illustrate the relevance of the function $L(h, S)$ to learning:

Consider an urn U containing a very large number of marbles possibly of many different types. Suppose we wish to "learn" what kind of marbles U contains by taking a small random sample X. If all the marbles in U are of distinct types, then it is clearly impossible to deduce much information about them. Suppose, however, that we know that there are only S types of marbles and that we wish to choose X so that it contains representatives of all but 1 percent of the marbles in U. Whatever the respective frequencies of the S types may be, the definition of $L(h, S)$ implies that if $|X| = L(100, S)$ then with probability greater than 99 percent we will have succeeded. To see why this implication holds, suppose that after $L(100, S)$ successive choices marble types representing between them at least 1 percent of the urn remain unchosen. In that case, we have made $L(100, S)$ Bernoulli trials each with a probability at least 1 percent of success (i.e., of choosing a previously unchosen type), and have succeeded fewer than S times (since some type remains unchosen). Note that what constitutes a success here will depend on previous choices. The probability of success, however, will always be at least 1 percent, independent of previous choices, and this is sufficient for our argument.

Based on the above Valiant's argument, I think that this is just the origin of what was later named *probably approximately correct* learning or *PAC* learning.

As we all know, human learning is not always 100% correct. Even if the score of a certain exam is 99, it is only limited to the result of that exam. It can be considered that human learning is a kind of probably approximately correct learning.

Here is a proposition.

For all real numbers $h > 1$, and all integers $S \geq 1$, that is, $1 < h \in \mathbb{R}$, $1 \leq S \in \mathbb{Z}_{+}$, satisfying $L(h, S) \leq 2h(S + \log h)$.

Where $2h(S + \log h)$ is the upper bound. The proof of the above proposition can be found in the reference (Valiant 1984b). This proposition applies to the entire range of S and h, indicating that $L(h, S)$ is essentially linear for both h and S.

3.3.1.5 Expressions

Valiant's paper discusses that there are three categories of concepts that can be learned in polynomial time through the above learning protocol. These categories can be characterized by defining the class of programs that recognize them. In different situations, these programs are special types of Boolean expressions.

These three expressions are i) conjunctive normal form (CNF) expressions, ii) disjunctive normal form (DNF) expressions, and iii) arbitrary expressions, where each variable appears only once.

Below is a simple explanation of conjunctive normal form expressions and disjunctive normal form expressions:

- A conjunctive normal form (CNF) expression is any product of clauses c_i, that is, $c_1 c_2 \cdots c_r$. Each clause is the sum of literals l_{ij}, that is, $l_{i1} + l_{i2} + \cdots + l_{is}$. A literal is a variable x_k or the negation of a variable $\bar{x}_k$. For example, $(x_1 + \bar{x}_2)(\bar{x}_1 + \bar{x}_2 + x_3)$ is a CNF expression. A k-CNF expression is also a CNF expression, but the number of literals in each clause does not exceed the positive integer k.

- A disjunctive normal form (DNF) expression is any sum of monomials m_i, that is, $m_1 + m_2 + \cdots + m_r$, where each monomial is the product of literals. For example, $x_1 \bar{x}_2 + x_1 x_3 \bar{x}_4$ is a DNF expression. If there are no negated variables in the disjunctive normal form expression, it is called a monotonic DNF expression. For example, $x_1 x_2 + x_1 x_3 x_4$ is a monotonic DNF expression.

3.3.2 *PAC Learning*

PAC learning is standing for probably approximately correct learning. PAC learning is a computational learning theory, which includes a PAC learning model.

In the learning model, a learner receives samples and needs to select an optimal generalization function from a class of candidate functions, called a hypothesis. Given any approximation ratio, success probability, or sample distribution, the learner must be able to learn its concept. The goal is to make this hypothesis to select a function with low generalization error (i.e., "approximately correct") with "high probability" (i.e., "probably").

3.3.2.1 Origin of the Term

As far as my investigation, I find that there is a developing process in the formation of the term "PAC learning."

In 1984, Leslie Valiant published a paper titled "A Theory of the Learnable" in the journal *Communications of the ACM*, proposing a theory of recognizing learnable concept classes from a probabilistic viewpoint.

In 1988, Dana Angluin of Yale University published a paper in the journal *Machine Learning*, in which Valiant's learnable theory was referred to as probably approximately correct identification or PAC identification for short (Angluin 1988).

In 1992, E. M. Oblow at Oak Ridge National Laboratory used the term probably approximately correct (PAC) learning in his paper (Oblow 1992).

In 2013, Leslie Valiant published a monograph titled *Probably Approximately Correct: Nature's Algorithms for Learning and Prospering in a Complex World* (Valiant 2013). That is to say, Valiant himself officially adopted the term *probably approximately correct*.

The important contribution of probably approximately correct learning is the introduction of the concept of computational theory into machine learning, especially proposing a computational learning model of how a learner finds an effective function and implements an effective process.

A variant of the PAC learning model is the weak learning model, and the problem of hypothesis boosting is proposed. The boosting algorithms in machine learning are developed based on this theory.

Below, before defining PAC learning, we will first give the definitions of learning model.

3.3.2.2 Learning Model

Let X denote the sample space and C represent the concept class. Let $x_i \in X$ be a sample and $c \in C$ denote a target concept. Without loss of generality, assume the target concept c is a binary function, so we have $c : X \to \{0, 1\}$. That is, if x_i is a

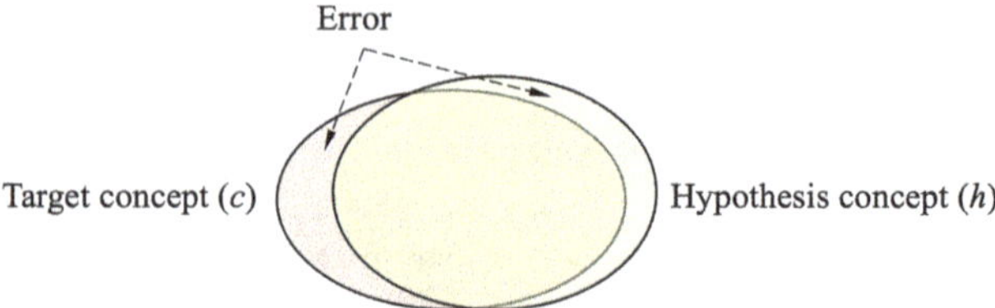

Fig. 3.1 Target concept, hypothesis concept, and their error

positive sample, then we have $c(x_i) = 1$; otherwise, x_i is a negative sample, and we have $c(x_i) = 0$.

According to the theory of learnability, let the sample x_i be randomly selected according to some probability distribution P, and it does not need to satisfy the condition of independent and identically distributed (i.i.d.).

Let the input of the learner be n observation data $(x_1, \ldots, x_n)$ and its annotation $(c(x_1), \ldots, c(x_n))$ used for training, Therefore, the training sample is represented as $S = \{(x_i, c(x_i)) \mid i = 1, \ldots, n\}$.

The learning process is as follows. First, find a hypothesis set that approximates the target concept, $H = \{h \mid h : X \to \{0, 1\}\}$, where h is called a hypothesis concept. The target concept c has a certain error with the hypothesis concept h, as shown in Fig. 3.1. Then, by using T for training, a hypothesis concept $h \in H$ is selected, so that the error between this hypothesis concept h and the target concept c is minimized, that is, h is approximately equal to c with high probability. This is another example of what is called "probably approximately correct."

3.3.2.3 PAC Learning Model

Here, we first give the definition of PAC learning and then discuss the characteristics of its learning model.

Definition 3.6 (PAC Learning) A concept class C is said to be PAC learnable, if there exists an algorithm $\mathcal{A}$ and a polynomial function poly $(\cdot, \cdot, \cdot, \cdot)$, such that for any $\epsilon > 0$ and $\delta > 0$, for all distributions D on X, and for any target concept $c \in C$, then for any sample size $m \geq$ poly $(1/\epsilon, 1/\delta, n, \text{size}(c))$, the following expression holds

$$P_{S \sim D^n}[R(h) \leq \epsilon] \geq 1 - \delta. \tag{3.26}$$

If $\mathcal{A}$ further runs in poly $(1/\epsilon, 1/\delta, n, \text{size}(c))$, then C is said to be efficiently PAC learnable. When such an algorithm $\mathcal{A}$ exists, it is called a PAC learning algorithm for C.

It should be emphasized that the definition of PAC learning has the following characteristics:

- PAC learning is a *distribution-free* model, that is, it does not make special assumptions about the distribution of the extracted data D; the idea of distribution-free comes from the learnability theory introduced in the previous section.
- The training samples and test samples used to define the error are obtained according to the same distribution D; in most cases, this is a necessary assumption to make generalization possible.
- The PAC learning framework is aimed at the learnability problem of the concept class C, not a specific concept. For this algorithm, the concept class C is known, but the target concept $c \in C$ is unknown.

For examples of PAC learning for specific learning problems, interested readers can refer to the literature (Mohri et al. 2012), which will not be elaborated here.

3.3.3 Occam Learning

Occam learning is also a learning theory model, the goal of which is to obtain a concise method based on training data. Occam learning is closely related to probably approximately correct (PAC) learning, and PAC evaluates learners based on their predictive ability on the test set.

The name of Occam learning was inspired by Occam's razor.

3.3.3.1 Occam's Razor

Occam's razor is a *law of parsimony*, that is,

> "simpler solutions are more likely to be correct than complex ones."

This rule has become a principle for problem-solving, that is, when faced with several solutions, the simplest solution should be chosen.

The idea is attributed to William of Occam (1287–1347), who was born in a small village called Occam in Surrey, southeast England, which is the origin of the name William of Occam. Later, he became a Franciscan Friar, scholastic philosopher, and theologian in England. William of Occam made drastic reforms to the obscure and lengthy explanations of his scholastic philosophy predecessors, so he was praised as "Occam's razor."

Similarly, in the scientific field, Occam's razor is used as an abductive heuristic rule for developing theoretical models, but it is not a strict arbitrator between candidate models, nor is it regarded as an indisputable scientific principle. However, simpler theories are more desirable than complex theories because the former are more testable.

In 1987, Anselm Blumer et al. published a paper titled "Occam's Razor" (Blumer et al. 1987) in the sixth issue of *Information Processing Letters*. This paper also applied Occam's razor to the problem of machine learning, as the principle of parsimony of Occam's razor has since been incorporated into the methodology of experimental science in the following form: given two explanations of the data, all other things being equal, the simpler explanation is preferable.

Since then, Occam's razor has also become a signboard in the field of machine learning.

3.3.3.2 Occam Learning Framework

Occam learning is named after Occam's razor. It is based on the law of parsimony of Occam's razor and applies it to the theory and mathematical proof of machine learning.

The conciseness of the concept c in the concept class C can be regarded as the length of the shortest bit string size(c) that can be represented in C. Occam learning links the simplicity of the output of the learning algorithm with its predictive ability for unknown data.

Definition 3.7 (Occam Learning) Let C be the concept class containing the target concept $c \in C$ and H be the hypothesis set. Then, for constants $\alpha \geq 0$ and $0 \leq \beta \leq 1$, the learning algorithm $\mathcal{A}$ is an alpha-beta Occam algorithm that uses H to learn C if and only if, given a set of n samples $S = \{x_i\}_{i=1}^{n}$ uses the concept $c(x)$ for labeling, we get n training samples $\{(x_i, c(x_i))\}_{i=1}^{n}$, which are used to train the learning algorithm $\mathcal{A}$ to get a hypothesis $h \in H$, so that

- h on S is consistent with c, that is, $\forall x \in S$, $h(x)$ is consistent with $c(x)$.
- size $(h) \leq (m \cdot \text{size}(c))^{\alpha} n^{\beta}$.

Where m is the maximum length for any sample $x \in S$.

For an Occam algorithm $\mathcal{A}$, if it runs in polynomial time in m, n, and size (c), it is called an efficient alpha-beta Occam algorithm. If there exists an efficient Occam algorithm $\mathcal{A}$ that uses H to learn C, then C is said to be an Occam learnable concept class of the hypothesis set H.

Occam learnability implies PAC learnability and the following theorem is given in the reference (Kearns and Vazirani 1994).

Theorem 3.1 (Occam Learning Implies PAC) *Let $\mathcal{A}$ be the alpha-beta Occam algorithm to use H to learn C, there exists a constant $\alpha \geq 0$, such that for any $0 < \epsilon$ and $\delta < 1$, given any distribution D, according to the concept of each bit length n in the sample labeled by $c \in C$:*

$$m > a \left(\frac{1}{\epsilon} \log \frac{1}{\delta} + \left(\frac{(n \cdot \text{size}(c))^{a}}{\epsilon} \right)^{\frac{1}{1-\beta}} \right), \tag{3.27}$$

The algorithm $\mathcal{A}$ will obtain a hypothesis $h \in H$, such that error $(h) \leq \epsilon$ *with a probability of at least $1 - \delta$.*

Here, the error (h) is associated with the concept c and distribution D, implying that the algorithm $\mathcal{A}$ also uses the hypothesis set H to learn the concept class C of PAC learners.

3.4 Bayesian Models

In the machine learning under probabilistic framework, one of the commonly used methods is Bayesian learning based on Bayes' theorem, also known as Bayesian reasoning or Bayesian inference.

The characteristic of the Bayesian model is that it obtains a hypothesis from data and then uses the hypothesis for prediction. In other words, the Bayesian model is used to infer unknown variable using known variables and to infer the conditional distribution of unknown variable given observable variables.

3.4.1 Bayes' Theorem

Let X and Y be discrete random variables, then Bayes' theorem is expressed as

$$P(Y|X) = \frac{P(X|Y)\,P(Y)}{P(X)}.$$
(3.28)

Where $P(X|Y)$ is the conditional probability known as the likelihood probability or likelihood, $P(Y)$ is the prior probability or prior, $P(X)$ is the evidence, and $P(Y|X)$ is the conditional probability known as the posterior probability or simply the posterior.

The proof of Bayes' theorem is not difficult:

From the joint probability equations introduced in Sect. 3.2.4 that are $P(X, Y) = P(X|Y)\,P(Y)$ and $P(X, Y) = P(Y|X)\,P(X)$, therefore $P(Y|X)\,P(X) = P(X|Y)\,P(Y)$, if $P(X) \neq 0$, the above theorem holds.

Bayes' theorem can be written as the following equation:

$$\text{Posterior} = \frac{\text{Likelihood} \times \text{Prior}}{\text{Evidence}}.$$
(3.29)

That is to say, if the Prior and Evidence are known, and the Likelihood is also known, then its Posterior can be obtained.

For continuous random variables, the Bayes' theorem also holds.

3.4.2 Bayesian Learning

The previous section introduced computational learning theory; let's consider the problem of human learning from a different viewpoint.

People often want to infer the cause based on the known effect. According to Bayes' theorem, it can be represented in the following form:

$$P\,(\text{Cause}|\text{Effect}) = \frac{P\,(\text{Effect}|\text{Cause})\,P\,(\text{Cause})}{P\,(\text{Effect})}. \tag{3.30}$$

Where $P\,(\text{Effect}|\text{Cause})$ is the likelihood of the causal relationship, $P\,(\text{Cause})$ is the prior probability of the cause, and $P\,(\text{Effect})$ is the evidence of the effect. If these conditions are known, the cause can be inferred from the effect, that is, the posterior probability $P\,(\text{Cause}|\text{Effect})$.

This is the theoretical basis of Bayesian learning.

For machine learning, let the hypothesis set be H, to get a hypothesis function $h \in H$, satisfying $h : X \to Y$, if expressed in conditional probability, it is $h \sim P\,(Y|X)$. Under the conditions that satisfy the Bayes' theorem, the corresponding results can be obtained.

Below is an example to further explain the importance of Bayesian learning.

Exercise 3.4 (Disease Inference) Inferring a disease from some symptoms based on Bayes' theorem.

Doctors often need to infer the probability of a disease based on certain symptoms. Suppose the likelihood of a disease causes a patient to have certain symptoms $P\,(\text{Symptoms}|\text{Disease})$ is 70%, the prior probability of a patient having the disease $P\,(\text{Disease})$ is 0.002%, and the evidence of any patient having the symptoms $P\,(\text{Symptoms})$ is 1%. Under these conditions, infer the posterior probability of the patient with the symptoms having the disease $P\,(\text{Disease}|\text{Symptoms})$.

Bayes' theorem can be rewritten in the following form and calculated:

$$P\,(\text{Disease}|\text{Symptoms}) = \frac{P\,(\text{Symptoms}|\text{Disease})\;P\,(\text{Disease})}{P\,(\text{Symptoms})}$$

$$= \frac{0.7 \times 0.00002}{0.01} = 0.0014.$$

The result of 0.0014 means that the probability of the disease in 700 people with the symptoms is less than one person.

For Eq. 3.30, if Cause is rewritten as Hypothesis and Effect is rewritten as Evidence, then we have

$$P\,(\text{Hypothesis}|\text{Evidence}) = \frac{P\,(\text{Evidence}|\text{Hypothesis})\,P\,(\text{Hypothesis})}{P\,(\text{Evidence})}. \tag{3.31}$$

Where P (Evidence|Hypothesis) is the likelihood of the evidence under the hypothesis, P (Hypothesis) is the prior probability of the hypothesis, P (Evidence) is evidence, and P (Hypothesis|Evidence) is the posterior probability of the hypothesis given the evidence.

3.4.3   PAC-Bayesian Learning

PAC-Bayesian learning, also known as PAC-Bayesian theory, is a product of combining Bayesian learning with PAC learning theory. It provides a convenient method of utilizing prior knowledge based on Bayesian learning while also providing strict and explicit generalization guarantees based on PAC learning theory.

The PAC-Bayesian theory has a process of formation and development.

In 1997, John Shawe-Taylor from the University of London and Robert C. Williamson from the Australian National University published a paper titled "A PAC Analysis of a Bayesian Estimator" at the *Conference on Computational Learning Theory* (COLT) of that year, applying PAC analysis to Bayesian classification function estimation for the first time (Shawe-Taylor and Williamson 1997).

In 1998, David A. McAllester from the Massachusetts Institute of Technology (MIT) published a paper titled "Some PAC-Bayesian Theorems" at the *Conference on Computational Learning Theory* (COLT) of that year, in which he presented two PAC-Bayesian theorems and proved that in all independent and identically distributed (i.i.d.) experiments, the learning algorithm can fully utilize the superiority of prior probabilities and also obtain PAC guarantees (McAllester 1998).

In 1999, David McAllester published a paper titled "PAC-Bayesian Model Averaging" at the *Conference on Computational Learning Theory* (COLT) of that year, further elaborating on the research approach of the PAC-Bayesian learning model (McAllester 1999).

In 2013, David McAllester et al. published Chapter 10 "PAC-Bayesian Theory" in the book *Empirical Inference*. Further clarification, the PAC-Bayesian framework is a frequentist method that replaces the finite VC dimension with a prior probability on a single hypothesis, which can have an infinite VC dimension (McAllester and Akinbiyi 2013).

In 2016, Pascal Germain et al. from the École Normale Supérieure in Paris, France, published a paper titled "PAC-Bayesian Theory Meets Bayesian Inference" at the *Conference on Neural Information Processing Systems* (NIPS) (Germain et al. 2016), establishing a direct link between Bayesian inference and PAC-Bayesian theory. That is, by using appropriate parameters and minimizing the PAC-Bayesian range, the marginal likelihood can be maximized.

The significance of PAC-Bayesian theory is that it provides PAC boundaries for general Bayesian algorithms.

Below is the formal definition of PAC-Bayesian learning by Germain et al. (2016).

Definition 3.8 (PAC-Bayesian Learning) Let X and Y denote the input and output variables, respectively, Θ is the parameter set, and h is the hypothesis that maps input X to output Y, i.e., $h : X \rightarrow Y$. Given a training set with n samples, $S = \{(X, Y)\} = \{(x_1, y_1), \ldots, (x_n, y_n)\}$, where $x_i \in X$ and $y_i \in Y$. Let $\theta \in \Theta$, $P(\theta)$ is the prior probability of the parameter set, and $P(Y|X, \theta)$ is the likelihood probability of the parameter set Θ, then according to Bayes' theorem, its posterior probability $P(\theta|X, Y)$ is equal to

$$P(\theta|X, Y) = \frac{P(\theta)\, P(Y|X, \theta)}{P(Y|X)} \propto P(\theta)\, P(Y|X, \theta). \tag{3.32}$$

Where $P(Y|X, \theta) = \prod_{i=1}^{n} P(y_i|x_i, \theta)$ and $P(Y|X) = \int_{\Theta} P(\theta)\, P(Y|X, \theta)\, d\theta$.

Let S be an independently and identically distributed training set according to the conditional probability $P(Y|X, \theta)$. Through the training of S, a hypothesis function $h_\theta \in \Theta$ is obtained, and its negative log-likelihood loss (nll) function $\mathcal{L}_{\text{nll}}$ is defined as follows:

$$\mathcal{L}_{\text{nll}}(h_\theta, X, Y) = -\ln P(Y|X, \theta). \tag{3.33}$$

Then, the empirical loss of the hypothesis $\hat{\mathcal{L}}$ can be related to the negative log-likelihood loss function $\mathcal{L}_{\text{nll}}$ as follows:

$$\hat{\mathcal{L}}(\theta, X, Y) = \frac{1}{n}\sum_{i=1}^{n} \mathcal{L}_{\text{nll}}(h_\theta, x_i, y_i) = -\frac{1}{n}\sum_{i=1}^{n} \ln P(y_i|x_i, \theta)$$

$$= -\frac{1}{n}\ln P(Y|X, \theta). \tag{3.34}$$

The above equation is written in exponential form:

$$P(Y|X, \theta) = e^{-n\hat{\mathcal{L}}(\theta, X, Y)}, \tag{3.35}$$

then

$$P(Y|X, \theta) \geq 1 - \delta = e^{-n\hat{\mathcal{L}}(\theta, X, Y)}. \tag{3.36}$$

3.5   Markov Models

A Markov model is a stochastic model used to model the systems with random changes. It is named after the Russian mathematician Andrey Markov.

The Markov model satisfies such a property, that is, its next state depends only on the current state and is independent of past states.

Table 3.1 Two types of
autonomous Markov models

System state	Fully observable	Partially observable
Autonomous	Markov chain	Hidden Markov model

We present two types of autonomous Markov models in Table 3.1, namely, the Markov chain and the hidden Markov model (HMM). The former is a fully observable state model, while the latter is a partially observable state model.

This section first introduces the concept of a stochastic process, then discusses the Markov property and Markov process, and finally discusses the Markov models of two types of autonomous decision-making methods, namely, the Markov chain and the hidden Markov model.

3.5.1 Stochastic Process

In probability theory and its related fields, a stochastic process is a mathematical object composed of a set of random variables and indexed by a set of numbers, which are usually considered as points in time.

There are many examples of stochastic processes, such as the growth of bacterial populations, current fluctuations caused by thermal noise, and the movement of gas molecules.

Definition 3.9 (Stochastic Process) A stochastic process (SP) is usually defined as a set of random variables on a probability space, that is,

$$SP = \{S(t) \mid t \in T\}.$$

Where T is an index set, also known as a parameter set; usually, T is interpreted as time, and each t is considered a point in time; $S(t)$ is a random variable at time t, also known as the state of the stochastic process at time t.

If the index set is a countable set, then SP is a discrete-time stochastic process, studied using probabilistic methods; when the index set is continuous, then SP is a continuous-time stochastic process, usually studied using analytical methods.

Stochastic processes are systematic mathematical models established for random phenomena and are widely used in many disciplines, including biology, chemistry, ecology, neuroscience, physics, finance, engineering, as well as technology fields such as image processing, signal processing, information theory, computer science, cryptography, and telecommunications.

Exercise 3.5 (Chinese Restaurant Process) The Chinese restaurant process is a process of discrete random partitioning, metaphorically referring to the process of customers in a Chinese restaurant randomly choosing tables.

Suppose there is a Chinese restaurant with an infinite number of tables, each of which can accommodate an infinite number of customers. Each customer chooses a table in the following random manner:

The first customer always chooses the first table. The next customer often sits at a table with people but may also choose an empty table. Repeat the above process.

The formal definition of the Chinese restaurant process is given below.

For a positive integer space $\mathbb{N}_+$, there exist $n, k \in \mathbb{N}_+$. Let t_n represent the table number chosen by the nth customer, α is the concentration parameter, then the probability distribution of the Chinese restaurant process is as follows:

(1) The 1st customer t_1 always chooses the 1st table, with a probability of $P(t_1 = 1) = 1$.

(2) When $n = N$ customer t_n comes to the restaurant, the number of people at the table is K ($1 \leq K \leq n - 1$), then:

- The probability of choosing an empty table is

$$P(t_n = K + 1 \mid t_1, t_2, \ldots, t_{n-1}) = \frac{\alpha}{n - 1 + \alpha}.$$

- The probability of choosing an occupied table $k \leq K$ is

$$P(t_n = k \mid t_1, t_2, \ldots, t_{n-1}) = \frac{c_k}{n - 1 + \alpha}.$$

Where $c_k = \sum_{n=1}^{N} \mathbb{I}(t_n = k)$ is the number of customers at the kth occupied table; $\mathbb{I}(\omega)$ is an indicator function, that is, $\mathbb{I}(\omega)$ equals 1 if ω is true; otherwise, it is 0.

In summary, the joint probability of N customers choosing tables is as follows:

$$P(t_1, t_2, \ldots, t_N) = \frac{\prod_{k=1}^{K}(c_k - 1)!}{\prod_{n=1}^{N}(n - 1 + \alpha)} \alpha^{K-1}.$$

The following example further illustrates the Chinese restaurant process.

There are six customers, as shown in Fig. 3.2; the chosen tables are: $(t_1, t_2, t_3, t_4, t_5, t_6) = (1,\ 1,\ 2,\ 1,\ 3,\ 2)$.

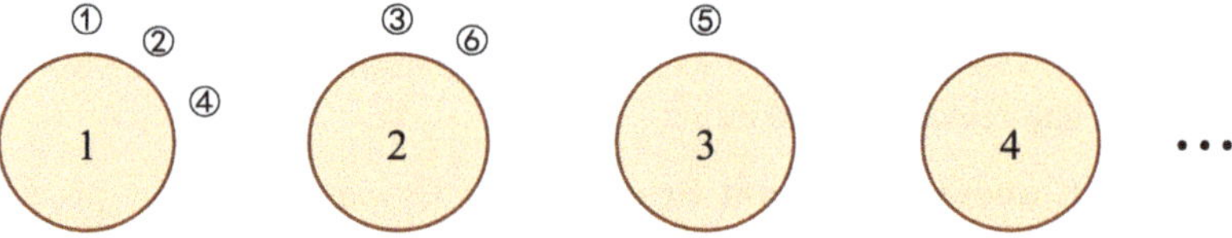

Fig. 3.2 Example of the Chinese restaurant process

Table 3.2 Probability of table selection in the Chinese restaurant process

	c_1	c_2	c_3	c_4
t_1	$\frac{\alpha}{\alpha}$	0	0	0
t_2	$\frac{1}{1+\alpha}$	$\frac{\alpha}{1+\alpha}$	$\frac{\alpha}{1+\alpha}$	$\frac{\alpha}{1+\alpha}$
t_3	$\frac{2}{2+\alpha}$	$\frac{\alpha}{2+\alpha}$	$\frac{\alpha}{2+\alpha}$	$\frac{\alpha}{2+\alpha}$
t_4	$\frac{2}{3+\alpha}$	$\frac{\alpha}{3+\alpha}$	$\frac{\alpha}{3+\alpha}$	$\frac{\alpha}{3+\alpha}$
t_5	$\frac{3}{4+\alpha}$	$\frac{\alpha}{4+\alpha}$	$\frac{\alpha}{4+\alpha}$	$\frac{\alpha}{4+\alpha}$
t_6	$\frac{3}{5+\alpha}$	$\frac{\alpha}{5+\alpha}$	$\frac{\alpha}{5+\alpha}$	$\frac{\alpha}{5+\alpha}$

That is, the number of customers who occupied the three tables is

$$\{c_1, c_2, c_3\} = \{3,\ 2,\ 1\}.$$

The relation between the number of customers and the number of occupied tables satisfies

$$\sum_{k=1}^{K} c_k = N.$$

According to the definition of the Chinese restaurant process, the probability of table selection for 1 to 6 customers is shown in Table 3.2.

The joint probability of choosing a table for these six customers is calculated as follows:

$$P(t_1, t_2, \ldots, t_6) = P(t_1)\, P(t_2|t_1)\, P(t_3|t_1, t_2)\, P(t_4|t_1, t_2, t_3)\, P(t_5|t_1, t_2, t_3, t_4)$$
$$P(t_6|t_1, t_2, t_3, t_4, t_5)$$
$$= \frac{\alpha}{\alpha} \cdot \frac{1}{1+\alpha} \cdot \frac{\alpha}{2+\alpha} \cdot \frac{2}{3+\alpha} \cdot \frac{\alpha}{4+\alpha} \cdot \frac{1}{5+\alpha}$$
$$= \frac{2}{(5+\alpha)!}\alpha^2.$$

The Chinese restaurant process has the following properties:

(1) Exchangeability: it means that rearranging the order of customers does not change the probability distribution of choosing tables.
(2) Consistency: it means that the partition rule obtained by removing the nth customer from the random partition is consistent with the random partition rule of the $n - 1$ customer.
(3) From the probability distribution of choosing empty tables $\frac{\alpha}{n-1+\alpha}$, it can be seen that under the same number of customers N, the larger the value of the concentration parameter α, the greater the probability value, resulting in a larger

number of tables with customers, and the fewer the number of customers at each table.

Mathematician David J. Aldous described the Chinese restaurant process in his 1983 book *Exchangeability and Related Topics* and credited it to Jim Pitman and Lester Dubins, who proposed the analogy of a Chinese restaurant for a discrete-time stochastic process after seeing many Chinese restaurants in San Francisco's Chinatown.[2]

These characteristics of the Chinese restaurant process are very useful and can simplify many problems in population genetics, language analysis, and image recognition. In machine learning, the Chinese restaurant process is closely related to Dirichlet processes and plays an important role in Bayesian nonparametric models (Gershman and Blei 2012). It is also used in clustering tasks (Lu et al. 2018).

3.5.2 Markov Property

The Markov property refers to the memory-less property of a stochastic process, that is, given the current state, the next state is only related to the current state and is conditionally independent of past states. The definition is given below.

Definition 3.10 (Markov Property) Given the current state of stochastic process $S(t)$ and past states $\{S(0), \ldots, S(t-1)\}$, if the conditional probability of the next state $S(t+1)$ only depends on the current state $S(t)$ and is irrelevant to the past states $\{S(0), \ldots, S(t-1)\}$, then the stochastic process is said to have Markov properties. It is expressed as

$$P(S(t+1) \mid S(0), \ldots, S(t-1), S(t)) = P(S(t+1) \mid S(t)). \qquad (3.37)$$

The above equation is a key element in determining whether a stochastic process has Markov property.

Below are two simple examples to explain the Markov property.

Exercise 3.6 (The Color of Eggs) There are three eggs in a jar, two with red shells and one with a white shell. If one was taken out yesterday and another one today, what will be the color of the egg to be taken out tomorrow?

If we only know the color of the egg taken out today and not the one taken out yesterday, then the probability of the color of the last egg to be taken out tomorrow is half red and half white; only if we know the color of the eggs taken out both yesterday and today, we can determine the color of the last egg to be taken out tomorrow. Obviously, this kind of stochastic process problem of judging the color of eggs does not have the Markov property.

---

[2] https://en.wikipedia.org/wiki/Chinese_restaurant_process

This is an example of "*sampling without replacement.*"

Exercise 3.7 (The Direction of a Toy Car) A toy car is controlled by a joystick, which can control the toy car to forward, backward, left, and right; the recorder can display the direction of the toy car. When the recorder shows that the direction of the toy car is north, the joystick is turned to the left. Asking: the direction of the toy car?

The answer to this question is self-evident. This is an example of "*sampling with replacement.*"

From this, the following conclusion can be drawn:

Corollary 3.1 (The Property of Sampling with Replacement) *For a stochastic process, sampling with replacement has the Markov property, while sampling without replacement does not have this property.*

Since the model with the Markov property only deals with the next state related to the current state, all Markov models have the Markov property. Some fields require the given model to be a Markov model, such as predictive modeling or probabilistic forecasting.

3.5.3 *Markov Process*

Andrey Markov studied the Markov process as early as around 1900 and published a paper on this topic in 1906. The theoretical basis of the Markov process is the stochastic process, and the Markov property is its important feature.

Definition 3.11 (Markov Process) A Markov process (MP) is a stochastic process with the Markov property, which can be represented as a 2-tuple: $\text{MP} = \langle S, P \rangle$. Where S is a state set and P is the transition probability of the current state transitioning to the next state.

The index set in a Markov process can be discrete or continuous. A Markov process with a discrete index set is expressed as

$$P\left(s_{t+1} | s_t\right) = \mathbb{P}\left[S\left(t+1\right) = s_{t+1} | S\left(t\right) = s_t\right]. \tag{3.38}$$

Where $\mathbb{P}[\cdot]$ denotes the transition matrix, that is,

$$\mathbb{P} = \begin{bmatrix} P_{00} & P_{01} & \cdots & P_{0t} \\ P_{10} & P_{11} & \cdots & P_{1t} \\ \vdots & \vdots & \ddots & \vdots \\ P_{t0} & P_{t1} & \cdots & P_{tt} \end{bmatrix}.$$

Random walks are an instance of the Markov process, which describes the path of a mathematical object in a space, which is a trajectory formed by a series of random steps in the space.

Other examples of random walks include the trajectory of molecules moving in a liquid or gas, the search path of foraging animals, and fluctuating stock prices.

Exercise 3.8 (Gambler's Ruin) A gambling-addicted gambler raises the next bet to a fixed proportion of the amount each time he wins but does not decrease it when he loses. Even if there is an expectation of winning money with each bet, the gambler will inevitably end up losing everything.

The integer random walk and the gambler's ruin problem mentioned above are both part of the discrete-time Markov process. The Brownian motion process and the one-dimensional Poisson process belong to the continuous-time Markov process.

The Markov process is an important type of stochastic process; it is the basis of general random simulation methods and is applied in many fields, for example, the cruise control system of motor vehicles, the queue of airport passengers, currency exchange rates, storage systems, the growth of certain species, and search engines.

3.5.4 Markov Chain

A Markov chain is a Markov process with a discrete state space or discrete index set, usually representing time.

The term Markov chain refers to a sequence of discrete random variables of a process, which defines its serial dependencies only in adjacent states and their change periods, as if in a "chain."

$\forall t$, $S(t) = s_t$ forms a countable set of values:

$$S(0) = s_0, S(1) = s_1, \cdots, S(t) = s_t, \tag{3.39}$$

that is referred to as a chain state space.

A specific Markov chain is defined by its transition probability $P\left(s_{t+1}|s_t\right)$, and the transition probability satisfies the Markov property, that is, it equals the probability of $S(t)$ under the condition of $S(t+1)$, and its calculation method is the same as Eq. 3.38.

Markov chains are usually represented by directed graphs, where the edges of the graph are marked as the probability from the state at time t to the state at time $t+1$; the same information is represented by the transition matrix from time t to time $t+1$.

Exercise 3.9 (Stock Market Trend Chart) Use a Markov chain to represent the bull market, bear market, and stagnant market trends exhibited by a stock market within a week.

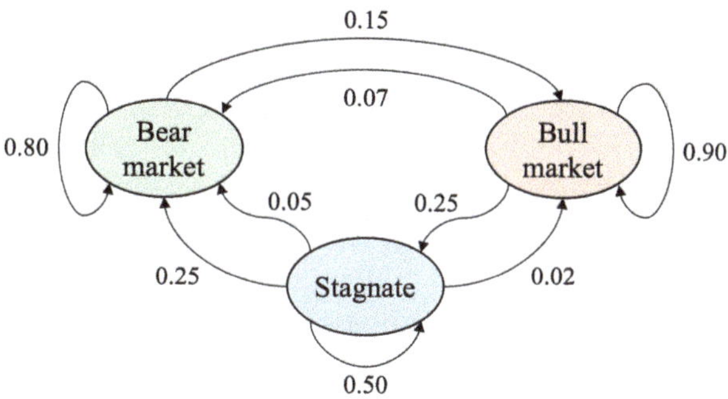

Fig. 3.3 Trend of a stock market within a week

As shown in Fig. 3.3, if the state space is represented as

$$\{1 = \text{Bull market}, 2 = \text{Bear market}, 3 = \text{Stagnant}\},$$

then the transition matrix is as follows:

$$\mathbb{P} = \begin{bmatrix} 0.90 & 0.07 & 0.02 \\ 0.15 & 0.80 & 0.05 \\ 0.25 & 0.25 & 0.50 \end{bmatrix}.$$

Markov chains have a wide range of applications, including physics, chemistry, medicine, music, gaming, and sports. In addition, the combination of Markov chains with the Monte Carlo method, known as Markov chain Monte Carlo (MCMC), is used to simulate random objects with specific probability distributions. This will be discussed in Sect. 3.8.

3.5.5 Hidden Markov Model

The hidden Markov model is used to describe a Markov process with unobservable unknown parameters. Its states can be characterized as a sequence of observation vectors, each of which is characterized as a probability distribution, forming a sequence of states with probability distribution. Therefore, the hidden Markov model is a double stochastic process, that is, a hidden Markov chain with a certain number of states and a set of explicit random functions.

Definition 3.12 (Hidden Markov Model) Let $\{Y_i\}_{i=1}^{n}$ and $\{X_i\}_{i=1}^{n}$ be discrete-time stochastic processes, if it satisfies the following two conditions:

- Y_i is a Markov process with hidden states.
- $P(X_i \mid Y_1 = y_1, \ldots, Y_{i-1} = y_{i-1}, Y_i = y_i) = P(X_i \mid Y_i = y_i)$.

Then, the set of pair $(\{Y_i\}, \{X_i\})_{i=1}^{n}$ is a hidden Markov model. $P(X_i \mid Y_i = y_i)$ is referred to as the output probability.

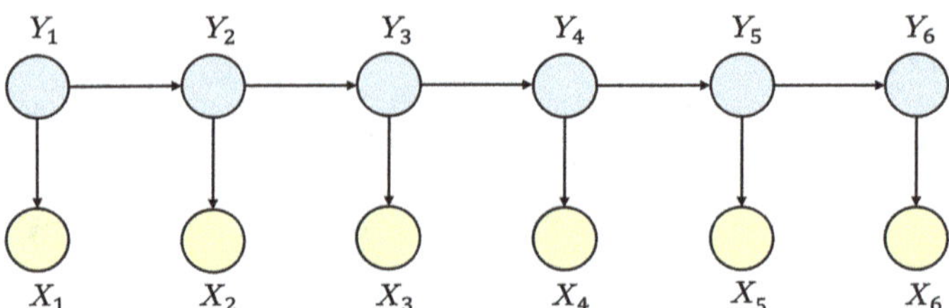

Fig. 3.4 Schematic diagram of the hidden Markov model

The schematic diagram of the hidden Markov model is shown in Fig. 3.4.

The hidden Markov model can be used to represent dynamic Bayesian networks (DBNs), which play an important role in machine learning. Dynamic Bayesian networks are Bayesian networks that unfold over time.

3.6 Probability Models

Probability models incorporate random variables and probability distributions into the model of events. For an event, a deterministic model gives a possible solution, while a probability model gives a probability distribution of a solution.

Early research on probability models mainly focused on the following two models, namely, the *discriminative model* and the *generative model*, also known as the discriminative approach and the generative approach.

Tony Jebara, in February 2002, discussed a general framework based on discriminative and generative models in his doctoral thesis titled *Discriminative, Generative and Imitative Learning* submitted to MIT, which can implement discriminative learning, generative learning, and imitative learning (Jebara 2002). In 2004, he published a book titled *Machine Learning: Discriminative and Generative*, which specifically discusses discriminative and generative models (Jebara 2004).

In 2012, Andrew Ng and Michael I. Jordan presented a paper at the annual *Conference on Neural Information Processing Systems* (NIPS), comparing discriminative and generative models using two representative algorithms as examples (Ng and Jordan 2002).

Some researchers have studied the *hybrid model* that combines discriminative and generative models for classification (Raina et al. 2004; Lester et al. 2005; Druck et al. 2007; Lasserre 2008), and a researcher has also studied the *mixed model* for same objective (Enzweiler and Gavrila 2008).

In addition to the above two probabilistic models, the team of Songchun Zhu, then a professor at the University of California, Los Angeles, introduced *descriptive model* in a paper titled "A Tale of Three Probabilistic Families: Discriminative, Descriptive and Generative Models" published in the second issue of the *Quarterly of Applied Mathematics* in 2019 (Wu et al. 2019).

Fig. 3.5 Schematic diagram
of the discriminative model

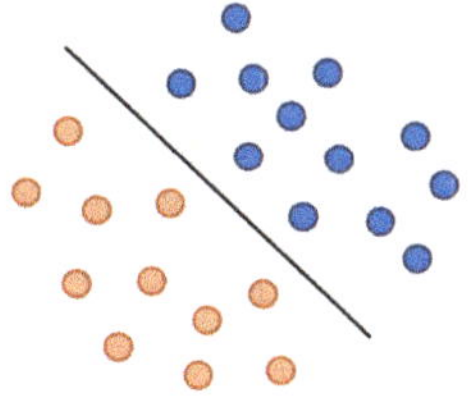

The following will mainly introduce the discriminative model and generative
model, followed by a brief comparison between them.

3.6.1 Discriminative Models

Discriminative models, also known as conditional models, are used to complete
classification or regression tasks in supervised learning. This model calculates the
conditional probability of its output result given the input data. Supervised learning
requires a labeled training dataset.

Let $x_1, \ldots, x_n \in X$ be the input variables, $y_1, \ldots, y_n \in Y$ be the target
output variables, and $S = \{(x_i, y_i) \mid i = 1, \ldots, n\}$ is the training dataset. The
characteristic of the discriminative model is that it models the decision boundary
directly using the conditional probability $P(Y|X)$ for calculation. Figure 3.5 is its
schematic diagram.

In addition, there is another type of discriminative model that learns the mapping
from input X to output Y based on training samples (Ng and Jordan 2002), but this
method does not belong to the category of probability models.

Typical algorithms of discriminative model include logistic regression and
conditional random field (CRF).

Taking the logistic regression method as an example, let $P(Y|X)$ be in parameter
form, $w_1, \ldots, w_n \in W$, and then its parameters are directly estimated from the
training data. The calculation method of logistic regression is as follows:

$$P(Y = 1|X, W) = \frac{1}{1 + \exp\left(w_0 + \sum_{i=1}^{n} w_i x_i\right)}, \tag{3.40}$$

$$P(Y = 0|X, W) = \frac{\exp\left(w_0 + \sum_{i=1}^{n} w_i x_i\right)}{1 + \exp\left(w_0 + \sum_{i=1}^{n} w_i x_i\right)}. \tag{3.41}$$

The above equations further explain the meaning of the statement that the
discriminative model directly calculates the conditional probability.

It should be pointed out that the latter equation is directly derived from the former
because the sum of these two probabilities must be equal to 1.

Fig. 3.6 Schematic diagram
of the generative model

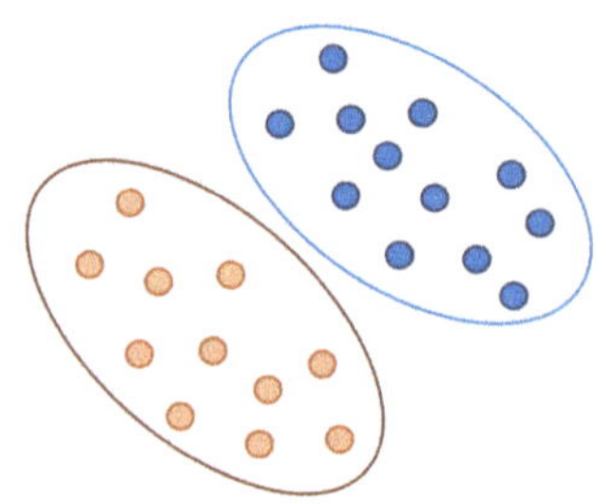

3.6.2 *Generative Models*

The generative model assumes that their results are produced by some latent variables through deterministic transformations. A generative model can be learned in an unsupervised manner, with datasets consisting only of input data and no corresponding output labels.

Let $x_1, \ldots, x_n \in X$ be the input variables and $y_1, \ldots, y_n \in Y$ be the target output variables.

The generative model has the following characteristics; their approach is to model the actual distribution, as shown in Fig. 3.6, which is manifested as directly calculating the joint distribution $P(X, Y)$ (Ng and Jordan 2002).

Therefore, the results of $P(X, Y)$ can be calculated through the joint probability distribution Eqs. 3.7 and 3.8, which is the key point of the generative model.

Classic algorithms of generative models include naive Bayes' classifiers and Gaussian mixture model (GMM).

For example, for $P(Y|X)$, from the joint distribution Eqs. 3.7 and 3.8, we can derive the following equation:

$$P(Y|X) = \frac{P(X, Y)}{P(X)} = \frac{P(X|Y)\,P(Y)}{P(X)}. \tag{3.42}$$

Therefore, we have

$$P(Y = 1 \mid X = x) = \frac{P(X = x, Y = 1)}{P(X = x)}$$

$$= \frac{P(X = x \mid Y = 1)\,P(Y = 1)}{P(X = x)}, \tag{3.43}$$

$$P(Y = 0 \mid X = x) = \frac{P(X = x, Y = 0)}{P(X = x)}$$

$$= \frac{P(X = x \mid Y = 0)\,P(Y = 0)}{P(X = x)}. \tag{3.44}$$

The above two equations not only establish the relationship between conditional distribution and joint distribution but also make the calculation of their joint distribution become the calculation of $P(X = x | Y = y)$ and $P(Y = y)$, which is also a formalized method for modeling actual distributions.

With the rise of deep learning, generative models combined with deep neural networks have formed a new class of methods, namely, deep generative model (DGM). The characteristic is that the deep neural network used as a generative model has many parameters, but these parameters are far less than the amount of data used to train it, so it must force the model to discover the essence of the internal data to generate these data.

Representative deep generative models include variational autoencoders (VAE), generative adversarial networks (GAN), and autoregressive models. These models will be introduced in Chap. 9.

In recent years, building large-scale deep generative models has become a trend. For example, the series of generative pre-trained transformers (GPT-n) launched by OpenAI are the autoregressive models that use deep learning methods to generate humanlike text (Brown et al. 2020).

3.6.3 Comparative Analysis

Discriminative models and generative models are widely used in machine learning. Taking the classification task of machine learning as an example, their differences are reflected in the following two aspects:

(1) The modeling methods are different: the discriminative model is to model the decision boundaries between classes, while the generative model is to model the actual distribution of each class.
(2) The learning probabilities are different: the discriminative model learns a conditional probability distribution, namely, $P(Y|X)$, whereas the generative model learns a joint probability distribution, i.e., $P(Y, X)$.

The representative algorithms based on discriminative model include logistic regression, perceptron, k-nearest neighbors, conditional random field, maximum entropy Markov model, and support vector machine.

And the representative algorithms based on generative model include naive Bayes' classifier, hidden Markov model, Gaussian mixture model, and Boltzmann machine.

3.7 Probabilistic Graphical Models

Probabilistic graphical models (PGMs), also known as graphical models or structured probabilistic models, are probability models that combine probability theory and graph theory.

Specifically, probabilistic graphical models are probability models that use graphs to represent the conditional dependency structure between random variables. They are mainly divided into directed graphical models, undirected graphical models, and factor graph models.

The development of probabilistic graphical models began in the 1980s.

Machine learning pioneer Michael Jordan was one of the early researchers of probabilistic graphical models. In 1998, he edited a book titled *Learning in Graphical Models* (Jordan 1998), which is divided into four main sections and covered articles by 33 authors. The titles of the four sections are Inference, Independence, Foundations for Learning, and Learning from Data.

In 2008, Martin J. Wainwright and Michael Jordan published *Graphical Models, Exponential Families, and Variational Inference* (Wainwright and Jordan 2008), stating: "The formalism of probabilistic graphical models provides a unifying framework for capturing complex dependencies among random variables, and building large-scale multivariate statistical models."

In 2009, Daphne Koller and Nir Friedman published a book titled *Probabilistic Graphical Models: Principles and Techniques* (Koller and Friedman 2009). The book is divided into four major parts with a total of 24 chapters. The titles of the four parts are Representation, Inference, Learning, Actions and Decisions.

In 2015, Luis Enrique Sucar published *Probabilistic Graphical Models: Principles and Applications* (Sucar 2015). The book is divided into four major parts with a total of 13 chapters. The titles of these four parts are Fundamentals, Probabilistic Models, Decision Models, Relational and Causal Models.

This section will first introduce the basic concepts of probabilistic graphs, then discuss the directed graphical model, undirected graphical model, and factor graphical model, and finally make a comparative analysis of the probabilistic graphical models.

3.7.1 Probabilistic Graphs

There are three types of probabilistic graphs (PGs), namely, directed graphs, undirected graphs, and factor graphs.

3.7.1.1 Directed Graphs

Definition 3.13 (Directed Graphs) A directed graph (DG) is an ordered 2-tuple, $DG = \left\langle X, \overrightarrow{E} \right\rangle$, where $X = \{X_i \mid i = 1, 2, \ldots, n\}$ is a set of random variables, X_i is a random variable, $\overrightarrow{E} = \{(X_i \to X_j) \mid X_i, X_j \in X\}$ is a set of directed edges, and $(X_i \to X_j)$ denotes a directed edge from X_i to X_j.

In a directed graph, if $\forall X_i$, $\exists (X_i \to X_i)$, it is called a directed cyclic graph (DCG). Otherwise, if $\forall X_i$, $\nexists (X_i \to X_i)$, it is called a directed acyclic graph (DAG).

3.7.1.2 Undirected Graphs

Definition 3.14 (Undirected Graphs) An undirected graph (UG) is an ordered 2-tuple, $\mathrm{UG} = \langle X, \overline{E} \rangle$, where $X = \{X_i \mid i = 1, 2, \ldots, n\}$ is a set of random variables, X_i is a random variable, $\overline{E} = \{(X_i - X_j) \mid X_i, X_j \in X\}$ is the set of undirected edges, $(X_i - X_j)$ denotes the undirected edge between X_i and X_j, and conversely, $(X_j - X_i)$ also holds.

3.7.1.3 Factor Graphs

Definition 3.15 (Factor Graphs) A factor graph (FG) is an ordered 2-tuple, $\mathrm{FG} = \langle N, \overline{E} \rangle$, where $N = \{X, F\}$ are two disjoint and independent sets of nodes, $X = \{X_i \mid i = 1, 2, \ldots, n\}$ is a set of random variables, $F = \{f_j \mid j = 1, 2, \ldots, m\}$ is a set of factors, and $\overline{E}$ is a set of undirected edges. The random variable is the parameter of the factor, denoted as $f_j\left(\hat{X}_j\right)$, where $\hat{X}_j = \{X_i \mid f_j - X_i \in \overline{E}\} \subseteq X$, $f_j - X_i$ indicates that there is an undirected edge between this factor and this random variable; conversely, $X_i - f_j$ also holds true.

Because the factor graph contains two disjoint and independent node sets, it is called a bipartite graph.

3.7.2 Directed Graphical Models

Definition 3.16 (Directed Graphical Models) A directed graphical model (DGM) is an ordered 2-tuple, $\mathrm{DGM} = \langle \mathrm{DG}, P_{\mathrm{DG}} \rangle$, where DG is a directed graph, P_{DG} is a set of directed graph probability distributions, and $P_{\mathrm{DG}} = \{P_{X_i} \mid X_i \in X\}$.

The directed graph model is suitable for modeling problems of unidirectional dependencies between random variables, represented by Bayesian networks (Pearl 1985).

The Bayesian network is a subset of directed graph model (DGM), which belongs to directed acyclic graph (DAG).

Definition 3.17 (Bayesian Networks) A Bayesian network (BN) is an ordered 2-tuple, $\mathrm{BN} = \langle \mathrm{DAG}, P_{\mathrm{BN}} \rangle$, where DAG is a directed acyclic graph and each random variable probability P_{X_i} in P_{BN} is a conditional probability distribution of the following form:

$$P_{X_i} = P(X_i \mid X_{i-1}, \ldots, X_1) = P(X_i \mid \mathrm{Parents}(X_i)). \tag{3.45}$$

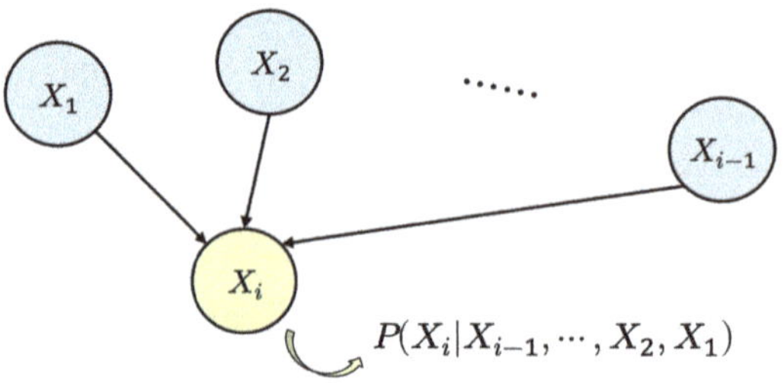

Fig. 3.7 Semantic relationship between nodes and parent nodes in Bayesian network

In the above equation, X_i is a random variable of any node in the directed graph, $X_{i-1}, \ldots, X_1$ are the parent nodes of X_i, represented as Parents (X_i), and its semantic relationship is shown in Fig. 3.7.

Therefore, the joint probability distribution of a Bayesian network P_{DPG} is calculated as follows:

$$P_{\mathrm{DPG}}(X_1, \ldots, X_n) = \prod_{i=1}^{n} P(X_i | \mathrm{Parents}(X_i)). \tag{3.46}$$

In summary, a Bayesian network is a network that uses a directed acyclic graph to represent random variables and their conditional correlations. It includes the following three elements:

(1) A set of nodes, each node corresponds to a random variable.
(2) Directed connections between these nodes.
(3) The conditional probability distribution of each node given its parent nodes, i.e., $P(X_i | \mathrm{Parents}(X_i))$.

The naive Bayes' classifier is the simplest Bayesian network, where samples of a specific category have different data representations on different feature attributes, and there is an independence assumption between features, that is, the features are uncorrelated.

In the directed graph model, there are also hidden Markov models (HMM), maximum entropy Markov models (MEMM), variational autoencoders (VAE) (Kingma and Welling 2014), etc.

The hidden Markov model (HMM) is used to describe a Markov process with hidden states, and the hidden states satisfy the Markov property. The hidden Markov model can be regarded as the simplest dynamic Bayesian network.

The maximum entropy Markov model (MEMM), also known as the conditional Markov model (CMM), is a probabilistic graphical model for sequence labeling that combines the feature of the hidden Markov model and the maximum entropy model. The maximum entropy Markov model extends the standard maximum entropy classifier by connecting the unknown values to be learned in the Markov chain, rather than being conditionally independent of each other.

The variational autoencoders (VAE) will be introduced in Chap. 9.

3.7.3 Undirected Graphical Models

In an undirected graph, there is no direction between any two nodes, so it is impossible to calculate the conditional probability of random variables like in a directed graph model.

Definition 3.18 (Undirected Graphical Models) An undirected graphical model (UGM) is an ordered 2-tuple, $\text{UGM} = \langle \text{UG}, P_{\text{UG}} \rangle$, where UG is an undirected graph, P_{UG} is a collection of probability distributions in the directed graph, and $P_{\text{UG}} = \left\{ \phi_{c_k} \left(X_{c_k} \right) \mid X_{c_k} \subseteq X \text{ and } k = 1, \ldots, K \right\}$, where X_{c_k} is a clique composed of several random variables, K is the maximum number of cliques, $\phi_{c_k} \left(X_{c_k} \right)$ is a nonnegative potential function, and $\phi_{c_k} : \text{Val} \left(X_{c_k} \right) \to \mathbb{R}_+$ is a mapping from the clique X_{c_k} mapping to the nonnegative real number space $\mathbb{R}_+$.

The undirected graph model is suitable for modeling problems where variables are interdependent, represented by Markov networks, also known as Markov random fields (MRFs).

Definition 3.19 (Markov Networks) A Markov network (MN) is an ordered 2-tuple, $\text{MN} = \langle \text{UG}, P_{\text{MN}} \rangle$, where UG is an undirected graph, P_{MN} is a joint probability distribution, and the calculation equation is as follows:

$$P_{\text{MN}} \left(X_1, \ldots, X_n \right) = \frac{1}{Z_{\text{MN}}} \prod_{k \in K} \phi_{c_k} \left(X_{c_k} \right). \tag{3.47}$$

In the above equation, it can be seen that for Markov networks, since there is no clear causal relationship between variables, their joint probability distribution often appears as a product of a series of potential functions. The results of these products are often not equal to 1, so they need to be normalized to form a valid probability distribution. The Z_{MN} in the above equation is a normalization constant of an undirected probability graph model, and its value is equal to

$$Z_{\text{MN}} = \sum_{X_i \in \mathcal{X}} \prod_{k \in K} \phi_{c_k} \left(X_{c_k} \right). \tag{3.48}$$

An example of a Markov network is shown in Fig. 3.8, in which solid circles represent nodes, the nodes contain random variables, and dashed lines represent node circles.

Therefore, the largest node circle of this Markov network is as follows:

$$P_{\text{MN}} \left(X_1, \ldots, X_5 \right) = \frac{1}{Z_{\text{MN}}} \phi_{c_1} \left(X_1, X_2, X_3 \right) \phi_{c_2} \left(X_1, X_3, X_4 \right) \phi_{c_3} \left(X_3, X_5 \right).$$

Fig. 3.8 Example of a
Markov network

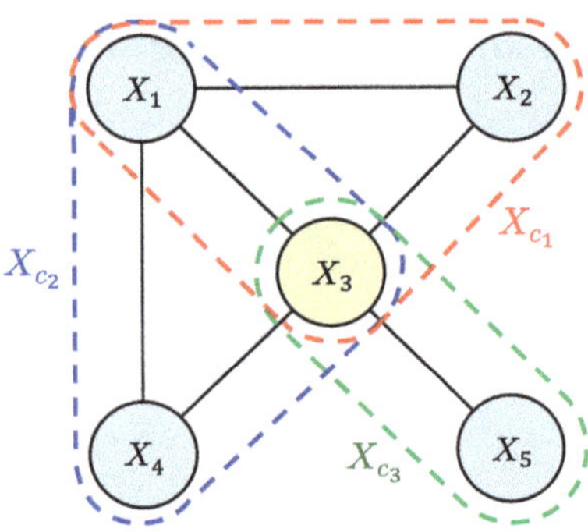

Where

$$X_{c_1} = \{X_1, X_2, X_3\},$$

$$X_{c_2} = \{X_1, X_3, X_4\},$$

$$X_{c_3} = \{X_3, X_5\}.$$

3.7.4 *Factor Graphical Models*

Definition 3.20 (Factor Graphical Models) A factor graphical model (FGM) is
an ordered 2-tuple, $\mathrm{FGM} = \langle \mathrm{FG}, P_{\mathrm{FG}} \rangle$, where FG is a factor graph and P_{FG} is its
joint probability distribution, which can be calculated by the following equation:

$$P_{\mathrm{FG}}(X_1, \ldots, X_n) = \frac{1}{Z_{\mathrm{FG}}} \prod_{j=1}^{m} f_j(X_j). \tag{3.49}$$

Like the undirected graph model, Z_{FG} is the normalization constant of the factor
graph model, and its calculation equation is as follows:

$$Z_{\mathrm{FG}} = \sum_{X_i \in \mathcal{X}} \prod_{j=1}^{m} f_j\left(\hat{X}_j\right). \tag{3.50}$$

Let the set of random variables $X = \{X_1, X_2, X_3, X_4\}$ and the set of factors
$F = \{f_1, f_2, f_3\}$, its factor graph is shown in Fig. 3.9.

According to Eq. 3.49, the corresponding joint probability distribution can be
obtained as follows:

$$P_{\mathrm{FG}}(X_1, X_2, X_3, X_4) = \frac{1}{Z_{\mathrm{FG}}} f_1(X_1) f_2(X_1, X_2, X_3) f_3(X_2, X_4). \tag{3.51}$$

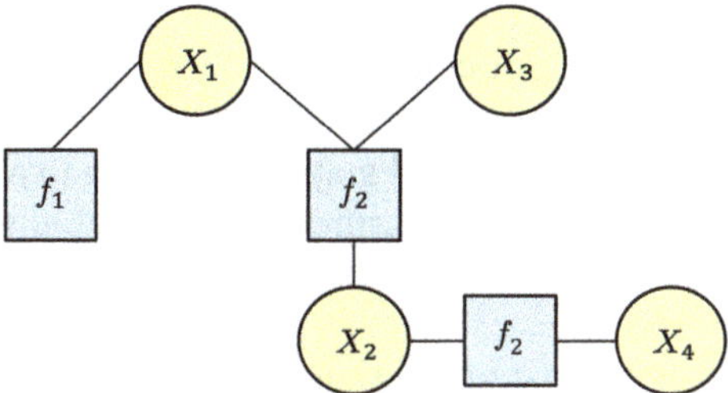

Fig. 3.9 Schematic diagram of the factor graph model

Where Z_{FG} is a normalization constant, which can be obtained from Eq. 3.50.

3.7.5 Comparative Analysis

3.7.5.1 Directed Graphs and Undirected Graphs

The representative of the directed graph model is the Bayesian network, which also includes the naive Bayes' classifier, hidden Markov model (HMM), and maximum entropy Markov model (MEMM). The naive Bayes' classifier is the simplest Bayesian network.

The undirected graph model is the Markov network, also known as the Markov random field (MRF), and the conditional random field (CRF) is its special case (Lafferty et al. 2001).

The schematic diagrams of four common probabilistic graphical models are shown in Fig. 3.10.

3.7.5.2 Discriminative and Generative

In previous section, we have introduced the discriminative model and the generative model.

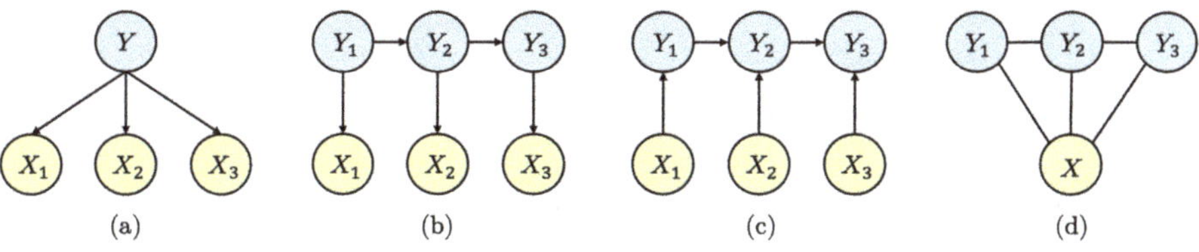

Fig. 3.10 Common probabilistic graphical models, where (**a**) naive Bayes' classifier, (**b**) hidden Markov model, (**c**) maximum entropy Markov model, and (**d**) conditional random field model

Since the form used by the generative model is $P(X, Y) = P(Y|X) P(X)$, therefore, it is natural for the generative model to be represented by directed graphs. Naive Bayes' classifiers, hidden Markov models (HMM), and maximum entropy Markov model (MEMM) in the directed graphs all belong to generative models.

On the other hand, since the discriminative model directly calculates the conditional probability $P(Y|X)$, a discriminative model can be represented by undirected graph. For example, conditional random fields (CRF), also known as Markov random fields, are a type of discriminative model.

However, there are exceptions as well, such as the maximum entropy Markov model (MEMM), which is a directed graph model, but it is a discriminative model.

In summary, in the probability graph models of Fig. 3.10a and b are generative models, while (c) and (d) are discriminative models.

3.8 Monte Carlo Methods

Monte Carlo methods are also known as Monte Carlo approximation or Monte Carlo simulation.

The basic idea of the Monte Carlo method is that when a problem is expressed as the probability of a certain random event or the expected value of a random variable, it is estimated by the frequency of the occurrence of this event through repeated sampling. This probability or the expected value of the random variable is used as an approximate solution to the problem.

Monte Carlo methods are a general term for a class of algorithms, including Monte Carlo integration, rejection sampling, adaptive rejection sampling, importance sampling, sampling importance resampling, Gibbs sampling, slice sampling, Markov chain Monte Carlo, Hamiltonian Monte Carlo, and Langevin Monte Carlo.

Monte Carlo methods have been widely used in physics, physical chemistry, operations research, financial engineering, macroeconomics, biomedical, and other fields and have also played an important role in the field of artificial intelligence, including machine learning.

This section first introduces where the name Monte Carlo method comes from and then explains Monte Carlo integration, importance sampling, Markov chain Monte Carlo, and Monte Carlo tree search methods.

3.8.1 Origin of the Name

The first Monte Carlo method was proposed in the late 1940s, in which the Monte Carlo is the code name for this method to keep it secret. The origin of this code name is related to a very important research project.

In the 1940s, the United States launched the Manhattan Project to develop atomic bombs. John von Neumann, Stanislaw Ulam, and others participated in the research of this project, which was located at the Los Alamos National Laboratory in the United States.

During the execution of the Manhattan Project, physicists needed to study radiation shielding and the distance that neutrons travel through various materials. Although most of the necessary data was already available at the time, such as the average distance that neutrons travel in matter before colliding with atomic nuclei, and the energy that neutrons might release after collision, it was still impossible to solve this problem using traditional deterministic mathematical methods.

Therefore, in 1946, Ulam first proposed the idea of random experiments, using the newly invented electronic computer to study neutron diffusion problems and other problems in mathematical physics. Specifically, it involves turning the process described by certain differential equations into a series of equivalent random operations. His idea was valued by von Neumann, who wrote the corresponding program with the world's first electronic general-purpose digital computer, *ENIAC*. Because using a truly random list of random numbers is very slow, von Neumann developed a program to generate pseudorandom numbers using the middle-square method. These works were carried out under extremely secret conditions and needed a code name, so their colleague Nicholas Metropolis suggested using the name of the Monte Carlo Casino in Monaco, because Ulam's uncle often borrowed money from relatives to gamble there. This is the origin of the name Monte Carlo.

In 1949, Nicholas Metropolis and Stanislaw Ulam published a paper titled "The Monte Carlo Method" in the *Journal of the American Statistical Association* (Metropolis and Ulam 1949).

After the Monte Carlo method was proposed, the RAND Corporation and the US Air Force promoted the Monte Carlo method, which has been widely used in many different fields.

In fact, similar random experiment work had appeared as early as the eighteenth century. In 1777, the French mathematician Comte de Buffon proposed using the "needle-throwing" experiment to estimate the value of π, which is the mathematical Buffon's needle problem and is considered the origin of the Monte Carlo method. Buffon's real name was Georges-Louis Leclerc, who was also a French natural scientist, cosmologist, and encyclopedist. In the 1930s, Enrico Fermi first used random sampling methods in his research on neutron diffusion.

3.8.2   *Monte Carlo Integration*

Usually, for a random variable X and its function $f(X)$, to calculate the expectation of the function, it can be obtained by calculating the integral of $f(x) p(x)$, that is,

$$\mathbb{E}\left[f_p\right] = \int_x f(x) p(x) \, dx, \tag{3.52}$$

where $p(x)$ is the probability distribution of x.

However, the calculation of the above integral is very difficult. The Monte Carlo method is a simple but effective alternative, and the specific method is as follows.

First, from the probability distribution $p(x)$ generates N independent samples, denoted as $x_1, \ldots, x_N$, then use the empirical distribution $\{f(x_i)\}_{i=1}^{N}$ to obtain an approximate $f(X)$ distribution, that is, use it to calculate the expected value of this random variable function, which is expressed in the following form:

$$\int_x f(x)\, p(x)\, dx \approx \frac{1}{N} \sum_{i=1}^{N} f(x_i). \tag{3.53}$$

The above equation is known as Monte Carlo integration. This method can not only solve the integral problems that cannot be directly solved but also has superiority over numerical integration, because the function is only evaluated on the corresponding probability distribution.

The Monte Carlo expectation is denoted as

$$\mathbb{E}\left[\hat{f}_p\right] = \frac{1}{N} \sum_{i=1}^{N} f(x_i). \tag{3.54}$$

Since x_i is a sample from the distribution $p(x)$, therefore we have $\mathbb{E}\left[\hat{f}_p\right] \approx \mathbb{E}\left[f_p\right]$; hence, the variance of $\hat{f}_p$ is

$$\mathrm{Var}\left(\hat{f}_p\right) \approx \frac{1}{N} \mathbb{E}\left[\left(f_p - \mathbb{E}\left[f_p\right]\right)^2\right]. \tag{3.55}$$

Below is an example to further illustrate the Monte Carlo methods.

Exercise 3.10 (Calculating π Using Monte Carlo Method) Here, we give a concept overview. First, consider a unit square with side length $r = 1.0$, and inscribe a circle with the radius r. Next, generate a large number of random points within the square. And then, count the number of points inside the quarter circle and unit square, respectively, the ratio of two numbers is an estimate of the ratio of the two areas, $\frac{\pi}{4}$. Finally, multiply the ratio by 4 to estimate π. The conceptual diagram is shown in Fig. 3.11.

Its Monte Carlo algorithm is as follows.

First, use the definite integral method to calculate the area of the quarter circle S', as shown below:

$$S' = \int_0^r \sqrt{r^2 - x^2}\, dx = \int_0^1 \sqrt{1 - x^2}\, dx.$$

Let $f(x) = \sqrt{1 - x^2}$, when the small bead is inside the circle, the random variable x is 0; otherwise, it is 1. And let $p(x)$ be $[0, r]$ on the uniform distribution,

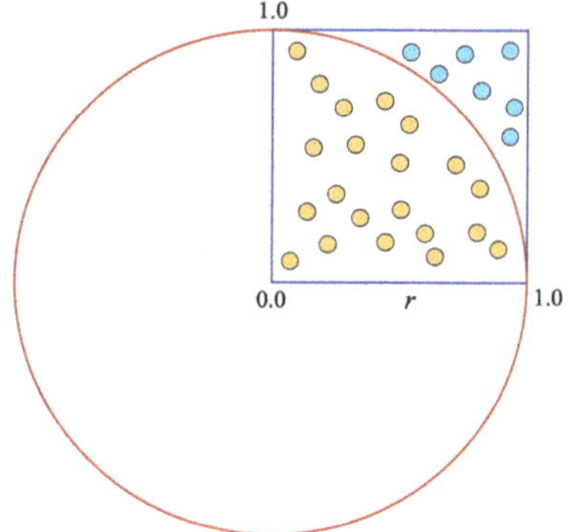

Fig. 3.11 Approximation of the value of π using the Monte Carlo method

hence $p(x) = \frac{1}{r} = 1$. Therefore, using Monte Carlo integration, we get the following result:

$$S' = \int_x f(x)\, p(x)\, dx \approx \frac{1}{N} \sum_{i=1}^{N} f(x_i).$$

The area of the circle is $S = \pi r^2 = 4S'$. Because the radius $r = 1$, therefore we have

$$\pi = 4S' = 4\frac{1}{N} \sum_{i=1}^{N} f(x_i).$$

Where N and $\sum_{i=1}^{N} f(x_i)$ are the numbers of small beads within the square and the quarter circle, respectively. From this, the approximate value of the circumference ratio π can be calculated.

When simulating a process using the Monte Carlo method, it is necessary to generate random variables of various probability distributions and then use methods such as Monte Carlo integration to estimate the numerical characteristics of the variables, thereby obtaining an approximate solution to the actual problem.

The reference (Murphy 2012) also provides another method for estimating the circumference ratio using Monte Carlo integration.

3.8.3 Importance Sampling

Importance sampling is a general technique for estimating the properties of a specific distribution. It does not generate samples from the current probability distribution $p(x)$ but samples from another distribution called the importance proposal distribution $q(x)$ and uses the ratio of the two as a weight. This is the so-called importance sampling.

The introduction of the importance proposal distribution in importance sampling is mainly for two reasons: first, the probability distribution $p(x)$ is often very complex, so another importance proposal distribution that is easy to estimate its importance properties is used for sampling. Second, it is used to reduce the variance brought by the Monte Carlo approximation, which is called variance reduction.

Importance sampling is based on Monte Carlo integration, and the integral of $f(x)p(x)$ can be calculated to obtain the expectation. An item $\frac{q(x)}{q(x)}$ is added to the equation, so that the original integral remains unchanged. Therefore, we have

$$\mathbb{E}\left[f_p\right] = \int_x f(x)\, p(x)\, dx = \int_x f(x)\, \frac{q(x)}{q(x)} p(x)\, dx = \int_x f(x)\, \frac{p(x)}{q(x)} q(x)\, dx. \tag{3.56}$$

Then, the Monte Carlo integration can be calculated using the following equation:

$$\int_x f(x)\, p(x)\, dx \approx \frac{1}{N} \sum_{i=1}^{N} \frac{p(x_i)}{q(x_i)} f(x_i). \tag{3.57}$$

The above equation constitutes the basis of importance sampling, where

$$w_i = \frac{p(x_i)}{q(x_i)}$$

known as importance weights. Therefore, Eq. 3.57 can also be expressed as

$$\int_x f(x)\, p(x)\, dx \approx \frac{\sum_{i=1}^{N} w_i\, f(x_i)}{\sum_{i=1}^{N} w_i}. \tag{3.58}$$

The Monte Carlo expectation of importance sampling is denoted as

$$\mathbb{E}\left[\hat{f}_q\right] = \frac{1}{N} \sum_{i=1}^{N} \frac{q(x_i)}{p(x_i)} f(x_i). \tag{3.59}$$

Then, the variance of the function $\hat{f}_q$ is

$$\mathrm{Var}\left(\hat{f}_q\right) \approx \frac{1}{N} \mathbb{E}\left[\left(f_p - \mathbb{E}\left[f_q\right]\right)^2\right]. \tag{3.60}$$

3.8.4 Markov Chain Monte Carlo

Sampling is an important part of the Monte Carlo method. However, for high-dimensional probability models, Monte Carlo sampling can be difficult to handle or even ineffective. Markov chain Monte Carlo (MCMC) provides a method for random sampling of high-dimensional probability distributions, where the next sample depends on the current sample. That is, by combining the Markov chain with Monte Carlo, and constructing a Markov chain that includes Monte Carlo samples, it can perform random sampling on high-dimensional probability distributions with sample-dependent probabilities.

The Markov chain Monte Carlo method includes several algorithms, the most common of which are *Gibbs sampling* (Andrieu et al. 2003; Bishop 2006; Goodfellow et al. 2016) and the *Metropolis-Hastings algorithm* (Andrieu et al. 2003; Bishop 2006).

Consider a strategy that constructs a finite state space $\mathcal{X} = \{x_1, x_2, \ldots, x_n\}$, uses the Markov chain to explore this state space $\mathcal{X}$, and generates samples $\{x_i\}_{i=1}^{n}$. The establishment of this strategy allows the chain to spend more time sampling in the most important areas, especially to make the sample x_i simulate from the probability distribution $p(x)$ obtained samples. In other words, when we cannot directly sample from $p(x)$ but can calculate $p(x)$ to obtain a normalization constant, use the Markov chain Monte Carlo method (Andrieu et al. 2003; Richey 2010).

The main feature of the Markov chain Monte Carlo method is that it constructs a Markov chain with the expected distribution as its equilibrium distribution and then records the state of the chain to obtain samples of the expected distribution. The more steps the method includes, the closer the sample distribution is to the actual expected distribution.

The Markov chain Monte Carlo method creates samples from a possible set of continuous random variables, whose probability density is proportional to a known function. These samples can be used to estimate the integral of the variable such as the expectation or variance.

In fact, the ensemble of a chain usually starts from a set of arbitrarily chosen points with certain intervals. This is a random walk process; they move randomly according to the algorithm, looking for places with higher contributions to the integral, and then move to the next point, assigning them a higher probability.

Section 3.5.4 of this chapter has introduced the knowledge about Markov chains. The MCMC method adopts the form of transition probability:

$$T_n(x_n, \ x_{n+1}) = p(x_{n+1}|x_n). \tag{3.61}$$

The Markov chain is specified by giving the conditional probability of subsequent variables. If the transition probabilities are the same for all n, it is called a *homogeneous Markov chain.*

The marginal probability of a particular variable can be expressed in the form of the marginal probability of previous variables in the chain:

$$p\left(x_{n+1}\right) = \sum_{x_t} p\left(x_{n+1}|x_n\right) p\left(x_n\right). \tag{3.62}$$

For a Markov chain, if each step in the chain maintains its distribution, then the distribution is said to be stationary. Therefore, for a homogeneous Markov chain with transition probability $T\left(x,\ x'\right)$, if the following condition is met

$$\hat{p}\left(x'\right) = \sum_{x} T\left(x,\ x'\right) \hat{p}\left(x\right), \tag{3.63}$$

then the distribution $\hat{p}\left(x'\right)$ remains unchanged. To ensure the expected distribution $p\left(x\right)$ remains unchanged, a sufficient but not necessary condition is to choose a transformation probability that satisfies the following condition:

$$\hat{p}\left(x\right) T\left(x,\ x'\right) = \hat{p}\left(x'\right) T\left(x',x\right), \tag{3.64}$$

to satisfy the property of detailed balance. It is easy to see that for a specific distribution $\hat{p}\left(x\right)$, the transition probability satisfying its detailed balance will keep this distribution unchanged, because

$$\sum_{x'} \hat{p}\left(x'\right) T\left(x',x\right) = \sum_{x} \hat{p}\left(x\right) T\left(x,\ x'\right) = \hat{p}\left(x\right) \sum_{x} p\left(x'|x\right) = \hat{p}\left(x\right). \tag{3.65}$$

Markov chains that obey detailed balance are reversible.

Our goal is to sample from a given distribution using a Markov chain. This can be achieved by constructing a Markov chain such that the expected distribution is invariant. However, for $n \to \infty$, it is also required that no matter what the initial distribution $p\left(x_0\right)$ is, the distribution $p\left(x_n\right)$ converges to the expected invariant distribution $\hat{p}\left(x\right)$. This property is called ergodicity, and the invariant distribution is called the equilibrium distribution. Obviously, an ergodic Markov chain can only have one equilibrium distribution. It can be seen that the homogeneous Markov chain is ergodic and it is only weakly constrained by the invariant distribution and transition probability (Neal 1993).

In practical applications, a set of transitions, called base transitions, are often constructed from $B_1, \ldots, B_k$ to build transition probabilities. The transition probability can be realized by the following form of mixed distribution:

$$T\left(x,\ x'\right) = \sum_{k=1}^{K} \alpha_k B_k\left(x,x'\right). \tag{3.66}$$

Where the mixture coefficients α_k $(k = 1, \ldots, K)$ satisfy $\alpha_k \geq 0$ and $\sum_k \alpha_k = 1$.

The Markov chain Monte Carlo method is mainly used for the numerical approximation of multiple integrals. Its applications include Bayesian statistics, computational physics, computational biology, computational linguistics, and the field of artificial intelligence including machine learning, such as addressing the search space problem in graph matching by constructing a Markov chain Monte Carlo based on unit algorithms (Zhang and Wang 2017).

3.8.5 Monte Carlo Tree Search

Monte Carlo tree search (MCTS) combines the random simulation of Monte Carlo with a search tree organically, forming a probabilistic heuristic search algorithm used for certain types of decision-making processes, especially for computer games. Important application examples include chess, Go, video game software, and AI poker bots.

Monte Carlo tree search is a best-first search method that does not require a position evaluation function. This algorithm can more accurately predict the most promising direction of movement in the search tree, making its evaluation more precise.

Each loop in the Monte Carlo tree search includes four steps: selection, expansion, simulation, and backpropagation. Monte Carlo tree search expands the search tree based on random sampling of the search space, focusing on analyzing the most promising branch nodes in the search tree. Each node in the Monte Carlo tree corresponds to a state of the game, and the evaluation of this state is based on the Monte Carlo method.

The state value of the Monte Carlo method is usually represented by W_i/N_i where W_i is the reward value of node i and N_i is the visit count of node i.

The steps of each loop of the Monte Carlo tree search are shown in Fig. 3.12, with specific explanations as follows:

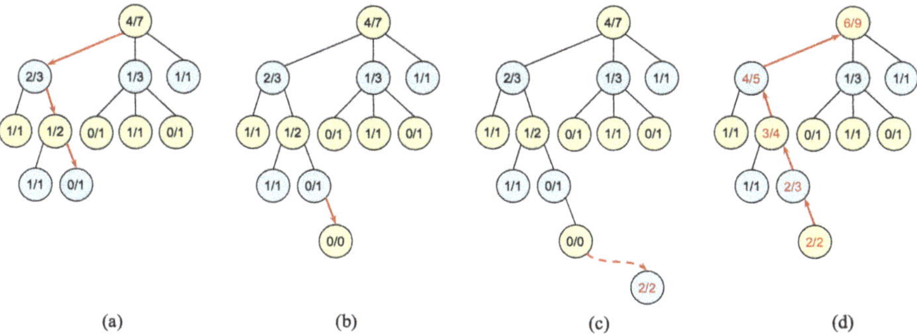

(a)   (b)   (c)   (d)

Fig. 3.12 Steps of Monte Carlo tree search

(a) Selection:

Starting from the root node, according to a certain strategy, a child node of a given node is selected. There are several strategies to choose from (Chaslot et al. 2008); here, we introduce the classic upper confidence bound (UCB) strategy, that is, $v_i + C\sqrt{\frac{\ln N}{n_i}}$, where v_i is the estimated value of node i, n_i is the number of visits to node i, while N is the total number of visits to its parent node and C is an adjustable parameter. Repeat the UCB strategy to get an optimal expansion node N, that is, node 0/1 in the fourth layer of Fig. 3.12a.

(b) Expansion:

Create a new node N_n in the search tree as a new child node of N. And initialize N_n as 0/0 in the fifth layer, as shown in Fig. 3.12b.

(c) Simulation:

In order for N_n to get an initial score, perform a Monte Carlo simulation until a result is obtained, which is used as the initial score of N_n. The Monte Carlo simulation can be purely random or weighted heuristic methods.

(d) Backpropagation:

After obtaining the simulation result of N_n, its parent node N and the state of each node on the backpropagation path will be updated according to the results of this simulation.

The advantage of Monte Carlo tree search is that the algorithm does not require any domain strategy or practical knowledge to make reasonable decisions, which is consistent with the random sampling of the Monte Carlo method. The case of calculating the value of π introduced in Exercise 3.10 is also the same. Monte Carlo tree search can work effectively without specific game knowledge beyond the basic rules of the game. This means that if Monte Carlo tree search is used for other games, only minor adjustments are needed, which makes Monte Carlo tree search a good method for solving game problems.

Monte Carlo tree search also has disadvantages. For example, in some not large-scale games, even within a tolerable time range, the optimal action mode cannot be found. This is mainly due to the small combinatorial space, and the key nodes cannot be visited many times to give a reasonable estimate. This shows that Monte Carlo tree search needs a sufficient number of iterations to converge to an optimal solution, and the Go program probably needs millions of battles to reach the level of a professional player (Munos 2014).

In recent years, many kinds of reinforcement techniques for Monte Carlo tree search have appeared, which can be divided into two categories: domain related and domain independent. The domain-related knowledge in specific game can be used to filter out unreasonable actions or to generate important games during the simulation process, which is closer to human game performance. This means that the result of the game will be close to reality rather than random simulation, so the nodes only need a few iterations to get a reasonable return value. Domain-related knowledge can produce a large performance improvement, but there will be some loss in speed, and it also lacks universality. Domain independent does not depend on a specific domain, so it has universality, which is also a point of concern in the academic world.

Further Reading

1. Zoubin Ghahramani. "Probabilistic Machine Learning and Artificial Intelligence." *Nature*, 521(7553), 2015.
   [Notes] This review paper points out that the probabilistic framework describes how to represent and manipulate uncertainty about models and predictions and has a central role in scientific data analysis, machine learning, robotics, cognitive science, and artificial intelligence.
2. Kevin P. Murphy. *Machine Learning: A Probabilistic Perspective*. MIT press, 2012.
   [Notes] This textbook offers a comprehensive and self-contained introduction to the field of machine learning, based on a unified, probabilistic approach.
3. Leslie G. Valiant. "A Theory of The Learnable." *Communications of the ACM*, 27 (11), 1984.
   [Notes] In this paper, Valiant regards learning as the phenomenon of knowledge acquisition in the absence of explicit programming. He gives a precise methodology for studying this phenomenon from a computational viewpoint. His work pioneered the field of computational learning theory.
4. Anselm Blumer, Andrzej Ehrenfeucht, David Haussler, and Manfred K. Warmuth. "Occam's razor." *Information Processing Letters*, 24 (6), 1987.
   [Notes] The paper applies Occam's razor to machine learning problems, suggesting that given two explanations for the data, all other things being equal, the simpler explanation is preferable. As a result, Occam's razor has become a signboard in the field of machine learning.
5. Michael J. Kearns, and Umesh V. Vazirani. *An Introduction to Computational Learning Theory*. MIT press, 1994.
   [Notes] The book is divided into eight chapters, which discuss the PAC learning model, Occam's razor, VC dimension, and other issues related to computational learning theory.
6. Leslie G. Valiant. *Probably Approximately Correct: Nature's Algorithms for Learning and Prospering in a Complex World*. Basic Books, 2013.
   [Notes] In this book, Valiant presents a masterful synthesis of learning and evolution to show how both individually and collectively we not only survive but prosper in a world as complex as our own. The key is probably approximately correct algorithms, a concept Valiant developed to explain how effective behavior can be learned.

References

Andrieu, C., N. De Freitas, A. Doucet, and M. I. Jordan. (2003). An introduction to MCMC for machine learning. *Machine Learning* 50(1–2): 5–43.

Angluin, D. (1988).Queries and concept learning. *Machine Learning* 2(4): 319–342.

Bishop, C. M. (2006). *Pattern recognition and machine learning*. London: Springer.

Blanchard, T., T. Lombrozo, and S. Nichols. (2018). Bayesian Occam's Razor is a razor of the people. *Cognitive Science* 42(4): 1345–1359.

Blumer, A., A. Ehrenfeucht, D. Haussler, and M. K. Warmuth. (1987).Occam's razor. *Information Processing Letters* 24(6): 377–380.

Brown, T., B. Mann, N. Ryder, M. Subbiah, J. D. Kaplan, P. Dhariwal, A. Neelakantan, P. Shyam, G. Sastry, and A. Askell. (2020). Language models are few-shot learners. In *Conference on Neural Information Processing Systems (NIPS)* 33: 1877–1901.

Chaslot, G., S. Bakkes, I. Szita, and P. Spronck. (2008). Monte-Carlo tree search: A new framework for game AI. In *Proceedings of the AAAI Conference on Artificial Intelligence and Interactive Digital Entertainment*.

Clark, A., and S. Lappin. (2011). *Probabilistic learning theory for language acquisition*. Linguistic Nativism and the Poverty of the Stimulus.

Druck, G., C. Pal, A. McCallum, and X. Zhu. (2007). *Semi-supervised classification with hybrid generative/discriminative methods*. New York: ACM.

Enzweiler, M., and D. M. Gavrila. (2008). A mixed generative-discriminative framework for pedestrian classification. In *IEEE Conference on Computer Vision and Pattern Recognition (CVPR)*. Piscataway: IEEE.

Germain, P., F. Bach, A. Lacoste, and S. Lacoste-Julien. (2016). PAC-Bayesian theory meets Bayesian inference. In *Conference on Neural Information Processing Systems (NIPS)*.

Gershman, S. J., and D. M. Blei. (2012). A tutorial on bayesian nonparametric models. *Journal of Mathematical Psychology* 56(1): 1–12.

Ghahramani, Z. (2015). Probabilistic machine learning and artificial intelligence. *Nature* 521: 452.

Goodfellow, I., Y. Bengio, and A. Courville. (2016). *Deep learning*. Cambridge: MIT Press.

Gordon, A. D., T. A. Henzinger, A. V. Nori, and S. K. Rajamani. (2014). *Probabilistic programming. proceedings of the on future of software engineering*. New York: ACM.

Guedj, B. (2019). A primer on PAC-Bayesian learning. arXiv preprint arXiv:1901.05353.

Jebara, T. (2002). Discriminative, generative and imitative learning, PhD thesis, Media laboratory, MIT.

Jebara, T. (2004). *Machine learning: Discriminative and generative*. Berlin: Springer.

Jordan, M. I. (1998). *Learning in graphical models*. Berlin: Springer.

Kearns, M. J., and U. V. Vazirani. (1994). *An introduction to computational learning theory*. Cambridge: MIT Press.

Kingma, D. P., and M. Welling. (2014). Auto-encoding variational Bayes. In *International Conference on Learning Representations (ICLR)*.

Koller, D., and N. Friedman. (2009). *Probabilistic graphical models: Principles and techniques*. Cambridge: MIT Press.

Korb, K. B., and A. E. Nicholson. (2011). *Bayesian Artificial Intelligence*, 2nd edn. Boca Raton: CRC Press.

Lafferty, J., A. McCallum, and F. Pereira. (2001). Conditional random fields: probabilistic models for segmenting and labeling sequence data. In *Proceedings of International Conference on Machine Learning (ICML)*.

Lasserre, J. A. (2008). *Hybrids of generative and discriminative methods for machine learning*. Cambridge: University of Cambridge.

Lester, J., T. Choudhury, N. Kern, G. Borriello, and B. Hannaford. (2005). A hybrid discriminative/generative approach for modeling human activities. In *International Joint Conferences on Artificial Intelligence (IJCAI)*.

Lu, J., M. Li, and D. Dunson. (2018). Reducing over-clustering via the powered Chinese restaurant process. arXiv preprint arXiv:1802.05392.

McAllester, D., and T. Akinbiyi. (2013). *PAC-Bayesian theory. Empirical inference*, 95–103. Berlin: Springer.

McAllester, D. A. (1998). Some PAC-Bayesian theorems. In *Conference on Computational Learning Theory*.

McAllester, D. A. (1999). *PAC-Bayesian model averaging*.

Metropolis, N., and S. Ulam. (1949). The Monte Carlo method. *Journal of the American Statistical Association* 44(247): 335–341.

Mohri, M., A. Rostamizadeh, and T. Ameet. (2012). *Foundations of machine learning*. Cambridge: MIT Press.

Munos, R. (2014). From Bandits to Monte-Carlo tree search: The optimistic principle applied to optimization and planning. *Foundations and Trends ® in Machine Learning* 7(1): 1–129.

Murphy, K. P. (2012). *Machine learning: A probabilistic perspective*. Cambridge: The MIT Press.

Murray, I., and Z. Ghahramani. (2005). A note on the evidence and Bayesian Occam's razor. Gatsby Computational Neuroscience Unit Tech Report.

Neal, R. M. (1993). *Probabilistic inference using Markov chain Monte Carlo methods*. Department of Computer Science, University of Toronto Toronto, Ontario, Canada.

Ng, A. Y., and M. I. Jordan. (2002). On discriminative vs. generative classifiers: A comparison of logistic regression and naive Bayes. In *Conference on Neural Information Processing Systems (NIPS)*.

Oblow, E. M. (1992). Implementing Valiant's learnability theory using random sets. *Machine Learning* 8(1): 45–73.

Pearl, J. (1985). Bayesian networks: A model of self-activated memory for evidential reasoning.

Raina, R., Y. Shen, A. Mccallum, and A. Y. Ng. (2004). Classification with hybrid generative/discriminative models. In *Conference on Neural Information Processing Systems (NIPS)*.

Richey, M. (2010). The evolution of Markov chain Monte Carlo methods. *The American Mathematical Monthly* 117(5): 383–413.

Shawe-Taylor, J., R. C. Williamson. (1997). A PAC analysis of a Bayesian estimator. In *Conference on Computational Learning Theory*.

Sucar, L. E. (2015). *Probabilistic graphical models: Principles and applications*. London: Springer.

Valiant, L. (1984a). A theory of the learnable. In *ACM Symposium on Theory of Computing*.

Valiant, L. (1984b). A theory of the learnable. *Communications of the ACM* 27: 1134–1142.

Valiant, L. (2013). *Probably approximately correct: Nature's algorithms for learning and prospering in a complex world*. Basic Books.

Vapnik, V. N. (1998). *Statistical learning theory*. New York: Wiley.

Vapnik, V. N. (1999). *The nature of statistical learning theory*, 2nd edn. New York: Springer.

Wainwright, M. J., and M. I. Jordan. (2008). Graphical models, exponential families, and variational inference. *Machine Learning* 1(1–2): 1–305.

Wu, Y. N., R. Gao, T. Han, and S.-C. Zhu. (2019). A tale of three probabilistic families: Discriminative, descriptive and generative models. *Quarterly of Applied Mathematics* 77(2): 423–465.

Zhang, R., and W. Wang. (2017). Second-and high-order graph matching for correspondence problems. *IEEE Transactions on Circuits and Systems for Video Technology* 28(10): 2978–2992.

Chapter 4
Statistical Framework

Abstract Statistics and probability theory complement each other, so do statistical framework and probabilistic framework. This chapter begins with an overview of statistical framework, including statistics and statistical learning. Secondly, we review the two component of classic statistics, namely, descriptive statistics and inferential statistics. We next introduce two statistical inference methods in mathematical statistics, namely, frequentist inference and Bayesian inference. This chapter then discusses statistical learning theory, where statistical models are the cornerstone of statistical learning theory, statistical learning models are the core of statistical learning theory, and growth functions, VC dimension, and Rademacher complexity are the components of statistical learning theory. Parametric and nonparametric models, as well as kernel methods, are also important components of statistical frameworks, and they are detailed in the last two sections of this chapter.

4.1 Overview

The foundation of the statistical framework is statistics and mathematical statistics, and the core of the statistical framework is statistical learning theory.

4.1.1 Statistics

Statistics is a discipline that collects, organizes, represents, analyzes, interprets, and makes predictions based on data. The main methods include descriptive statistics and inferential statistics.

Mathematical statistics, also known as modern statistics, is the application of mathematics in statistics. Its mathematical foundation is probability theory, and it also includes mathematical analysis, linear algebra, stochastic analysis, and differential equations. The main content includes probability distributions, statistical inference, regression analysis, and nonparametric statistics.

Among them, the probability distribution is a function, such as the normal distribution, or binomial distribution, which provides accurate and effective data representation means for random experiments, statistical inference, machine learning, etc. Statistical inference is the process of inferring the characteristics of the probability distribution of random variables using probability theory methods. Regression analysis is a statistical process for estimating the relationship between dependent variables and independent variables, which is also used by machine learning to solve regression tasks, while nonparametric statistics is a type of statistical analysis that does not rely on the assumption of a specific underlying distribution or any other specific assumptions about the population parameters.

Statistics and probability theory have a natural connection, both are research fields focused on analyzing the relative frequency of events, but their focuses are different: probability theory is used to predict the possibility of future events, while statistics is used to analyze the frequency of past events.

Someone made an analogy to explain the relationship between probability and statistics: given a bucket of balls, two colors each occupy a certain proportion, and a certain number of balls are drawn from it; probability theory is to infer the color of the balls in hand, given the color of the balls in the bucket; in contrast, statistics is to estimate the color of the balls in the bucket, given the color of the balls in hand. This analogy vividly explains the relationship between the population and the sample in probability and statistics.

From the perspective of methodology, probability theory is inference, statistics is induction; probability theory is the foundation, and statistics is the application.

4.1.2 Statistical Learning

Statistical learning is a theoretical analysis framework of machine learning, originating from statistics and functional analysis.

Vladimir N. Vapnik's *Statistical Learning Theory* (Vapnik 1998) and *The Nature of Statistical Learning Theory* (Vapnik 1999) are classic books on statistical learning theory. The two books (Bousquet et al. 2003; James et al. 2013) are introductory books on statistical learning.

The most direct result of statistical learning theory is the support vector machine (SVM) algorithm in machine learning (Boser et al. 1992; Cortes and Vapnik 1995). We have in Sect. 1.3.4 of this book reviewed the history of the development of support vector machines.

Parametric models and nonparametric models are important models in statistics, especially in statistical learning. From the viewpoint of parameters, the parametric models refer to the models that have a fixed number of parameters, while the nonparametric models are not parameter-less but refer to the models whose number of parameters grows with the amount of data.

Kernel methods also play an important role in statistical learning. The "kernel" refers to a window function, and "kernel methods" are a class of algorithms based on kernel functions, including support vector machines (SVM); many machine learning algorithms use kernel methods.

It is worth pointing out that statistical learning theory is a framework of computational learning theory and an important part of machine learning theory. However, statistical learning is not equivalent to machine learning, and the view that machine learning is modern statistics is somewhat biased.

4.2 Descriptive Statistics

Descriptive statistics is the summary statistics that quantitatively calculates and describes data through numerical calculations or visualization. It aims to summarize sample datasets, rather than using these sample data to obtain information about the population. Therefore, descriptive statistics is also called descriptive statistical analysis.

Specifically, when describing a sample in descriptive statistics, only the sample of interest is obtained, the data in the sample is recorded, and then the summary statistical information and charts are used to display the attributes of this group of samples. Descriptive statistics only describe the actual measured samples, not trying to infer the properties of a complete population.

In descriptive statistics, it is necessary to describe the distribution form and distribution characteristics of random variables and to perform numerical calculations on random variables.

The distribution form of random variables can be described by indicators such as skewness and kurtosis. Its distribution form is determined by the number of peaks it contains, which can be divided into three types: unimodal, bimodal, and multimodal.

The distribution characteristics of random variables can be represented in graphical or tabular form, including histograms, boxplots, and stem-and-leaf plots.

The numerical calculations of descriptive statistics mainly include measures of central tendency and measures of dispersion, the content of which is shown in Fig. 4.1.

Next, we will introduce these two types of numerical calculation methods in descriptive statistics.

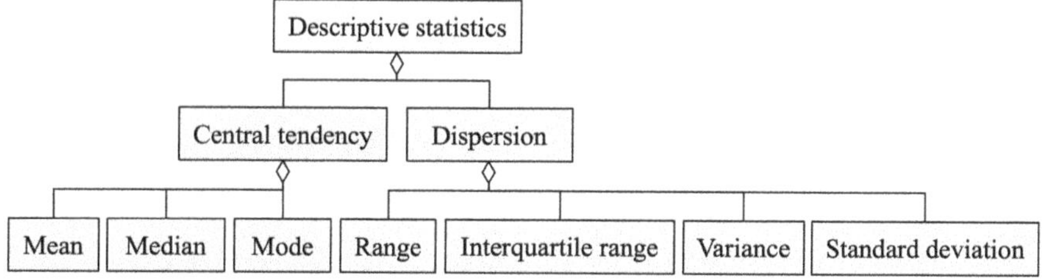

Fig. 4.1 Content of numerical calculations in descriptive statistics

4.2.1   *Central Tendency*

Central tendency refers to the central value or typical value of a probability distribution, which can also be called the center or location of the probability distribution. In plain speaking, the measure of central tendency is referred to as the measure of the average.

The central tendency mainly includes the mean, median, and mode.

4.2.1.1   Mean

The mean, also known as the arithmetic mean, refers to the average, which is the sum of the data items in the dataset divided by the number of data items.

Let $x_1, x_2, \ldots, x_n$ is a finite dataset, $\bar{x}_n$ represents its mean, and its calculation equation is as follows:

$$\bar{x}_n = \frac{1}{n} \sum_{i=1}^{n} x_i. \tag{4.1}$$

4.2.1.2   Median

Here, we discuss the issue of the median for the following two situations.

Firstly, for the finite ordered dataset $x_1, x_2, \ldots, x_n$, let m denote the median, and its calculation equation is as follows:

$$m = \frac{1}{2} \left(x_{\lfloor (n+1)/2 \rfloor} + x_{\lceil (n+1)/2 \rceil} \right). \tag{4.2}$$

Where $\lfloor \cdot \rfloor$ and $\lceil \cdot \rceil$ represent the floor function and the ceiling function, respectively.

As can be seen from the above equation, the median is the "middle" value: if the number of items in the ordered dataset is odd, the median is the middle data item; if the number of items is even, the median is the sum of the two adjacent middle numbers divided by 2.

Secondly, for any probability distribution, a median of a real number m can be found.

If the probability distribution is discrete, the following inequality is satisfied:

$$P(X \leq m) \geq \frac{1}{2} \text{ and } P(X \geq m) \geq \frac{1}{2}. \tag{4.3}$$

If the probability distribution is continuous, the inequality is as follows:

$$\int_{(-\infty,m]} p\,(x)\,dx \geq \frac{1}{2} \quad \text{and} \quad \int_{[m,\infty]} p\,(x)\,dx \geq \frac{1}{2}. \tag{4.4}$$

4.2.1.3 Mode

For a dataset with a finite number of items, the mode is the number that appears most frequently in the dataset.

If X is a discrete random variable, then the mode is the value of x at which the probability mass function takes its maximum value, that is, $X = x$. In other words, it is the value most likely to be sampled. For a given discrete distribution, the mode is not necessarily unique, as the probability mass function may take its maximum value at several points such as x_1, x_2. The extreme case is when there is a uniform distribution, where all values are equal.

When the probability density function of a continuous distribution has multiple local maxima, it is common to treat all the local maxima as the mode of the distribution. Such a continuous distribution is called a multi-peak distribution. The mode of a continuous probability distribution is often taken to be any value of x for which its probability density function has a local maximum, so that any peak is a mode.

4.2.1.4 Comparative Analysis

The central tendency has the following properties: if a given dataset is symmetrically distributed, the mean, median, and mode are the same; otherwise, the median and mean are sequentially biased to the skewed side to the mode. The schematic diagram of the above properties is shown in Fig. 4.2.

Let's look at following simple exercise to see how the above three measures relate to each other.

Exercise 4.1 (Comparing the Central Tendency of the Following Datasets) Given dataset 1 $= \{1, 2, 3, 4, 5, 5, 6, 7, 8, 9\}$ and dataset 2 $= \{1, 2, 2, 3, 4, 5, 6, 7,$

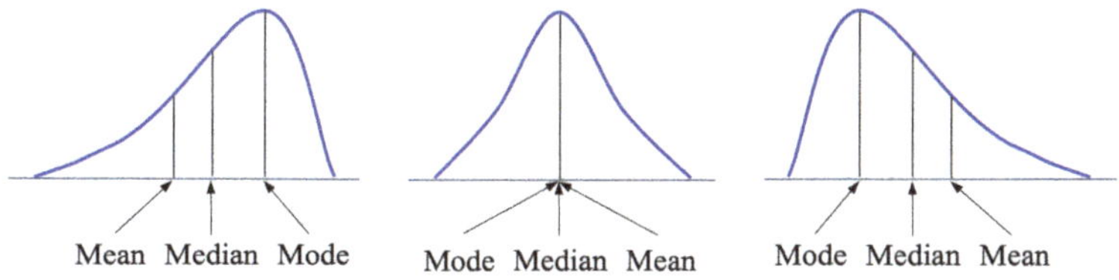

Fig. 4.2 Central tendency under different data distribution

Table 4.1 Comparison of mean, median, and mode

	Dataset	Calculate	Result
Mean	Dataset 1	$(1+2+3+4+5+5+6+7+8+9)/10$	5
	Dataset 2	$(1+2+2+3+4+5+6+7+8+9)/10$	4.7
Median	Dataset 1	$(5+5)/2$	5
	Dataset 2	$(4+5)/2$	4.5
Mode	Dataset 1	$\cdots, 5, 5, \cdots$	5
	Dataset 2	$\cdots, 2, 2, \cdots$	2

8, 9}, then the instances and results of their mean, median, and mode are shown in Table 4.1, respectively.

4.2.2 Dispersion

The dispersion also known as spread or variability measures the dispersion or change in the distribution of a dataset. It consists of range, interquartile range, variance, and standard deviation.

4.2.2.1 Range

The range is the difference between the largest and smallest values in an ordered dataset.

Because its calculation is very simple, it can reflect the dispersion of the ordered datasets from one side. However, when there are outliers in the dataset, it is difficult to reflect the actual dispersion of the dataset.

4.2.2.2 Interquartile Range

Given a dataset whose items are in ascending order, as shown in Fig. 4.3, the interquartile range (IQR) is the difference between the upper quartile value Q_3 and the lower quartile value Q_1, that is, $IQR = Q_3 - Q_1$. In other words, if the dataset

Fig. 4.3 Schematic diagram of the interquartile range

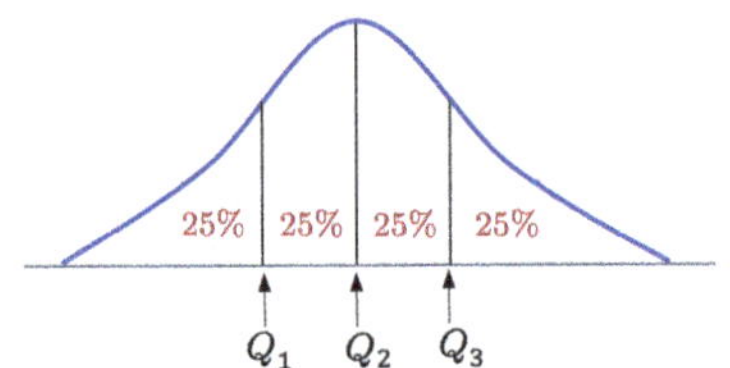

is divided into four equal parts, the interquartile range is the value at the position of the three dividing points.

Therefore, the interquartile range is also known as quartiles, mid-spread, or middle 50%, where the term "middle 50%" is more illustrative.

4.2.2.3 Variance and Standard Deviation

The calculation of variance and standard deviation is an important part of statistical analysis.

Let $x_1, x_2, \ldots, x_n$ is a finite set of data, and σ^2 represents the variance, then its calculation equation is as follows:

$$\sigma^2 = \frac{1}{n} \sum_{i=1}^{n} (x_i - \overline{x}_n)^2 . \tag{4.5}$$

Where $\overline{x}_n$ is the mean.

Careful readers will find that the variance has already been introduced in Sect. 3.2 "Basics of Probability Theory." The mathematical representations of variance in statistics and probability theory are not exactly the same. If interested, you can compare the two to further understand the similarities and differences between statistics and probability theory.

The standard deviation is defined as the square root of the variance, represented by the symbol σ, that is,

$$\sigma = \sqrt{\sigma^2}. \tag{4.6}$$

4.3 Inferential Statistics

Inferential statistics, also known as inductive statistics, is the process of drawing a sample from the population, inferring the distribution characteristics of the sample through statistical analysis, and then obtaining the properties of the population.

In inferential statistics, the observed dataset is considered to be a sample obtained from the population, so it is necessary to ensure that the sample accurately reflects the population, which is an important factor for inferential statistics. Generally, the following points must be achieved:

(1) Define the population under study.
(2) Draw a representative sample from this population.
(3) Perform an analysis that includes sampling errors.

Inferential statistics is based on descriptive statistics and probability theory and uses the regularities inferred from sample data to infer, predict, and estimate

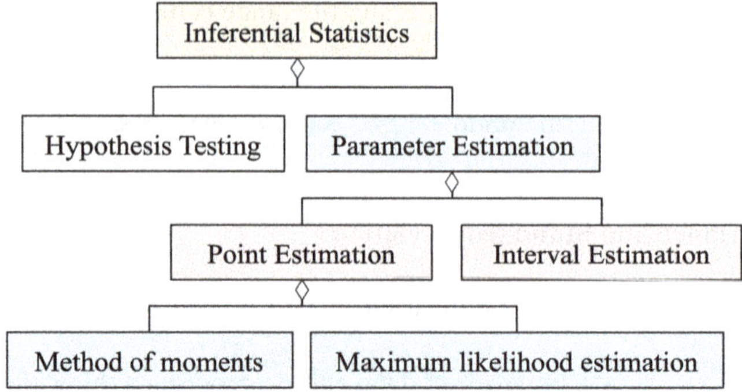

Fig. 4.4 Main types covered by inferential statistics

the properties of the population. Inferential statistics is suitable for inferring the population from a large sample data.

The main types covered by inferential statistics are shown in Fig. 4.4. The difference between it and descriptive statistics is that the latter only focuses on the properties of the observed data, not on the assumption that the observed dataset comes from the population.

Next, we will introduce the two main types of inferential statistics, namely, parameter estimation and hypothesis testing.

4.3.1 Parameter Estimation

Parameter estimation is the process of statistical data based on samples drawn from the population, and this statistical data is used as an approximation of the population parameters.

There are mainly two subtypes of parameter estimation, namely, point estimation and interval estimation. Their relationship is shown in Fig. 4.5.

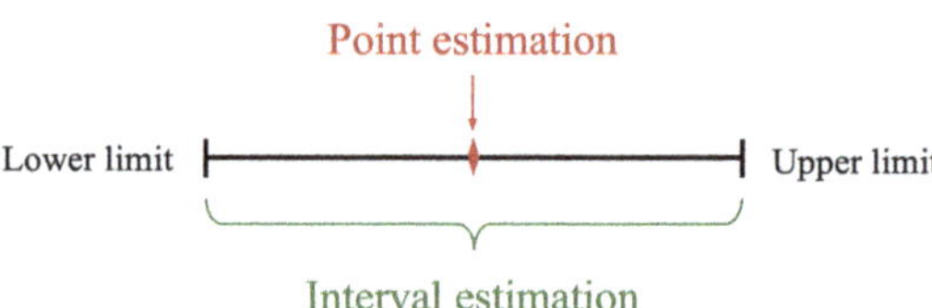

Fig. 4.5 Relationship between point estimation and interval estimation

4.3.1.1 Point Estimation

Point estimation is to estimate the parameters in the population distribution based on a certain characteristic value of the sample. Common methods of point estimation include the method of moments, maximum likelihood estimation, Bayesian estimation, and expectation-maximization (EM) algorithm. The method of moments and maximum likelihood estimation are introduced in the following.

The method of moments is a method for estimating population parameters, which equates (observable) sample moments with (unobservable) population moments and then solves this equation to estimate its value.

The kth moment of the random variable X is denoted as $E\left(X^k\right)$. Let the distribution of the random variable X be $f_X(w; \theta)$, the task is to estimate k unknown parameters $\theta_1, \theta_2, \ldots, \theta_k$, and the kth moment (population moment) of the true distribution can be represented as θ function, namely,

$$\mu_1 \equiv E\left(X^1\right) = g_1\left(\theta_1, \theta_2, \ldots, \theta_k\right),$$

$$\mu_2 \equiv E\left(X^2\right) = g_2\left(\theta_1, \theta_2, \ldots, \theta_k\right),$$

$$\vdots$$

$$\mu_k \equiv E\left(X^k\right) = g_k\left(\theta_1, \theta_2, \ldots, \theta_k\right). \tag{4.7}$$

Suppose a sample of length n is obtained, the result is $w_1, w_2, \ldots, w_n$. Let

$$\hat{\mu}_j = \frac{1}{n}\sum_{i=1}^{n} w_i^j, \quad j = 1, 2, \ldots, k, \tag{4.8}$$

to estimate the jth sample moment of μ_j. The method of moments denoted by $\hat{\theta}_1, \hat{\theta}_2, \ldots, \hat{\theta}_k$ is defined as the solution to the following equation:

$$\hat{\mu}_1 = g_1\left(\hat{\theta}_1, \hat{\theta}_2, \ldots, \hat{\theta}_k\right),$$

$$\hat{\mu}_2 = g_2\left(\hat{\theta}_1, \hat{\theta}_2, \ldots, \hat{\theta}_k\right),$$

$$\vdots$$

$$\hat{\mu}_k = g_k\left(\hat{\theta}_1, \hat{\theta}_2, \ldots, \hat{\theta}_k\right). \tag{4.9}$$

Maximum likelihood estimation (MLE) is also a commonly used statistical method for fitting statistical models to data and estimating the parameters of the model.

Typically, for a set of fixed data and potential statistical models, the maximum likelihood method will choose the model parameters, that is, the parameters that maximize the likelihood function. These parameter values will generate the distribution of the observed data. Maximum likelihood estimation (MLE) provides a unified estimation method, which is well-defined in the case of normal distribution and many other problems.

Let $\theta \in \Theta$ be the unknown parameters appearing in the distribution of X. For an observed dataset of size n, $X = x_1, x_2, \ldots, x_n$, the maximum likelihood estimation (MLE) method is used to estimate θ.

First, construct the likelihood function of n observed dataset $L_n(X|\theta) = L_n(x_1, x_2, \ldots, x_n|\theta)$. Since the observed dataset X is a discrete random variable, its probability mass function is $P\{X = x_i\} = p(x_i|\theta)$, and then its likelihood function is

$$L_n(x_1, x_2, \ldots, x_n|\theta) = p(x_1|\theta)p(x_2|\theta)\cdots p(x_n|\theta) = \prod_{i=1}^{n} p(x_i|\theta). \qquad (4.10)$$

Where the estimated value of θ is

$$\hat{\theta}_{\text{MLE}} = \arg\max_{\theta \in \Theta} L_n(X|\theta) = \arg\max_{\theta \in \Theta} L_n(x_1, x_2, \ldots, x_n|\theta). \qquad (4.11)$$

Intuitively, the above equation will choose the parameter value that makes the observed dataset most likely. The $L_n(X|\theta)$ that maximizes the likelihood function $\hat{\theta}_{\text{MLE}} \in \Theta$ is known as the maximum likelihood estimation. In practice, to turn the product of the likelihood function into addition, the natural logarithm of the likelihood function is usually used, which is called the log-likelihood function, i.e.,

$$\ln L_n(X|\theta) = \ln L_n(x_1, x_2, \ldots, x_n|\theta) = \sum_{i=1}^{n} \ln p(x_i|\theta). \qquad (4.12)$$

The above equation is calculated as follows:

$$\frac{\partial \ln L_n(X|\theta)}{\partial \theta} = 0,$$

to obtain its maximum value.

Since the logarithmic function is monotonic, the log-likelihood function $\ln L_n(X|\theta)$ and the likelihood function $L_n(X|\theta)$ reach their maximum values at the same point.

Exercise 4.2 (Estimating Parameters Using Maximum Likelihood Estimation)
Let X be an exponential distribution with parameter λ. Based on a sample with size $= n$, using the maximum likelihood estimation (MLE) method to estimate the parameter λ.

Since $X : \mathcal{E}(\lambda)$, the likelihood function is

$$L_n(x_1, x_2, \ldots, x_n, \lambda) = \lambda^n \exp\left\{-\lambda \sum_{i=1}^{n} x_i\right\},$$

and

$$\ln L_n(X|\lambda) = n \ln \lambda - \lambda \sum_{i=1}^{n} x_i.$$

Therefore

$$\frac{\partial \ln L_n(X|\lambda)}{\partial \lambda} = \frac{n}{\lambda} - \sum_{i=1}^{n} x_i.$$

By solving the following equation

$$\frac{n}{\lambda} - \sum_{i=1}^{n} x_i = 0,$$

we get

$$\hat{\lambda}_{\mathrm{MLE}} = \frac{1}{\frac{1}{n}\sum_{i=1}^{n} x_i} = \frac{1}{\overline{x}_n}.$$

In the equation above, $\hat{\lambda}_{\mathrm{MLE}}$ is the estimate of the parameter λ.

It is worth noting that the parameter λ estimated by the above method is the same as that obtained using the method of moments.

4.3.1.2 Interval Estimation

In many cases, what is of interest is not point estimation, but the construction of a confidence interval, and the method of constructing such an interval is called interval estimation.

Interval estimation is to take an appropriate interval from the sample to estimate the parameters of the population distribution. Specifically, a confidence interval is a type of interval estimate derived from the observed dataset that may contain the true value of an unknown population parameter, with an associated confidence level. The confidence level represents the proportion of possible confidence intervals that contain the true value of the population parameter. In other words, if confidence intervals are constructed at a given confidence level using statistics from an infinite number of independent samples, the proportion of these intervals that contain the true parameter value will be equal to the confidence level. The formal definition of a confidence interval is given below.

Let $X = x_1, x_2, \ldots, x_n$ be a random sample with a probability distribution with statistical parameter θ. The confidence interval for parameter θ is determined by a pair of random variables $u(X)$ and $v(X)$, then $\forall \theta$, and the confidence interval is expressed as

$$P_\theta \left(u\left(X\right) < \theta < v\left(X\right) \right) = \gamma. \tag{4.13}$$

Where the γ represents the confidence level and $u(X)$ and $v(X)$ are also known as the random endpoint of this interval. The confidence level γ is close to but not greater than 1, which can be expressed as $\gamma = 1 - \alpha$, where α is a small nonnegative number close to 0. $P_\theta(\cdot)$ represents the probability distribution dependent on the statistical parameter θ.

4.3.2 Hypothesis Testing

Hypothesis testing is the process of making a statistical hypothesis about the population parameters and then testing whether this statistical hypothesis holds based on sample data.

Statistical hypotheses are sometimes referred to as confirmatory data analysis, which is the result of modeling based on a set of samples and is a testable hypothesis.

The method used for hypothesis testing depends on the nature of the sample data and the path of analysis. Hypothesis testing is used to infer the results of hypotheses from larger sample data.

In statistics, the hypothesis about parameters is the statistical hypothesis about one or more parameters. The hypothesis to be tested for its correctness is called the null hypothesis, also known as the original hypothesis, denoted as H_0. The null hypothesis is usually determined by the analyst and reflects the analyst's view of the unknown parameters.

The other hypothesis relative to the null hypothesis is called the alternative hypothesis, denoted as H_1; it usually reflects an alternative hypothesis about the possible value of the parameter.

To test the above null hypothesis and alternative hypothesis, two more terms are needed: one is the significance level, denoted as α, which represents the probability of rejecting the null hypothesis; the other is the p-value, which represents the probability of the null hypothesis being true.

The following is an example to illustrate the process of hypothesis testing.

Exercise 4.3 (Vitamin C and the Flu) To verify whether vitamin C can prevent the flu, the following hypothesis test is conducted.

- Null hypothesis (H_0): Taking vitamin C cannot prevent the flu.
- Alternative hypothesis (H_1): Taking vitamin C can prevent the flu.

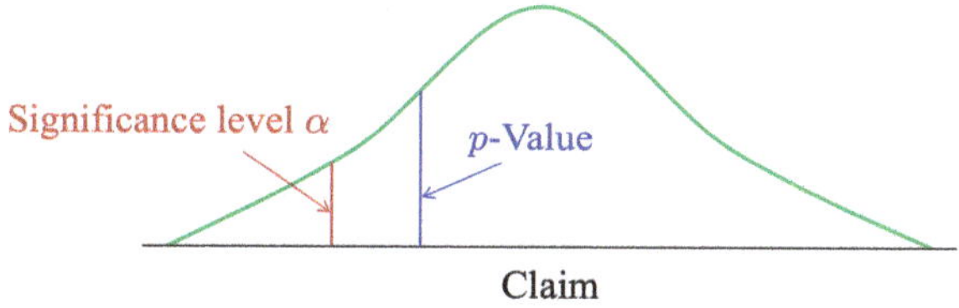

Fig. 4.6 Example of hypothesis testing

- Significance level (α): Setting to 0.04.

The experiment process is as follows: during the flu season, the experimental group is given vitamin C, while the control group is given placebos, and whether the participants in these two groups get sick during the flu season is recorded. Through statistical analysis, the probability of the null hypothesis being true is that $p = 0.18$.

Since $p = 0.18 \gg \alpha = 0.04$, it does not meet the condition to reject the null hypothesis H_0; in other words, there is no evidence to support the alternative hypothesis H_1 is valid. The conclusion drawn from the above hypothesis test is that vitamin C does not have the effect of preventing the flu. Therefore, the null hypothesis is valid. Its schematic diagram is shown in Fig. 4.6.

4.4 Statistical Inference

The inferential statistics introduced in the previous section is a statistical analysis method based on induction, that is, the process of inferring the properties of the population by statistics on the sample.

Statistical inference[1] is the process of using probability theory methods to infer the probability distribution characteristics of random variables, which is one of the important contents of mathematical statistics.

In machine learning, the meaning of "inference" is to make predictions with a trained model. Therefore, deriving and adjusting model parameters is called training, and using the prediction model is called inference.

This section mainly discusses two statistical inference methods based on probability theory, namely, frequentist inference and Bayesian inference.

4.4.1 Frequentist Inference

Frequentist inference is one of the mathematical methods of inferential statistics, which obtains the properties of the population by analyzing the frequency or proportion in the sample data.

---

[1] https://en.wikipedia.org/wiki/Statistical_inference

The process of frequentist inference is as follows. For the probability distribution of the population, corresponding sample data is generated by repeated sampling, and then the frequency characteristics or proportion of this sample data are quantified to obtain its confidence interval.

Frequentist inference originates from the frequentist interpretation of probability. Specifically, any given experiment can be considered as one in an infinite sequence of identical experiments that could potentially be repeated; each repetition can produce statistically independent results.

Therefore, frequentist inference can be represented as $P(D|H)$, i.e., the probability of the data D given the hypothesis H. That is to say, frequentist inference treats the hypothesis as fixed and the data as random. The so-called treating hypothesis as fix means that the probability of the hypothesis is either 1 or 0, that is, $P(H) \rightarrow \{1, 0\}$. The so-called treating data as random means that if the same experiment is repeated, the data will change randomly.

Coin flipping is a classic frequentist experiment, often modeled using the Bernoulli distribution, and its probability mass function can be expressed as follows:

$$f(x; p) = p^x (1 - p)^{1-x}. \tag{4.14}$$

Where x is a random variable that represents the result of a coin flip, let 1 be the head and 0 be the tail. And p is a parameter that represents the probability of the head's appearance.

For Eq. 4.14, there are the following two cases:

- If x is equal to 1, then $f(1; p) = p$.
- Otherwise, x is equal to 0, $f(0; p) = 1 - p$.

Therefore, the Bernoulli distribution perfectly describes the probability of the outcome of the binary experiment, that is, when the probability of the heads of the coin is equal to p, the probability of the tails is $1 - p$. This is the simplest example of a discrete probability distribution.

The maximum likelihood estimation (MLE) method is often used to determine the parameters of the model based on sample data.

Let $D = x_1, x_2, \ldots, x_n$ be the sample data and θ be the parameters of the model. Using the likelihood maximization method of the parameter to obtain its likelihood function

$$L(D; \theta) = P(D|\theta) = f(x_1, x_2, \ldots, x_n|\theta) = \prod_{i=1}^{n} f(x_i|\theta). \tag{4.15}$$

The meaning of the above equation is that the likelihood is the probability of the sample data D given the parameter θ.

Based on the above likelihood function, we can obtain its maximum likelihood estimate (MLE) and find the parameter values that maximize the likelihood function:

$$\hat{\theta} = \arg\max_{\theta} L\left(D; \theta\right). \tag{4.16}$$

4.4.2   Bayesian Inference

The so-called Bayesian inference is based on Bayes' theorem to infer the maximum posterior probability of parameters.

Therefore, Bayesian inference can be represented as $P\left(H|D\right)$, i.e., the probability of hypothesis H given the data D. That is to say, Bayesian inference regards the data as fixed and the hypothesis as random. The so-called regarding data as fix refers to the data we have. The so-called hypothesis as random refers to the probability of the hypothesis being a value between 0 and 1, that is, $0 \le P\left(H\right) \le 1$.

Let $D = x_1, x_2, \ldots, x_n$ be the sample data and θ be the model parameters. Then, the posterior probability of parameter θ given sample data D can be calculated by

$$P\left(\theta|D\right) = \frac{P\left(\theta\right) P\left(D|\theta\right)}{P\left(D\right)}.$$

Where $P\left(\theta\right)$ is the prior, $P\left(D|\theta\right)$ is the likelihood, and $P\left(D\right)$ is the evidence, also known as the normalization constant. Therefore, we have

$$P\left(\theta|D\right) \propto P\left(\theta\right) P\left(D|\theta\right). \tag{4.17}$$

Consequently, the maximum parameter value of this posterior probability is

$$\hat{\theta} = \arg\max_{\theta} P\left(\theta|D\right). \tag{4.18}$$

4.4.3   Comparative Analysis

We can compare and analyze the frequentist inference and Bayesian inference from the following two viewpoints.

(1) From the viewpoint of theoretical foundations, the differences between these two types of inference are as follows:

- Frequentist inference: its theoretical foundation is the law of large numbers (LLN) and the central limit theorem (CLT) in probability theory.
- Bayesian inference: its theoretical foundation is the Bayesian theorem in probability theory.

(2) From the viewpoint of conditional probability, the differences between these
two types of inference are as follows:

- Frequentist inference: for the hypothesis H and sample data D, the conditional probability of being interested is $P(D|H)$. That is to say, this method treats data as random and the hypothesis as fixed. In other words, if the experiment is repeated, the data will vary, but the hypothesis will not change. This method is called frequentist inference because it focuses on the frequency of expected observational data given a certain hypothesis about the field. In hypothesis testing, the value of probability P is the value of $P(D|H)$, where H is usually a null hypothesis.
- Bayesian inference: for the sample data D and the hypothesis H, the conditional probability of being interested is $P(H|D)$. That is to say, this method treats data as fixed and hypotheses as random. In other words, the data is already given, and the probability of the hypothesis being true or false is between 0 and 1. The reason it is called Bayesian inference refers to the uses of Bayes' theorem for calculating this conditional probability.

4.5   Statistical Learning Theory

Statistical learning theory (Vapnik 1998, 1999) originates from statistics and functional analysis. It is a theoretical framework for supervised learning and can be seen as a theory of inductive learning.

However, before the 1990s, it was just a purely theoretical analysis method. The support vector machine (SVM) algorithm based on statistical learning theory that emerged in the mid-1990s made this theory not only a tool for theoretical analysis but also a practical method for finding predictive functions or estimation functions based on sample data.

In the following section, we first present a formal definition of the statistical model, next give a formal definition of the statistical learning model, and then discuss several issues related to model capabilities, namely, growth function, VC dimension, and Rademacher complexity.

4.5.1   Statistical Model

Definition 4.1 (Statistical Model) A statistical model (SM) can be represented as a 3-tuple $\mathrm{SM} = \langle \mathcal{X}, \mathcal{Y}, P \rangle$, where $\mathcal{X}$ is the input space of the data, $\mathcal{Y}$ is the output space, also known as the target space of the data, and P denotes the probability function.

Without loss of generality, the parameterization problem is omitted in the above model.

Using a random data generator, represented as probability $P(x)$, generates a set of observation data D, that is,

$$D = \{x_1, x_2, \ldots, x_n\} \subseteq \mathcal{X}.$$

Where $x_i \in \mathcal{X}$ and n belong to the set of nonnegative natural numbers $\mathbb{N}_+$, that is, $n \in \mathbb{N}_+$.

And, through a statistical function, that is, the conditional probability $P(y|x)$ gets the probability of y given x, and $y \in \mathcal{Y}$.

The above definition implies such an assumption: $P(x)$ and $P(y|x)$ are two "true but unknown" probability distributions. It should be mentioned that the term "true" means that two probability distributions are objectively existing; and the so-called unknown refers to that they cannot be calculated accurately.

For example, to calculate the average weight of all 5-year-old boys in a certain area, we can randomly select a certain number of 5-year-old boys from that area as samples, measure their weight separately, and then infer the overall result based on these sample data. In this case, the process of selecting boys can be seen as sampling with $P(x)$, and the average weight y of boys in that area can be seen as the statistical result obtained with $P(y|x)$. In this case, it can be said that these two probability distributions are true but unknown.

A model is a simplification or approximation of reality, but it is not a 100% accurate representation of reality.

There is a common aphorism in the field of statistics:

All models are wrong, but some are useful.

This aphorism is said to come from the British statistician George E. P. Box (1919–2013). However, some people believe that this saying is not original to Box, and similar sayings have existed for a long time. But at least it can be said that Box made it a famous aphorism.

This aphorism is considered not only applicable to statistical models but also often applicable to other scientific models.

4.5.2 Statistical Learning Model

The statistical learning model is a learning model built on the statistical model and is a part of statistical learning theory.

Definition 4.2 (Statistical Learning Model) A statistical learning model (SLM) can be seen as a 4-tuple SLM $= \langle \mathcal{X}, \mathcal{Y}, P, H \rangle$, where $\langle \mathcal{X}, \mathcal{Y}, P \rangle$ is the same as the statistical model defined in Definition 4.1 and H is the hypothesis set, also known as the hypothesis function set.

Let $x \in \mathcal{X}$ and $y \in \mathcal{Y}$, a training sample set S is generated in the following way:

(1) A random observation data x generator, represented as the probability distribution $P(x)$, generates a set of observation data D, that is,

$$D = \{x_1, x_2, \ldots, x_n\} \subseteq \mathcal{X}.$$

(2) An annotator for input data x to label its target data y, represented as a conditional probability distribution $P(y|x)$, used to label each $x_i \in D$ with its corresponding y_i.

(3) According to the equation of joint probability distribution, $P(x, y) = P(x)P(y|x)$, we can obtain the following training samples:

$$S = \{(x_1, y_1), (x_2, y_2), \ldots, (x_n, y_n)\} \subseteq \mathcal{X} \times \mathcal{Y}.$$

Where n belongs to the set of nonnegative natural numbers $\mathbb{N}_+$, that is, $n \in \mathbb{N}_+$.

Similar to the statistical model, the above definition also implies such an assumption: there exist two "true but unknown" probability distributions $P(x)$ and $P(y|x)$ in the process of generating training sample S; their meaning is similar to the statistical model defined in previous subsection.

Use the above training sample S to train the hypothesis set H, and obtain an optimal hypothesis function $h \in H$, so that the predicted output obtained by hypothesis function $\hat{y} = h(x)$ has a smallest error with the target output annotated in the training sample y, that is, the loss function $\mathcal{L}(\hat{y}, y)$ is minimized, expressed as

$$\underset{\hat{y}=h(x)}{\arg\min} \, \mathcal{L}(\hat{y}, y) = \underset{h \in H}{\arg\min} \, \mathcal{L}(h(x), y). \tag{4.19}$$

Let $\hat{y} = h(x)$ be denoted as $h(\hat{y}|x)$, then the process of minimizing the loss function is to make $h(\hat{y}|x)$ sufficiently approximate to the target conditional probability $P(y|x)$; this process is learning.

The above statistical learning model belongs to the supervised learning paradigm, which will be further discussed in Chap. 8 of Part III.

4.5.3 Growth Function

The growth function, also known as the shatter coefficient or shattering number, is used to measure the complexity of the hypothesis set H in statistical learning models.

Before defining the growth function, we first describe the constraint set H_D and then introduce the concept of shattering.

Definition 4.3 (Constraint Set) Let $H : \mathcal{X} \rightarrow \{0, 1\}$ be the hypothesis set, $D \subseteq \mathcal{X}$ be a finite dataset, then the constraint set H_D is defined as the set of H acting on D, that is,

$$H_D = \{(h(x_1), \ldots, h(x_n)) \mid x_i \in D, h(x_i) \in H, n \in \mathbb{N}_+\}. \tag{4.20}$$

Where $\mathbb{N}_+$ denotes the set of nonnegative natural numbers.

The $H : X \to \{0, 1\}$, which means using the hypothesis h to perform binary classification on the data x_i, such as wrong and correct answers, negative and positive results, spam and non-spam emails. Since the number of elements in the dataset D is n, there can be at most $\{0, 1\}^n$ results, that is, $|H_D| \leq 2^n$, where "$|H_D|$" denotes the number of elements in H_D.

Definition 4.4 (Growth Function) The growth function $G_H : \mathbb{N} \to \mathbb{N}$ is defined as

$$G_H(n) = \max_{n=|D|} |H_D|. \tag{4.21}$$

That is to say, the growth function G_H is the maximum number of elements in the constraint set H_D.

Definition 4.5 (Shattering) The dataset D can be shattered by the hypothesis set H, if there exists a hypothesis set H and a dataset D, such that $|H_D| = 2^n$.

From Definitions 4.4 and 4.5, it can be concluded that when the shattering condition is met, $G_H(n) = 2^n$.

Below is a simple exercise to understand the meaning of the growth function.

Exercise 4.4 (Data Shattering Problem in One-Dimensional Space) Given a dataset with three data points. $D = x_1, x_2, x_3$ in one-dimensional space $X \subseteq \mathbb{R}^1$, where $x_i \in X$. And there exists a hypothesis $h \in H$, and $h(x_i) \to \{0, 1\}$, so that the number of combinations is $2^3 = 8$, each combination is shown in the table below:

	(0)	(1)	(2)	(3)	(4)	(5)	(6)	(7)
$h(x_1)$	0	0	0	0	1	1	1	1
$h(x_2)$	0	0	1	1	0	0	1	1
$h(x_3)$	0	1	0	1	0	1	0	1

The combination of column (2) and column (5) in the above table:

$$\{h(x_1) = 0, h(x_2) = 1, h(x_3) = 0\},$$

$$\{h(x_1) = 1, h(x_2) = 0, h(x_3) = 1\},$$

they cannot be shattered, that is, they are unable to perform binary classification, so their growth function $G_H(3) = 8 - 2 = 6$.

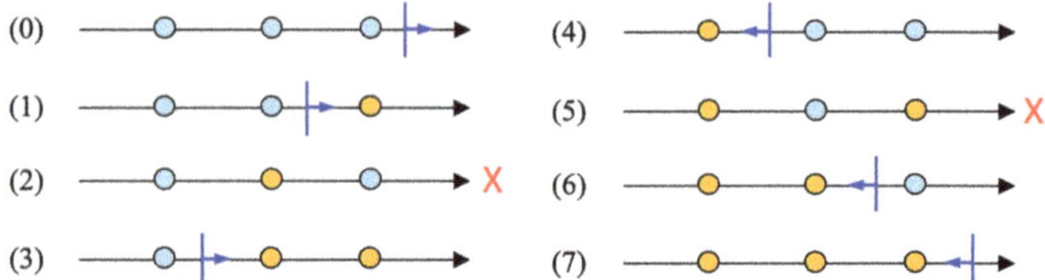

Fig. 4.7 Example of shattering three data in one-dimensional space

The schematic diagram of the above exercise is shown in Fig. 4.7.

The growth function is a purely combinatorial method used to measure the complexity of the hypothesis set, also known as the capacity, richness, expressive power, or flexibility of the hypothesis set.

The calculation of the growth function is not easy, because according to its definition, it is necessary to calculate $G_H(n)$ for all $h \in H$.

Next, we will introduce another method to measure the complexity of the hypothesis set H, namely, the VC dimension. This method is based on a scalar that is closely related to the growth function.

4.5.4 VC Dimension

VC dimension stands for Vapnik-Chervonenkis (VC) dimension, which was proposed in 1971 by Vladimir N. Vapnik and Alexey Chervonenkis in the paper "On the Uniform Convergence of Relative Frequencies of Events to Their Probabilities" that is written in Russian (Vapnik and Chervonenkis 1971).

The VC dimension is the core content of VC theory which stands for "Vapnik-Chervonenkis theory" and was proposed and improved from the late 1960s to the 1990s (Vapnik 1998, 1999). VC theory attempts to explain the learning process from a statistical point of view, and this theory gave birth to the support vector machine (SVM).

Vladimir Vapnik detailed the VC dimension and its VC theory in his books *Statistical Learning Theory* (Vapnik 1998) and *The Nature of Statistical Learning Theory* (Vapnik 1999).

The VC dimension is used to measure the complexity of the hypothesis set, becoming an important concept in statistical learning theory and directly related to the growth function. The VC dimension is also a purely combinatorial method, but it is usually easier to calculate than the growth function.

To facilitate the definition of the VC dimension of the hypothesis set H, we introduce the concept of linear binary partition, that is, for a dataset $D \subseteq X$ with n samples and output space $\mathcal{Y} = \{0, 1\}^n$, the way to shatter it is to do a linear binary partition on it.

Given a hypothesis set H, if H can perform all possible linear binary classification on the dataset D, then it is said that D is linearly shattered by H. Obviously, for linear binary partition, $|(h(x_1), h(x_2), \ldots, h(x_n))| \leq 2^n$, therefore, $G_H(n) = 2^n$.

Definition 4.6 (VC Dimension) The VC dimension of a hypothesis set H, denoted as VCdim (H), is the maximum value of n in the growth function that can be shattered by the hypothesis set H, that is,

$$\text{VCdim}(H) = \max\left\{n \mid G_H(n) = 2^n\right\}. \tag{4.22}$$

That is to say, any dataset with n maximum samples can be linearly binary shattered by the hypothesis set H, if and only if its growth function $G_H(n)$ equals 2^n.

It is worth noting that another meaning of n as the maximum number of samples is that when the number of samples is greater than n, for example, a dataset with $n + 1$ samples, it cannot be linearly binary partitioned. In addition, if for any n, it satisfies $G_H(n) = 2^n$, then VCdim $(H) = \infty$. That is, if all sample numbers n can be shattered by the hypothesis set H, then its VC dimension is infinite.

To further understand the concept of VC dimension, it is related to the dimension of the data space, so the following theorem and corollary hold (Burges 1998).

Theorem 4.1 *Consider some set of n data points in $\mathbb{R}^m$. Choose any one of the data points as the origin, and then these n data points can be shattered by oriented hyperplanes, if and only if the position vectors of the remaining data points are linearly independent.*

Corollary 4.1 *The VC dimension of the set of oriented hyperplanes in $\mathbb{R}^m$ is $m + 1$.*

Since we can always choose $m + 1$ data points in $\mathbb{R}^m$, and then choose one of the data points as the origin, such that the position vectors of the remaining m data points are linearly independent, but can never choose $m+2$ such data points, because no $m + 1$ position vectors in $\mathbb{R}^m$ can be linearly independent.

Exercise 4.5 (Data Shattering Problem in Two-Dimensional Space) As shown in Fig. 4.8, for the dataset located in the two-dimensional real vector set $\mathbb{R}^2$, the VC dimension to perform linear binary partition is 3, but it is impossible to perform linear binary partition on 4 or more data points in two-dimensional space.

Why study VC dimension? Because its role is to analyze the complexity of classification functions.

It needs to be emphasized that the VC dimension is defined for binary functions, that is, for the function space corresponding to $\{0, 1\}$. For function spaces with more than two values, some researchers have proposed several generalization methods. For example, for multivalued functions, that is, function to $\{0, \ldots, n\}$, the Natarajan dimension can be used (Natarajan 1989); for real-valued functions, such as the real number interval $[0, 1]$, the pseudo-dimension can be used (Morgenstern and Roughgarden 2015).

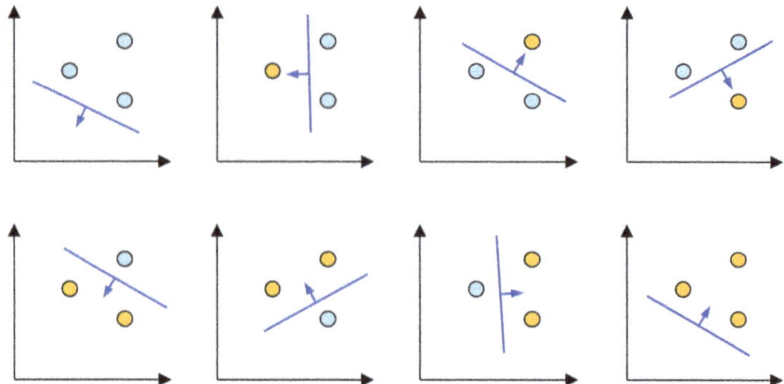

Fig. 4.8 Example of shattering three data in two-dimensional space

4.5.5 *Rademacher Complexity*

Rademacher complexity is named after the German-American mathematician Hans Rademacher and is used to measure the ability of a class of real-valued functions relative to a probability distribution.

Unlike the VC dimension introduced in the previous subsection, Rademacher complexity is not limited to binary functions and has been proven that it can also be useful in analyzing other learning algorithms.

Below, we first propose the consistency issues to facilitate the understanding of the Rademacher complexity, next give the definition of the Rademacher average, and then introduce the Rademacher complexity based on it.

4.5.5.1 Consistency Issues

Let $\mathcal{X}$ be input space, $D = \{x_1, x_2, \ldots, x_n\} \subseteq \mathcal{X}$ be the input data, and $\{y_1, y_2, \ldots, y_n\}$ be the corresponding labels, where $y_i \in \{-1, +1\}$. And let H be the hypothesis set, there exists a hypothesis $h \in H$, with $h : \mathcal{X} \to \{-1, +1\}$, i.e., $\forall x_i \in D, h(x_i) \to \{-1, +1\}$.

Based on the above assumptions, we may find such a phenomenon: $y_i h(x_i)$ equals to $+1$ if the prediction of $h(x_i)$ is consistent with the label y_i; but $y_i h(x_i)$ equals to -1 if the prediction of $h(x_i)$ is inconsistent with the label y_i. That is to say, $\forall i \in \{1, 2, \ldots, n\}$

$$y_i h(x_i) = \begin{cases} +1 & \text{if } \langle y_i, h(x_i) \rangle = \langle -1, -1 \rangle \text{ or } \langle +1, +1 \rangle \\ -1 & \text{if } \langle y_i, h(x_i) \rangle = \langle -1, +1 \rangle \text{ or } \langle +1, -1 \rangle \end{cases}$$

We call the above phenomenon the consistency issue over label prediction.

For the n data points in D, if the consistency over label prediction is accumulated and scored, one point will be added when label prediction is consistent, and one point will be deducted if they are inconsistent. Furthermore, it is easy to calculate the consistency average (CA) over label prediction:

$$\mathrm{CA}_{\mathrm{lp}}(H) = \arg\max_{h \in H} \left(\frac{1}{n} \sum_{i=1}^{n} y_i h\left(\boldsymbol{x}_i\right) \right). \tag{4.23}$$

From the above expression, we can see that when a hypothesis h in the hypothesis set H is found that have maximized consistency average, it means that the hypothesis has the highest prediction accuracy and the smallest error relative to the label.

Next, we make a substitution for Eq. 4.23 by using σ_i to replace y_i, where the σ_i is an independent random noise belonging to Rademacher distribution, which selects one from $\{-1, +1\}$ randomly, and $P(\sigma_i = +1) = P(\sigma_i = -1) = 0.5$. That is, $\forall i \in \{1, 2, \ldots, n\}$:

$$\sigma_i = \begin{cases} +1 & \text{with probability } 0.5 \\ -1 & \text{with probability } 0.5 \end{cases}.$$

At this point, what we need to consider is the consistency issue over noise prediction, i.e., for $h \in H$, make $h : D \to \{\sigma_i\}_{i=1}^{n}$.

Similarly, we calculate the consistency average over noise prediction as below:

$$\mathrm{CA}_{\mathrm{np}}(H) = \arg\max_{h \in H} \left(\frac{1}{n} \sum_{i=1}^{n} \sigma_i h\left(\boldsymbol{x}_i\right) \right). \tag{4.24}$$

Due to the consistency of the prediction by $h\left(\boldsymbol{x}_i\right)$ relying on random noise σ_i, so that we calculate the expectation for Eq. 4.24, that is,

$$\mathrm{CA}_{\mathrm{np}}(H) = \mathbb{E}_{\sigma} \left[\max_{h \in H} \left(\frac{1}{n} \sum_{i=1}^{n} \sigma_i h\left(\boldsymbol{x}_i\right) \right) \right]. \tag{4.25}$$

4.5.5.2   Rademacher Average and Complexity

On the basis of consistency issues, it will be easy to explain the Rademacher average and its Rademacher complexity.

Definition 4.7 (Rademacher Average) Let $\mathcal{Z}$ be input space, $S = \{z_1, z_2, \ldots, z_n\}$ $\subseteq \mathcal{Z}$ be a sample data, F be a real-valued function class, and $F : \mathcal{Z} \to \{\sigma_i\}_{i=1}^{n}$. The

Rademacher average of the function class F over the sample data S is as follows:

$$\text{Red}_S (F) = \mathbb{E}_\sigma \left[\sup_{f \in F} \left(\frac{1}{n} \sum_{i=1}^{n} \sigma_i f (z_i) \right) \right]. \tag{4.26}$$

Where σ_i denotes independent random noise, $\sigma_i \in \{-1, +1\}$, and $P (\sigma_i = +1) = P (\sigma_i = -1) = 0.5$.

According to the definition of Rademacher average, we can infer that it quantifies the consistency between the function class F and the random noise $\{\sigma_i\}_{i=1}^{n}$ which can be used to measure the fitting ability between them.

When a function f is found in function class F that maximizes the Rademacher average, it means that the function has the highest consistency and the smallest error that relatives to the random noise $\{\sigma_i\}_{i=1}^{n}$ on sample S.

Definition 4.8 (Rademacher Complexity) Let P be a Rademacher distribution over sample data S, the Rademacher complexity of the function class F over the distribution P is as follows:

$$\text{Red}_P (F) = \mathbb{E}_{S \sim P} [\text{Red}_S (F)]. \tag{4.27}$$

From the above definition, it is known that the Rademacher complexity $\text{Red}_P (F)$ is the expectation of Rademacher average $\text{Red}_S (F)$. This complexity depends on the distribution P rather than the dataset S.

Taking a simple comparison of the capability measure between the Rademacher complexity and the VC dimension, the conclusion is as follows:

- Rademacher complexity is different from the VC dimension, as the former is not limited to binary functions and can also be used to analyze other types of learning algorithms.
- For the constraint of generalization error, the Rademacher complexity is more stringent than the VC dimension, and it can not only be used for classification like the VC dimension but also regression.
- Rademacher complexity is related to the distribution of data, while the VC dimension is a combination that is independent of distribution.

4.5.5.3 Data-Dependent Bounds

Rademacher complexity can be used to derive data-dependent bounds on the learnability of function classes. Intuitively, a function class with smaller Rademacher complexity is easier to learn. Here, we introduce two of these bounds (Shalev-Shwartz and Ben-David 2014).

First, the bound of representativeness. In machine learning, the desire to have a training set that represents the true distribution of a sample data S can be quantified by the concept of representativeness.

Let P denote the probability distribution used to draw samples, H denote the set of hypotheses, and F denote the corresponding error functions. For each hypothesis $h \in H$, there exists an error function $f \in F$ that reflects the prediction error of h on sample z. For example, let h be a binary classifier that returns 1 if h classifies a sample correctly and 0 otherwise. Define

$$L_P(f) = \mathbb{E}_{z \sim P}[f(z)], \tag{4.28}$$

$$L_S(f) = \frac{1}{n} \sum_{i=1}^{n} f(z_i). \tag{4.29}$$

Where Eq. 4.28 represents the expected error of the error function $f \in F$ on the actual distribution P and Eq. 4.29 represents the estimated error of the error function $f \in F$ on the sample data S.

The representativeness of the sample data S to the probability distribution P and the error functions F is then defined as follows:

$$\text{Rep}_P(F, S) = \sup_{f \in F}(L_P(f) - L_S(f)). \tag{4.30}$$

Smaller representativeness is better, because it provides a way to avoid overfitting: this means that the expected error of a classifier is not much higher than its estimated error, so choosing a classifier with a low estimated error will ensure its low expected error. However, the notion of representativeness is relative and thus cannot be compared across different samples.

The expected representativeness of a sample can be bounded by the Rademacher complexity of the function class:

$$\mathbb{E}_{S \sim P}\left[\text{Rep}_P(F, S)\right] \le 2 \cdot \mathbb{E}_{S \sim P}[\text{Red}_S(F)]. \tag{4.31}$$

Second, the bound of generalization error. When the Rademacher complexity is small, the hypothesis set H can be learned by using minimized generalization error.

For each $\delta > 0$, there exists a probability of at least $1 - \delta$; then for sample data S and each hypothesis $h \in H$, the following bound of generalization error holds

$$L_P(h) - L_S(h) \le 2 \cdot \text{Red}_S(F) + 4\sqrt{\frac{2 \ln(4/\delta)}{n}}, \tag{4.32}$$

where the generalization error of hypothesis h is minimized.

The generalization error is also called expected error or expected risk. We'll cover this in more detail in Chap. 8.

4.6   Parametric and Nonparametric

For a statistical model (SM), from the point of view of the parameters of the model, it can be divided into the parametric model and nonparametric model, and there is also the semiparametric model.

This section mainly introduces parametric models and nonparametric models.

4.6.1   Parametric Models

Definition 4.9 (Parametric Models) For a parameterized statistical model, $\text{PSM} = \langle \mathcal{X}, \mathcal{Y}, P, \Theta \rangle$, it is called a parametric model, if the Θ be a set of parameters with a finite number of parameters.

In other words, given a parameter set $\Theta = \{\theta_1, \theta_2, \ldots, \theta_k\}$, and an observed dataset $D \subseteq \mathcal{X}$, the model $\text{PSM} = \langle \mathcal{X}, \mathcal{Y}, P, \Theta \rangle$ is called a parametric model, if under the condition of this parameter set Θ, the probability of feature predictions x is independent of the dataset D. Among them, the parametric model of discrete distribution is expressed as $P(x|\theta, D) = P(x|\theta)$, and the parametric model of a continuous distribution is expressed as $p_{\theta, D}(x) = p_\theta(x)$.

The parametric model has a fixed number of parameters and makes strict assumptions about the probability distribution of random variables, such as following a normal distribution. Since the random variable x is independent of the observed dataset D, even if the amount of observed dataset is infinite, the complexity of this model is bounded. The characteristic of the parametric model is that it is easy to use. After training with the given sample data, it can directly predict and learn from unknown data.

A typical parametric model is the Gaussian mixed model (GMM). It is composed of k parts, where the ith part has a mean of μ_i and a variance of σ_i. Namely,

$$P(x|\theta) = \sum_{i=1}^{k} \phi_i \mathcal{N}(x|\mu_i, \sigma_i), \tag{4.33}$$

$$\mathcal{N}(x|\mu_i, \sigma_i) = \frac{1}{\sqrt{(2\pi)^k |\sigma_i|}} \exp\left(-\frac{1}{2}(x - \mu_i)^\mathsf{T} \sigma^{-1}(x - \mu)\right). \tag{4.34}$$

The parameters in the above equation are the mean μ_i and variance σ_i, that is, $\theta = (\mu, \sigma)$. By learning the parameters, it is possible to fit the training data as much as possible, thereby obtaining a Gaussian mixture model that can describe this set of data.

In addition, there are several other parametric models, as shown in Table 4.2.

It lists several representative machine learning algorithms based on parametric models in Table 4.3.

Table 4.2 Examples of other parametric models

Model name	Parameter	Expression
Poisson distribution	$\theta = \lambda$	$P\left(j \vert \theta\right) = \frac{\lambda^j}{j!} \exp\left(-\lambda\right)$
Gaussian distribution	$\theta = (\mu, \sigma)$	$p_\theta\left(x\right) = \frac{1}{\sqrt{2\pi}\sigma} \exp\left(-\frac{(x-\mu)^2}{2\sigma^2}\right)$
Weibull distribution	$\theta = (\lambda, \beta, \mu)$	$p_\theta\left(x\right) = \frac{\beta}{\lambda}\left(\frac{x-\mu}{\lambda}\right)^{\beta-1} \exp\left(-\left(\frac{x-\mu}{\lambda}\right)^\beta\right) \mathbb{I}\left(x > \mu\right)$
Binomial distribution	$\theta = (n, p)$	$P\left(k \vert \theta\right) = \frac{n!}{k!(n-k!)} p^k \left(1 - p\right)^{n-k}$

Table 4.3 Common machine learning algorithms based on parametric models

Algorithm	Brief introduction
Linear regression	A supervised learning algorithm that calculates the linear relationship between independent and dependent variables is used for regression task
Logistic regression	A supervised learning algorithm that predicts the probability that input data belongs to a given category is used for classification task
Linear discriminant analysis	A supervised learning algorithm that finds the linear combination of features in the input dataset is used for classification task
k-means	An unsupervised learning algorithm that groups input data into k clusters is used for clustering task
Naive Bayes	A supervised learning algorithm that determines the category of input data based on Bayes' rule is used for classification task

4.6.2 Nonparametric Models

Definition 4.10 (Nonparametric Models) For a parameterized statistical model $\mathrm{PSM} = \langle \mathcal{X}, \mathcal{Y}, P, \Theta \rangle$, it is called a nonparametric model if the Θ is a parameter set without fixed number of parameters.

In other words, the number of the parameter set Θ in the nonparametric model is not fixed but increases or decreases with the change in the amount of data in the dataset D.

It needs to be emphasized that nonparametric models are not parameter-free, but are parameterized without fixed dimensions, and are therefore also known as variable parametric models.

Nonparametric models are useful when there is a lot of data but a lack of prior knowledge, as there is no need to focus on parameter selection, making them very flexible. However, nonparametric models often struggle computationally with large-scale datasets. The distribution of such data is usually impossible to simulate with prior knowledge. Simply put, nonparametric models do not make assumptions about the overall distribution of the sample but directly analyze the sample. Therefore, it is

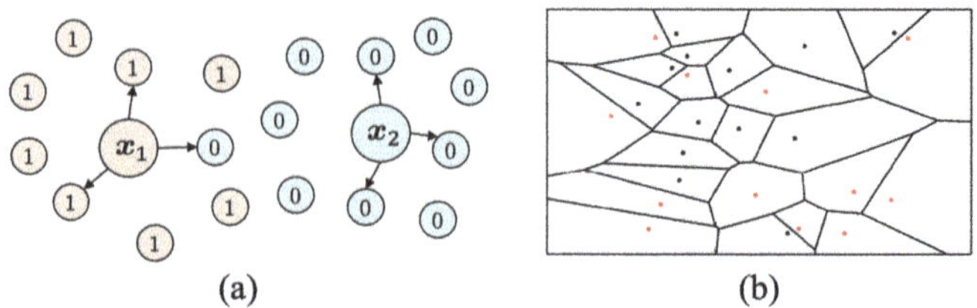

Fig. 4.9 Schematic diagram of classification using k-nearest neighbors

necessary to adjust the parameters during the learning process, usually continuously increasing the scale of parameters, such as k-nearest neighbor method.

In machine learning based on nonparametric models, the effective number of parameters increases with the number of sample instances; hence, it is called instance-based learning. In addition, because nonparametric models store all data instances in memory during processing, they are also known as memory-based learning (Russell and Norvig 2009; Alpaydin 2009).

The one of typical machine learning algorithms, k-nearest neighbors, is based on nonparametric models, which can be used for classification or regression. It simply looks at the k closest points in the observed dataset to a test data point, calculates how many members are around that data point, and returns an empirical score as an estimate. For example, for the classification problem of k-nearest neighbors, the following equation can be used to calculate the estimated value of neighboring points:

$$p\left(y = c \mid \boldsymbol{x}, D, k\right) = \frac{1}{k} \sum_{i \in N_k(\boldsymbol{x}, D)} \mathbb{I}\left(y_i = c\right), \tag{4.35}$$

where $N_k\left(\boldsymbol{x}, D\right)$ is the index of the k-nearest neighbors of data point $\boldsymbol{x}$ in the observed dataset D and $\mathbb{I}\left(\omega\right)$ is the indicator function of the expression ω, $\mathbb{I}\left(\omega\right)$ equals 1 if ω is true, and 0 otherwise.

Figure 4.9a is a schematic diagram of classifying two-dimensional observed data D using the method of k-nearest neighbors. Let $k = 3$, the labels of the 3 nearest neighbors of the test data point $\boldsymbol{x}_1$ are 1, 1, and 0, and the labels of the 3 nearest neighbors of the data point $\boldsymbol{x}_2$ are 0, 0, and 0. Therefore, according to Eq. 4.35, we have

$$p\left(y = 1 \mid \boldsymbol{x}_1, D, k = 3\right) = 2/3,$$

$$p\left(y = 1 \mid \boldsymbol{x}_2, D, k = 3\right) = 0/3.$$

Table 4.4 Common machine learning algorithms based on nonparametric models

Algorithm	Brief introduction
k-nearest neighbors	A supervised learning algorithm that divides input data into k groups based on distance, which can be used for classification and regression tasks
Kernel density estimation	The application of kernel smoothing in probability density estimation, i.e., estimating the probability density function of a random variable based on the kernel as a weight
Gaussian process regression	A supervised learning algorithm aimed at solving classification and regression problems based on Gaussian processes
Decision trees	A type of supervised learning algorithm that uses a treelike model of decisions for classification and regression tasks
Kernel support vector machine	A supervised learning algorithm that uses a Gaussian kernel and kernel trick to map nonlinearly separable datasets to a high-dimensional space where they become linearly separable, used for nonlinear classification tasks

Figure 4.9b is the Voronoi diagram, also known as Voronoi tessellation, when $k = 1$ (Murphy 2012).

It lists several nonparametric machine learning algorithms to Table 4.4.

4.6.3   Comparative Analysis

The main difference between parametric and nonparametric models lies in whether there is a fixed number of parameters, in other words, whether the model's parameters increase or decrease with changes in the amount of training data.

Parametric models have a fixed number of parameters and make strict assumptions about the probability distribution of random variables.

Nonparametric models are not parameter-free but rather cannot be parameterized with a fixed number of parameters; hence, they are also known as variable parametric models (Table 4.5).

4.7   Kernel Methods

Many tasks in machine learning, such as classification, clustering, ranking, and dimensionality reduction, require discovering and analyzing the relationships between data in a dataset. There are two types of solutions: one is to transform the original data into a feature vector representation through a specific feature mapping; the other is to map the similarity between data pairs in nonlinear data in the original space to a linearly separable space through a specific kernel function.

Table 4.5 Differences between parametric and nonparametric models

Parametric models	Nonparametric models
Use a fixed number of parameters to build the model	Use a non-fixed number of parameters to build the model
Parameter analysis is to test means	Nonparametric analysis is to test medians
Only applicable to variables	Applicable to both variables and attributes
Always make strict assumptions about the data	Usually make fewer assumptions about the data
Require fewer data	Require more data
Assume a certain probability distribution	No assumption of the probability distribution
Handle intervals data or ratio data	Handle raw data
Faster computation speed	Slower computation speed

Kernel methods originated in the 1960s (Aizerman et al. 1964). Its name comes from the kernel functions they use, which can calculate the inner product of all data pairs in a high-dimensional, implicit feature space without calculating the coordinates of the data in that space. Since inner product operations are much easier than coordinate calculations, this is known as the kernel trick.

Many machine learning algorithms use kernel methods, such as kernel perceptron, kernel SVM, Gaussian processes, principal component analysis (PCA), canonical correlation analysis (CCA), ridge regression, spectral clustering, and linear adaptive filters.

Due to their effectiveness, kernel methods are not only introduced in most machine learning textbooks, but there are also several monographs and papers dedicated to kernel methods, for example, the monograph *Learning with Kernels: Support Vector Machines, Regularization, Optimization, and Beyond* by Bernhard Schölkopf and Alexander J. Smola published in 2002 (Schölkopf and Smola 2002), the monograph *Kernel Methods for Pattern Analysis* by John Shawe-Taylor and Nello Cristianini published in 2004 (Shawe-Taylor and Cristianini 2004), the review paper "Kernel Methods in Machine Learning" by Thomas Hofmann et al. published in the *Annals of Statistics* in 2008 (Hofmann et al. 2008), and "Kernel Methods for Deep Learning" by Youngmin Cho and his advisor Lawrence K. Saul presented at the *Conference on Neural Information Processing Systems* (NIPS) 2009 (Cho and Saul 2009).

This section mainly discusses the basic principles of kernel methods, while machine learning algorithms that use kernel methods will be discussed separately in the corresponding chapters.

Fig. 4.10 Mapping from input space to feature space

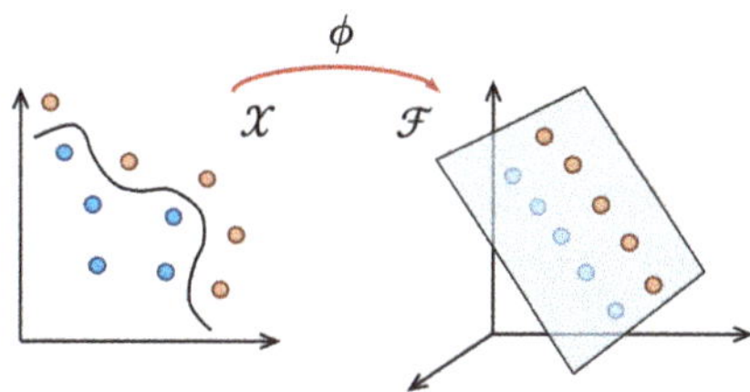

4.7.1   Kernel Trick

For a parameterized statistical learning model PSLM $= \langle \mathcal{X}, \mathcal{Y}, P, H, \Theta \rangle$, there exists the following labeled sample data:

$$S = \{(x_i, y_i) \mid x_i \in \mathcal{X}, \ y_i \in \mathcal{Y} \text{ and } i = 1, \ldots, n\}.$$

The goal is to predict unknown data after the training by this sample.

In machine learning, we often encounter nonlinear problems, that is, the sample data is linearly inseparable. For this, we can consider the following two treatment methods.

One is based on the sample dataset S, to get a nonlinear hypothesis $h \in H$ closest to the target probability function $P(y|x)$, such that $h : \mathcal{X} \to \mathcal{Y}$.

Another way, as shown in Fig. 4.10, is to map the input space to a linearly separable high-dimensional feature space $\mathcal{F}$, that is, to find a feature mapping ϕ, satisfying

$$\phi : \mathcal{X} \to \mathcal{F}, \tag{4.36}$$

where $\dim(\mathcal{F}) \gg \dim(\mathcal{X})$.

And find a hypothesis $h \in H$, such that $h : \mathcal{F} \to \mathcal{Y}$.

Among the above two methods, the former is often difficult to find a suitable nonlinear model for the nonlinearly separable input space $\mathcal{X}$; the latter requires finding an appropriate feature mapping function to construct a linearly separable feature space $\mathcal{F}$, but the dimension of the feature space is often greater than the dimension of the original input space and may even be ultrahigh dimensional.

For example, for a text classification task that needs to handle phrases composed of three consecutive words, the 3-gram model is used as the feature sequence. For a text with only one hundred thousand words, the dimension of its feature space $\mathcal{F}$ reaches $\dim(\mathcal{F}) = 10^{15}$.

Therefore, in such a high-dimensional feature space, to obtain a linear classification hyperplane, a large amount of coordinate calculation is required, which is very costly. An effective way to solve this problem is the kernel trick, which is based on the kernel function, and its feature space belongs to the Hilbert space.

4.7.2   Hilbert Space

The feature space $\mathcal{F}$ in kernel methods is a Hilbert space, which is closely related to the inner product space.

In linear algebra, an inner product space is a vector space with an additional inner product structure, which associates each pair of vectors in the space with a scalar. The inner product can be explained geometrically, such as by calculating the length of vectors and the angle between vectors, and it also provides a method to define the orthogonality between vectors, that is, zero inner product.

The inner product is usually represented by the symbol $\langle \cdot, \cdot \rangle$. Given two vectors $x, x' \in \mathbb{R}^m$, their inner product satisfies the following equation:

$$\langle x, x' \rangle = \sum_{i=1}^{m} [x]_i \, [x]'_i. \tag{4.37}$$

The length of the vector x is represented by its norm as follows:

$$\| x \| = \sqrt{\langle x, x' \rangle}. \tag{4.38}$$

In Euclidean space, the inner product is equal to the dot product, also known as the scalar product. However, the inner product space is generalized from Euclidean space to any- or infinite-dimensional vector space, where functional analysis comes into play.

Definition 4.11 (Hilbert Space) A Hilbert space $\mathcal{F}$ is an inner product space with separability and completeness.

Let an element of this space be represented as $\varphi \in \mathcal{F}$. The completeness refers to every Cauchy sequence φ_n ($n \geq 1$) converges to an element φ; the Cauchy sequence satisfies the following property: when $\varphi \to \infty$

$$\sup_{m > n} \| \varphi_n - \varphi_m \| \to 0. \tag{4.39}$$

$\mathcal{F}$ is called separable, only if it satisfies the following condition: for any $\epsilon > 0$, there exists a finite set of elements $\varphi_1, \ldots, \varphi_n \in \mathcal{F}$, such that for all $\varphi \in F$

$$\min_i \| \varphi_i - \varphi \| < \epsilon. \tag{4.40}$$

4.7.3 *Kernel Functions*

Definition 4.12 (Kernel Function) Let $\mathcal{X}$ be a non-empty input space, the symmetric function

$$\kappa : \mathcal{X} \times \mathcal{X} \to \mathbb{R} , \qquad (4.41)$$

is known as the kernel function, also simply called the kernel, where $\mathbb{R}$ is a set of real numbers.

The so-called symmetric function refers to the property as below:

$$\forall x, x' \in \mathcal{X}, \quad \kappa\left(x, x'\right) = \kappa\left(x', x\right).$$

Firstly, the question is what the essence of kernel function is. From Eq. 4.41, it can be seen that $\mathcal{X} \times \mathcal{X}$ is the Cartesian product of the input spaces and any data pair $\left(x, x'\right)$ in the input space is mapped to a real number in the set of real numbers $\mathbb{R}$ by the kernel function κ. So that the purpose of the kernel function κ is to compare the data pair $\left(x, x'\right)$ in the input space, that is, the so-called similarity measure, the result is a real number. Therefore, the kernel function is a similarity function between any data pair in the input space.

The next question is how to measure the similarity between any data pair in the input space. The kernel function further introduces the inner product operation of Hilbert space. The specific method is to perform feature mapping on all data pairs in the input space to obtain feature vectors located in Hilbert space $\varphi\left(x\right)$ and $\varphi\left(x'\right)$ and then calculate their inner product. Therefore, the kernel function becomes the following equation:

$$\kappa\left(x, x'\right) = \left\langle \varphi\left(x\right), \varphi\left(x'\right)\right\rangle. \qquad (4.42)$$

Equation 4.42 indicates that for some nonlinearly separable input spaces, such a kernel function can be used, that is, a linearly separable Hilbert space is constructed through feature mapping $\mathcal{F}$, and then the inner product between all data pairs in this space is calculated, without the need to calculate the coordinates of each data in this space.

The list of some commonly used kernel functions is shown in Table 4.6.

The advantage of kernel functions is that they significantly reduce the cost of computation. Because calculating the inner product in a high-dimensional feature space is much simpler than calculating coordinates, the complexity of calculating the inner product is usually $O\left(m\right)$, while the complexity of calculating its coordinates is about $O\left(\dim\left(\mathcal{F}\right)\right)$, where $\dim\left(\mathcal{F}\right) \gg m$, that is, the dimension of the feature space is much larger than the dimension of the original input space m.

Table 4.6 Commonly used kernel functions

Name	Kernel function
Polynomial kernel	$\kappa\left(\boldsymbol{x}, \boldsymbol{x}'\right) = \left(\langle \boldsymbol{x}, \boldsymbol{x}'\rangle + c\right)^d$
Hyperbolic tangent kernel	$\kappa\left(\boldsymbol{x}, \boldsymbol{x}'\right) = \tanh\left(\kappa\langle \boldsymbol{x}, \boldsymbol{x}'\rangle + \Theta\right)$
Gaussian kernel	$\kappa\left(\boldsymbol{x}, \boldsymbol{x}'\right) = \exp\left(-\frac{\|\boldsymbol{x}-\boldsymbol{x}'\|^2}{2\sigma^2}\right)$
Laplacian kernel	$\kappa\left(\boldsymbol{x}, \boldsymbol{x}'\right) = \exp\left(-\frac{\|\boldsymbol{x}-\boldsymbol{x}'\|}{\sigma}\right)$
Power kernel	$\kappa\left(\boldsymbol{x}, \boldsymbol{x}'\right) = \|\boldsymbol{x}-\boldsymbol{x}'\|^d$

4.7.4   Positive-Definite Kernel

Before giving the definition of a positive-definite kernel, we introduce the concepts of the Gram matrix and positive-definite matrix.

Definition 4.13 (Gram Matrix) Given a kernel function $\kappa : \mathcal{X} \times \mathcal{X} \to \mathbb{R}$ and inputs $\boldsymbol{x}_1, \boldsymbol{x}_2, \ldots, \boldsymbol{x}_n \in \mathcal{X}$, then the following $n \times n$ matrix

$$\mathbf{K} = \left[\kappa\left(\boldsymbol{x}_i, \boldsymbol{x}_j\right)\right]_{i,j} \in \mathbb{R}^{n \times n} \tag{4.43}$$

is known as the Gram matrix (or kernel matrix) corresponding to $\boldsymbol{x}_1, \boldsymbol{x}_2, \ldots, \boldsymbol{x}_n$.

Definition 4.14 (Positive-Definite Matrix) Let the column vector be $\boldsymbol{c} = (c_1, \ldots, c_n)^{\mathsf{T}} \in \mathbb{R}^{n \times 1}$, $\mathbf{K}$ is a $n \times n$ Gram matrix, then it is called a positive-definite matrix if the following $n \times n$ matrix

$$\boldsymbol{c}^{\mathsf{T}}\mathbf{K}\boldsymbol{c} = \sum_{i=1}^{n}\sum_{j=1}^{n} c_i c_j \kappa\left(\boldsymbol{x}_i, \boldsymbol{x}_j\right) \geq 0, \tag{4.44}$$

is satisfied.

Definition 4.15 (Positive-Definite Kernel) For a kernel function $\kappa : \mathcal{X} \times \mathcal{X} \to \mathbb{R}$ and inputs $\boldsymbol{x}_1, \boldsymbol{x}_2, \ldots, \boldsymbol{x}_n \in \mathcal{X}$, its kernel function is called a positive-definite kernel if it satisfies the condition of a positive-definite matrix.

It can be proven that the kernel function performing inner product operation in Hilbert space is a positive-definite kernel. From Eqs. 4.42 and 4.44, we can conclude

$$\sum_{i=1}^{n}\sum_{j=1}^{n} c_i c_j \kappa\left(\boldsymbol{x}_i, \boldsymbol{x}_j\right) = \sum_{i=1}^{n}\sum_{j=1}^{n} c_i c_j \langle \varphi\left(\boldsymbol{x}_i\right), \varphi\left(\boldsymbol{x}_j\right)\rangle$$

$$= \sum_{i=1}^{n}\sum_{j=1}^{n} \langle c_i \varphi\left(\boldsymbol{x}_i\right), c_j \varphi\left(\boldsymbol{x}_j\right)\rangle$$

$$= \left\|\sum_{i=1}^{n} c_i \varphi\left(\boldsymbol{x}_i\right)\right\|^2 \geq 0.$$

The above result comes from the nonnegativity property of the norm.

Many types of kernel functions satisfy the condition of positive-definite kernel, such as polynomial kernels, Gaussian kernels, and variance kernels.

The book *Kernel Methods for Pattern Analysis* discusses the basic kernel functions and their types, as well as the three main ways to simplify kernel function calculations, namely, using closed-form analytical expressions, recursive relationships, and sampling methods (Shawe-Taylor and Cristianini 2004). It also provides a list of various kernel functions in Appendix D of the book for easy reference.

4.7.5   *Reproducing Kernel*

Let $\mathcal{F}_X$ be a Hilbert space composed of functions $f \in \mathcal{F}_X$, that is,

$$f : X \to \mathbb{R}. \tag{4.45}$$

Where the function f is expressed in the following form

$$f(\cdot) = \sum_{i=1}^{n} c_i \kappa(x_i, \cdot).$$

Also, let

$$f'(\cdot) = \sum_{j=1}^{n'} c_j' \kappa\left(\cdot, x_j'\right).$$

Then, we have

$$\langle f, f' \rangle = \sum_{i=1}^{n} \sum_{j=1}^{n'} c_i c_j' \kappa\left(x_i, x_j'\right) = \sum_{i=1}^{n} c_i f'(x_i) = \sum_{j=1}^{n'} c_j' f\left(x_j'\right). \tag{4.46}$$

From the above equation, we can further derive

$$\langle f, f \rangle = \sum_{i=1}^{n} \sum_{j=1}^{n} c_i c_j \kappa\left(x_i, x_j\right) = \sum_{i=1}^{n} c_i f(x_i) \tag{4.47}$$

The above equation can be derived from the following properties if we let $f = \kappa(x, \cdot)$, and

$$\langle f, \kappa(x_i, \cdot) \rangle = \sum_{i=1}^{n} c_i \kappa(x_i, \cdot) = f(x). \tag{4.48}$$

Therefore, the following reproducing kernel definition holds.

Definition 4.16 (Reproducing Kernel) A kernel function is known as a reproducing kernel function if the following condition is satisfied:

$$f(x) = \langle f, \kappa(x, \cdot) \rangle. \tag{4.49}$$

The reproducing kernel function is also called reproducing kernel for short.

Similarly, it can be concluded that the reproducing kernel also satisfies the property:

$$\langle \kappa(x, \cdot), \kappa(\cdot, x') \rangle = \kappa(x, x'). \tag{4.50}$$

Based on the reproducing kernel, the following definition can be derived.

Definition 4.17 (Reproducing Kernel Hilbert Space) The Hilbert space dealing with the reproducing kernel function $\mathcal{F}_\chi$ is called the reproducing kernel Hilbert space.

It's worth emphasizing that the kernel function is the similarity measure for any data pairs in the input space (x, x') and reproducing kernel function is the regularity measure for the kernel function.

Further Reading

1. Vladimir N. Vapnik. *The Nature of Statistical Learning Theory*, Springer science & business media, 1995.
   [Notes] This book is to discuss the fundamental ideas which lie behind the statistical theory of learning and generalization, considered to be the cornerstone of statistical learning theory.
2. Vladimir N. Vapnik. "An Overview of Statistical Learning Theory." *IEEE Transactions on Neural Networks*. 10(5), 1999.
   [Notes] This review paper was written after his book of *The Nature of Statistical Learning Theory* in 1995, which encapsulates the main points of statistical learning theory.
3. Trevor Hastie, Robert Tibshirani, Jerome Friedman. *The Elements of Statistical Learning: Data Mining, Inference, and Prediction* (2nd Edition). New York: Springer, 2009.

[Notes] This book covers the methods from supervised learning to unsupervised learning, including neural networks, support vector machines, classification trees, and boosting methods. It also introduces graphical models, random forests, ensemble methods, regression, and clustering methods.

4. Gareth James, Daniela Witten, Trevor Hastie, and Robert Tibshirani. *An Introduction to Statistical Learning with Applications in R* (2nd Edition). New York, Springer, 2021.

[Notes] This book introduces some statistical modeling, prediction techniques, and related applications. It includes linear regression, classification, resampling methods, tree-based methods, support vector machines, deep learning, and unsupervised learning and also provides some application examples based on the *R* language.

References

Aizerman, M. A., E.M. Braverman, and L.I. Rozonoer. (1964). Theoretical foundations of the potential function method in pattern recognition learning. *Automation and Remote Control* 25: 821–837.

Alpaydin, E. (2009). *Introduction to machine learning* (2nd Edn.). Cambridge: MIT Press.

Boser, B. E., I. M. Guyon, and V. N. Vapnik. (1992). A training algorithm for optimal margin classifiers. In *Fifth Annual Workshop on Computational Learning Theory*.

Bousquet, O., S. Boucheron, and G. Lugosi. (2003). *Introduction to statistical learning theory. Summer school on machine learning*. Berlin: Springer.

Burges, C. J. (1998). A tutorial on support vector machines for pattern recognition. *Data Mining and Knowledge Discovery* 2(2): 121–167.

Cho, Y., and L. K. Saul. (2009). Kernel methods for deep learning. In *Conference on Neural Information Processing Systems (NIPS)*.

Cortes, C., and V. Vapnik. (1995). Support vector networks. *Machine Learning* 20(3): 273–297.

Hofmann, T., B. Schölkopf, and A. J. Smola. (2008). Kernel methods in machine learning. *The Annals of Statistics* 1171–1220.

James, G., D. Witten, T. Hastie, and R. Tibshirani. (2013). *An introduction to statistical learning: With applications in R*. New York: Springer.

Morgenstern, J. H., and T. Roughgarden. (2015). On the pseudo-dimension of near-optimal auctions. In *Conference on Neural Information Processing Systems (NIPS)*.

Murphy, K. P. (2012). *Machine learning: A probabilistic perspective*. Cambridge: The MIT Press.

Natarajan, B. K. (1989). On learning sets and functions. *Machine Learning* 4(1): 67–97.

Russell, S., and P. Norvig. (2009). *Artificial intelligence: A modern approach*, 3rd ed. Pearson Education.

Schölkopf, B., and A. J. Smola. (2002). *Learning with kernels: Support vector machines, regularization, optimization, and beyond*. Cambridge: MIT Press.

Shalev-Shwartz, S., and S. Ben-David. (2014). *Understanding machine learning: From theory to algorithms*. Cambridge: Cambridge University Press.

Shawe-Taylor, J., and N. Cristianini. (2004). *Kernel methods for pattern analysis*. Cambridge: Cambridge University Press.

Vapnik, V. N. (1998). *Statistical learning theory*. New York: Wiley.

Vapnik, V. N. (1999). *The nature of statistical learning theory*, 2nd ed. New York: Springer.

Vapnik, V. N., and A. Chervonenkis. (1971). On the uniform convergence of relative frequencies of events to their probabilities. *Theory of Probability and Its Applications* 16(2): 264–279.

Chapter 5
Connectionist Framework

Abstract The foundation of the connectionist framework stems from connectionism, which simulates the information processing and learning process of the brain through artificial neural networks. Took an overview on the framework, we introduce connectionist learning theories, including Hebb theory, parallel distributed processing, connectionist models, and neural network theory. Next, we explain some units in neural networks, including biological neuron, artificial neuron, spike neuron, long short-term memory (LSTM) unit, gated recurrent unit (GRU), and capsule unit. Then, we discuss the structures of neural networks, namely, feedforward, feedback, and symmetric neural networks. After that, we introduce the optimizations of neural networks, such as backpropagation, evolutionary methods, and weight-agnostic methods without training the neural networks. Finally, the single- and multiagent networks are also explained with two case studies.

5.1 Overview

The connectionist framework refers to the machine learning framework based on connectionist approaches and connectionist models. Its manifestation is known as artificial neural networks, an artificial representation of the brain, used to simulate the brain's information processing and learning process. This school of thought is known as connectionism.

Connectionism, also known as bionicsism or physiologism, believes that the learning process of the brain lies in the connection mechanism of the brain's neural network.

The history of the development of neural networks has been introduced in Chap. 1 of this book. Connectionism cored with artificial neural network was pioneered in the 1940s, gradually attracting the attention of some researchers and even becoming a theory and method in the fields of artificial intelligence, cognitive psychology, cognitive science, neuroscience, and philosophical psychology. Donald Hebb's connectionist view is known as Hebbian theory, an interpretation of the brain's learning process, and is therefore also known as Hebbian learning.

In the early 1980s, the field of cognitive science referred to it as connectionist models tried to explain mental phenomena with connectionist models or neural network models. It is believed that from the perspective of knowledge processing, there should be higher-level knowledge processing units than neurons.

The core of connectionism is the theory of connectionist learning, also known as the theory of connectionist learning.

The connectionist framework is one of the important frameworks of machine learning. Neural networks based on the connectionist framework have become a hot spot in machine learning research and its applications in recent years.

5.2 Connectionist Learning Theories

It can be considered that connectionist learning theories are originated from Hebbian theory, flourished in parallel distributed processing, and developed into connection models and neural network theories. The following four sections introduce these stages of development and their main content.

5.2.1 Hebbian Theory

Hebbian theory originated from Donald Hebb's book titled *The Organization of Behaviour: A Neuropsychological Theory* published in 1949. The book attempts to interpret the adaptation process of brain neurons (cells) during the learning process from a neuropsychological perspective (Hebb 1949).

Hebb described the relationship between neural reflex activity and nerve cells in his book as follows:

> Let us assume that the persistence or repetition of a reverberatory activity (or 'trace') tends to induce lasting cellular changes that add to its stability.

He then made the following precise statement about this assumption:

> When an axon of cell A is near enough to excite a cell B and repeatedly or persistently takes part in firing it, some growth process or metabolic change takes place in one or both cells such that A's efficiency, as one of the cells firing B, is increased.

The above paragraph is known as Hebb's postulate, also known as Hebb's rule. Specifically, Hebb's postulate includes the following two rules:

(1) If two neurons on either side of a connection are activated synchronously, then the weight of that connection is increased.

(2) Conversely, if two neurons on either side of a connection are activated asynchronously, then the weight of that connection is decreased.

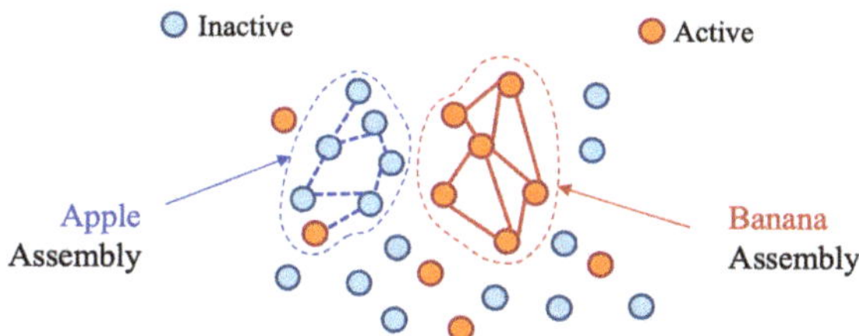

Fig. 5.1 Illustration of Hebb's cell assembly theory

Hebb's rule provides a theoretical basis for learning based on neural networks, that is, learning is a local phenomenon without environmental feedback.

Hebb also proposed the concept of "cell assembly" in his book:

Any frequently repeated, particular stimulation will lead to the slow development of a 'cell-assembly,' a diffuse structure comprising cells in the cortex and diencephalon (and also, perhaps in the basal ganglia of the cerebrum), capable of acting briefly as a closed system, delivering facilitation to other such systems and usually having a specific motor facilitation. A series of such events constitutes a 'phase sequence' – the thought process. Each assembly action may be amused by a preceding assembly, by a sensory event, or-normally-by both. The central facilitation from one of these activities on the next is the prototype of 'attention'.

Therefore, Hebb's theory is also known as cell assembly theory. Figure 5.1 is an example of the cell assembly theory: in the process of perceiving banana information, the relevant nerve cells are excited, while the nerve cells related to the apple are not excited. Donald Hebb is known as the father of neuropsychology and neural networks, and models established following Hebb's theory are known as Hebbian learning.

5.2.2 *Parallel Distributed Processing*

Parallel distributed processing (PDP) was once synonymous with connectionism, which began in the 1970s and made significant progress in the 1980s. PDP emphasizes the parallel processing nature of neural networks and the distributed nature of neurons.

Looking back, the 1970s to 1980s coincided with the boom in parallel computing and distributed computing research, and the term parallel distributed processing may have been born out of this.

Among the publications on PDP, the most influential is *Parallel Distributed Processing: Explorations in the Microstructure of Cognition* edited by David E. Rumelhart, James L. McClelland, and the PDP Research Group. Those publications consist of two volumes: "Volume 1: Foundations" (Rumelhart et al. 1986c) and "Volume 2: Psychological and Biological Models" (McClelland et al. 1987). These two volumes are considered pioneering works in connectionism, leading people to equate parallel distributed processing with connectionism, even though the term connectionism is not used in the volumes.

The second article in Volume 1 is titled "A General Framework for Parallel Distributed Processing" by David Rumelhart, Geoffrey Hinton, and James McClelland (Rumelhart et al. 1986b). The article lists eight important attributes of the parallel distributed processing (PDP) model, namely:

(1) A set of processing units
(2) A state of activation
(3) An output function for each unit
(4) A pattern of connectivity among units
(5) A propagation rule for propagating patterns of activities through the network of connectivities
(6) An activation rule for combining the inputs impinging on a unit with the current state of that unit to produce a new level of activation for the unit
(7) A learning rule whereby patterns of connectivity are modified by experience
(8) An environment within which the system must operate

In addition, the seventh article in Volume 1, "Learning and Relearning in Boltzmann Machines," was written by Geoffrey Hinton and Terrence Sejnowski (Hinton and Sejnowski 1986). More than twenty years later, Geoffrey Hinton and his student Ruslan Salakhutdinov published a paper titled "Deep Boltzmann Machines" at the *International Conference on Artificial Intelligence and Statistics* in 2009 (Salakhutdinov and Hinton 2009). The continuous deep work on Boltzmann machines by Hinton and his team is very admirable.

In 2014, the journal *Cognitive Science* of the Cognitive Science Society published a special article "Parallel Distributed Processing at 25: Further Explorations in the Microstructure of Cognition" by Timothy T. Rogers and James McClelland. This paper not only commemorates the 25th anniversary of the publication of the two volumes of Parallel Distributed Processing (PDP) but also discusses how the PDP model can play a greater role within the framework of cognitive science (Rogers and McClelland 2014).

5.2.3 Connectionist Models

Connectionism attempts to explain mental phenomena using artificial neural networks. Its main argument is that mental phenomena can be simulated with networks, which consist of many simple and unified units that are interconnected. The form of the units and connections can vary depending on the model. The units in the network can represent neurons in the brain, and the connections can represent synapses of neurons.

Connectionism also proposes a cognitive theory, which is based on simulating concurrent distributed signal activity through digitizable representations of connections, and the function of learning can be achieved by adjusting the weights of the connections based on experience.

The core of connectionism is referred to as connectionist models. In 1982, Jerome A. Feldman and Dana H. Ballard published a paper titled "Connectionist Models and Their Properties" in the journal *Cognitive Science* (Feldman and Ballard 1982). The paper proposed a general connectionist model and discussed its application in cognitive science. The paper also discussed the stability and noise sensitivity of the model, distributed decision-making, issues of time and sequence, and the representation of complex concepts.

In 1988, 1990, and 1993, three sessions of the *Connectionist Models Summer School* were held. The first two were held at Carnegie Mellon University in the United States, and the third was at the University of California, San Diego. In addition, three proceedings of the Connectionist Models Summer School were published.

In the *Oxford Companion to Consciousness* published in 2009, James McClelland and Axel Cleeremans annotated the term "Connectionist Models" as follows (McClelland and Cleeremans 2009):

> Connectionist models take inspiration from the manner in which information processing occurs in the brain. Processing involves the propagation of activation among simple units (artificial neurons) organized in networks, that is, linked to each other through weighted connections representing synapses or groups thereof. Each unit then transmits its activation level to other units in the network by means of its connections to those units. The activation function, that is, the function that describes how each unit computes its activation based on its inputs, may be a simple linear function, but is more typically non-linear.

5.2.4 Neural Network Theory

The concept of neural networks can be traced back to a classic paper titled "A Logical Calculus of the Ideas Immanent in Nervous Activity" published in *The Bulletin of Mathematical Biophysics* in 1943 by Warren S. McCulloch and Walter Pitts (McCulloch and Pitts 1943). The authors of the paper attempted to understand how the brain generates highly complex patterns by using many interconnected basic cells; these basic brain cells are referred to as neurons. The paper presented a highly simplified model of a neuron, which was later referred to as the McCulloch-Pitts neuron model.

The computing system, inspired by biological neurons in the brain, is also known as a connectionist system, named after the concept of connectionism. Artificial neural networks are an implementation of machine learning. In recent years, especially with the continuous development of deep learning, artificial neural networks have become a hot research topic and have played an important role in many application fields.

In 1997, Philippe de Wilde published a book titled *Neural Network Models: Theory and Projects*, which deeply explained neural network models, and each chapter also provided a research topic. In June 2013, Philippe de Wilde released the second edition of the book (De Wilde 2013).

The perceptron, composed of single-layer neurons, is the earliest artificial neural network. Based on the single-layer perceptron, a multilayer perceptron (MLP) appeared: it consists of an input layer, an output layer, and multiple hidden layers, also known as a multilayer artificial neural network.

From the view of hierarchy, artificial neural networks can be divided into shallow neural networks and deep neural networks. There is no strict definition of the boundary between the number of layers in shallow and deep, but neural networks with two or fewer hidden layers are usually called shallow networks, and neural networks with multiple hidden layers are called deep networks.

Machine learning based on deep neural networks is called deep learning.

In August 2007, Alexander I. Galushkin published a monograph titled *Neural Networks Theory* (Galushkin 2007), elevating neural networks to a theoretical level.

5.3 Units of Neural Networks

The unit in artificial neural networks was initially just a simple artificial neuron, proposed in 1943. Spiking neuron was proposed in 1952. With the improvement of the functions of artificial neural networks, the long short-term memory (LSTM) unit was proposed in 1997, the gated recurrent unit (GRU) was introduced in 2014, and the Capsule was proposed as the unit of the capsule neural network in 2017.

5.3.1 Biological Neuron

Neuroscience starts from studying the nervous system and researching how the brain processes information. It organically combines behavior, cognition, and brain mechanisms, trying to comprehensively reveal the information processing process and its neural mechanisms when humans and animals perceive objects, form representations, use language, memorize information, and make reasoning decisions.

Brain science is one of the main research contents of neuroscience. The brain is very superior in rational decision-making, although it is not perfect. The brain has decision-making functions, and prediction and inference are the keys to decision-making.

The basic unit of the brain's nervous system is the biological neuron that consists of dendrite and axon. The dendrite contains the cell body and nucleus. On the long axon are nerve fibers composed of the myelin sheath and its Schwann cell, node of Ranvier, and the tail which is called the axon terminals.

Each neuron's dendrites can receive incoming electrical pulse signals and transmit them into the cell body. After being transmitted through the axon, the signal is then sent to another neuron or other tissues in the body, such as muscles or glands, through the end of the axon.

5.3.2   Artificial Neuron

Below, we first present the structure of an artificial neuron and then introduce several commonly used activation functions.

5.3.2.1   Neuron Structure

The structure of an artificial neuron is shown in Fig. 5.2.

Its inputs are $x_1, x_2, x_3, \ldots$ and the constant representing the bias weight $+1$, then multiplied by the weights $w_1, w_2, w_3, \ldots$ and the scalar b, then calculate their linear accumulation sum:

$$u = \sum_i w_i x_i + b. \tag{5.1}$$

Where each weight w_i determines the contribution of the input x_i to the output. Then, through the activation function $f(u)$ the output y is obtained, that is,

$$y = f(u). \tag{5.2}$$

It should be noted that the inputs and outputs of artificial neurons are all scalars.

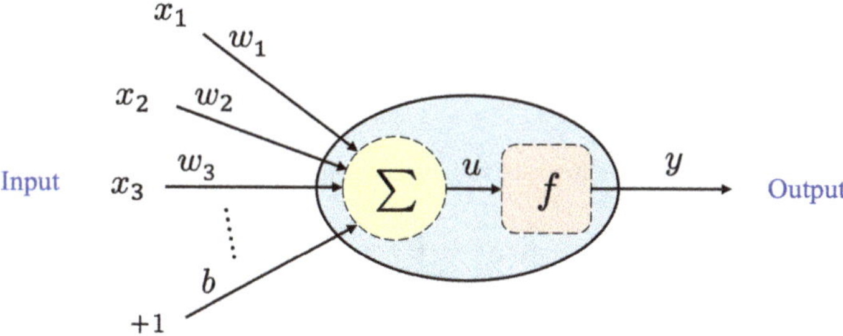

Fig. 5.2 Structure of an artificial neuron

5.3.2.2   Activation Function

The activation function $f(u)$ in the neuron is used to transform the weighted input sum u into the corresponding output y. Here, it introduces several commonly used activation functions, namely, binary threshold, Sigmoid, TanH, and ReLU.

(1) Binary threshold function

The function of the binary threshold is as follows:

$$f(u) = \begin{cases} 0 & \text{if } u < \theta \\ 1 & \text{otherwise} \end{cases}. \tag{5.3}$$

Where θ is referred to as the threshold.

The binary threshold function is shown in Fig. 5.3a, with outputs of only 0 and 1, while the binary step function is a special case of the binary threshold function, where $\theta = 0$.

Neurons using the binary threshold function can be referred to as binary neurons, and neural networks using binary neurons can only be used for binary logic operations or linear classification. McCulloch-Pitts neurons are binary neurons, and the single-layer perceptron also uses binary neurons.

The single-layer perceptron using binary neurons has limitations, that is, it can only learn linearly separable patterns and cannot be used to handle the XOR problem (Minsky 1969).

The limitations are illustrated in Fig. 5.4. Suppose the single-layer perceptron has only two input values, i.e., $x_1, x_2 \in \{0, 1\}$, and the output value is $y \in \{0, 1\}$. For

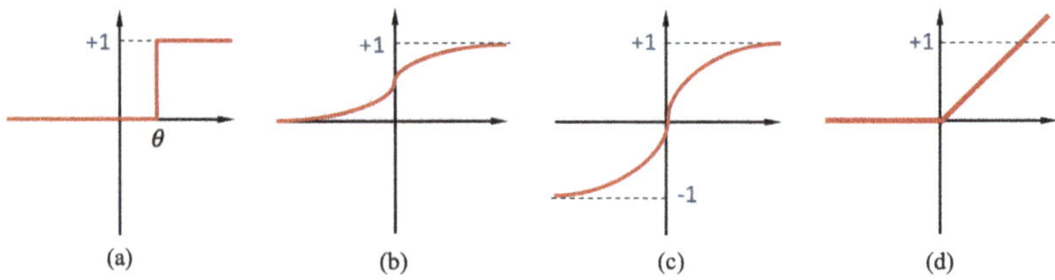

Fig. 5.3 Curves of four representative activation functions

x_1	x_2	θ	y
0	0	2	0
0	1	2	0
1	0	2	0
1	1	2	1

AND

x_1	x_2	θ	y
0	0	1	0
0	1	1	1
1	0	1	1
1	1	1	1

OR

x_1	x_2	θ	y
0	0	?	0
0	1	?	1
1	0	?	1
1	1	?	0

XOR

Fig. 5.4 Logical AND, OR, and XOR problems for single-layer perceptron

the logical AND problem, if the threshold is set to $\theta = 2$, the desired result can be obtained; for the logical OR problem, the threshold only needs to be set to $\theta = 1$; it can also get the desired results; however, for the logical XOR problem, it cannot be solved by setting a threshold.

With the continuous progress of research on artificial neural networks, nonlinear activation functions have emerged, whose outputs are differentiable functions of the inputs. The three activation functions introduced below, namely, Sigmoid, TanH, and ReLU, are all nonlinear activation functions.

(2) Sigmoid function

It is a bounded, differentiable real-valued function:

$$f(u) = \sigma(u) = \frac{1}{1 + e^{-u}}. \tag{5.4}$$

Where $\sigma(\cdot)$ represents the Sigmoid function, also known as the soft step function. Its output increases monotonically with the input, and the range is $(0, 1)$, as shown in Fig. 5.3b.

The term "Sigmoid" means "S shaped," so the Sigmoid function can also be directly called as the "S-shaped function." This function has an important feature, that is, its derivative can be expressed by its output, that is,

$$\frac{d\sigma(u)}{du} = \sigma(u) \cdot (1 - \sigma(u)). \tag{5.5}$$

This is an important feature used by the backpropagation (BP) algorithm. And the backpropagation algorithm will be further discussed in Sect. 5.6 of this chapter.

(3) Hyperbolic tangent function

The hyperbolic tangent function, denoted as TanH, is also a differentiable function with a range of $(-1, 1)$. The expressions for the function and its derivative are as follows:

$$f(u) = \tanh(u) = \frac{\left(e^u - e^{-u}\right)}{\left(e^u + e^{-u}\right)}, \tag{5.6}$$

$$\frac{d\tanh(u)}{du} = 1 - f(u)^2. \tag{5.7}$$

Figure 5.3c is the curve of the TanH function.

(4) Rectified linear unit

The rectified linear unit (ReLU) is also a differentiable function, as shown in Fig. 5.3d; its range is $[0, \infty)$. Equations 5.8 and 5.9 are its function and derivative, respectively:

$$f(u) = \max(0, u) = \begin{cases} 0, & u \le 0 \\ u, & u > 0 \end{cases}, \tag{5.8}$$

$$\frac{df(u)}{du} = \begin{cases} 0, & u \le 0 \\ 1, & u > 0 \end{cases}. \tag{5.9}$$

The activation function ReLU is adopted by most convolutional neural networks.

It should be noted that the characteristics of the above activation functions are single output, and the activation function Softmax with multiple outputs will be detailed in Sect. 12.5.

In addition, for readers who want to further understand activation functions, you can refer to the review article titled "Activation Functions: Comparison of Trends in Practice and Research for Deep Learning" (Nwankpa et al. 2018).

5.3.3 Spiking Neuron

In history, the first spiking neuron called the Hodgkin-Huxley model was proposed in 1952 (Hodgkin et al. 1952). Based on it, a spiking neural network (SNN) was presented (Maass 1994), and then some similar networks emerged such as pulse-coupled neural network (PCNN) Ranganath et al. (1995), pulsed neural network (PNN) (Humphries and Gurney 2001), and impulsive neural network (INN) (Stamova and Stamov 2016).

The spiking neuron is inspired by the biological neuron. In neuroscience, the "spike" is used to represent the electrical pulse signal received by the dendrites of biological neurons, also known as the action potential, as shown in Fig. 5.5a. Its waveform is extremely sharp, and the peak duration is very short (about one millisecond).

5.3.3.1 Spiking Neuron Model

The computational unit in a spiking neural network is a spiking neuron, which is based on spiking neuron models (Gerstner and Kistler 2002), a mathematical description of the characteristics of biological spiking neurons.

Compared with classical artificial neurons, spiking neurons are closer to biological neurons. Wolfgang Maass, in his paper titled "Networks of Spiking Neurons: The Third Generation of Neural Network Models" (Maass 1997), divided neural

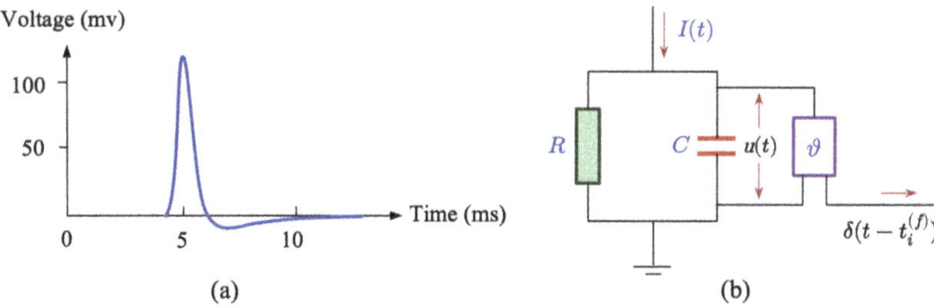

Fig. 5.5 The schematic diagram of (**a**) the action potential and (**b**) the leaky integrate-and-fire model

networks into three generations based on the computational units in the neural network models: the first generation is based on McCulloch-Pitts neurons, with perceptrons as its representative; the second generation is based on neurons with activation functions, such as convolutional neural networks; and the third generation is the spiking neural network based on spiking neurons.

Representative spiking neuron models include the Hodgkin-Huxley model (Hodgkin et al. 1952), the FitzHugh-Nagumo model (FitzHugh 1961; Nagumo et al. 1962), the Hindmarsh-Rose model (Hindmarsh and Rose 1984), the spike response model (SPM) (Gerstner and van Hemmen 1992), and the integrate-and-fire model (Gerstner and Kistler 2002). Among them, Alan L. Hodgkin and Andrew F. Huxley, who proposed the Hodgkin-Huxley model, won the Nobel Prize in Physiology or Medicine in 1963.[1]

The spiking neuron models have the following common characteristics:

(1) It transmits information through a discrete spike train.
(2) Its firing probability to generate spikes increases with excitatory input and decreases with inhibitory input.
(3) When the internal state reaches a given threshold, one or more output spikes will be generated.

5.3.3.2   Leaky Integrate-and-Fire Model

The leaky integrate-and-fire model is a famous spiking neuron model, which establishes the relationship between the membrane currents at the input and the membrane voltage at the output of neurons. The basic schematic of this model is shown in Fig. 5.5b.

The leaky integrate-and-fire model is composed of parallel resistors R and capacitors C, and ϑ is the threshold. The input driving current $I(t)$ is charging

---

[1] https://www.nobelprize.org/prizes/medicine/1963/summary/

the RC circuit, causing the voltage across the capacitor $u(t)$ to continuously rise. If it makes $u(t) = \vartheta$ at $t_i^{(f)}$ moment, then it generates output pulse $\delta\left(t - t_i^{(f)}\right)$.

The driving current can be divided into two parts, $I(t) = I_R + I_C$, where I_R is the current passing through the linear resistor R, which can be calculated using Ohm's law, that is, $I_R = \frac{u(t)}{R}$, where $u(t)$ is the voltage across the resistor. And I_C is charging the capacitor C, according to the definition of capacitance, $C = \frac{q}{u}(t)$, where q is the charge; hence, $I_C = C\frac{du}{dt}$. Therefore, the following equation holds

$$I(t) = \frac{u(t)}{R} + C\frac{du}{dt}. \tag{5.10}$$

Multiply both sides of the above equation by R and then introduce the time constant of the leaky integrator $\tau_m = RC$, thus deriving the standard equation of the leaky integrator excitation model:

$$\tau_m \frac{du}{dt} = -u(t) + RI(t). \tag{5.11}$$

5.3.4 Long Short-Term Memory Unit

A long short-term memory (LSTM) unit typically consists of a cell, an input gate, an output gate, and a forget gate.

The prototype of the long short-term memory unit was proposed in 1997 by Sepp Hochreiter and Jürgen Schmidhuber (Hochreiter and Schmidhuber 1997). In 1999, Felix Gers and his mentor Jürgen Schmidhuber, along with Fred Cummins, introduced the forget gate into the LSTM architecture, allowing the LSTM to reset its state. In 2000, Gers, Schmidhuber, and Cummins introduced peephole connections into the LSTM architecture.

The long short-term memory unit can serve as a unit of a recurrent neural network (RNN), and a recurrent neural network composed of LSTM units is called an LSTM network. It is worth noting that although some researchers simply refer to the LSTM network as LSTM, do not confuse the LSTM network with the LSTM unit.

There are several types of LSTM units, with the typical LSTM unit shown in Fig. 5.6.

Fig. 5.6 Structure of a typical LSTM unit

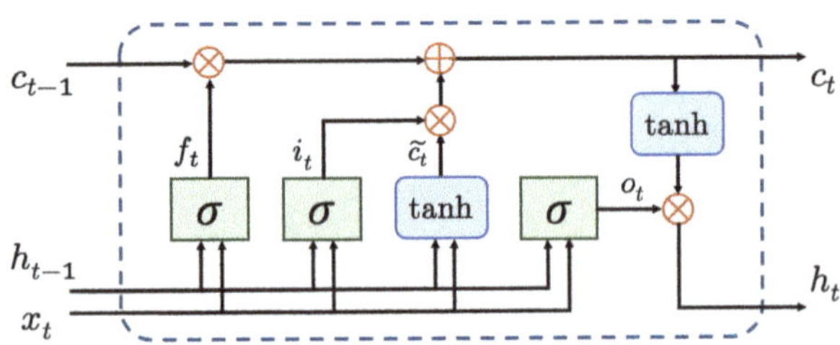

In Fig. 5.6, t is the time step, c_t is the cell, i_t is the input gate, o_t is the output gate, f_t is the forget gate, x_t is the input, and h_t is the output. The expressions for these parameters are as follows:

$$f_t = \sigma\left(W_f x_t + U_f h_{t-1} + b_f\right),$$

$$i_t = \sigma\left(W_i x_i + U_i h_{t-1} + b_i\right),$$

$$\widetilde{c}_t = \tanh\left(W_c x_t + U_c h_{t-1} + b_c\right),$$

$$c_t = f_t * c_{t-1} + i_t * \widetilde{c}_t,$$

$$o_t = \sigma\left(W_o x_t + U_o h_{t-1} + b_o\right),$$

$$h_t = o_t * \tanh\left(c_t\right). \tag{5.12}$$

Where W and b are a parameter matrix and vector.

The memory cell is responsible for remembering information over any time interval, while the input, output, and forget gates are responsible for regulating the information entering and leaving the memory cell. These characteristics of the LSTM unit make the LSTM network based on it capable of handling sequential information.

If you want to learn more about LSTM networks, you may first take a look at the blog titled "Understanding LSTM Networks" by Christopher Olah.[2]

5.3.5   *Gated Recurrent Unit*

The gated recurrent unit (GRU) is a gating mechanism for recurrent neural networks (Cho et al. 2014), which can also be seen as a variant of the long short-term memory (LSTM) unit. The GRU was proposed by Kyunghyun Cho et al. in 2014 when Cho was doing postdoctoral research in the Yoshua Bengio team at University of Montreal, Canada.

Figure 5.7 shows the structure of the gated recurrent unit. It is similar to an LSTM unit with a forget gate, but it has one less output gate than the LSTM. In addition, it combines the forget gate and the input gate into a single update gate. It also merges the memory cell and the hidden state, resulting in a model that is simpler than the standard LSTM model.

---

[2] https://colah.github.io/posts/2015-08-Understanding-LSTMs/

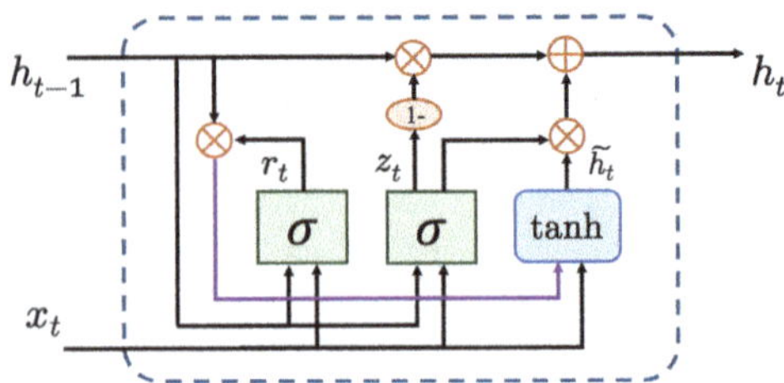

Fig. 5.7 Structure of the gated recurrent unit (GRU)

In Fig. 5.7, t is time step, x_t is input, z_t is update gate, r_t is reset gate, $\widetilde{h}_t$ is memory cell, and h_t is output. The expressions of gated recurrent unit (GRU) are as follows:

$$z_t = \sigma \left(W_z \cdot [h_{t-1}, x_t] \right),$$

$$r_t = \sigma \left(W_r \cdot [h_{t-1}, x_t] \right),$$

$$\widetilde{h}_t = \tanh \left(W \cdot [r_t * h_{t-1}, x_t] \right),$$

$$h_t = (1 - z_t) * h_{t-1} + z_t * \widetilde{h}_t. \tag{5.13}$$

Where W is the parameter matrix.

GRU is more concise and performs similarly to LSTM in tasks such as polyphonic music modeling, speech signal modeling, and natural language processing and shows better performance on some smaller datasets (Chung et al. 2014).

However, some studies suggest that LSTM is more powerful than GRU because the former can easily perform unbounded counting, which the latter cannot do (Weiss et al. 2018).

5.3.6　Capsule Unit

In 2017, Professor Geoffrey Hinton et al. proposed the capsule neural network (CapsNet), with the capsule as its basic unit (Sabour et al. 2017), where the capsule is defined as follows:

> A capsule is a group of neurons whose activity vector represents the instantiation parameters of a specific type of entity such as an object or an object part.

The idea behind a capsule is inspired by cortical mini-columns in the brain. Some scientists who engaged in neuroscience research have found that there are numerous cortical mini-columns in the cerebral cortex. A cortical mini-column is a vertical columnar body in the cerebral cortex, composed of 80 to 120 interconnected neurons. All neurons in the cortical mini-column have a common receptive field and common output. Therefore, these scientists believe that cortical mini-columns are likely to be the basic computational units that make up the cerebral cortex.

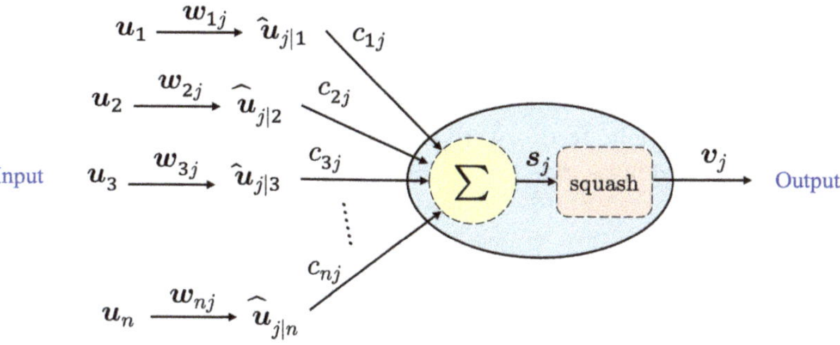

Fig. 5.8 Structure of the capsule unit

Professor Geoffrey Hinton and his team use capsules to represent various parameters of objects such as position, size, and color. Formally, a capsule is represented as an activity vector, each element of which accommodates an instantiated value of a parameter.

The structure of the capsule unit is shown in Fig. 5.8.

By comparing the capsule with the artificial neuron depicted in Fig. 5.2, it can be seen that the input and output of the capsule are vectors; each input u_i comes from the previous layer of neural capsules.

The processing steps of a capsule can be divided into the following three steps:

First, the input u_i needs to be processed.

An affine transformation is performed, i.e., multiplied by a transformation matrix w_{ij}; the purpose is to transform the input vector into the output vector of the next layer. This is an operation unique to neural capsules:

$$\hat{u}_{j|i} = w_{ij}u_i. \tag{5.14}$$

Next, calculate the weighted sum of $\hat{u}_{j|i}$, where c_{ij} is the coupling coefficient:

$$s_j = \sum_i c_{ij}\hat{u}_{j|i}. \tag{5.15}$$

Finally, the output is obtained through a nonlinear activation function for vectors. The activation function in the neural capsule is referred to as the squashing function, and its expression is as follows:

$$v_j = \frac{\| s_j \|^2}{1 + \| s_j \|^2} \frac{s_j}{\| s_j \|}. \tag{5.16}$$

5.4 Structures of Neural Networks

From the viewpoints on the structures of neural networks, they can be divided into three main forms, namely, feedforward neural networks, feedback neural networks, and symmetric neural networks.

5.4.1 Feedforward Neural Networks

The feedforward neural networks can be traced back to single-layer perceptron proposed in 1958, but modern feedforward neural networks trained using the backpropagation method originated in the early 1970s.

As shown in Fig. 5.9, feedforward neural networks allow information to be transmitted in one direction only, from the input layer, through several hidden layers, to the output layer, with no feedback between layers and no self-loop in the neurons themselves.

For example, if a learner is needed, $y = f(x)$, the goal is to map the input data x to y. Design a neural network and define a mapping $y = f(x; \theta)$, and adjust the network parameters θ through training to obtain an optimal function f^*, so that $y = f^*(x; \theta)$.

The $y = f(x; \theta)$ is referred to as a feedforward model, because the process of information flow starts from the input x, passes through the intermediate function f for calculation, and finally gets the output y. The information flow in this process is unidirectional, with no feedback. The aforementioned feedforward model implemented by neural networks is a single-layer feedforward neural network.

If $f = f^{(N)} \circ \cdots \circ f^{(2)} \circ f^{(1)}$ is composed of N functions, then

$$y = f^{(N)} \circ \cdots \circ f^{(2)} \circ f^{(1)}\left(x; \theta^{(1)}\right). \tag{5.17}$$

It is called a multilayer feedforward model, where $f^{(i)}\left(x; \theta^{(i)}\right)$ is the function of the ith layer. Similarly, a multilayer feedforward model implemented with neural networks is a multilayer feedforward neural network. This chain structure is very

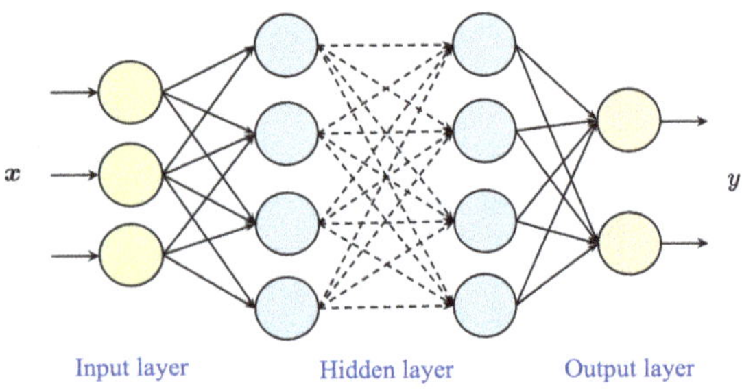

Fig. 5.9 The schematic diagram of a feedforward neural network

Fig. 5.10 Chain structure of feedforward neural network

suitable for building multilayer feedforward neural networks. Figure 5.10 is an abstract diagram of this chain structure, $f^{(1)}$ is the function of the first layer, $f^{(2)}$ is the function of the second layer, and so on.

In a narrow sense, perceptrons and multilayer perceptrons belong to feedforward neural networks; in a broad sense, convolutional neural networks, autoencoders, etc. can all be considered as feedforward neural networks, because these neural networks meet the definition of feedforward neural networks.

Feedforward neural networks are often suitable for dealing with spatially related information, for example, convolutional neural networks are suitable for image classification, and autoencoders are suitable for image dimensionality reduction.

5.4.2 Feedback Neural Networks

Feedback neural networks can be further divided into recurrent neural networks and recursive neural networks.

5.4.2.1 Recurrent Neural Networks

Recurrent neural networks (RNN) originated from the content-addressable memory systems with binary threshold units proposed by John Hopfield in 1982, which was later called Hopfield network (Hopfield 1982). From 1985 to 1986, David Rumelhart, Geoffrey Hinton, and Ronald Williams published papers in which they discussed recurrent neural networks and their structure (Rumelhart et al. 1985, 1986a).

As shown in Fig. 5.11, unlike feedforward neural networks, recurrent neural networks also include backward connections between their neurons or layers, forming directed cycles, which can retain the internal state of the network and enable it to show dynamic time characteristics.

Let t denote the moment and $\boldsymbol{x}^{(t)}$ is the input data, if the output $y^{(t+1)}$ at time $t+1$ can be represented as follows:

$$y^{(t+1)} = f\left(y^{(t)}, \boldsymbol{x}^{(t)}; \boldsymbol{\theta}\right) = f\left(f\left(y^{(t-1)}, \boldsymbol{x}^{(t-1)}\right), \boldsymbol{x}^{(t)}; \boldsymbol{\theta}\right); \boldsymbol{\theta}). \tag{5.18}$$

This is referred to as a recurrent method. Because the definition of $y^{(t+1)} = f\left(\cdot\right)$ at time $t+1$ is same as the definition of $y^{(t)} = f\left(\cdot\right)$ at time t.

Figure 5.12a is a loop node in the recurrent neural network, and Fig. 5.12b is the logical structure of the node after it is expanded.

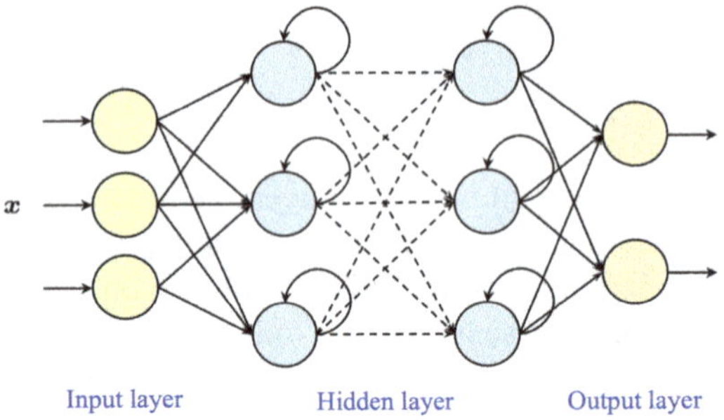

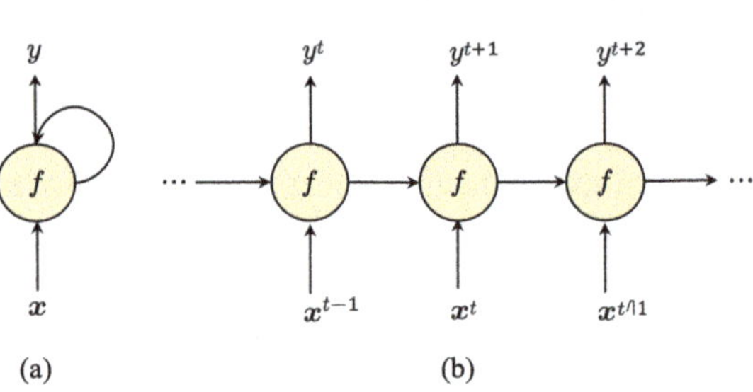

Fig. 5.11 The schematic diagram of recurrent neural network

Fig. 5.12 Logical structure of nodes in a recurrent neural network

A unit of a recurrent neural network can be a neuron or another one of units. A recurrent neural network composed of LSTM unit described in Sect. 5.3.4 is called an LSTM network, or simply LSTM, which was proposed by Sepp Hochreiter and Jürgen Schmidhuber in 1997. In addition, the gated recurrent unit (GRU) described in Sect. 5.3.5 can also be used, which was proposed in 2014 and is a gating mechanism of recurrent neural networks. The Hopfield network is also a form of recurrent neural network.

Recurrent neural networks are often suitable for dealing with dynamic time behaviors related to time series, such as unconstrained handwriting recognition, speech recognition, and video analysis.

5.4.2.2 Recursive Neural Network

Recursive neural networks are a type of treelike network that applies the same set of weights recursively and traverses the network in topological order to get the corresponding results.

It should be emphasized that recursive neural networks are different from the previously mentioned recurrent neural networks, and they should not be confused:

first, the meanings of recursive and recurrent are different. In mathematics and computer science, recursion has a clear definition. The explanation on *WikiDiff* is "Recurrent is recurring time after time while recursive is drawing upon itself, referring back."

Second, recursive neural networks share weights between different nodes, while recurrent neural networks share weights within the same node.

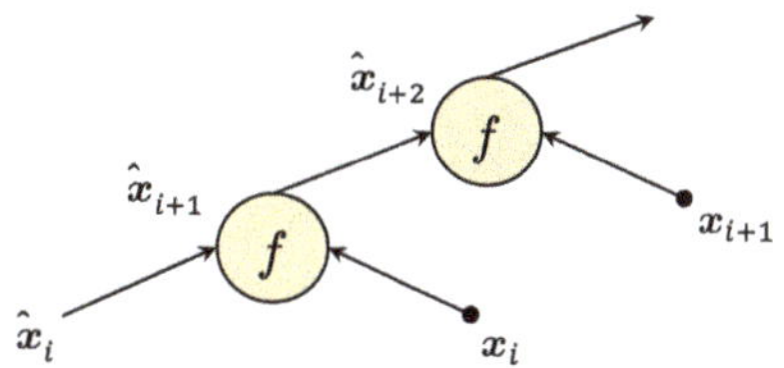

Fig. 5.13 The schematic diagram of a recursive neural network

Third, recursive neural networks are structurally recursive, while recurrent neural networks are cyclic in time.

Below, we first give a formal definition of the recursive neural network and then further discuss the differences between the recursive neural network and the recurrent neural network.

Without loss of generality, a part of the recursive neural network can be represented as a binary tree as shown in Fig. 5.13, where x_i represents the value of the leaf nodes of the tree and $\hat{x}_i$ represents the output value of non-leaf nodes or as the input value of the previous layer nodes. For any non-leaf node in the binary tree, its left and right child nodes can both be leaf nodes, both can be non-leaf nodes, or even only one can be a leaf node.

The following expression can be used to explain its mathematical meaning:

$$\hat{x}_{i+2} = f\left(\hat{x}_{i+1}, x_{i+1}; \theta\right) = f\left(f\left(\hat{x}_i, x_i; \theta\right), x_{i+1}; \theta\right). \tag{5.19}$$

The above expression reflects the essence of recursive neural networks: similar to recurrent neural networks, recursive neural networks also call $f(\cdot)$ within $f(\cdot)$; unlike recurrent neural networks, recursive neural networks only depend on the level of the tree and do not consider time t.

From the structures of recurrent neural networks and recursive neural networks, the former is shown in Fig. 5.12b as a chain-like neural network structure, while the latter is shown in Fig. 5.13 as a treelike neural network structure.

Some researchers believe that recursive neural networks are a generalization of recurrent neural networks or that recurrent neural networks are a special type of recursive neural network (Irsoy and Cardie 2014; Goodfellow et al. 2016).

Recursive neural networks are used more often in natural language processing, for example, the paper titled "Deep Recursive Neural Networks for Compositionality in Language" published at the 2014 *Conference on Neural Information Processing Systems* (NIPS) (Irsoy and Cardie 2014) and the paper "Template-based Math Word Problem Solvers with Recursive Neural Networks" published at the 33rd *AAAI Conference on Artificial Intelligence* (AAAI) (Wang et al. 2019).

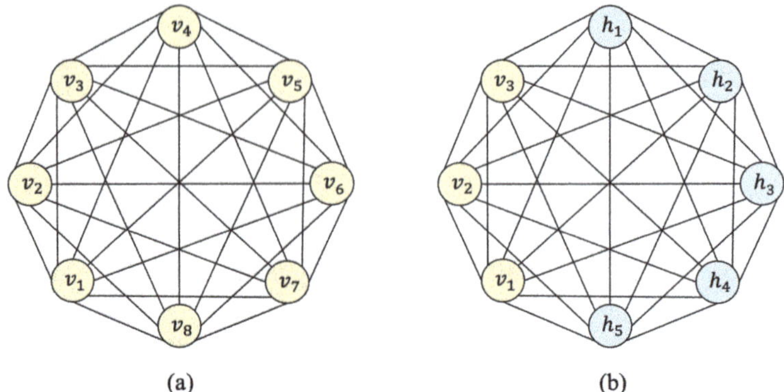

(a)   (b)

Fig. 5.14 Two types of symmetric neural networks

5.4.3 Symmetric Neural Networks

Symmetric neural networks, also known as symmetrically connected neural networks or symmetrically connected networks, are similar to recurrent neural networks, but the connections between units are symmetric, that is, the weights in both directions are the same.

The symmetric neural networks are easier to analyze than recurrent neural networks, but their functionality is also more restricted because the network obeys an energy function. Symmetric connection networks without hidden units are called Hopfield networks, and symmetric connection networks with hidden units are called Boltzmann machines.

The nodes in the Hopfield network are called binary threshold units, which have only two output values, forming a content-addressable memory system, and ensuring convergence to local minima (Hopfield 1982). Figure 5.14a is a schematic diagram of a Hopfield network with 8 units.

The nodes in the Boltzmann machine are called neuron-like units, which also have only two output values, but can randomly decide whether their output is 0 or 1. The random sampling function uses the Boltzmann distribution of statistical mechanics, hence the name Boltzmann machine (Hinton and Sejnowski 1983; Ackley et al. 1985). Figure 5.14b is a schematic diagram of it, with 3 visible units and 5 hidden units.

5.5 Optimizations of Neural Networks

Multilayer artificial neural networks are composed of many artificial neurons connected in a certain way and layer, each with a weight between two layers of neurons, which can be adjusted in some way, and then get the corresponding output value through the activation function, so that the neural network can adapt to the input and learn.

This is the optimization of neural networks and is also often referred to as the training of neural networks.

Next, we will introduce three optimization methods, which are the backpropagation algorithm, evolutionary algorithms, and weight-agnostic method.

5.5.1   Backpropagation

Backpropagation is short for backward propagation of errors. In this subsection, we will overview the algorithm of backpropagation first and then introduce some research about backpropagation and human brain.

5.5.1.1   Backpropagation Algorithm

The backpropagation algorithm is a common method for training multilayer neural networks and adjusting network weight errors, usually combined with the gradient descent optimization method. The blue dashed line in Fig. 5.15 is the path of backpropagation in a multilayer neural network.

The backpropagation algorithm originated from the automatic differentiation (AD) method of nested differentiable function discrete connected networks proposed by Seppo Linnainmaa in 1970. Paul J. Werbos proposed the possibility of using Seppo Linnainmaa's method for the training of artificial neural networks in 1974.

From the 1980s onward, some neural network researchers have published results using the backpropagation algorithm to train multilayer neural networks (Rumelhart et al. 1985). In the paper "Learning Representations by Back-propagating Errors" published in *Nature* magazine in 1986 by David Rumelhart, Geoffrey Hinton, and Ronald Williams, it was pointed out that the backpropagation algorithm can repeatedly adjust the weights of the network, making the error between the actual output vector of the network and the expected output vector minimal, and can

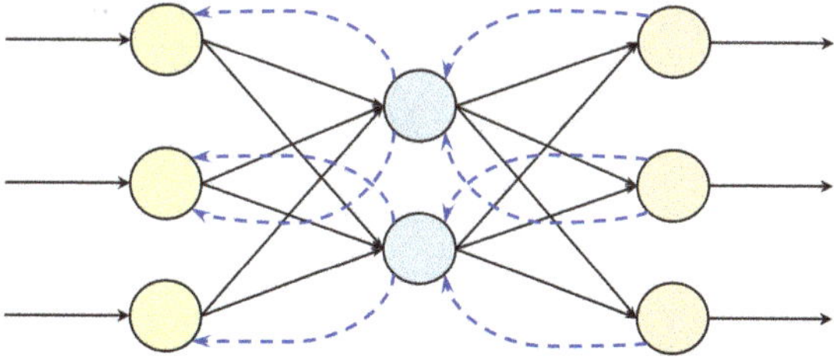

Fig. 5.15 Path of backpropagation in a multilayer neural network

also make the data in the hidden layer of the neural network form useful learning representations (Rumelhart et al. 1986b).

Multilayer neural networks can be seen as implementing a composite function from input space to output space, called the network function, denoted as φ. The learning problem in it is to find the best combination of weights, so that this network function φ tries to approximate the target function f. However, since the target function f cannot be obtained directly, the network function φ can only be trained through some samples to approximate the target function f.

Consider a feedforward neural network with n dimensional input and m dimensional output, which contains any number of hidden layers. Each neuron's activation function in the network is continuous and differentiable, and the weights are randomly selected real values. Given a training set composed of p ordered pairs of n dimensional and m dimensional

$$\left\{ (x_1, t_1), (x_2, t_2), \ldots, (x_p, t_p) \right\}.$$

When the x_i of the training set is submitted as input to the network, it produces an actual output o_i. However, this actual output is usually not equal to the target output t_i. We hope that for all $i = 1, \ldots, p$, the o_i and t_i should be as similar as possible. More precisely, it is to minimize the error function of the network. That is,

$$E = \frac{1}{2} \sum_{i=1}^{p} \| o_i - t_i \|^2. \tag{5.20}$$

The role of the backpropagation algorithm is to find the local minimum of the error function. The process is as follows: first, initialize the neural network with randomly selected weights; then calculate the gradient of the error function and use it to correct the initial weights. The process of calculating the gradient is recursive.

The first step in the minimization process is to expand the neural network so that it can automatically calculate its error function. Figure 5.16 is a schematic diagram of it.

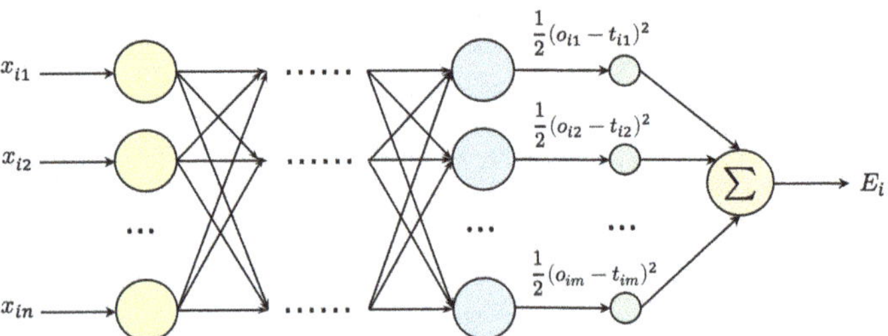

Fig. 5.16 The expanded network for calculating the error function

The output of each unit in the network o_{ij} is connected to a node, and the function $\frac{1}{2}\left(o_{ij} - t_{ij}\right)^2$ is calculated, where o_{ij} represents the jth component of the output vector o_i and t_{ij} is the jth component of the target vector t_i, $j = 1, \ldots, m$. Then, calculate the ith error function:

$$E_i = \frac{1}{2} \sum_{j=1}^{m} \left(o_{ij} - t_{ij}\right)^2 . \tag{5.21}$$

A network extension needs to be built for each t_i. A computing unit collects all quadratic errors and outputs their sum $\sum_{i=1}^{p} E_i$. Therefore, the output of this extended network, that is, the error function E, can be obtained:

$$E = \sum_{i=1}^{p} E_i . \tag{5.22}$$

In this way, a neural network is obtained that can calculate the total error of a given training set. The weights in the network become the only parameters that can be modified to make the quadratic error E as low as possible. Because E is specifically calculated by the combination of expanding network node functions, it is a continuous differentiable function of the l weights $W_1, W_2, \ldots, W_l$ in the network. In this way, it is possible to minimize E by using the iterative process of gradient descent, for which it is necessary to calculate its gradient in the following way:

$$\nabla E = \left\{ \frac{\partial E}{\partial W_1}, \frac{\partial E}{\partial W_2}, \cdots , \frac{\partial E}{\partial W_l} \right\} . \tag{5.23}$$

Each weight is updated using the following increment:

$$\Delta W_i = -\gamma \frac{\partial E}{\partial W_i} \quad \text{for } i = 1, \ldots, l. \tag{5.24}$$

Where γ represents the learning constant, which is a proportionality parameter. It defines the step length in the negative gradient direction for each iteration.

As the network expands, the entire learning problem is simplified to calculate the gradient of the network function relative to its weight. Once a method for calculating this gradient is obtained, the network weights can be adjusted iteratively. This will find the minimum value of the error function, where $\nabla E = 0$.

5.5.1.2   Backpropagation and the Brain

The backpropagation algorithm is an effective method for training artificial neural networks. However, the problem of whether there is a backpropagation mechanism in the biological brain is worth thinking about and researching. Geoffrey Hinton has had an invited talk at the *NIPS 2007 Deep Learning Workshop*, titled "How to do Backpropagation in a Brain."

It is worth mentioning that on April 17, 2020, *Nature Reviews Neuroscience*, a sub-journal of Nature, published a paper titled "Backpropagation and The Brain." The co-first authors are Timothy P. Lillicrap from DeepMind and University College London and Adam Santoro from DeepMind, and the co-corresponding authors are Timothy P. Lillicrap and Geoffrey Hinton (Lillicrap et al. 2020). The paper in its abstract proposes the following view:

> Nonetheless, recent developments in neuroscience and the successes of artificial neural networks have reinvigorated interest in whether backpropagation offers insights for understanding learning in the cortex. The backpropagation algorithm learns quickly by computing synaptic updates using feedback connections to deliver error signals. Although feedback connections are ubiquitous in the cortex, it is difficult to see how they could deliver the error signals required by strict equationtions of backpropagation. Here we build on past and recent developments to argue that feedback connections may instead induce neural activities whose differences can be used to locally approximate these signals and hence drive effective learning in deep networks in the brain.

About whether there is a backpropagation mechanism in the brain, the paper writes:

> There is no direct evidence that the brain uses a backpropagation-like algorithm for learning. However, past research has shown that models trained with backpropagation can explain observed neural responses, such as the response characteristics of neurons in the posterior parietal cortex and primary motor cortex. A new wave of evidence from neuroscience modeling of the visual cortex is driving this trend. This work shows that multilayer models trained with backpropagation for object classification often perform better than other models in matching the visual ventral stream representations in primates. Models not trained with backpropagation (such as bio-inspired models using Gabor filters, or networks using non-backpropagation optimization) do not perform as well as networks optimized with backpropagation, and their representations do not match those in the inferior temporal cortex as well as the representations discovered by models trained with backpropagation.

The paper concludes:

> With the advent of greater computing power, larger datasets, and some technical improvements, backpropagation can now train multilayer neural networks to compete with human abilities. We believe that backpropagation provides a conceptual framework for understanding how the cortex learns, but many mysteries remain about how the brain could approximate it.

With the continuous development of neuroscience, the mystery between backpropagation and the brain will eventually be unraveled. However, as far as my opinion, artificial neural networks working on electronic computers are different from biological neural networks in the biological brain. Electronic computers are hardware, artificial neural networks, and their backpropagation algorithms are software, and biological neural networks in the brain are called wetware.[3] Hardware

---

[3] https://en.wikipedia.org/wiki/Wetware_(brain)

and software can be inspired by wetware to build, but it may not work to explain wetware with successful cases of hardware and software.

Wetware is a term created to mimic hardware, software, and firmware, referring to the life forms of organisms, used to describe the parts found in organisms equivalent to hardware and software, especially the central nervous system and the mind of humans. The hardware part of wetware is the neurons in the brain, and the software part is considered to be the pulse sequence passing through various neurons. The term wetware is believed to have become popular from the science fiction novel *Wetware* published in 1988 by Rudy Rucker, who is an American mathematician, computer scientist, and science fiction writer. The term wetware is also commonly used in some academic articles (Kalnikait and Whittaker 2007; Bray 2009).

5.5.2   *Evolutionary Methods*

Evolutionary methods are another approach to neural network optimization, mainly including evolutionary algorithms and neuroevolution.

5.5.2.1   Evolutionary Algorithms

Evolutionary algorithms are population-based general metaheuristic optimization algorithms, also the subset of evolutionary computing in artificial intelligence.

Evolutionary algorithms originate from biological evolution mechanisms, such as reproduction, mutation, recombination, and selection. The candidate solutions to optimization problems play the role of individuals in the population, and the fitness function determines the quality of the candidate solutions. With the repeated action of the above evolution mechanisms, the population gradually evolves.

There are several types of evolutionary algorithms, including genetic algorithms, genetic programming, evolutionary programming, gene expression programming, differential evolution, and evolution strategy. Since the 1990s, some researchers have tried to use genetic algorithms, genetic programming, evolutionary programming, and evolution strategies for the optimization of artificial neural networks (Ronald and Schoenauer 1994; Gruau 1994; Angeline et al. 1994; Yao and Liu 1997).

5.5.2.2   Neuroevolution

From the 2000s to the present, neuroevolution has attracted the attention of some researchers, who are committed to researching how to use neuroevolution to develop neural networks (Stanley et al. 2005).

Neuroevolution is a method that combines evolutionary algorithms with artificial neural networks, that is, using evolutionary algorithms to generate artificial neural networks, including their parameters, topology, and rules.

In 2017, researchers from Uber AI Labs published their research results: compared with complex gradient-based backpropagation algorithms, simple gradient-free genetic algorithms and neuroevolution algorithms are superior (Such et al. 2017).

Nature Machine Intelligence, an online Nature sub-journal, published a review article titled "Designing Neural Networks through Neuroevolution" in its first issue of 2019 (Stanley et al. 2019). The first author, Kenneth O. Stanley, is a professor at the University of Central Florida and a senior research scientist at Uber AI Labs. The article begins by stating:

> Much of recent machine learning has focused on deep learning, in which neural network weights are trained through variants of stochastic gradient descent. An alternative approach comes from the field of neuroevolution, which harnesses evolutionary algorithms to optimize neural networks, inspired by the fact that natural brains themselves are the products of an evolutionary process.

This review article further elaborates:

> Neuroevolution enables important capabilities that are typically unavailable to gradient-based approaches, including learning neural network building blocks (for example, activation functions), hyperparameters, architectures, and even the algorithms for learning themselves. Neuroevolution also differs from deep learning (and deep reinforcement learning) by maintaining a population of solutions during the search, enabling extreme exploration and massive parallelization. Finally, because neuroevolution research has (until recently) developed largely in isolation from gradient-based neural network research, it has developed many unique and effective techniques that should be effective in other machine learning areas too.

This review article, while revisiting the foundations of classic neuroevolution, focuses on several key aspects of modern neuroevolution, including large-scale computation, the benefits of novelty and diversity, the capabilities of indirect encoding, and the field's contributions to meta-learning and neural architecture search.

In looking forward to the development prospects of neuroevolution, the review article writes:

> Although neuroevolution has long existed as an independent research area, a trend is emerging wherein the line between neuroevolution and deep learning, both of which concern neural networks, gradually blurs. Ideas from neuroevolution infuse deep learning, both through hybridization of evolution and gradient descent, and through conceptual insights transferring from one paradigm to the other. This trend is likely to accelerate especially as computational resources expand, opening up evolutionary experiments that once focused on small models to large-scale discovery. Just as evolution and neural networks proved a potent combination in nature, so does neuroevolution offer an intriguing path to recapitulating some of the fruits of evolving brains.

It is worth pointing out that the core idea of evolution theory is "survival of the fittest," which includes mutual constraints between species and between species and nature. Therefore, neuroevolution, inspired by evolution theory, should also set appropriate constraints when designing its evolutionary algorithms.

5.5.3   Weight-Agnostic Method

In June 2019, Adam Gaier from the Bonn-Rhein-Sieg University of Applied Sciences in Germany and David Ha from Google Brain submitted a paper to the electronic preprint repository *arXiv*, titled "Weight Agnostic Neural Networks." The paper was later published at the *Conference on Neural Information Processing Systems* (NeurIPS) held in December 2019 (Gaier and Ha 2019).

It should be mentioned that the original abbreviation of the *Conference on Neural Information Processing Systems*, NIPS, was considered by some to provoke inappropriate sexual associations and suspected of gender discrimination. Under their opposition, it was changed to NeurIPS on November 17, 2018. Therefore, when this book refers to the abbreviation of the *Conference on Neural Information Processing Systems* after 2018, we follow this change.

Back to the main topic of this subsection, Adam Gaier and David Ha wrote in the introduction of the paper:

> In biology, precocial species are those whose young already possess certain abilities from the moment of birth. There is evidence to show that lizard and snake hatchlings already possess behaviors to escape from predators. Shortly after hatching, ducks are able to swim and eat on their own, and turkeys can visually recognize predators.

This paper is inspired by the aforementioned precocial behavior in nature to conduct research on weight-agnostic neural networks. The so-called weight agnostic refers to such neural network architectures with strong inductive biases, which can perform various tasks with randomly given initial weights, without the need for weight parameter training like traditional neural networks. Therefore, the key technology is how to explore this kind of weight-agnostic neural network, called neural architecture search (NAS). A conceptual diagram is given in the paper.

The exploration algorithm of this neural network is as follows:

Initialize: an initial population of the minimal neural network topologies is created.

Evaluate: each network is evaluated over multiple rollouts, with a different shared weight value assigned at each rollout.

Rank: the networks are ranked according to their performance and complexity.

Vary: a new population is created by varying the highest-ranked network topologies, chosen probabilistically through tournament selection.

Then, the algorithm repeats from the step 2, Evaluate, generating weight-agnostic topologies of gradually increasing complexity, which perform better after several iterations.

The paper also provides two examples of weight-agnostic neural networks.

5.6 Agents of Neural Networks

Looking at deep neural networks from the view of intelligent agents, they can be divided into single- and multiagent neural networks.

5.6.1 Single- and Multiagent Networks

A single-agent neural network refers to a network that has the logical function of only one agent. Most shallow or deep networks are single-agent networks, for example, perceptrons, multilayer perceptrons, convolutional neural networks (CNN), long short-term memory (LSTM) networks, gated recurrent unit (GRU) networks, deep Boltzmann machines (DBM), deep belief networks (DBN), residual networks, and spiking neural networks (SNN).

It is worth emphasizing that the single-agent network is unrelated to the depth of the network. Even an artificial neural network with thousands of layers, if it only has the function of a single intelligent agent, is still a single-agent network.

Multiagent neural networks refer to networks that have the logical functions of at least two agents, with the logical relationships between those agents. These are some neural network models developed in recent years, for handling more complex tasks. Representative multiagent networks include autoencoder (AE), generative adversarial network (GAN), collaborative neural network, and knowledge distillation network.

Below, we introduce to the convolutional neural networks in single agents, as well as knowledge distillation network with two agents. Several other multiagent neural networks will be introduced in Chap. 9.

5.6.2 Case Study of Single-Agent Networks

In the early artificial neural networks, the neurons of every two layers were fully connected, and the parameter number of the network appeared a combinatorial growth. Therefore, the number of layers in a fully connected neural network cannot be too many, the model's generalization ability is relatively weak, and it is prone to overfitting problems, making it difficult to train.

Convolutional neural network (CNN) is a successful case that breaks through the full connection method.

Convolutional neural network (CNN) is a deep network. As shown in Fig. 5.17, in addition to the input layer and output layer, it includes several convolutional layers, pooling layers, and fully connected layers.

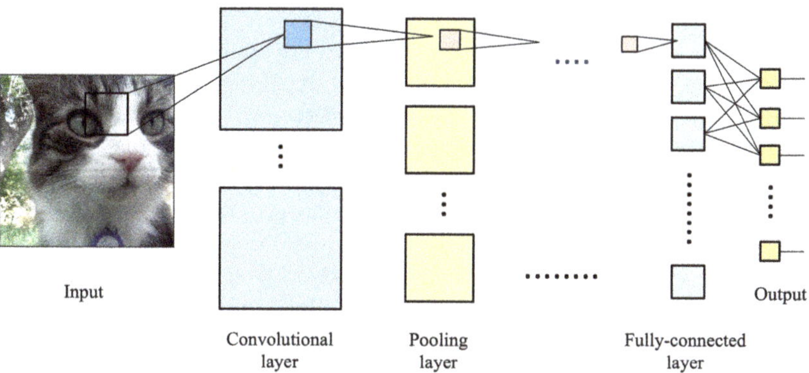

Fig. 5.17   The schematic diagram of convolutional neural network

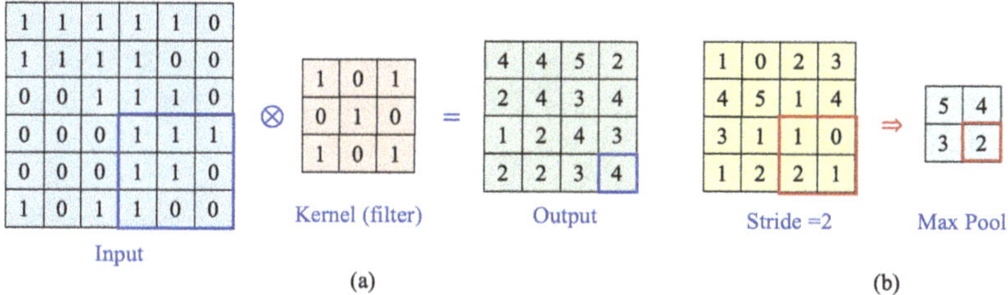

Fig. 5.18   Convolution and pooling operations

For input image data, CNN mainly uses the following four strategies to reduce the number of parameters: (i) local connection, (ii) shared weights, (iii) pooling, and (iv) multilayer.

For an image with $1,000 \times 1,000$ pixels, let the hidden unit be 10^6, for a fully connected neural network, the number of parameters is 10^{12}, that is, one trillion parameters!

However, in the convolutional neural network, the same is $1,000 \times 1,000$ pixel image, using 100 filters for convolution, each filter is 10×10, and the number of parameters is only 10^4. Figure 5.18a shows that on a 6×6 image, the size of the filter is 3×3, and the same filter moves across the image in a sliding window manner, obtaining different local area features.

Pooling in convolutional neural networks is a kind of aggregation statistical operation, i.e., sampling, taking the maximum or average value from the features in a certain local area. Pooling can not only reduce the parameters of the model but also prevent overfitting to a certain extent, improving the results. Because the pooling operation has translation invariance, that is, the features obtained by pooling remain unchanged after the image is translated. Common pooling operations include max pooling and average pooling. Figure 5.18b shows the four features obtained after applying pooling to four nonoverlapping areas of the image.

The seven-layer convolutional neural network LeNet-5 proposed by Yann LeCun et al. in 1998 is a pioneer of convolutional neural networks (LeCun et al. 1998). The convolutional neural network AlexNet proposed by Alex Krizhevsky et al. in 2012 is a successful model of convolutional neural networks (Krizhevsky et al. 2012), which won first place in the ImageNet ILSVRC-2012 competition, and the top-5 detection error rate was 10.9% points higher than the second place.

AlexNet has five convolutional layers, with a max pooling layer following the 1st, 2nd, and 5th convolutional layers, followed by two fully connected layers, and the final layer is a softmax classification layer. AlexNet has a total of 650,000 neurons and 60 million parameters. AlexNet uses the ReLU activation function, multi-GPU training method, local response normalization, overlapping pooling strategy, two data augmentation methods of translating images and changing RGB channel values, and a dropout strategy.

ReLU is an activation function, its full name is rectified linear unit, and the expression is $f(x) = \max(0, x)$. This unsaturated nonlinear activation function improves the training speed of the model, making the training speed of the convolutional neural network on the CIFAR dataset six times that of the hyperbolic tangent function TanH.

AlexNet uses a dropout strategy on both fully connected layers, that is, the neurons on these two fully connected layers stop working with a probability of 0.5. The dropout strategy makes the model not overly dependent on certain specific neurons, reduces overfitting, enhances generalization, and is more robust.

Since then, with the heat of deep learning, various variants of convolutional neural networks have emerged. Representative ones include VGGNet (Simonyan and Zisserman 2014) and GoogLeNet/Inception V1 (Szegedy et al. 2015) launched in 2014, ResNet (He et al. 2015) published in 2015, and DenseNet (Huang et al. 2017) proposed in 2017.

5.6.3 Case Study of Multiagent Networks

Knowledge distillation network (KDN) first appeared in the paper "Distilling the Knowledge in a Neural Network" by Geoffrey Hinton et al. at the *NIPS 2014 Deep Learning and Representation Learning Workshop* and then uploaded the paper to the electronic preprint repository *arXiv* (Hinton et al. 2015).

This network uses the knowledge distillation method to transfer the knowledge of the pretrained large model to the small model, as shown in Fig. 5.19. According to this principle, the literature (Gou et al. 2021) regards the large model as the teacher model and the small model as the student model.

Since then, there have been some variants of knowledge distillation. For instance, the paper "Learning Efficient Object Detection Models with Knowledge Distillation" published at the 2017 *Conference on Neural Information Processing Systems* (NIPS) (Chen et al. 2017), a paper titled "Apprentice: Using Knowledge Distillation Techniques to Improve Low-Precision Network Accuracy" was published at

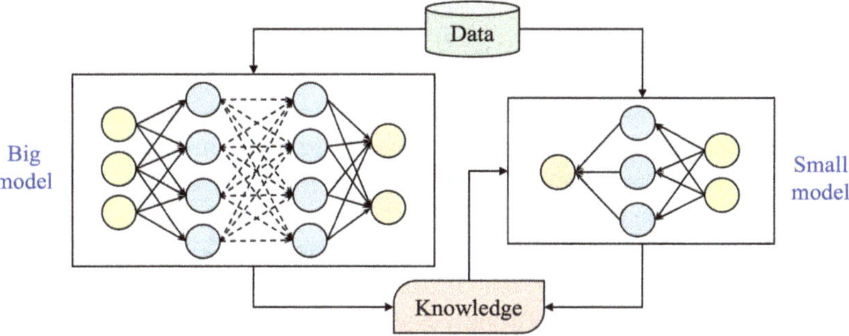

Fig. 5.19 The schematic diagram of knowledge distillation neural network

the 2018 *International Conference on Learning Representations* (ICLR) (Mishra 2018), the "Progressive Teacher-student Learning for Early Action Prediction" was published at the 2019 *IEEE/CVF Conference on Computer Vision and Pattern Recognition* (CVPR) (Wang et al. 2019), and the "Distilling Knowledge from Graph Convolutional Networks" was published at the 2020 *IEEE/CVF Conference on Computer Vision and Pattern Recognition* (CVPR) (Yang et al. 2020).

Further Reading

1. David E. Rumelhart, Geoffrey E. Hinton, and Ronald J. Williams. "Learning Representations by Back-Propagating Errors." *Nature*, 323(6088), 1986.
   [Notes] This is the earliest paper that uses the backpropagation algorithm to train multilayer artificial neural networks. The backpropagation algorithm repeatedly adjusts the weights of the connections in the network, so as to minimize a measure of the difference between the actual output and the desired output of the neural network. As a result, its internal hidden units come to represent important features of the task domain.
2. Yann LeCun, Léon Bottou, Yoshua Bengio, and Patrick Haffner. "Gradient-based learning applied to document recognition." *Proceedings of the IEEE*, 86(11), 1998.
   [Notes] This paper introduced the pioneering seven-layer convolutional neural network called LeNet-5, used by American NCR company for a bank check recognition system, to recognize 32 × 32 pixel handwritten Arabic numerals.
3. Timothy T. Rogers and James L. McClelland. "Parallel distributed processing at 25: Further explorations in the microstructure of cognition." *Cognitive science*, 38(6), 2014.
   [Notes] This is not only a commemorative paper on the 25th anniversary of the publication on parallel distributed processing but also explores how the model can play a greater role in the framework of cognitive science.

4. Kaiming He, Xiangyu Zhang, Shaoqing Ren, and Jian Sun. "Deep Residual Learning for Image Recognition." *IEEE Conference on Computer Vision and Pattern Recognition* (CVPR), 2016.

   [Notes] This paper presents a residual learning framework to ease the training of networks that are substantially deeper than those used previously. An ensemble of these residual nets achieves 3.57% error on the ImageNet test set. This result won the 1st place on the ILSVRC 2015 classification task.

5. Yoshua Bengio, Yann Lecun, and Geoffrey Hinton. "Deep Learning for AI." *Communications of the ACM*, 64(7), 2021.

   [Notes] Received the 2018 Turing Award, the three pioneers of deep learning published the *Turing Lecture* on Communications of the ACM, in which they retrospected the origins of deep learning, described a few of the more recent advances, and discussed some of the future challenges.

References

Ackley, D. H., G. E. Hinton, and T. J. Sejnowski. (1985). A learning algorithm for boltzmann machines. *Cognitive Science* 9(1): 147–169.

Angeline, P. J., G. M. Saunders, and J. B. Pollack. (1994). An evolutionary algorithm that constructs recurrent neural networks. *IEEE Transactions on Neural Networks* 5(1): 54–65.

Bray, D. (2009). *Wetware: a computer in every living cell*. New Haven: Yale University Press.

Chen, G., W. Choi, X. Yu, T. Han, and M. Chandraker. (2017). Learning efficient object detection models with knowledge distillation. In *Conference on Neural Information Processing Systems (NIPS) 30*.

Cho, K., B. Van Merriënboer, C. Gulcehre, D. Bahdanau, F. Bougares, H. Schwenk, and Y. Bengio. (2014). Learning Phrase Representations Using RNN Encoder-Decoder for Statistical Machine Translation. arXiv preprint arXiv:1406.1078.

Chung, J., C. Gulcehre, K. Cho, and Y. Bengio. (2014). Empirical Evaluation of Gated Recurrent Neural Networks on Sequence Modeling. arXiv preprint arXiv:1412.3555.

De Wilde, P. (2013). *Neural network models: Theory and projects*. Berlin: Springer Science & Business Media.

Feldman, J. A., and D. H. Ballard. (1982). Connectionist models and their properties. *Cognitive Science* 6(3): 205–254.

FitzHugh, R. (1961). IImpulses and physiological states in theoretical models of nerve membrane. *Biophysical Journal* 1(6): 445–466.

Gaier, A., and D. Ha. (2019). Weight agnostic neural. In *Conference on Neural Information Processing Systems (IPS)*

Galushkin, A. I. (2007). *Neural networks theory*. Berlin: Springer Science & Business Media.

Gerstner, W., and W. M. Kistler. (2002). *Spiking neuron models: Single neurons, populations, plasticity*. Cambridge: Cambridge University Press.

Gerstner, W., and J. L. van Hemmen. (1992). Associative memory in a network of 'spiking' neurons. *Network: Computation in Neural Systems* 3(2): 139–164.

Goodfellow, I., Y. Bengio, and A. Courville. (2016). *Deep learning*. Cambridge: MIT Press.

Gou, J., B. Yu, S. J. Maybank, and D. Tao. (2021). Knowledge distillation: A survey. *International Journal of Computer Vision* 129(6): 1789–1819.

Gruau, F. (1994). Neural Network Synthesis Using Cellular Encoding and the Genetic Algorithm, Laboratoire de l'Informatique du Parallilisme, Ecole Normale Supirieure de Lyon.

He, K., X. Zhang, S. Ren, and J. Sun. (2015). Deep Residual Learning for Image Recognition. arXiv e-prints: arXiv: 1512.03385.

Hebb, D. O. (1949). The organization of behaviour. *Journal of Applied Behavior Analysis* 25(3): 575–577.

Hindmarsh, J. L., and R. Rose. (1984). A model of neuronal bursting using three coupled first order differential equations. *Proceedings of the Royal Society of London. Series B. Biological Sciences* 221(1222): 87–102.

Hinton, G. E., and T. J. Sejnowski. (1983). *Optimal Perceptual Inference.* Citeseer.

Hinton, G. E., and T. J. Sejnowski. (1986). Learning and relearning in boltzmann machines. *Parallel Distributed Processing: Explorations in the Microstructure of Cognition* 1: 2.

Hinton, G., O. Vinyals, and J. Dean. (2015). Distilling the Knowledge in a Neural Network. arXiv preprint arXiv:1503.02531.

Hochreiter, S., and J. Schmidhuber. (1997). Long short-term memory. *Neural Computation* 9(8): 1735–1780.

Hodgkin, A. L., and A. F. Huxley. (1952). A quantitative description of membrane current and its application to conduction and excitation in nerve. *The Journal of Physiology* 117(4): 500–544.

Hopfield, J. J. (1982). Neural networks and physical systems with emergent collective computational abilities. *Proceedings of the National Academy of Sciences* 79(8): 2554–2558.

Huang, G., Z. Liu, L. Van Der Maaten, and K. Q. Weinberger. (2017). Densely connected convolutional networks. In *Proceedings of the IEEE Conference on Computer Vision And Pattern Recognition.*

Humphries, M. D., and K. N. Gurney. (2001). A pulsed neural network model of bursting in the Basal Ganglia. *Neural Networks* 14(6–7): 845–863.

Irsoy, O., and C. Cardie. (2014). Deep recursive neural networks for compositionality in language. In *Conference on Neural Information Processing Systems (NIPS).*

Kalnikaité, V., and S. Whittaker. (2007). Software or wetware? Discovering when and why people use digital prosthetic memory. In *Proceedings of the SIGCHI Conference on Human Factors in Computing Systems.*

Krizhevsky, A., I. Sutskever, and G. E. Hinton. (2012). Imagenet classification with deep convolutional neural networks. In *Conference on Neural Information Processing Systems (NIPS).*

LeCun, Y., L. Bottou, Y. Bengio, and P. Haffner. (1998). Gradient-based learning applied to document recognition. *Proceedings of the IEEE* 86(11): 2278–2324.

Lillicrap, T. P., A. Santoro, L. Marris, C. J. Akerman, and G. Hinton. (2020). Backpropagation and the brain. *Nature Reviews Neuroscience* 21(6): 335–346.

Maass, W. (1994). On the computational complexity of networks of spiking neurons. In *Conference on Neural Information Processing Systems (NIPS).*

Maass, W. (1997). Networks of spiking neurons: The third generation of neural network models. *Neural Networks* 10(9): 1659–1671.

McClelland, J. L., and A. Cleeremans. (2009). *Connectionist models. Oxford companion to consciousness.* New York: Oxford University Press.

McClelland, J. L., D. E. Rumelhart, and P. R. Group. (1987). *Parallel distributed processing, volume 2: Explorations in the microstructure of cognition: Psychological and biological models.* Cambridge: MIT Press.

McCulloch, W. S., and W. Pitts. (1943). A logical calculus of the ideas immanent in nervous activity. *The Bulletin of Mathematical Biophysics* 5(4): 115–133.

Minsky, M., and S. Papert. (1969). *Perceptrons: An introduction to computational geometry.* Cambridge: The MIT Press.

Mishra, A., and D. Marr. (2018). Apprentice: Using knowledge distillation techniques to improve low-precision network accuracy. In *International Conference on Learning Representations.*

Nagumo, J., S. Arimoto, and S. Yoshizawa. (1962). An active pulse transmission line simulating nerve axon. *Proceedings of the IRE* 50(10): 2061–2070.

Nwankpa, C., W. Ijomah, A. Gachagan, and S. Marshall. (2018). Activation Functions: Comparison of Trends in Practice and Research for Deep Learning. arXiv preprint arXiv:1811.03378.

Ranganath, H., G. Kuntimad, and J. Johnson. (1995). Pulse coupled neural networks for image processing. In *Proceedings IEEE Southeastcon '95. Visualize the Future*, Piscataway: IEEE.

Rogers, T. T., and J. L. McClelland. (2014). Parallel distributed processing at 25: Further explorations in the microstructure of cognition. *Cognitive Science* 38(6): 1024–1077.

Ronald, E., and M. Schoenauer. (1994). Genetic lander: An experiment in accurate neuro-genetic control. In *International Conference on Parallel Problem Solving from Nature*. Berlin: Springer.

Rumelhart, D. E., G. E. Hinton, and R. J. Williams. (1985). *Learning internal representations by error propagation*. California: California University San Diego La Jolla Institue for Cognitive Science.

Rumelhart, D. E., G. E. Hinton, and J. L. McClelland. (1986a). A general framework for parallel distributed processing. *Parallel Distributed Processing: Explorations in the Microstructure of Cognition* 1(45–76): 26

Rumelhart, D. E., G. E. Hinton, and R. J. Williams. (1986b). Learning representations by back-propagating errors. *Nature* 323(6088): 533–536.

Rumelhart, D. E., J. L. McClelland, and P. R. Group. (1986c). *Parallel distributed processing, volume 1: Explorations in the microstructure of cognition: Foundations*.

Sabour, S., N. Frosst, and G. E. Hinton. (2017). Dynamic routing between capsules. In *Conference on Neural Information Processing Systems (NIPS)*.

Salakhutdinov, R., and G. Hinton. (2009). Deep boltzmann machines. In *Proceedings of International Conference on Artificial Intelligence and Statistics*, PMLR 5: 448–455.

Simonyan, K., and A. Zisserman. (2014). Very Deep Convolutional Networks for Large-Scale Image Recognition. arXiv preprint arXiv:1409.1556.

Stamova, I., and G. Stamov (2016). Impulsive neural networks. *Applied Impulsive Mathematical Models*, Springer: 207–269.

Stanley, K. O., B. D. Bryant, and R. Miikkulainen. (2005). Real-time neuroevolution in the NERO video game. *IEEE Transactions on Evolutionary Computation* 9(6): 653–668.

Stanley, K. O., J. Clune, J. Lehman, and R. Miikkulainen. (2019). Designing neural networks through neuroevolution. *Nature Machine Intelligence* 1(1): 24–35.

Such, F. P., V. Madhavan, E. Conti, J. Lehman, K. O. Stanley, and J. Clune. (2017). Deep Neuroevolution: Genetic Algorithms Are a Competitive Alternative for Training Deep Neural Networks for Reinforcement Learning. arXiv preprint arXiv:1712.06567.

Szegedy, C., W. Liu, Y. Jia, P. Sermanet, S. E. Reed, D. Anguelov, D. Erhan, V. Vanhoucke, and A. Rabinovich. (2015). Going deeper with convolutions. In *Conference on Computer Vision and Pattern Recognition*.

Wang, X., J. F. Hu, J. H. Lai, J. Zhang, and W. S. Zheng. (2019). Progressive teacher-student learning for early action prediction. In *Proceedings of the IEEE Conference on Computer Vision and Pattern Recognition (CVPR)*.

Wang, L., D. Zhang, J. Zhang, X. Xu, L. Gao, B. Dai, and H. T. Shen. (2019). Template-based math word problem solvers with recursive neural networks. In *The Thirty-Third AAAI Conference on Artificial Intelligence (AAAI)*.

Weiss, G., Y. Goldberg, and E. Yahav. (2018). On the Practical Computational Power of Finite Precision RNNs for Language Recognition. arXiv preprint arXiv:1805.04908.

Yang, Y., J. Qiu, M. Song, D. Tao, and X. Wang. (2020). Distilling knowledge from graph convolutional networks. In *Proceedings of the IEEE/CVF Conference on Computer Vision and Pattern Recognition*.

Yao, X., and Y. Liu. (1997). A new evolutionary system for evolving artificial neural networks. *IEEE Transactions on Neural Networks* 8(3): 694–713.

Chapter 6
Symbolic Framework

Abstract Symbolic framework is one of the learning frameworks based on symbol system and its symbolic logic. We first overview symbolism, symbolic learning theory, physical symbol systems, and GOFAI. Second, we discuss formal logic consisting of symbolic logic and mathematical logic. Next, we explain logic learning, especially logic learning theory, the logical learning modes, and logic programming methods, where the logical learning modes consist of deductive learning, inductive learning, and abductive learning. We then introduce rule learning represented by association rules, including association rules discovery and synthesis. After that, we learn causal learning, specifically causal relations, causal inference, and two approaches to causal learning. After reviewing Moravec's paradox, thinking fast and slow, we finally introduce some approaches of neuro-symbolic learning that emerged in recent years.

6.1 Overview

Symbols refer to all advanced symbols invented and understood by humans, including signs, numbers, characters, and words.

Based on symbols, one has also created many systems composed of symbols, logical relationships between symbols, and their processing methods, referred to as symbolic systems.

Symbolic systems are the advanced form of representation, thinking, communication, and dissemination. Various natural languages have their symbolic systems, and many disciplines, such as mathematics, physics, chemistry, logic, and computer science, have also established different symbolic systems.

Similar to symbolic artificial intelligence in the field of artificial intelligence, symbolic machine learning has also emerged in the field of machine learning.

The symbolic framework is another theoretical framework of machine learning based on symbolic systems.

This section first discusses symbolism, then introduces symbolic learning theory, physical symbol systems, and GOFAI.

W. Wang, *Principles of Machine Learning*,
https://doi.org/10.1007/978-981-97-5333-8_6

6.1.1 Symbolism

Symbolism is also known as logicism, the school of psychology, or computerism.

Symbolism is based on the principles of physical symbology and the principle of limited rationality. It views human thinking as a combination of symbols, uses symbols and their relationships to represent information and knowledge, and completes human cognition and learning through the storage, extraction, reasoning, and transformation of symbols. Symbolists invent specific algorithms to process these symbols, solve problems, and derive new knowledge.

The symbolic method is the earliest technical approach in the field of artificial intelligence. As early as 1956, the year of the birth of artificial intelligence, Allen Newell, Herbert Simon, and John C. Shaw of Carnegie Mellon University in the United States developed a program named "Logic Theorist." This is the world's first artificial intelligence program, used to mimic human logical reasoning and problem-solving abilities, also known as a logic deductive system. The first successful case of symbolic artificial intelligence was the use of "Logic Theorist" to prove 38 of the first 52 theorems in Alfred N. Whitehead and Bertrand Russell's monograph *Principia Mathematica*, and to find new, more concise proofs for some theorems.

From the late 1950s to the late 1980s, symbolic methods dominated the fields of artificial intelligence and machine learning research. Logic programming represented by the Prolog language, rule-oriented production systems, decision trees, and decision forests in machine learning are a few milestones of symbolic artificial intelligence and symbolic machine learning. Heuristic algorithms, expert systems, and knowledge engineering theories and techniques have also made significant progress.

6.1.2 Symbolic Learning Theory

Symbolic learning theory originates from psychology. This is why it is also known as psychological symbolism.

In *APA Dictionary of Psychology*, offered by American Psychological Association, the symbolic learning theory[1] is defined as follows:

> A theory that attempts to explain how imagery works in performance enhancement. It suggests that imagery develops and enhances a coding system that creates a mental blueprint of what has to be done to complete an action.

Symbolic learning theory originated from the purposive behaviorism proposed by American psychologist Edward C. Tolman (1886–1959). He published his monograph *Purposive Behavior in Animals and Men* in 1932. Tolman believed that learning develops from knowledge of the environment and how to interact with

---

[1] https://dictionary.apa.org/symbolic-learning-theory.

it. Tolman's goal was to discover the complex cognitive mechanisms that guide behavior and their purpose.

Purposive behaviorism is a branch of psychology, considered a bridge between behaviorism and cognitive theory. It combines objective research on behavior while also considering the purpose or goal of behavior. Purposive behaviorism believes that behavior has a potential goal, and all action goals will guide its behavior until the goal is achieved.

Purposive behaviorism is different from behaviorism based on the stimulus-response (S-R) theory of psychologists like Edward Thorndike. The latent learning theory based on purposive behaviorism is also different from the reinforcement learning theory based on classical behaviorism.

6.1.3 Physical Symbol System

In 1976, the Association for Computing Machinery (ACM) awarded the 1975 Turing Award to two pioneers of artificial intelligence —Allen Newell and Herbert Simon. In March 1976, *Communication of the ACM* published their *Turing Award Lecture* titled "Computer Science as Empirical Inquiry: Symbols and Search" (Newell and Simon 1976), in which they described the physical symbol system as follows:

> A physical symbol system consists of a set of entities, called symbols, which are physical patterns that can occur as components of another type of entity called an expression (or symbol structure). Thus, a symbol structure is composed of several instances (or tokens) of symbols related in some physical way (such as one token being next to another). At any instant in time, the system will contain a collection of these symbol structures. Besides these structures, the system also contains a collection of processes that operate on expressions to produce other expressions: processes of creation, modification, reproduction, and destruction. A physical symbol system is a machine that produces through time an evolving collection of symbol structures. Such a system exists in a world of objects wider than just these symbolic expressions themselves.

In their Turing award lecture, Newell and Simon proposed two core concepts of symbols, objects, and expression structures, namely:

- Designation:
  Given an expression, if the expression can affect an object or appear to depend on the object, the expression is said to "designate" the object.

  The essence of "designation" is that the system can access its objects through expressions.
- Interpretation:
  If an expression designates a process, and the system can execute the process when given the expression, the system is said to be able to "interpret" the expression.

  "Interpretation" implies a special form of dependent behavior: given an expression, the system can execute the designated process. That is, the process can be evoked and executed from the designated expression.

Newell and Simon also proposed the following physical symbol system hypothesis, namely:

> A physical symbol system has the necessary and sufficient means for general intelligent action.

They further elaborated: The so-called "necessary" means that any system that exhibits general intelligence will be proven to be a physical symbol system upon analysis; the so-called "sufficient" means that any physical symbol system of sufficient size can be organized to exhibit general intelligence; the so-called "general intelligent behavior" is the hope that in any actual situation, intelligent behavior suitable for the system's end and adapted to environmental needs can be generated within a certain speed and complexity range.

Newell and Simon also pointed out:

> This is an empirical hypothesis. ... The hypothesis could indeed be false. Intelligent behavior is not so easy to produce that any system will exhibit it willy-nilly. Indeed, there are people whose analyses lead them to conclude either on philosophical or scientific grounds that the hypothesis is false. Scientifically, one can attack or defend it only by bringing forth empirical evidence about the natural world.

In the collected papers titled *50 Years of Artificial Intelligence: Essays Dedicated to the 50th Anniversary of Artificial Intelligence* published in 2007, there is a paper titled "The Physical Symbol System Hypothesis: Status and Prospects" (Nilsson 2007) which was published by Nils J. Nilsson of Stanford University in the United States. The paper analyzes some negative views on the physical symbol system hypothesis, pointing out that these negative views are mainly based on the non-symbol hypothesis of intelligent behavior and the non-computational and unconscious hypothesis of the brain. Notably, Nilsson believes that artificial intelligence systems that reach the level of human intelligence will be a combination of symbol processing and non-symbol processing.

6.1.4 GOFAI

The GOFAI stands for "good old-fashioned artificial intelligence," which first appeared in the book titled *Artificial Intelligence: The Very Idea* published in 1985 by John Haugeland, a philosophy professor at the University of Pittsburgh (Haugeland 1985).

By the way, upon investigation, the meaning of "good old-fashioned" is "even though it's old-fashioned, it's still good." John Haugeland conducts research in the philosophy of mind, cognitive science, phenomenology, etc., and this book is a treatise on the philosophy of artificial intelligence.

The period when Haugeland published this book was a time of rapid development of large-scale integrated circuits and microprocessors and a rapid increase in computer processing power, which provided good computational power for statistical learning and multilayer neural networks. It was also from this period that symbolic artificial intelligence and symbolic machine learning, which do not require high computational power, gradually declined. The reason is that symbolic systems can only handle problems that can be represented as physical symbols, while vision, hearing, and speech are difficult to represent with symbols.

Since the 2010s, deep learning has gradually become popular worldwide. Complex tasks in computer vision, computer speech, natural language processing, computer games, medicine, and biology, once deep learning is adopted, or existing methods are combined with deep learning, often achieve astonishing results.

However, it is worth noting that after more than a decade of popularity, the limitations of deep learning have gradually emerged. Some researchers have begun to consider the complementarity of deep learning and symbolic systems, proposing neuro-symbolic learning methods that combine neural networks with symbolic AI. This will be further discussed in Sect. 6.6 of this chapter.

The creation of symbols and symbolic systems is one of the important signs that distinguish humans from other animals and is a cognitive tool above perception. With symbols and symbolic systems, human knowledge can be recorded, learned, communicated, and passed on. Therefore, in the theoretical framework of machine learning, the symbolic framework is indispensable.

6.2 Formal Logic

Formal logic abstracts logic into a symbolic language, that is, formal languages, giving logic certain mathematical properties and allowing mathematical operations.

In this section, we will introduce the types of formal logic first and explain symbolic logic and mathematical logic respectively.

6.2.1 Types of Formal Logic

Formal logic can be divided into symbolic logic and mathematical logic. Among them, symbolic logic includes propositional logic and first-order logic, while mathematical logic includes modal logic, fuzzy logic, and description logic.

For easy memory, the types of formal logic are represented in Unified Modeling Language (UML), as depicted in Fig. 6.1.

6.2.2 Symbolic Logic

Symbolic logic, also known as classical logic, is a development of formal logic employing a special symbols capable of manipulation in accordance with precise rules, which consists of propositional logic and first-order logic.

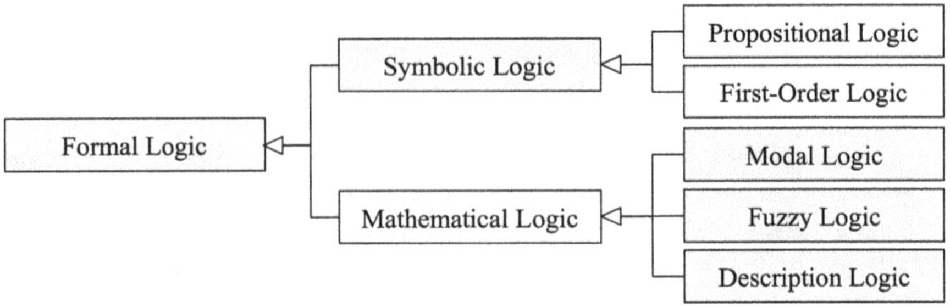

Fig. 6.1 Types of formal logic

Table 6.1 Comparison of propositional logic and first-order logic in symbolic logic

Formal language	Logic scope	Logical value
Propositional logic	Facts	True/False/Unknown
First-order logic	Facts, objects, relations	True/False/Unknown

Table 6.2 Main symbols in logic

Category	Symbol	Brief explanation
Connectives	$\neg$	Logical negation
	$\wedge$	Logical conjunction
	$\vee$	Logical disjunction
	$\Rightarrow$	Implication
	$\Leftrightarrow$	Material equivalence
	$\models$	Entails
	$\not\models$	Not valid
Quantifiers	$\forall$	Universal quantifier
	$\exists$	Existential quantifier
Equality	$=$	Equal to

The comparison of the two is listed in Table 6.1.

The middle column of Table 6.1 depicts the logical scope studied by symbolic logic, such as propositional logic only studying facts, and first-order logic also studying objects and relations. The right column is the logical value, that is, the result of logical determination, such as true and false. The symbols often used in symbolic logic are shown in Table 6.2. These symbols are divided into three categories: connective symbols, quantifier symbols, and equality symbol.

6.2.2.1 Propositional Logic

Definition 6.1 (Propositional Logic) Propositional logic is a formal system for dealing with statements that have a truth value of true or false and for constructing rules for proving theorems.

The propositional logic is also known as propositional calculus.

$$
\begin{array}{rcl}
\textit{Sentence} & \rightarrow & \textit{AtomicSentence} \mid \textit{ComplexSentence} \\
\textit{AtomicSentence} & \rightarrow & \textit{True} \mid \textit{False} \mid P \mid Q \mid R \mid \dots \\
\textit{ComplexSentence} & \rightarrow & <\textit{Sentence}> \mid [\,\textit{Sentence}\,] \\
 & \mid & \neg\ \textit{Sentence} \\
 & \mid & \textit{Sentence} \wedge \textit{Sentence} \\
 & \mid & \textit{Sentence} \vee \textit{Sentence} \\
 & \mid & \textit{Sentence} \Rightarrow \textit{Sentence} \\
 & \mid & \textit{Sentence} \Leftrightarrow \textit{Sentence} \\
\textit{OPERATOR PRECEDENCE} & : & \neg,\ \wedge,\ \vee,\ \Rightarrow,\ \Leftrightarrow
\end{array}
$$

Fig. 6.2 The BNF representation for the syntax rules of propositional logic

The sentences of propositional logic consist of atomic sentences and complex sentences. Atomic sentences include variables or logical values, while complex sentences are depicted in Table 6.2 composed of atomic sentences and connective symbols. Figure 6.2 is a description of the syntax rules of propositional logic using the Backus–Naur form (BNF).

Propositional logic is also known as zeroth-order logic, a term that facilitates comparison with first-order logic.

6.2.2.2   First-Order Logic

Definition 6.2 (First-Order Logic) First-order logic is a formal system built on top of propositional logic, which can use quantified variables of non-logical objects and statements containing variables.

The first-order logic is also known as predicate logic or first-order predicate calculus.

Figure 6.3 is a BNF description of the syntax rules of first-order logic. It can be seen that it has added the quantifier symbols, equality symbols, and predicates from Table 6.2. Comparing it with the syntax rules of propositional logic, one can see the similarities and differences between the two.

Definition 6.3 (Terms) The terms of first-order logic intuitively represent objects.

The formation rules for terms in first-order logic are depicted in Table 6.3.

Definition 6.4 (Formulas) The formulas of first-order logic intuitively express statements that can be true or false.

The formation rules of formulas in first-order logic are shown in Table 6.4.

The first-order logic expressions, based on the syntax, terms, and formula rules of first-order logic, are shown in Fig. 6.4.

$$
\begin{aligned}
\textit{Sentence} \;&\rightarrow\; \textit{AtomicSentence} \mid \textit{ComplexSentence} \\
\textit{AtomicSentence} \;&\rightarrow\; \textit{Predicate} \mid \textit{Predicate(Term, ...)} \mid \textit{Term} = \textit{Term} \\
\textit{ComplexSentence} \;&\rightarrow\; < \textit{Sentence} > \mid [\, \textit{Sentence} \,] \mid \neg\, \textit{Sentence} \\
&\mid\; \textit{Sentence} \wedge \textit{Sentence} \mid \textit{Sentence} \vee \textit{Sentence} \\
&\mid\; \textit{Sentence} \Rightarrow \textit{Sentence} \mid \textit{Sentence} \Leftrightarrow \textit{Sentence} \\
&\mid\; \textit{Quantifier Variable, ... Sentence} \\
\textit{Term} \;&\rightarrow\; \textit{Function(Term, ...)} \mid \textit{Constant} \mid \textit{Variable} \\
\textit{Quantifier} \;&\rightarrow\; \forall \mid \exists \\
\textit{Constant} \;&\rightarrow\; A \mid X_1 \mid \textit{John} \mid ... \\
\textit{Variable} \;&\rightarrow\; a \mid x \mid s \mid ... \\
\textit{Predicate} \;&\rightarrow\; \textit{True} \mid \textit{False} \mid \textit{After} \mid \textit{Loves} \mid \textit{Raining} \mid ... \\
\textit{Function} \;&\rightarrow\; \textit{Mother} \mid \textit{LeftLeg} \mid ... \\
\textit{OPERATOR PRECEDENCE} \;&:\; \neg, =, \wedge, \vee, \Rightarrow, \Leftrightarrow
\end{aligned}
$$

Fig. 6.3 The BNF representation for the syntax rules of first-order logic

Table 6.3 Formation rules for terms in first-order logic

Term	Formation rules
Variables	Any variable symbol is a term
Constants	Any constant symbol is also a term
Functions	Any expression composed of n parameters $f\,(t_1, \ldots, t_n)$ is a term, where each parameter t_i is a term, and f is an n-ary function symbol. The symbol representing an individual constant is a 0-ary function symbol, so it is also a term

Table 6.4 Formation rules for formulas in first-order logic

Formula	Formation rules
Predicate symbols	If P is an n-ary predicate symbol and $t_1, \ldots, t_n$ are terms, then $P\,(t_1, \ldots, t_n)$ is a formula
Equality	If t_1 and t_2 are terms, then $t_1 = t_2$ is a formula
Negation	If φ is a formula, then $\neg\varphi$ is a formula
Binary connectives	If φ and ψ is a formula, then $\varphi \Rightarrow \psi$ is a formula. Similar rules can be applied to other binary logical connections
Quantifiers	If φ is a formula and x is a variable, then $\forall x\varphi$ and $\exists x\varphi$ are formulas

6.2.3 *Mathematical Logic*

Mathematical logic, also known as non-classical logic or alternative logic, is formed by extending, differentiating, and varying the symbolic logic (classical logic).

Mathematical logic mainly includes modal logic, fuzzy logic, and description logic.

$$
\begin{aligned}
&Mary \in FemalePersons \\
&John \in MalePersons \\
&SisterOf(Mary, John) \\
&FemalePersons \subset Persons \\
&MalePersons \subset Persons \\
&\forall x, x \in Persons \Rightarrow [\forall y, HasMother(x, y) \Rightarrow y \in FemalePersons]
\end{aligned}
$$

Fig. 6.4 Examples of first-order predicate logic expressions

6.2.3.1 Modal Logic

Modal logic began in 1910. Its semantics were established in 1959. And its rapid development was in the period of 1960s.

Modal logic involves the concept of possible worlds in logic, which includes the real world and the imaginary world, used to express modal claims.

In any given possible world, a proposition can be understood in terms of the modality of the world it is in. Propositions in possible worlds include:

- true propositions,
- false propositions,
- possible propositions,
- impossible propositions,
- necessarily propositions,
- contingent propositions.

The logic in possible worlds is modal logic.

Modal logic is a form of formal logic that extends classical propositional logic and first-order logic by adding the ability to express modality.

A modality is a word that expresses a mode state, used to qualify a statement. For example, in the sentence "it might rain tomorrow," the word "might" plays the role of a modality. In addition, words like "perhaps," "can," "certain," "inevitable," are words that qualify statements and are used for modal processing.

Formalized modal logic uses modal operators to represent modality. There are two basic unary modal operators, namely: "$\square$" represents "necessarily", and "$\lozenge$" represents "possibly". In classical modal logic, each can be expressed in its negated form. For example:

$$\lozenge P = \neg\square\neg P.$$

$$\square P = \neg\lozenge\neg P.$$

> *if $v(w, P)$ then $w \vDash P$*
>
> $w \vDash \neg P$ *if and only if $w \nvDash P$*
>
> $w \vDash (P \wedge Q)$ *if and only if $w \vDash P$ and $w \vDash Q$*
>
> $w \vDash \Box P$ *if and only if for every element u of W, if wRu then $u \vDash P$*
>
> $u \vDash \Diamond P$ *if and only if for some element u of W, it holds that wRu and $u \vDash P$*
>
> $\vDash P$ *if and only if $w^* \vDash P$*

Fig. 6.5 Recursive definition of the truth of a formula at a world

In the aforementioned two modal logic statements, if P represents "rain", the meanings are as follows:

> Possibly rain, equivalent to, not necessarily does not rain.
>
> Necessarily rain, equivalent to, not possibly does not rain.

Definition 6.5 (Semantics of Modal Logic) The semantics of modal logic can be formalized as a 3-tuple $\mathcal{M} = \langle W, R, v \rangle$, where: the W is a set of possible worlds; R is a binary relation called accessibility on W; and v is a valuation function, which assigns a truth value to each pair of an atomic formula and a world.

For instance, if $w, u \in W$, the wRu means that the world u is accessible from the world w; and if there exists $v(w, P)$, then the atomic formula P is true in w. And w^* represents the actual world.

The recursive definition of the truth of a formula at a world is shown in Fig. 6.5.

The modality dealing with necessity, possibility, and contingency is called alethic modality. Modal logic was originally proposed to deal with these concepts, so some people call alethic modality classical or traditional modal logic.

In alethic modality, there are three properties: possibility, necessity, and contingency. The definitions are as follows:

- Possibility   If it is true in at least one possible world.
- Necessity   If it is true in all possible worlds.
- Contingency   If it is true in some possible worlds and false in others.

Later, some researchers extended modal logic and proposed several other modalities of modal logic, including:

- Temporal modality   To express propositions related to time.
- Deontic modality   To express obligations, permissions, and related concepts.
- Epistemic modality   To represent modality related to knowledge inference.
- Doxastic modality   To represents modality related to belief inference.

Each logical function and its example of those modalities are shown in Table 6.5.

Table 6.5 Other four modalities of modal logic

Name	Logical function	Example
Temporal modality	Represents propositions related to time	Past, always, future, forever
Deontic modality	Represents obligations, permissions, and related concepts	Mandatory, permissible
Epistemic modality	Related to the inference of knowledge	As my knowledge that …
Doxastic modality	Related to reasoning with beliefs	It is believed that …

How modal logic can be combined with the field of machine learning is a worthwhile research area. Some researchers have combined modal logic with neural networks and proposed connectionist modal logic (Garcez et al. 2007, 2009).

6.2.3.2   Fuzzy Logic

Fuzzy logic is a form of many-valued logic that is formed by assigning truth values to propositions. The standard set of truth values is [0, 1]. Where: 0 represents "totally false", 1 represents "totally true", and any real value between 0 and 1 represents "partially true".

The term fuzzy logic was named in 1965 by Lotfi A. Zadeh of the University of California, Berkeley, when he proposed the theory of fuzzy sets (Zadeh 1965, 1988). The predecessor of fuzzy logic is infinite-valued logic, which began in the 1920s.

Fuzzy logic and fuzzy sets are mathematical methods for representing fuzzy and imprecise information, with the ability to recognize, characterize, operate, interpret, and utilize data and information that are fuzzy and lack certainty.

Fuzzy logic and fuzzy sets have been widely used in the field of machine learning, known as fuzzy machine learning, such as fuzzy clustering (Ruspini, 2019), fuzzy neural networks (Buckley and Hayashi 1994), and fuzzy deep neural networks (Das et al. 2020).

6.2.3.3   Description Logic

Description logic is a set of formalized knowledge representation languages, a decidable fragment of first-order logic (Baader et al. 2003; Krötzsch et al. 2012), and a subset of first-order logic (Nardi and Brachman 2003).

The logical expressive power of description logic lies between propositional logic and first-order logic in classical logic. However, unlike first-order logic, the core inference problems of description logic are usually decidable, and effective decision processes have been designed and implemented for these problems.

The study of description logic began with terminological systems. Later, the focus was on the concept formation structure in the language, and the term concept

Table 6.6 Synonyms in description logic and first-order logic

Description logic	First-order logic
Individual	Constant
Concept	Unary predicate
Role	Binary predicate

Table 6.7 Main symbols of description logic

Symbol	Description	Example
$\top$	Universal concept	$\top$
$\bot$	Empty concept	$\bot$
$\sqcap$	Concept conjunction	$C \sqcap D$
$\sqcup$	Concept disjunction	$C \sqcup D$
$\neg$	Concept negation	$\neg C$
$\forall$	Universal quantifier	$\forall R, C$
$\exists$	Existential quantifier	$\exists R, C$
$\sqsubseteq$	Concept inclusion	$C \sqsubseteq D$
$\equiv$	Concept equivalence	$C \equiv D$
$\doteq$	Concept definition	$C \doteq D$
:	Concept assertion	$a : C$
:	Role assertion	$(a, b) : R$

language was introduced. In the 1980s, the term description logic became popular after the focus shifted further to the characteristics of the underlying logical system (Baader et al. 2003).

The synonyms and their corresponding relationships in description logic and first-order logic are shown in Table 6.6.

The main symbols of description logic are shown in Table 6.7.

The following is an example.

Exercise 6.1 (A University's Courses) Describing a university's courses (Course), setting the minimum and maximum number of students who enroll in each course: the number of students in each course cannot be less than 5, the course is taught by one teacher; advanced courses (AdvCourse) are for graduate students (GradStudent), and the number of students cannot exceed 20; while basic courses (BasCourse) can only be taught by professors, and the number of students cannot exceed 50.

The result is shown in Fig. 6.6 using description logic.

In addition to the aforementioned concepts of description logic, there are several extended description logics, namely: spatial description logic, temporal description logic, spatiotemporal description logic, and fuzzy description logic. Each type of description logic shows a different balance between expressiveness and reasoning complexity by supporting different mathematical constructors.

Description logic is used in the field of artificial intelligence to describe and interpret concepts known as terminological knowledge in the application domain. It provides a logical formalization method for ontology and the semantic web.

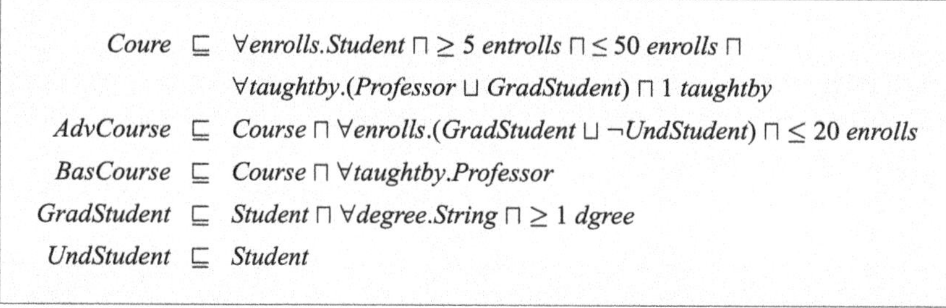

$$
\begin{aligned}
\textit{Coure} &\sqsubseteq \forall enrolls.Student \sqcap\ \geq 5\ entrolls \sqcap\ \leq 50\ enrolls \sqcap \\
&\quad \forall taughtby.(Professor \sqcup GradStudent) \sqcap 1\ taughtby \\
\textit{AdvCourse} &\sqsubseteq Course \sqcap \forall enrolls.(GradStudent \sqcup \neg UndStudent) \sqcap\ \leq 20\ enrolls \\
\textit{BasCourse} &\sqsubseteq Course \sqcap \forall taughtby.Professor \\
\textit{GradStudent} &\sqsubseteq Student \sqcap \forall degree.String \sqcap\ \geq 1\ dgree \\
\textit{UndStudent} &\sqsubseteq Student
\end{aligned}
$$

Fig. 6.6 A university's courses described by description logic

Some researchers are also engaged in research combining description logic with neural symbols (Singh et al. 2023).

6.3 Logic Learning

The theoretical basis of logical learning is the theory of logical learning. The main patterns of logical learning include deductive learning, inductive learning, and abductive learning. Logical programming is an effective way to implement logical learning.

6.3.1 Logical Learning Theory

The theory of logical learning was proposed by American psychologist Joseph F. Rychlak in 1980s. His work can be roughly divided into two aspects: theoretical and experimental. His theoretical work mainly focuses on exploring and understanding the theoretical and philosophical foundations of psychology, while his experimental work focuses on scientific experiments, aiming to test his theory of logical learning through experiments.

In 1981, Joseph F. Rychlak published a paper titled "Logical Learning Theory: Propositions, Corollaries, and Research Evidence" in the *Journal of Personality and Social Psychology*. He proposed that the theory of logical learning, as a teleological learning theory, is used to replace the non-teleological learning theory that dominates psychology (Rychlak 1981). In 1994, he published a book titled *Logical Learning Theory: A Human Teleology and its Empirical Support* (Rychlak 1994). In 2005, he published an autobiographical article in the *Journal of Personality Assessment* titled "In Search and Proof of Human Beings, Not Machines", in which he discussed his research and proposal of the theory of logical learning, although he was already 76 years old this year.

As aforementioned, Joseph Rychlak's theory of logical learning mainly studies psychology from a teleological perspective rather than machine learning. However, his view that the cognitive process is a logical process rather than a mechanical one is worth pondering and exploring for researchers engaged in artificial intelligence, especially machine learning.

Completeness, consistency, decidability, and expressiveness are the main features of logical learning.

If human thinking is divided into two types, concrete thinking and abstract thinking, then logical learning is most suitable for the second type of thinking, but most other methods in machine learning belong to the first type. Therefore, logical learning still has a lot of room for development.

The core concept of logical learning is logical form and its argumentation. Logical form refers to the manifestation of logic, and argumentation is carried out by using different types of logical reasoning.

In terms of perception, such as visual, auditory, and language perception, various methods and algorithms based on probability learning theory, statistical learning theory, and connectionist learning theory have achieved considerable success. But for cognition, since the brain's cognition is a high-level mental activity, the theory of logical learning should and will inevitably play an important role.

6.3.2 Logical Learning Modes

There are mainly three modes of logical reasoning: deductive reasoning, inductive reasoning, and abductive reasoning.

The aforementioned three modes of reasoning are the consensus of many philosophers and logicians. Other modes of reasoning, such as analogical reasoning and fallacious reasoning, are also considered to be derived from the aforementioned three reasoning modes and therefore cannot be considered as separate reasoning modes.

Based on the aforementioned three reasoning modes, three learning modes have been formed, namely: deductive learning, inductive learning, and abductive learning. These are three important learning modes.

6.3.2.1 Deductive Learning

Deductive learning (Mitchell 1997) originates from deductive reasoning (Johnson 1999).

Deductive reasoning, also known as deductive logic, or logical deduction, is a method of reasoning that draws individual conclusions through reasoning based on known general or universal premises. In simple terms, it is reasoning from general to specific.

There are mainly three types of deductive reasoning, namely: affirming the antecedent (modus ponens), denying the consequent (modus tollens), and syllogism.

The classic example of deductive reasoning by the ancient Greek philosopher Aristotle is the method of syllogism.

Exercise 6.2 (The Death of Socrates) All humans are mortal; Socrates is human; therefore Socrates will also die.

The validity of the aforementioned reasoning lies in: "All human are mortal" is a well-known objective law, called the major premise; and "Socrates is a human" is a minor premise under the condition of the major premise, from which the specific conclusion "Socrates will also die" is derived in accordance with the objective law.

Here, P represents the hypothesis, Q represents the conclusion, $P \rightarrow Q$ represents the conditional statement, then the formalized description of the three types of deductive reasoning is shown in Table 6.8.

Deductive reasoning can infer the conclusion Q from the hypothesis P, only if Q is the formal logic conclusion of P. The truth of the hypothesis P can guarantee the truth of the conclusion Q via effective deduction.

Therefore, based on the aforementioned deductive reasoning, the following definition is given for deductive learning.

Definition 6.6 (Deductive Learning) Deductive learning is the learning that proposes hypotheses based on some existing facts or rules, and then confirms them through observation.

Figure 6.7a is a schematic diagram of deductive learning, which depicts the process of deductive learning.

Table 6.8 Three types of deductive reasoning

	Modus ponens	Modus tollens	Syllogism
Major premise	$P \rightarrow Q$	$P \rightarrow Q$	$P \rightarrow Q$
Minor premise	P	$\neg Q$	$Q \rightarrow R$
Conclusion	Q	$\neg P$	$P \rightarrow R$

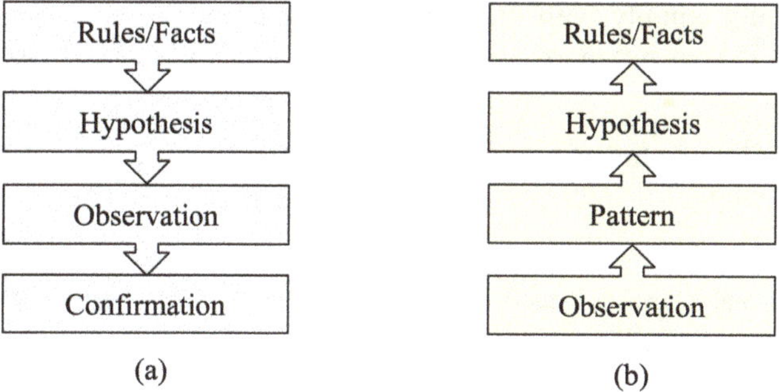

Fig. 6.7 Deductive learning vs. inductive learning

In the *Encyclopedia of Machine Learning* (Sammut and Webb 2011), deductive learning is described as follows:

> Deductive learning is a subclass of machine learning that studies algorithms for learning provably correct knowledge.

Interestingly, the *Encyclopedia of Machine Learning* also provides two synonyms for deductive learning, namely: analytical learning and explanation-based learning.

In the book *Machine Learning* published by Tom M. Mitchell in 1997, there is a section specifically introducing deductive learning (Mitchell 1997).

6.3.2.2   Inductive Learning

Inductive learning (Mitchell 1997) originates from inductive reasoning (Hayes et al. 2010).

Inductive reasoning is a form of reasoning that uses individual instances as premises to derive general conclusions. In simple terms, it is reasoning from specific to general. If described in statistical terms, induction is the process of deriving the population from samples.

Exercise 6.3 (Area of a Triangle) The area of a right triangle is half the product of the base and the height, and the area of an acute triangle and an obtuse triangle is also half the product of the base and the height, so the area of all triangles is half the product of the base and the height.

As we all know, the problem of triangle area is elementary geometry knowledge. This is an example of correct inductive reasoning. However, inductive reasoning cannot guarantee that every conclusion is correct.

The major types of inductive reasoning include generalization and prediction. Their formalized descriptions are shown in Table 6.9.

Inductive reasoning allows for inference from P to Q, but within Q does not necessarily fully comply with P. That is to say, P may give us good reason to acknowledge Q, but it cannot ensure that Q is 100% correct. Please see a well-known example.

Exercise 6.4 (The Color of Swans) All swans we have seen are white; therefore, all swans are white.

Table 6.9 Two major types of inductive reasoning and their formalized descriptions

	Premise	Conclusion
Generalization	Proportion Q of the sample has attribute a	So, the proportion Q of the population has attribute a
Prediction	Proportion Q of observed members of group G have had attribute a	So, there is a probability corresponding to Q that other members of group G will have attribute a when next observed

Obviously, the truth of the aforementioned reasoning is not impeccable, because there are exceptions: black swans are also found in the wetlands of Australia and New Zealand. The so-called black swan events refer to unpredictable, unusual, and even sudden events, but they are not impossible.

Unlike deductive reasoning, inductive reasoning does not draw universal conclusions on a closed domain, so it is applicable to situations of cognitive uncertainty. That is to say, inductive reasoning essentially supports uncertain reasoning. Based on some uncertain knowledge, the credibility of the conclusion can be obtained through inductive reasoning under given premises.

Just as deductive reasoning derived deductive learning, inductive reasoning also gave birth to inductive learning. The following definition is given here.

Definition 6.7 (Inductive Learning) Inductive learning is the learning of general facts or rules by observing certain patterns, forming hypotheses based on these patterns, and then deriving from these hypotheses.

Figure 6.7b is a schematic diagram of inductive learning, which depicts the process of inductive learning.

Inductive learning is also another entry in the *Encyclopedia of Machine Learning* (Sammut and Webb 2011), defined as follows:

Inductive learning is a subset of machine learning that studies algorithms for learning knowledge based on statistical regularities. The learned knowledge typically has no deductive guarantees of correctness, though there may be statistical forms of guarantees.

It is worth noting that the aforementioned definition defines inductive learning as learning based on statistical regularities. From another perspective, it can be understood that most of the learning based on statistics belongs to inductive learning.

Tom Mitchell's book *Machine Learning* also discusses inductive learning in many places (Mitchell 1997).

6.3.2.3 Abductive Learning

Abductive learning (Thagard and Shelley 1997; Dai et al. 2019) originates from abductive reasoning (Thagard and Shelley 1997).

Before discussing abductive reasoning, let's first comprehend the meaning of the term "abduction".

The *Encyclopedia of Machine Learning* (Sammut and Webb 2011) explains the origin and meaning of the term "abduction" as follows:

The word abduction comes from the Latin word abducere, which means 'to lead away from.' It is sometimes used interchangeably with the word retroduction, which comes from the Latin words retro, meaning 'backwards', and ducere meaning 'to lead.' The term abduction is most commonly used to describe forms of reasoning that are concerned with the generation and evaluation of explanatory hypotheses.

> - The surprising fact, C, is observed.
>
> - But if A were true, C would be a matter of course.
>
> - Hence, there is reason to suspect that A is true.

Fig. 6.8 Argument schema of abductive reasoning

Table 6.10 Abductive reasoning and formalized description

Abduction	Reasoning rule
Fact	Q
Hypothesis	$P \rightarrow Q$
Premise	P

Abductive reasoning, also called retroduction, is a form of reasoning proposed by American philosopher and logician Charles S. Peirce (1839–1914), which is one of the three indispensable forms of reasoning alongside deductive and inductive reasoning.

Peirce explicitly presents abduction as a logical form, and Fig. 6.8 is its argument schema (Haig 2005).

In the aforementioned schema, fact C can be a specific event or a generalization of experience. The A is a premise, which can be understood as an explanatory hypothesis or theory. And C occurs following it, but not from an isolated A, but from an A related to background knowledge. Finally, A should not be understood as correct, but as credible and worth further study.

Abductive reasoning starts from observed facts and seeks their most likely premises. In short, as shown in Table 6.10, it is a reverse reasoning process from conclusion to premise.

Abductive reasoning allows the hypothesis P to be the premise of Q; as a result of reasoning, it traces back the cause from the conclusion Q, proposes the hypothesis P, and then obtains the premise P through argumentation and verification. Scientific research often starts from observed surprising phenomena, proposes hypotheses about the causes of these phenomena through reverse inference, and then further synthesizes arguments and verifies these hypotheses to obtain their most reasonable explanation.

Unlike deductive reasoning, the premise derived from the conclusion in abductive reasoning is only a proposed, best explanation. Although abductive reasoning is reasoning for the best explanation, not all abductive reasoning can obtain the best explanation. In the view of Charles Peirce, deduction and induction are not forms of reasoning to generate new knowledge. The task of generating new knowledge can only be accomplished through abductive reasoning. In scientific activities, the process of seeking the best explanation and analyzing the cause is the process of scientific discovery.

Following the basic idea of abductive reasoning, abductive learning has also emerged. The *Encyclopedia of Machine Learning* (Sammut and Webb 2011) gives the following definition of abductive learning:

> Abductive reasoning, then, is often portrayed as explanatory reasoning that leads back from facts to a proposed explanation of those facts. It is different from inductive reasoning, which is commonly concerned with descriptive inference that results in generalizations. The phrase abductive learning can be taken to cover a wide range of concerns where learning outcomes result from the employment of abductive reasoning. Abductive learning occurs widely in scientific, professional, lay and educational endeavors.

Based on the aforementioned definition of abductive reasoning, here we also provide a definition of abductive learning.

Definition 6.8 (Abductive Learning) Abductive learning is the learning that starts from observed facts (events, experiences, etc.), seeks the most likely or optimal premises (hypotheses, theories, etc.), and generates new knowledge from this learning.

Abductive learning takes many forms, with the following three being the most common.

(1) Existential abduction.

That is, there are explanatory hypotheses or theoretical assumptions used to explain the premises of related facts, not their essence. For example, some people can use viral infection to explain some symptoms of the common cold, but they cannot explain the nature of the virus itself. In science, the multivariate statistical method of exploratory factor analysis is sometimes used to hypothesize the existence of latent causal factors and to explain related effect indicators. For example, IQ test scores can be used to measure intelligence levels (Haig 2005). Therefore, exploratory factor analysis is a learning method that helps generate basic reasonable hypotheses that can explain variable positive correlations.

(2) Analogical abduction.

That is, abductive learning is achieved through analogical modeling, which is valuable for understanding the essence of causal mechanisms inherent in science and daily life. Expanding the understanding of the essence of unknown theoretical causal mechanisms based on well-known and understood knowledge is an important means. Using the strategy of analogical modeling, appropriate analogies can be drawn from known sources, known properties, and known behaviors to establish models of unknown subjective or causal mechanisms. Two examples illustrate the emergence of this strategy: one is the molecular model of gas based on the analogy with balls stored in a container; the other is the computational model of thinking based on the analogy with a computer.

(3) Best explanation.

That is, from the perspective of explanatory value, the rationality of knowledge claims is proven. When a hypothesis or theory is judged to be the best explanation, it must be accepted. Best explanation learning has a wide range of applications in human affairs. For example, Darwin argued for the superiority of his theory of evolution because it provides an explanation closer to the facts than creationism.

Abductive learning has attracted the attention of some machine learning researchers, and corresponding papers have been published (Crowder and Carbone 2017; Zhou 2019; Dai et al. 2019).

6.3.2.4 Comparative Analysis

Here, we make a comparative analysis for the aforementioned three patterns of logical learning.

Charles Peirce deeply studied the intrinsic connection between argumentation methods and reasoning patterns. Argumentation is a series of arguments, presented as argument outlines, usually used to persuade someone or provide reasons for them to accept its conclusion.

Arguments can be divided into analytical arguments and synthetic arguments, where analytical arguments are deductive, and synthetic arguments are inductive or abductive. The intrinsic connection between them can be seen in Fig. 6.9.

Charles Peirce, drawing on the syllogism method, used "beans" as an example to illustrate these three different reasoning patterns, as detailed in Fig. 6.10. Where the rule represents the major premise, the case represents the minor premise, and the result represents the conclusion. Different reasoning patterns show different orders.

From Figs. 6.9 and 6.10, it can be seen that: deductive reasoning and inductive reasoning are almost inverse, deductive reasoning goes from general to specific conclusion, while inductive reasoning goes from specific to general conclusion; inductive reasoning and abductive reasoning are inverse, inductive reasoning is to find out the rules, while abductive reasoning is to make the best explanation of the phenomenon.

Therefore, it can be concluded that the two learning modes of deductive learning and inductive learning are also inverse: deductive learning starts from facts or rules, goes through hypotheses, observations, and then confirmation, which can be regarded as a top-down learning process; inductive learning starts from observations, discovers its patterns, forms hypotheses, and then obtains facts or

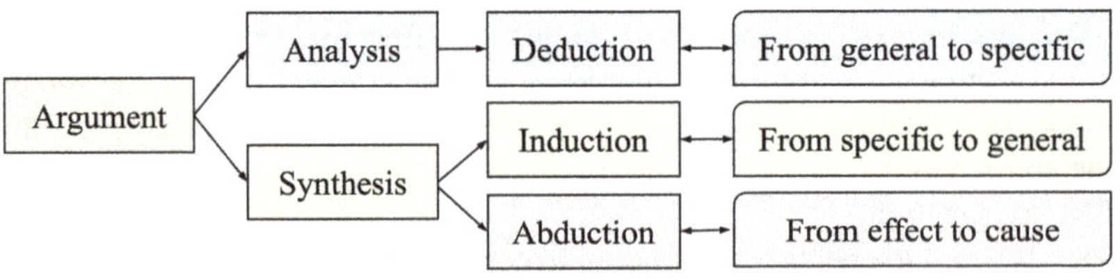

Fig. 6.9 The intrinsic connection between argumentation and three major reasoning modes

Deduction	Rule	All beans in this bag are white	Specific conclusion (always true)
	Case	These beans are from this bag	
	Result	These beans are white	
Induction	Case	These beans are from this bag	General conclusion (maybe true)
	Result	These beans are white	
	Rule	All beans in this bag are white	
Abduction	Rule	All beans in this bag are white	Best explanation (maybe true)
	Result	These beans are white	
	Case	These beans are from this bag	

Fig. 6.10   Examples and characteristics of the three major reasoning modes

rules learning, that is, a bottom-up learning process. Abductive learning, on the other hand, starts from observed facts, seeks the premise with the best explanation, and generates new knowledge from this. The three learning modes based on the three reasoning modes have become the theoretical basis of machine learning.

6.3.3   *Logic Programming Methods*

Logic programming (LP) is a programming paradigm based on formal logic. This subsection first introduces the logic programming language and then discusses two extended logic programming methods in machine learning, namely: inductive logic programming and probabilistic logic programming.

6.3.3.1   Logic Programming Language

The representative of logic programming languages is Prolog, whose name is formed by combining the first three letters of the two nouns in "programming in logic." Prolog language originates from first-order logic in formal logic.

Unlike other procedural programming languages (such as C language), or object-oriented programming languages (such as C++, and Java), Prolog is a declarative programming language. Its program logic is expressed as relations, representing facts or rules, and its reasoning function is initiated by running queries on these relations.

The representation of rules in Prolog language is as follows:

$$H : -B_1, \ldots, B_n.$$

$$(6.1)$$

```
likes(bill, car).

animal(X)  :- cat(X).

bird(X)    :- animal(X), has(X, feather).
```

Fig. 6.11 Three simple Prolog clauses

Its meaning is:

$$H \Leftarrow B_1 \wedge \cdots \wedge B_n.$$

Where, "$B_1 \wedge \cdots \wedge B_n$" is referred to as the antecedents, the symbol "$\wedge$" represents logical conjunction, and "H" is referred to as the consequent.

Prolog's reasoning adopts the reasoning method of first-order logic, namely forward chaining and backward chaining.

The Prolog language has been widely used in theorem proving, expert systems, natural language processing, machine learning (Zelle and Mooney 1994), etc. Figure 6.11 shows three simple Prolog clauses: the first one is "Bill likes cars", describing a fact; the second one is "Cats are animals", describing a rule; the last one is "Animals with feathers are birds", which is also a rule.

Logic programming (LP) can be represented as:

$$LP = Logic + Computational\ Control. \tag{6.2}$$

From the aforementioned formula, it can be seen that: Logic programming is the combination of logic and computational control, where the logic appears in the form of a program.

6.3.3.2 Inductive Logic Programming

Inductive logic programming (ILP) is a supervised learning method that uses a logic programming form similar to the Prolog language to represent background knowledge, samples, and hypotheses uniformly. Compared with other machine learning methods, ILP is easy to understand and can include additional information in the learning problem.

Inductive logic programming can be represented as:

$$ILP = Logic + Computational\ Control + Statistics. \tag{6.3}$$

From the aforementioned formula, it can be seen that: inductive logic programming is a combination of logic, statistics, and computational control. It can also be considered as adding a statistical induction mechanism to the logic programming Eq. 6.2.

Inductive logic programming can be traced back to the doctoral thesis titled *"Automatic Methods of Inductive Inference"* published in 1972 by Gordon Plotkin of the University of Edinburgh Plotkin (1972).

In 1981, Ehud Shapiro wrote a research report titled "Inductive Inference of Theories from Facts", in which he mainly elaborated on the model inference problem and its algorithms (Shapiro 1981). Ehud Shapiro et al. also built the first model inference system, a program written in Prolog language, which can perform inductive inference from positive examples and negative examples.

In 1991, Stephen Muggleton published a paper titled "Inductive Logic Programming" (Muggleton 1991), the term inductive logic programming was used for the first time. Muggleton also founded the *International Conference on Inductive Logic Programming*, which has been held annually since 1991.

Induction can be divided into two types, one is philosophical induction, and the other is mathematical induction. The former refers to proposing a theory to explain observed facts, while the latter is to prove the properties of all members in an ordered set. The induction in inductive logic programming belongs to philosophical induction.

Inductive logic programming is a framework of inductive machine learning. It is a method of inferring and obtaining hypotheses from positive examples, negative examples, and background knowledge.

The formal definition of inductive logic programming is as follows:

Let B be the given background knowledge and serve as a logical inference, in the form of Horn clauses used in logic programming. Let E^+ and E^- be positive and negative examples. A correct hypothesis h is a logical proposition in Table 6.11.

The schema of inductive logic programming is as follows:

Positive examples + Negative examples + Background knowledge $\Rightarrow$ Hypothesis.

For example, given the following background knowledge:

$$parent(\text{Helen, Mary}) \wedge parent(\text{Tom, Eve}) \wedge female(\text{Helen}) \wedge$$
$$female(\text{Mary}) \wedge female(\text{Eve}).$$

A simple positive example can be inferred:

$$daughter(\text{Mary, Helen}) \wedge daughter(\text{Eve, Tom}).$$

Table 6.11 Conditions of logical propositions

Proposition	Expression
Necessity	$B \not\models E^+$
Sufficiency	$B \wedge h \models E^+$
Weak consistency	$B \wedge h \not\models \text{false}$
Strong consistency	$B \wedge h \wedge E^- \not\models \text{false}$

6.3.3.3   Probabilistic Logic Programming

Many applications of machine learning require the ability to reason with uncertain knowledge, so finding the right logical induction is particularly important. Combining probability theory, which has the ability to handle uncertainty, with first-order logic, which has formal reasoning capabilities, forms a richer and more expressive probabilistic logic.

Nils J. Nilsson from Stanford University in the United States first used this term in his 1986 paper titled "Probabilistic Logic" (Nilsson 1986).

Probabilistic logic extends the truth table of first-order logic, the results of which can be obtained through probabilistic expressions. That is, it generalizes the semantics of first-order logic, where the truth value of a statement is a probability value between 0 and 1. When the probability of all sentences is 0 or 1, its semantics can be simplified to traditional logical implication. This generalization applies to any logical system because it can ensure the consistency of a finite set of statements.

Probabilistic logic programming (PLP) is a method of combining the logic programming paradigm with probability, suitable for modeling in domains with complex and uncertain relationships between elements (Ng and Subrahmanian 1992; Lukasiewicz 1998).

The development of probabilistic logic programming has also promoted the study of probabilistic logic learning. As shown in Fig. 6.12, probabilistic logic learning is the combination of probability, logic, and learning (De Raedt and Kersting 2003).

In the context of probabilistic logic learning, the meanings of probability, logic, and learning are as follows: "probability" refers to the use of probability inference mechanisms based on probability theory, such as Bayesian networks, hidden Markov models, and stochastic grammars; "logic" refers to first-order logic and its relational representation, that is, using the existing research results in the logical domain; "learning" refers to machine learning. The purpose of probabilistic logic learning is to organically combine these three components.

Probabilistic logic programming is developed based on logic programming, represented as:

$$PLP = Logic + Computational\ Control + Probability \qquad (6.4)$$

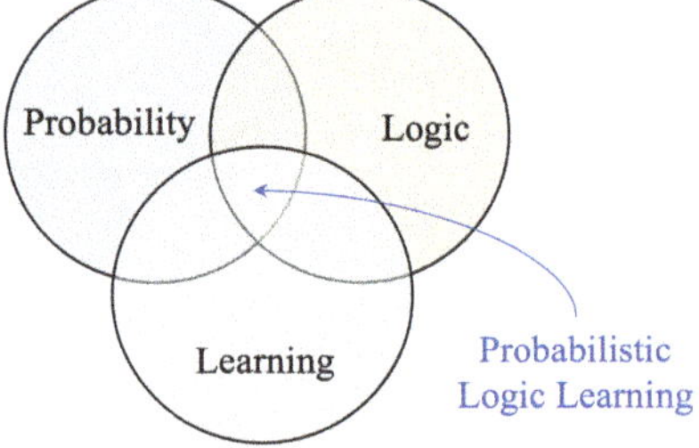

Fig. 6.12 Probabilistic logic learning is the intersection of probability, logic, and learning

It can be considered that probabilistic logic programming is based on the logic programming Eq. 6.2, with the addition of a probability inference mechanism.

Probabilistic logic programming was proposed in the early 1990s, and many approaches have emerged so far (De Raedt and Kimmig 2015). Here, ProbLog is used as an example (De Raedt et al. 2007) to introduce probabilistic logic programming.

ProbLog was released in the form of a paper titled "ProbLog: A Probabilistic Prolog and its Application in Link Discovery" at *International Joint Conferences on Artificial Intelligence* (IJCAI), one of the top international conferences in artificial intelligence. It can be seen that ProbLog is formed by extending the probability function based on the existing logic programming language Prolog and is a probabilistic Prolog language.

A ProbLog program consists of a Prolog program and a set of probability values. A Prolog program is composed of a set of statements, which can be expressed as rules or facts. The set of facts in Prolog is represented as: $f_1, f_2, \ldots, f_n$, where f_i is a fact. Also, a set of probability values is represented as: $p_1, p_2, \ldots, p_n$, where p_i is a probability annotation, $p_i \in [0, 1]$. Therefore, the set of facts in ProbLog can be denoted as:

$$F = \{p_1 :: f_1, p_2 :: f_2, \ldots, p_n :: f_n\}. \tag{6.5}$$

The aforementioned formula means that each fact of ProbLog f_i is equivalent to a Boolean random variable, the probability of its Boolean value being "true" is p_i, and the probability of being "false" is $1 - p_i$.

For a fact set of a ProbLog logic program $F' \subseteq F$, its probability distribution is:

$$P_F\left(F'\right) = \prod_{f_i \in F'} p_i \prod_{f_i \in F \setminus F'} (1 - p_i). \tag{6.6}$$

Probabilistic logic programming is constantly developing and evolving. Eight years after the release of ProbLog, the second generation ProbLog2 has also been launched (Dries et al. 2015). In addition, at the 2018 *Conference on Neural Information Processing Systems* (NeurIPS), a paper titled "DeepProbLog: Neural Probabilistic Logic Programming" (Manhaeve et al. 2018) proposed a framework integrating neural networks and probabilistic logic programming. DeepProbLog fully utilizes the expressive power and advantages of both and can be trained end-to-end based on instances.

By the way, it is just from the 2018 *Conference on Neural Information Processing Systems* that the abbreviation of the conference name was changed to NeurIPS. Because a few commentators criticized the original abbreviation, NIPS, as encouraging sexism due to its association with the word nipples, and as being a slur against Japanese.

After decades of development, probabilistic logic programming has become an important branch of the logic programming field. Since 2014, the annual *Workshop on Probabilistic Logic Programming*[2] has been held during the *International Conference on Logic Programming*.

6.4   Rule Learning

Rule learning is also known as rule-based learning, or rule-based machine learning (Weiss and Indurkhya 1995), including association rule learning, decision trees, and random forests constructed from multiple decision trees.

Here, we take association rule learning as an example to introduce rule-based machine learning methods.

Association rule learning, also known as association analysis, is used to discover relationships between variables in large transaction datasets. The method used is to discover strong rules based on different measures of interestingness, and new rules will be generated as the analysis increases. The ultimate goal is to enable the machine to simulate the feature extraction of the human brain and improve the ability to abstract its associations from new unclassified data.

It should be pointed out that association rule learning focuses on the association between variables, not the causal relationship.

In 1993, Rakesh Agrawal et al. proposed a method to discover all important association rules in large retail sales databases (Agrawal et al. 1993), namely association rule learning. This is also the origin of the famous *Market Basket Analysis*. Michael Hahsler discussed this in his 2015 paper titled "A Probabilistic Comparison of Commonly Used Interest Measures for Association Rules" (Hahsler, 2015). Pang-Ning Tan et al. also gave a detailed explanation of association rule analysis in Chap. 6 "Association Analysis: Basic Concepts and Algorithms" of the book *Introduction to Data Mining* published in 2018 (Tan et al. 2018).

Here, association rule learning is introduced from the following three aspects: association rule, association rule discovery, and association rule synthesis.

6.4.1   Association Rule Learning

Definition 6.9 (Association Rule) The association rule (AR) can be formally defined as a 3-tuple, AR $= \langle I, T, R \rangle$. Where: $I = \{i_1, i_2, \ldots, i_m\}$ is an itemset, each item is presented as a binary attribute; $T = \{t_1, t_2, \ldots, t_n\}$ is a transaction dataset, each transaction $t_i \subseteq I$ is a subset of items in I and has a unique transaction identity; $R : X \Rightarrow Y$ is a rule, X and Y are respectively referred to as the antecedent and the consequent of the rule, $X, Y \subseteq I$, and $X \cap Y = \emptyset$.

---

[2] https://stoics.org.uk/plp/.

Table 6.12 Example in the transaction dataset T

Transaction ID	*butter*	*bread*	*egg*	*milk*	*beer*	*diaper*
1	0	1	1	1	0	0
2	1	0	1	1	0	0
3	1	1	0	1	0	0
4	0	0	0	0	1	1

A simple example is shown in Table 6.12.

The description is as follows.

A supermarket's transaction dataset T contains the data of six items and four transactions, the itemset is

$$I = \{butter, bread, egg, milk, beer, diaper\}.$$

Where, the value "1" for each transaction indicates the corresponding item exists, and the value "0" indicates the corresponding item does not exist.

An instance rule derived from the aforementioned transaction dataset is

$$\{butter, \ egg, \ bread\} \Rightarrow \{milk\},$$

which means that customers who buy *butter*, *egg*, and *bread* often also buy *milk*. In addition, another instance rule can be derived, although there is only one example in the four transactions, namely

$$\{diaper\} \Rightarrow \{beer\},$$

this rule coincidentally aligns with the "Beer and Diapers" story.

6.4.2   *Association Rule Discovery*

Association rule discovery is the process of selecting useful rules from all possible association rules. The rules can be discovered by some measurement methods, among which the commonly used are "support" and "confidence."

6.4.2.1   Support

Support, denoted as supp($\cdot$), includes the support of the antecedent and the support of the association rule.

The support of the antecedent X is the frequency of this itemset appearing in the transaction dataset, that is:

$$\text{supp}(X) = \frac{|t \in T; X \subseteq t|}{|T|} = P(X). \tag{6.7}$$

For example, according to the aforementioned equation, the support rate of the itemset $X = \{beer, diaper\}$ in Table 6.12 is $1/5 = 0.2$.

The itemset with a support rate greater than the support threshold σ, that is $\text{supp}(X) > \sigma$, is called a frequent itemset.

The support rate of the association rule $X \Rightarrow Y$ is the support rate of all itemset in the rule, denoted as:

$$\text{supp}\,(X \Rightarrow Y) = \text{supp}\,(X \cup Y) = P\,(X \cap Y). \tag{6.8}$$

6.4.2.2   Confidence

Confidence, denoted as $\text{conf}(\cdot)$, only measures for association rules.

The confidence of the association rule $X \Rightarrow Y$ is the probability of its consequent given its antecedent, that is:

$$\text{conf}\,(X \Rightarrow Y) = \frac{\text{supp}\,(X \Rightarrow Y)}{\text{supp}\,(X)} = \frac{\text{supp}\,(X \cup Y)}{\text{supp}\,(X)} = \frac{P\,(X \cap Y)}{P\,(X)} = P\,(Y|X).$$

$$\tag{6.9}$$

For example, for the rule extracted from Table 6.12, $\{butter, bread\} \Rightarrow \{milk\}$, its confidence is $0.2/0.2 = 1$.

The association rules that are not less than the confidence threshold γ, that is, $\text{conf}(X \Rightarrow Y) \geq \gamma$, are called strong rules.

6.4.3   *Association Rule Synthesis*

There are many methods for association rule synthesis, and the task of rule synthesis is usually decomposed into the following two subtasks:

(1) Frequent itemset synthesis, that is, discover frequent itemset from many transactions in the transaction dataset.
(2) Strong rules synthesis, extract strong rules based on the discovered frequent itemset.

The computational cost required to synthesize frequent itemset is much greater than the computational overhead required to synthesize strong rules. The famous Apriori algorithm is a method that greatly reduces the amount of computation (Agrawal and Srikant 1994), and it is listed as one of the "Top 10 Algorithms in Data Mining" (Wu et al. 2008).

The Apriori algorithm uses support rate to prune candidate frequent itemset, thereby reducing the computational complexity of frequent itemset synthesis, and reducing the complexity of generating association rules. The pruning principle of

Apriori is as follows: if an itemset is an infrequent itemset, all subsets containing this itemset are pruned.

After the Apriori algorithm, some researchers have proposed several methods to improve efficiency, including Eclat, FP-Tree, FP-growth, AprioriTID, AprioriHybrid and other algorithms.

6.5 Causal Learning

There are several terms related to causal learning: causality, cause and effect, and causal relation.

In 1999, Avi Sion published a book titled *The Logic of Causation: Definition, Induction and Deduction of Deterministic Causality*, and the third edition was released in 2010 (Sion 2010).

Avi Sion made a highly summarize on the terms related to causal in this book:

Causality refers to causal relations, i.e. the relations between causes and effects.

He is a doctor of philosophy, has been engaged in the study of logic, philosophy, and spirituality. Avi Sion has published 27 books since 1990, covering logic, phenomenology, epistemology, causality, psychology, ethics, etc.

Below, the main discussion is on causal relation, causal inference, and causal learning that are closely related to machine learning.

6.5.1 Causal Relations

Some researchers have proposed the following four types of causal relations through observation and analysis, that are: common-cause relations, common-effect relations, causal chains, and causal steady states.

(1) Common-cause relations: One cause leads to several effects. As shown in Fig. 6.13a, cause A leads to effects B and C respectively. For example, the flu can lead fever, cough, headache, and the common cause of these effects is the flu virus.

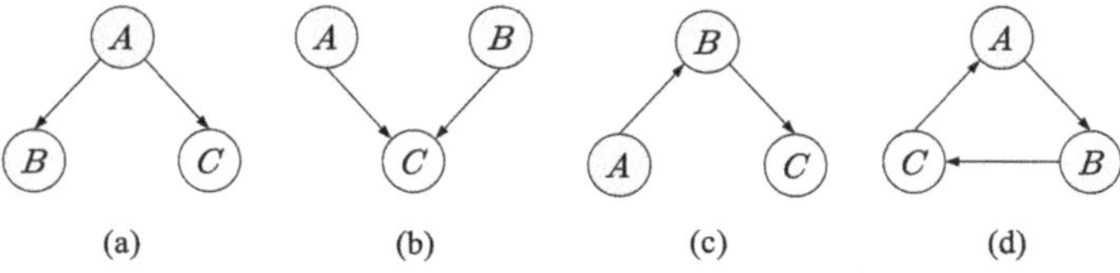

Fig. 6.13 Four types of causal relations

(2) Common-effect relationships: Several causes lead to one effect. As shown in Fig. 6.13b, causes A and B both lead to the same effect C. For example, untreated waste smoke, exhaust gas, wastewater, waste residue, and waste materials lead environmental pollution.

(3) Causal chains: One cause leads to one effect, and that effect is the cause of another effect. As shown in Fig. 6.13c, cause A leads to effect B, and B in turn causes effect C. For example, high housing prices reduce the purchasing power of homebuyers, and sluggish consumption can lead to economic imbalance.

(4) Causal homeostasis: Several causal relations form a cycle, constituting a stable state. As shown in Fig. 6.13d, A, B, and C are interrelated, forming a stable cyclical state. For example, in natural ecosystems, the ecological balance established through the interaction between organisms and the environment, and between organisms, can be considered a form of causal homeostasis.

With the deep development of machine learning, causal relations have attracted great importance to some researchers.

This is because causal relations are very important for machine learning. For example: if a robot knows that a glass vase will break when it falls on the ground, there is no need to throw many glass vases on the ground to train the robot to understand what will happen; for autonomous vehicles, if a causal prediction function is pre-installed in its intelligent control system, it can avoid traffic accidents to the greatest extent possible, without having to experience numerous traffic accidents to learn safe driving.

One of the three giants of deep learning, Yoshua Bengio, pointed out in October 2019 that deep learning is good at discovering patterns in large amounts of data, but cannot explain the connections between them. He believes that artificial intelligence can only fully realize its potential by understanding causal relations beyond the application field of pattern recognition, thus making artificial intelligence systems smarter and more efficient.

Yoshua Bengio and his team tried to turn the above views into reality. In January 2019, the team submitted a paper titled "A Meta-Transfer Objective for Learning to Disentangle Causal Mechanisms" to the electronic preprint repository *arXiv*, which was accepted and published by the *International Conference on Learning Representations* (ICLR) in 2020 (Bengio et al. 2020). This paper proposes Meta-Learn Causal Structures, which allows learners to quickly adapt to sparse distribution changes caused by interventions, subject actions, and other unstable sources. In October 2019, the team submitted another paper to the electronic preprint repository *arXiv*, titled "Learning Neural Causal Models from Unknown Interventions" (Ke et al. 2019), proposing a new framework for meta-learning causal models, where the relationship between each variable and its parent variable is modeled by a neural network and adjusted by structured meta-parameters that capture the overall topology of the directed graphical model.

Researchers engaged in machine learning and causal relation research are not uncommon, and there has been a clear upward trend in recent years. In November 2019, Bernhard Schölkopf, as the single author, submitted a paper titled "Causality in Machine Learning" to the electronic preprint repository *arXiv* (Schölkopf 2019). Schölkopf is the director of the Max Planck Institute for Intelligent Systems in Germany, which is affiliated with the Max Planck Society, funded by the German federal and state governments.

In the aforementioned paper, Schölkopf points out that, statistical learning is a "method driven by independent and identically distributed data," a well-known issue. Current machine learning often assumes that the dataset satisfies the independent and identically distributed premise, that is, all samples come from the same generation process, and the samples are independent of each other. When the data meet the independent and identically distributed conditions, machine learning often works well, and some even surpass humans. Otherwise, machine learning will produce inexplicable results. For example, when shopping online to buy a laptop bag for an existing laptop, the recommendation system of the online shop also recommends a laptop.

Therefore, the paper proposes to develop from statistical models to causal models.

Regarding the causal relationship in machine learning, the paper discusses several aspects in detail, including: the hierarchy of causal modeling, independent causal mechanisms, the discovery of cause and effect, and causal representation learning.

Judea Pearl, a computer scientist and philosopher at the University of California, Los Angeles, is a pioneer and advocate of artificial intelligence causal inference. However, for a long time, almost no one thought about the relationship between causal inference and machine learning. Bernhard Schölkopf's paper discusses the established and should be established connections between the two, which was appreciated by Judea Pearl and tweeted, considering the paper to be:

A very comprehensive, delightful and inspiring paper. Recommended to ALL, not just MANY ML/AL folks.

6.5.2 *Causal Inference*

Causal inference is the judgment of the impact of cause on effect, especially the inference process of what effect the change of cause will lead to.

Some researchers in the field of machine learning have realized that machine learning is not only predicting the results of data, but also analyzing the cause that affect its effect, and inferring its effect from this cause.

Judea Pearl is a pioneer and advocate of causal inference. He published in 2000 the book *Causality: Models, Reasoning and Inference* (Pearl 2000), which detailed and analyzed the causal relations. He described a structured causal model (SCM) in the book, which includes and integrates other causal relation methods, and provides a unified mathematical basis for causal inference.

Judea Pearl wrote in the postscript of the book:

Causality is not mystical or metaphysical. It can be understood in terms of simple processes, and it can be expressed in a friendly mathematical language, ready for computer analysis.

In 2009, Judea Pearl published a paper titled "Causal Inference in Statistics: A Review" in American *Statistics Surveys* magazine (Pearl 2009), and pointed out that:

The aim of standard statistical analysis, typified by regression, estimation, and hypothesis testing techniques, is to assess parameters of a distribution from samples drawn of that distribution. With the help of such parameters, one can infer associations among variables, estimate beliefs or probabilities of past and future events, as well as update those probabilities in light of new evidence or new measurements. These tasks are managed well by standard statistical analysis so long as experimental conditions remain the same.

He also pointed out,

Causal analysis goes one step further; its aim is to infer not only beliefs or probabilities under static conditions, but also the dynamics of beliefs under changing conditions, for example, changes induced by treatments or external interventions. $\cdots$ This distinction implies that causal and associational concepts do not mix. $\cdots$ Correlation does not imply causation.

In 2010, Judea Pearl published a paper titled "An Introduction to Causal Inference" in the *International Journal of Biostatistics* (Pearl 2010), further elaborating on his views on causal inference.

In recognition of Judea Pearl's outstanding contributions to causal inference and probabilistic methods in artificial intelligence, he was awarded the Turing Award in 2011.

In 2017, Jonas Peters, Dominik Janzing, and Bernhard Schölkopf co-authored a book titled *Elements of Causal Inference: Foundations and Learning Algorithms*, published by the MIT Press (Peters et al. 2017).

In May 2018, at the age of nearly 82, Judea Pearl coauthored *The Book of Why: The New Science of Cause and Effect* with Dana Mackenzie (Pearl and Mackenzie 2018). The book gives this new science a very concise name: causal inference. It is clear that the author has elevated causal inference to a new science.

Causal inference is mainly divided into three levels: association, intervention, and counterfactuals. Judea Pearl depicted the relationship between these three levels in his 2019 paper (Pearl 2019), which is shown in Fig. 6.14.

The three levels hierarchy of causal inference in Fig. 6.14 are from low to high.

Level 1   is called association, the expression is $P(y|x)$, to infer the causal relations in statistical conditions. For example, the occurrence of certain symptoms will inform what kind of disease it is. This association can be inferred directly using conditional probability. Since this level does not require causal information, it is placed at the lowest level of the three-layer hierarchy. The commonality of current machine learning methods is to answer such questions.

	Level	Activity	Typical Questions	Examples
1	Association $P(y|x)$	Seeing	• What is? • How would seeing X change my belief in Y?	• What does a symptom tell me about a disease? • What does a survey tell us about the election results?
2	Intervention $P(y|do(x),z)$	Doing, Intervening	• What if? • What if I do X?	• What if I take aspirin, will my headache be cured? • What if we ban cigarettes?
3	Counterfactuals $P(y_x|x',y')$	Imagining, Retrospection	• Why? • Was it X that caused Y? • What if I had acted differently?	• Was it the aspirin that stopped my headache? • Would Kennedy be alive had Oswald not shot him? • What if I had not been smoking the past two years?

Fig. 6.14 Three-level hierarchy of causal inference. Questions at level i (i = 1,2,3) can be answered only if information from level j ($j > i$) is available

Level 2   is known as intervention, the expression is $P(y|do(x),z)$, to infer the causal relations in factual conditions. Its level is higher than association. It not only involves what has been observed but also what should be obtained after the change. At this level, a typical question is: If I take aspirin, will my headache be cured? Such questions cannot be answered solely from existing data, because they involve the reaction after intervention.

Level 3   is referred to as counterfactual, the expression is $P(y_x|x',y')$, to infer the causal relations in counterfactual conditions. The reason for arranging it at the highest level of the three-layer hierarchy is that the counterfactual contains issues of intervention and association. A typical question is: What would happen if different actions were taken? This reasoning pattern of counterfactual can be traced back to the views of philosophers David Hume and John Stuart Mill. Common phrases composed of adjectives of counterfactual are counterfactual thinking, counterfactual logic, counterfactual reasoning, etc.

Causal inference is an important theoretical framework of machine learning. Humans often think from a causal viewpoint, if we understand the reasons for certain things happening, we can improve their outcomes by changing that reason.

Causal inference is a tool that allows machine learning algorithms to reason in a way similar to humans.

6.5.3   Causal Learning Approaches

Causal learning is learning the relationship between causes and effects or determining the causal status between two or more events.

There are two approaches to causal learning: one is bottom-up, judging the causal relationship through the common information between accidental events; the other is top-down, inferring the causal relationship from observations and verifying its correctness. The latter is closely connected with the causal inference mentioned in the previous section.

In 2005, Cambridge University Press published *The Cambridge Handbook of Thinking and Reasoning*, a total of 858 pages. The title of Chap. 7 is "Causal Learning", authored by Marc J. Buehner of the School of Psychology at Cardiff University, UK, and Patricia W. Cheng of the Department of Psychology at the University of California, Los Angeles (Buehner and Cheng 2005).

In 2012, Oxford University Press also published an 864-page *The Oxford Handbook of Thinking and Reasoning*, for which Patricia Cheng and Marc Buehner wrote Chapter 12, also titled "Causal Learning" (Cheng and Buehner 2012).

Psychology is studying causal learning from a computational perspective, focusing on causal discovery, mainly using two approaches, namely the associative approach and the causal approach. The associative approach is an intuitive, purely statistical method, while the causal approach adds causal hypotheses to the inference process.

Causal learning is gradually gaining attention in the field of machine learning. Many researchers compare machine intelligence with natural intelligence. Natural intelligence refers to the intelligence exhibited by biological entities. They realize that machine intelligence is quite limited in some problems, but natural intelligence excels at it. So far, the data that machine learning can handle often assume independent and identically distributed (i.i.d.) conditions. However, the challenge faced by machine learning is how to deal with non-independent and identically distributed (non-i.i.d.) data. Machine learning under i.i.d. conditions is called in-distribution generalization, while machine learning under non-i.i.d. conditions is called out-of-distribution generalization. Despite both involving generalization, out-of-distribution generalization is a highly challenging form of generalization.

In May 2021, a review paper titled "Toward Causal Representation Learning" (Schölkopf et al. 2021) was published in the *Proceedings of the IEEE*. The paper has seven authors, with Bernhard Schölkopf as the first author. This is another paper by him following "Causality for Machine Learning" in 2019, and Yoshua Bengio is a coauthor of this paper.

The paper provides a systematic review of causal representation learning. The term causal representation learning encompasses causal learning and representation learning. The former refers to automatically identifying potential causal variables and their relationships from raw data, while the latter refers to automatically extracting required features from raw data. Causal representation learning provides a new research approach for some unresolved issues in machine learning, such as the robustness, reusability, and out-of-distribution generalization of learning models.

The main sections are as follows: levels of causal modeling, causal models and inference, independent causal mechanisms, causal discovery and machine learning, learning causal variables, and implications for machine learning.

Table 6.13 Basic categories of models

Model	Mechanistic/physical	Structural causal	Causal graphical	Statistical
Predict in i.i.d. setting	Yes	Yes	Yes	Yes
Predict under distribution shift/intervention	Yes	Yes	Yes	No
Answer counterfactual questions	Yes	Yes	No	No
Obtain physical insight	Yes	?	?	No
Learn from data	?	?	?	Yes

In Table 6.13, the first row represents four types of models, namely mechanistic/physical, structural causal, causal graphical, and statistical models. The first column represents five tasks, namely predicting in i.i.d. setting, predicting under distribution shifts/intervention, answering counterfactual questions, obtaining physical insight, and learning from data.

Among the four types of models, the first type is the mechanistic/physical model, usually characterized by differential equations, which provides a comprehensive description of the system and can complete the first four tasks. The second type is the structural causal model abstracted from physical realism, composed of a set of causal variables and a set of causal conditions, supporting the first three tasks. The third type, the causal graphical model, is a graphical model that uses nodes and edges in the graph to represent causal relations, supporting the first two tasks. The fourth type is the statistical model, which, besides being able to predict in an i.i.d. environment and learn from data, does not support the other three tasks.

The paper uses a graphical approach to illustrate the differences between statistical models and causal models. Given three sets of variables, a statistical model specifies a single probability distribution, while a causal model represents a set of distributions, each corresponding to a possible intervention.

The conceptual foundation of statistical learning is the joint probability distribution, which can only infer data that satisfy the i.i.d. condition; whereas causal learning allows inference of data through intervention and can provide understanding and prediction of intervention effects, thus it does not need to satisfy the prerequisite condition of data being i.i.d. This is the biggest difference between statistical learning and causal learning.

The difference between statistical and causal models can be expressed by the principle of independent causal mechanisms (ICMs), where the causal generation process of system variables is composed of independent modules that do not affect each other. Under this premise, it means that the conditional distribution of each variable given its cause mechanism will not affect other mechanisms.

The paper extends the ICM principle and takes the sparse mechanism shift (SMS) hypothesis as a result of the ICM principle, where small distribution changes often appear sparsely or locally in causal decomposition, that is, they should not usually affect all factors simultaneously.

It also studied the connection between causal representation learning and disentangled representations in deep learning, and discussed semi-supervised learning, domain generalization and robustness, and how machine learning models can benefit from causal learning.

In May 2021, Jovana Mitrovic et al. from Google's DeepMind company published a paper titled "Representation Learning via Invariant Causal Mechanisms" (Mitrovic et al. 2020) at the *International Conference on Learning Representations* (ICLR). What the title of the paper reveals is that the authors propose a new self-supervised objective, which enforces invariant prediction of proxy targets across augmentations through an invariance regularizer, which yields improved generalization guarantees.

6.6 Neuro-Symbolic Learning

In recent years, hybrid machine learning has attracted the attention of some researchers and become a new research approach.

The so-called hybrid mainly combines neural networks and symbolic methods, represented by neuro-symbolic learning (which is also written as neural-symbolic learning). In addition, some similar terms have appeared, such as neuro-symbolic computing, neuro-symbolic artificial intelligence, neuro-symbolic machine learning (Mao et al. 2019), and neuro-symbolic method.

With the rapid development of artificial intelligence and machine learning, some very complex intelligent tasks, such as Chess, Go, Texas Hold'em, medical diagnosis, and protein structure prediction, artificial intelligence has surpassed human intelligence. However, there are some simple tasks that even toddlers, chicks, or cubs can complete, but artificial intelligence is hard to handle. In addition, with the rapid development of deep learning, breakthrough achievements have been made in dealing with vision, speech, and natural language perception, but artificial intelligence and machine learning are lacking in cognitive processing.

Below, we first introduce Moravec's paradox and the theory of thinking fast and slow and then discuss the neuro-symbolic method.

6.6.1 Moravec's Paradox

Hans P. Moravec is a famous scholar in robotics and artificial intelligence in the United States. He published a book in 1988 titled *Mind Children: The Future of Robot and Human Intelligence* (Moravec 1988), in which he points out:

> It is comparatively easy to make computers exhibit adult level performance on intelligence tests or playing checkers, and difficult or impossible to give them the skills of a one year old when it comes to perception and mobility.

This statement by Moravec is known as Moravec's paradox.

Moravec explained this paradox based on evolution. The perception and action skills that humans possess are all biological instincts, which have evolved in the process of natural selection. In this process, natural selection tends to evolve and optimize these skills. The older the skill, the longer the evolution time. The evolution time of human abstract thinking is relatively short. He concluded from this that the difficulty of reverse engineering human skills is roughly proportional to the time of evolution of the skill. The oldest human skills are largely unconscious, and seem effortless, but are difficult to implement with computers. However, human abstract thinking, logical reasoning, etc., although they are advanced human intelligence, reverse engineering with computers is not impossible.

About Moravec's paradox, I think we can analyze it from other viewpoints:

> Human advanced cognitive activities are complex cognitive processes, which are often formalized and symbolized for ease of communication and dissemination. Therefore, these formalized and symbolized advanced cognitive activities of humans are easy for computers to represent and reason. Chess, Go, poker, etc., are competitive games created by humans, and artificial intelligence is also created by humans using computer software and hardware to simulate human intelligence. Therefore, artificial intelligence can defeat human intelligence in competitive games. Medical diagnosis and protein structure prediction are also formalization of human diagnosis and determination methods. In these areas, artificial intelligence could surpass human intelligence.
>
> However, the perception and action skills of toddlers rely on intuition and instinct. Intuition is a direct perception or feeling without conscious analysis and reasoning, while instinct is an innate ability, including both innate and acquired factors. The common feature of intuition and instinct is that they are difficult to formalize and symbolize, so it is difficult to represent knowledge and process with computers. Therefore, it is not surprising that the most effective method of using artificial intelligence to process intuition and instinct has not yet been found.

6.6.2   Thinking Fast and Slow: System 1 and System 2

Daniel Kahneman published a book in 2011 titled *Thinking, Fast and Slow* (Kahneman 2011).

Kahneman was born in Tel Aviv, British Mandate of Palestine (now Israel) in 1934, received a bachelor's degree from the Hebrew University of Jerusalem, and master's and doctorate degrees from the University of California, Berkeley.

Table 6.14 System 1 vs. System 2

System 1	System 2
Belong to fast thinking, relying on intuition, memory, and habits, and making judgments or reactions quickly	Belong to slow thinking, requiring focused attention, deep rational and logical thinking, and making decisions after careful consideration
Be often unconscious, fast but not pursuing accuracy, but sometimes it is influenced by the rule of thumb that seeing is believing, and the illusions it produces often lead to wrong choices	Be conscious, relying on long-term learned knowledge and accumulated experience, analyzing and solving problems through attention and logical thinking, and making complex decisions. It is not prone to errors, but it will be influenced by the intuitive judgment of system 1

He is considered the most famous psychologist after Freud, and his research covers psychology, cognitive science, economics, and other fields. He was awarded the 2002 Nobel Memorial Prize in Economic Sciences, in recognition of his outstanding contributions to integrating insights from psychological research into economic science, especially concerning human judgment and decision-making under uncertainty.

"Thinking, Fast and Slow" is Daniel Kahneman's representative work in psychology. He believes that the brain has two ways of thinking, fast and slow, and people often switch between fast and slow thinking. He associates this with the two systems proposed by psychologists Keith Stanovich and Richard West, namely system 1 and system 2 depicted in Table 6.14.

At this point, some readers may ask: What is the relationship between Kahneman's fast and slow thinking and machine learning?

Please see the following news.

At the 2020 *AAAI Conference on Artificial Intelligence* (AAAI), a unique "Fireside Chat with Daniel Kahneman" was held, discussing the current state and future of artificial intelligence and human decision-making. On the stage, besides Daniel Kahneman, there were the three giants of deep learning, the 2018 Turing Award winners Joshua Bengio, Geoffrey Hinton, and Yann LeCun, and the host was Francesca Rossi, IBM's global head of AI ethics.

Daniel believes that today's deep learning is more like a product of System 1, but not System 2. Deep learning can match patterns and make predictions, but what deep learning lacks is the ability to reason. Many people think this is crucial.

Some researchers have begun to delve into the issue of how deep learning supports fast and slow thinking (i.e., System 1 and System 2).

At the 2017 *Conference on Neural Information Processing Systems* (NIPS), three researchers from University College London published a paper titled "Thinking Fast and Slow with Deep Learning and Tree Search" (Anthony et al. 2017).

Joshua Bengio gave two academic talks separately at the *Conference on Neural Information Processing Systems* (NeurIPS) in December 2019 and the *AAAI Conference on Artificial Intelligence* (AAAI) in February 2020, titled "From System 1 Deep Learning to System 2 Deep Learning" and "Deep Learning for System 2 Processing". Both of his academic talks discussed how deep learning solves the problem of fast and slow thinking, believing that current deep learning belongs to System 1, and the future should extend deep learning to System 2 to achieve higher-level cognition. Specific approaches include out-of-distribution generalization and transfer, high-level semantic representations, compositionality, and causality.

In December 2020, 11 researchers from IBM and four universities, including Francesca Rossi, submitted a paper to the electronic preprint repository *arXiv* titled "Thinking Fast and Slow in AI" (Booch et al. 2020), after then which was officially published at the 35th *AAAI Conference on Artificial Intelligence* (AAAI) in February 2021. The paper argues that if we have a deep understanding of the reasons for some human abilities still lacking in artificial intelligence (such as adaptability, generalization ability, common sense, and causal reasoning), we can obtain similar abilities by embedding these causal components in artificial intelligence systems. Drawing inspiration from Daniel Kahneman's cognitive theory, the paper proposed ten research questions in the fast and slow thinking of artificial intelligence.

6.6.3 Neuro-Symbolic Approaches

The connectionist framework represented by neural networks and the symbolic framework with logical reasoning capabilities are the two main theoretical frameworks of machine learning. Deep neural networks are very effective for certain unstructured data such as vision, speech, and natural language, but their ability to handle abstract problems such as logic and reasoning is very limited. Symbolic artificial intelligence is powerful in logical reasoning, but it is clumsy in handling large amounts of unstructured data. Neural-symbolic learning methods are the product of combining the strengths of these two theoretical frameworks.

In 2009, Artur S. d'Avila Garcez, Luis C. Lamb, and Dov M. Gabbay from the UK published a book titled *Neural-Symbolic Cognitive Reasoning* (Garcez et al. 2009). The book proposes to integrate neural network models with two of the most basic cognitive abilities of humans, i.e., learning from experience and reasoning based on knowledge.

In 2012, Artur Garcez, Krysia B. Broda, and Dov Gabbay published a book titled *Neural-Symbolic Learning Systems: Foundations and Applications* (Garcez et al. 2012). The book points out that neuro-symbolic learning systems play to the strengths of neural and symbolic artificial intelligence and will play a core role in artificial intelligence. The book is divided into three parts, covering the main topics of neuro-symbolic integration, namely: theoretical advances in knowledge representation and learning, knowledge extraction from pre-trained neural networks, and inconsistency handling in neuro-symbolic systems.

In February 2020, Gary Marcus, a professor of psychology at New York University, submitted a 59-page paper to the electronic preprint repository *arXiv*, titled "The Next Decade in AI: Four Steps Towards Robust Artificial Intelligence" (Marcus 2020).

It is worth pointing out that, Gary Marcus, before his above paper, participated in the "AI Debate[3]" held in Montreal, Canada, in December 2019. He and Joshua Bengio were the debaters, and Vincent Boucher was the moderator of the debate. Joshua Bengio and Gary Marcus debated the direction of artificial intelligence development. Bengio believes that sequential reasoning can be performed while staying in the deep learning framework, while Marcus believes that it is unrealistic to handle abstraction and reasoning with a single structure, emphasizing the value of hybrid models.

The aforementioned paper by Gary Marcus was organized and published after the debate. In the paper, he outlined the following four steps: the development of hybrid neuro-symbolic architectures; the construction of cognitive frameworks and large-scale knowledge bases; the development of abstract reasoning tools; and the complex mechanisms of representation and induction in cognitive models.

By the way, Gary Marcus also coauthored *Rebooting AI: Building Artificial Intelligence We Can Trust* with Ernest Davis, which was published in 2019 (Marcus and Davis 2019).

In September 2020, Zhang Bo, a fellow of Chinese Academy of Sciences, et al. from the Artificial Intelligence Research Institute of Tsinghua University published an article titled "Towards the Third Generation of Artificial Intelligence" in the 70th anniversary special issue of *China Science* (Zhang et al. 2020), elaborating on the concept of the third generation of artificial intelligence. They pointed out that since the birth of artificial intelligence, there has always been a dispute between symbolism and connectionism. Today, it seems that these are just different ways of simulating human minds (or brains). To establish a new interpretable and robust AI theory and method, and to develop safe, trustworthy, reliable, and scalable AI technology, it is necessary to combine these two, which is the inevitable path for the development of AI.

In December 2020, two researchers from City, University of London, UK, and Federal University of Rio Grande do Sul, Brazil, submitted a review article titled "Neurosymbolic AI: The 3rd Wave" to the electronic preprint repository *arXiv* (Garcez and Lamb 2020). They believe that symbolic artificial intelligence, represented by knowledge-driven, is the first wave, and deep neural networks characterized by data-driven are the second wave. The organic combination of these two, i.e., neuro-symbolic artificial intelligence, is the third wave.

In January 2022, Pascal Hitzler, a professor at Kansas State University, and Md Kamruzzaman Sarker, an assistant professor at University of Hartford, edited and published a proceeding titled *Neuro-Symbolic Artificial Intelligence: The State of the Art*, which includes 17 articles on neuro-symbolic methods, reflecting the breadth and depth of the latest developments in this field (Hitzler and Sarker 2022), which is one of the series *Frontiers in Artificial Intelligence and Applications*.

---

[3] https://www.quebecartificialintelligence.com/aidebate/.

Further Reading

1. Robin Manhaeve, Sebastijan Duman, Angelika Kimmig, Thomas Demeester, and Luc De Raedt. "DeepProbLog: Neural Probabilistic Logic Programming." *Conference on Neural Information Processing Systems* (NeurIPS), 2018.
[Notes] This paper proposes a probabilistic logic programming language that incorporates deep learning by means of neural predicates, referred to as Deep-ProbLog, that supports (i) both symbolic and subsymbolic representations and inference, (ii) program induction, (iii) probabilistic logic programming, and (iv) deep learning from examples.
2. Peter Kontschieder, Madalina Fiterau, Antonio Criminisi, and Samuel Rota Bulo. "Deep Neural Decision Forests." *IEEE International Conference on Computer Vision* (ICCV), 2015.
[Notes] The paper presents a novel approach that unifies classification trees with the representation learning functionality known from deep convolutional networks, by training them in an end-to-end manner. And it won the Marr Prize at the ICCV 2015.
3. Yoshua Bengio, Tristan Deleu, Nasim Rahaman, Rosemary Ke, Sébastien Lachapelle, Olexa Bilaniuk, Anirudh Goyal, and Christopher Pal. "A Meta-Transfer Objective for Learning to Disentangle Causal Mechanisms." *International Conference on Learning Representations* (ICLR), 2019.
[Notes] The paper proposes to use a meta-learning objective that maximizes the speed of transfer on a modified distribution to learn how to modularize acquired knowledge and discover causal dependencies.
4. Bernhard Schölkopf. "Causality for machine learning." *arXiv preprint* arXiv:1911.10500, 2019.
[Notes] This article discusses where links have been and should be established between graphical causal inference and machine learning.
5. Bernhard Schölkopf, Francesco Locatello, Stefan Bauer, Nan Rosemary Ke, Nal Kalchbrenner, Anirudh Goyal, and Yoshua Bengio. "Toward Causal Representation Learning." *Proceedings of the IEEE*, 2021.
[Notes] This is a review paper on causal learning, also known as causal representation learning. The first author is Bernhard Schölkopf, and it is another academic contribution following his 2019 paper "Causality for Machine Learning".
6. Grady Booch, Francesco Fabiano, Lior Horesh, Kiran Kate, Jonathan Lenchner, Nick Linck, Andreas Loreggia, et al. "Thinking Fast and Slow in AI." *AAAI Conference on Artificial Intelligence*, 2021.
[Notes] This paper proposes a research direction to advance AI, which draws inspiration from cognitive theories of human decision making.
7. Artur d'Avila Garcez, and Luis C. Lamb. "Neurosymbolic AI: the 3rd Wave." *arXiv preprint* arXiv:2012.05876, 2020.
[Notes] In this paper, they relate recent and early research results in neuro-symbolic AI with the objective of identifying the key ingredients of the next wave of AI systems. This paper focuses on research that integrates in a principled

way neural network-based learning with symbolic knowledge representation and logical reasoning.

References

Agrawal, R., T. Imielinski, and A. Swami. (1993). Mining association rules between sets of items in large databases. In *ACM SIGMOD International Conference on Management of Data.*

Agrawal, R., and R. Srikant. (1994). Fast algorithms for mining association rules. In *20th International Conference on Very Large Data Bases (VLDB)*, Santiago, Chile.

Anthony, T., Z. Tian, and D. Barber. (2017). Thinking fast and slow with deep learning and tree search. In *Conference on Neural Information Processing Systems (NIPS).*

Baader, F., D. Calvanese, D. L. Mcguinness, D. Nardi, and P. F. Patel-Schneider. (2003). *The description logic handbook: theory, implementation, and applications.* Cambridge University Press.

Bengio, Y., T. Deleu, N. Rahaman, R. Ke, S. Lachapelle, O. Bilaniuk, A. Goyal, and C. Pal. (2020). A meta-transfer objective for learning to disentangle causal mechanisms. In *International Conference on Learning Representations (ICLR).*

Booch, G., F. Fabiano, L. Horesh, K. Kate, J. Lenchner, N. Linck, A. Loreggia, K. Murugesan, N. Mattei, and F. Rossi. (2020). Thinking Fast and Slow in AI. arXiv preprint arXiv:2010.06002.

Buckley, J. J., and Y. Hayashi. (1994). Fuzzy neural networks: A survey. *Fuzzy Sets and Systems* 66(1): 1–13.

Buehner, M. J., and P. W. Cheng. (2005). Causal learning. *The Cambridge Handbook of Thinking and Reasoning*: 143–168.

Cheng, P. W., and M. J. Buehner. (2012). Causal learning. *The Oxford Handbook of Thinking and Reasoning*: 210–233.

Crowder, J. A., and J. N. Carbone. (2017). Abductive Artificial Intelligence Learning Models.

Dai, W.-Z., Q. Xu, Y. Yu, and Z.-H. Zhou. (2019). Bridging machine learning and logical reasoning by abductive learning. In *Conference on Neural Information Processing Systems (NeurIPS).*

Das, R., S. Sen, and U. Maulik. (2020). A survey on fuzzy deep neural networks. *ACM Computing Surveys (CSUR)* 53(3): 1–25.

De Raedt, L., and K. Kersting. (2003). Probabilistic logic learning. *ACM SIGKDD Explorations Newsletter* 5(1): 31–48.

De Raedt, L., and A. Kimmig. (2015). Probabilistic (logic) programming concepts. *Machine Learning* 100(1): 5–47.

De Raedt, L., A. Kimmig, and H. Toivonen. (2007). ProbLog: A probabilistic prolog and its application in link discovery. In *International Joint Conferences on Artificial Intelligence (IJCAI)*, Hyderabad.

Dries, A., A. Kimmig, W. Meert, J. Renkens, G. Van den Broeck, J. Vlasselaer, and L. De Raedt. (2015). Problog2: Probabilistic logic programming. In *Joint European Conference on Machine Learning and Knowledge Discovery in Databases.* Springer.

Garcez, A. d. A., and L. C. Lamb. (2020). Neurosymbolic AI: The 3rd Wave. arXiv preprint arXiv:2012.05876.

Garcez, A. S. A., L. C. Lamb, and D. M. Gabbay. (2009). *Neural-symbolic cognitive reasoning.* New York: Springer Science & Business Media.

Garcez, A. S. d. A., K. B. Broda, and D. M. Gabbay. (2012). *Neural-symbolic learning systems: Foundations and applications.* New York: Springer Science & Business Media.

Garcez, A. S. d. A., L. C. Lamb, and D. M. Gabbay. (2007). Connectionist modal logic: representing modalities in neural networks. *Theoretical Computer Science* 371(1–2): 34–53.

Garcez, A. S. d. A., L. C. Lamb, and D. M. Gabbay. (2009). Connectionist modal logic. *Neural-Symbolic Cognitive Reasoning* 55–74.

Haig, B. D. (2005). Exploratory factor analysis, theory generation, and scientific method. *Multivariate Behavioral Research* 40(3): 303–329.

Haugeland, J. (1985). *Artificial intelligence: The very idea.* MIT Press.

Hayes, B. K., E. Heit, and H. Swendsen. (2010). Inductive reasoning. *Wiley Interdisciplinary Reviews Cognitive Science* 1(2): 278–292.

Hitzler, P., and M. K. Sarker. (2022). *Neuro-symbolic artificial intelligence: the state of the art.* IOS Press.

Johnson-Laird, P. (1999). Deductive reasoning. *Annual Review of Psychology* 50(1): 109–135.

Kahneman, D. (2011). *Thinking, fast and slow.* Macmillan.

Ke, N. R., O. Bilaniuk, A. Goyal, S. Bauer, H. Larochelle, C. Pal, and Y. Bengio. (2019). Learning Neural Causal Models from Unknown Interventions. arXiv preprint arXiv:1910.01075.

Krötzsch, M., F. Simancik, and I. Horrocks. (2012). A Description Logic Primer. arXiv Preprint arXiv:1201.4089v3.

Lukasiewicz, T. (1998). Probabilistic logic programming. In *European Conference on Artificial Intelligence.*

Manhaeve, R., S. Dumancic, A. Kimmig, T. Demeester, and L. De Raedt. (2018). DeepProbLog: neural probabilistic logic programming. In *Conference on Neural Information Processing Systems (NIPS).*

Mao, J., C. Gan, P. Kohli, J. B. Tenenbaum, and J. Wu. (2019). The neuro-symbolic concept learner: interpreting scenes, words, and sentences from natural supervision. In *International Conference on Learning Representations (ICLR).*

Marcus, G. (2020). The Next Decade in AI: Four Steps Towards Robust Artificial Intelligence. arXiv preprint arXiv:2002.06177.

Marcus, G., and E. Davis. (2019). Rebooting AI: Building Artificial Intelligence We Can Trust, Vintage.

Mitchell, T. M. (1997). *Machine learning.* McGraw-Hill Science.

Mitrovic, J., B. McWilliams, J. C. Walker, L. H. Buesing, and C. Blundell. (2020). Representation learning via invariant causal mechanisms. In *International Conference on Learning Representations.*

Moravec, H. (1988). *Mind Children: The Future of Robot and Human Intelligence.* Harvard University Press.

Muggleton, S. (1991). Inductive logic programming. *New Generation Computing* (8).

Nardi, D., and R. J. Brachman. (2003). An introduction to description logics. *Description Logic Handbook* 1: 40.

Newell, A., and H. A. Simon. (1976). Computer science as empirical inquiry: symbols and search. *Communication of the ACM* 19(3).

Ng, R., and V. S. Subrahmanian. (1992). Probabilistic logic programming. *Information and Computation* 101(2): 150–201.

Nilsson, N. J. (1986). Probabilistic logic. *Artificial Intelligence* 28(1): 71–87.

Nilsson, N. J. (2007). The physical symbol system hypothesis: status and prospects. *50 Years of Artificial Intelligence* 9–17.

Pearl, J. (2000). *Causality: models, reasoning and inference.* New York: Springer.

Pearl, J. (2009). Causal inference in statistics: an overview. *Statistics Surveys* 3: 96–146.

Pearl, J. (2010). An introduction to causal inference. *The International Journal of Biostatistics* 6(2).

Pearl, J. (2019). The seven tools of causal inference, with reflections on machine learning. *Communications of the ACM* 62(3): 54–60.

Pearl, J., and D. Mackenzie. (2018). *The book of why: the new science of cause and effect.* Basic Books.

Peters, J., D. Janzing, and B. Schölkopf. (2017). *Elements of causal inference: Foundations and learning algorithms.* Cambridge, MA: The MIT Press.

Plotkin, G. (1972). *Automatic methods of inductive inference.* University of Edinburgh.

Rychlak, J. F. (1981). Logical learning theory: propositions, corollaries, and research evidence. *Journal of Personality and Social Psychology* 40(4): 731.

Rychlak, J. F. (1994). *Logical learning theory: a human teleology and its empirical support*. U of Nebraska Press.

Rychlak, J. F. (2005). In search and proof of human beings, not machines. *Journal of Personality Assessment* 85(3): 239–256.

Sammut, C., and G. I. Webb. (2011). *Encyclopedia of machine learning*. New York: Springer Science & Business Media.

Schölkopf, B. (2019). Causality for Machine Learning. arXiv preprint arXiv:1911.10500.

Schölkopf, B., F. Locatello, S. Bauer, N. R. Ke, N. Kalchbrenner, A. Goyal, and Y. Bengio. (2021). Toward causal representation learning. *Proceedings of the IEEE*.

Shapiro, E. (1981). Inductive Inference of Theories From Facts.

Singh, G., S. Bhatia, and R. Mutharaju. (2023). Neuro-Symbolic RDF and Description Logic Reasoners: The State-Of-The-Art and Challenges. arXiv preprint arXiv:2308.04814.

Sion, A. (2010). *The logic of causation: definition, induction and deduction of deterministic causality*. CreateSpace Independent Publishing Platform.

Tan, P.-N., M. Steinbach, A. Karpatne, and V. Kumar. (2018). *Introduction to data mining (2nd edition)*. London: Pearson Education, Inc.

Tan, P.-N., M. Steinbach, and V. Kumar. (2014). *Introduction to data mining*. London: Pearson Education Limited.

Thagard, P., and C. Shelley. (1997). *Abductive reasoning: logic, visual thinking, and coherence*. Dordrecht: Springer Netherlands.

Weiss, S. M., and N. Indurkhya. (1995). Rule-based machine learning methods for functional prediction. *Journal of Artificial Intelligence Research* 3: 383–403.

Wu, X., V. Kumar, J. R. Quinlan, J. Ghosh, Q. Yang, H. Motoda, G. J. Mclachlan, A. Ng, B. Liu, and P. S. Yu. (2008). Top 10 algorithms in data mining. *Knowledge and Information Systems* 14(1): 1–37.

Zadeh, L. A. (1965). Fuzzy sets. *Information & Control* 8(3): 338–353.

Zadeh, L. A. (1988). Fuzzy logic. *Computer* 21(4): 83–93.

Zelle, J. M., and R. J. Mooney. (1994). Inducing deterministic prolog parsers from treebanks: a machine learning approach. In *Twelfth National Conference on Artificial Intelligence (AAAI-94)*.

Zhang, B., J. Zhu, and H. Su. (2020). Toward the third generation of artificial intelligence (in Chinese). *Scientia Sinica Informationis* 50: 1281–1302.

Zhou, Z. H. (2019). Abductive learning: towards bridging machine learning and logical reasoning. *Science China Information Sciences* 62(7): 76101.

Chapter 7
Behavioral Framework

Abstract The behavioral framework is also one of the important theoretical frameworks for machine learning, especially the learning from the environment. In this chapter, we first overview behaviorism doctrine and behavioral psychology. Next, we introduce behavioral learning theory, including respondent conditioning and operant conditioning, and the later propose the concept of reinforcement learning. We then introduce behavioral decision theory and the behavior decision process and expected utility. After that, we explain two types of decision models: Bayesian decision models and Markov decision models. Finally, we present three types of behavior decisions, including single-stage decision, multistage decision, as well as sequential decision, where the sequential decision is made during the interactive learning with the environment.

7.1 Overview

In this section, we will briefly introduce behaviorism doctrine and behavioral psychology, which are the theoretical foundations of behavioral framework.

7.1.1 Behaviorism Doctrine

Behaviorism doctrine is a systematic theory for studying the behavior of intelligent agents, focusing on observable, quantifiable behavior, and then conducting psychological research and analysis.

The main point of behaviorism is that psychology should study observable, measurable psychological behavior and verify and explain it through experiments, rather than merely based on some psychological speculations. This led to the proposal of behavioral psychology.

The intelligent activities and behavioral characteristics of humans or other advanced animals are formed in the process of self-adaptation and self-optimization in a given environment, learning through interaction with the environment, and

    241
W. Wang, *Principles of Machine Learning*,
https://doi.org/10.1007/978-981-97-5333-8_7

accumulating learning effects from the rewards and punishments of environmental feedback. Behaviorism views the behavior of intelligent agents as the external manifestation of their intelligent activities. If a machine's behavior appears to be intelligent or intellectual, it should be considered an intelligent or intellectual machine.

John B. Watson is a famous American psychologist and is considered the founder of behaviorism. In 1913, he published a paper titled "Psychology as The Behaviorist Views It" in the journal *Psychological Review* (Watson 1913), stating:

> The behaviorist views psychology as a purely directive experimental branch of natural science. Its theoretical goal is the prediction and control of behavior. So far, human psychology has been unsuccessful due to the mistaken notion that introspection is the only method available to psychology, and that it is the study of consciousness. Actually, psychology is the study of behavior and therefore need not take recourse to conscious phenomena. Hence, animal psychology is as valid a field of study as human psychology. The laws of behavior of animals must be determined and evaluated in and for themselves, regardless of their generalizability to other animals or humans. This suggested elimination of states of consciousness as the objects of investigation will remove the barrier that exists between psychology and other natural sciences, without neglecting the essential problems of introspective psychology.

John Watson's paper originated from a lecture he gave at Columbia University in 1913 and is considered *The Behaviorist Manifesto*. John Watson is also known as the father of behaviorism.

Watson also has a frequently quoted view:

> Give me a dozen healthy infants, well-formed, and my own specified world to bring them up in and I'll guarantee to take any one at random and train him to become any type of specialist I might select – doctor, lawyer, artist, merchant-chief and, yes, even beggar-man and thief, regardless of his talents, penchants, tendencies, abilities, vocations, and race of his ancestors.[1]

The aforementioned quote comes from John Watson's book *Behaviorism* published in 1924. The basic view is that the ultimate determinant of behavior is the way the infant grows and the environment in which it lives, not the genes of its parents. That is, behavior can be measured, trained, and changed.

Behaviorists define learning as an observable change in behavior and believe that it is always impossible to know what is happening in the mind. Trying to guess or infer things that cannot be observed through experience is inappropriate. We should understand what is being learned by observing obvious changes in behavior. Behavioral psychology only focuses on observable stimulus-response behaviors because they can be studied in a systematic and observable way, regardless of their internal psychological state.

The view of behaviorist theory is that only observable behavior can be called cognition, and emotions and feelings are too subjective. Behaviorists also believe that all behavior is the result of experience. Anyone, regardless of their background, can be trained to act in a specific way under appropriate conditions.

---

[1] https://dandradebehaviorism.weebly.com/john-watson.html.

7.1.2 Behavioral Psychology

From the 1920s to the mid-1950s, behavioral psychology gradually became the mainstream school of psychology. Some people believe that the popularity of behavioral psychology lies in establishing psychology as an objective, measurable science.

Behaviorist psychology has several branches, the representative branches mainly include the following three.

(1) Methodological behaviorism:
It believes that only public events, that is, the dynamic behavior of intelligent agents, can be objectively observed. Although thoughts and feelings are still recognized as existing, they are not considered part of behavioral science.
(2) Psychological behaviorism:
It introduces new principles of human learning, that is, humans not only follow the learning principles of animals but also follow the unique learning principles of humans.
(3) Purposive behaviorism:
It combines objective research on behavior and also considers the purpose or goal of behavior.

7.2 Behavioral Learning Theory

Early research showed that animals can through trial and error learn certain behaviors until satisfactory results are achieved. After decades of research and experimentation, some researchers have gradually discovered the mechanism of behavioral learning and formed a behavioral learning theory.

Reinforcement learning in machine learning has a deep historical connection with behavioral learning theory.

This section briefly describes the basics of behavioral learning and introduces several representative research results, namely: respondent conditioning, law of effect, operant conditioning, and reinforcement schedules.

7.2.1 Behavioral Learning

Behavioral learning refers to all behaviors being learned through interaction with the environment. Learning is the result of responding to external events, and innate or genetic factors have little influence on behavioral learning.

The response made by intelligent agents to changes in environmental conditions is known as conditioning, also known as conditioned response.

The study of behavioral learning focuses on the observable external behavior of intelligent agents, not the difficult-to-observe mental activities of intelligent agents. Because observable external behavior can be objectively measured and studied systematically. Behaviorism emphasizes the importance of environmental factors in behavioral learning.

The view of behavioral learning believes that the behavioral learning has no essential difference between humans and animals. Therefore, research and experiments on behavioral learning can be conducted not only on humans but also on animals. However, due to the ease of implementation and control of the experimental environment of small animals, such experiments have become the main experimental means and data sources for behaviorists.

7.2.2 *Respondent Conditioning*

Respondent conditioning originated from the conditioned response (CR) proposed by Russian physiologist Ivan Pavlov, so it is also known as classical conditioning[2] theory.

Pavlov published his theory of conditioned response in 1897. While studying the eating process of dogs, he noticed that some dogs would salivate before the food was served. He tried to explain this strange behavior. He originally thought that the phenomenon was due to a physical process, such as the smell of food causing a physiological response in the dog. However, he found through experiments that the bell ringing before the food was served was the cause of the dog's salivation, because at this time the dog could not smell any food. This behavior can be considered a phenomenon, indicating that the dog has learned that food will be delivered with the bell. This is a process where a psychological response (hearing the bell) leads to a physiological response (salivation).

Pavlov published the aforementioned research results in 1897.

In addition to Pavlov's study of respondent conditioning through dogs, some researchers have conducted research on the nictitating membrane of rabbit eyes (Gormezano et al. 1962).

The so-called nictitating membrane is a transparent or semitransparent third eyelid unique to animals like rabbits, which is closed under certain circumstances to protect their eyes.

Typically, when a rabbit is restrained and its eyes are blown on or given a slight electric shock, which is an unconditioned stimulus (US), it will cause the rabbit's nictitating membrane to close, a situation also known as an unconditioned response. Some conditioned stimuli, such as sound or light, will not cause the rabbit's nictitating membrane to close, a situation also known as a conditioned response. They conducted an experiment in which the rabbit was repeatedly alternated

---

[2] https://en.wikipedia.org/wiki/Classical_conditioning.

between the conditioned stimulus and the unconditioned stimulus, and the result was that the conditioned stimulus also caused the rabbit's nictitating membrane to close. Because this conditioned response is produced by the repeated alternation of the conditioned stimulus and the unconditioned stimulus, the unconditioned stimulus is often referred to as a reinforcing stimulus. The strength of the conditioned response is measured by the speed of the nictitating membrane closure or the likelihood of closing before the unconditioned stimulus, which is a measure of the strength of the conditioned stimulus.

The aforementioned experiment shows that learning takes place in the alternation process of the conditioned stimulus and the unconditioned stimulus, thus giving birth to the Rescorla–Wagner model of reactive conditioning.

The Rescorla–Wagner model was proposed by Robert A. Rescorla and Allan R. Wagner, psychologists at Yale University, in 1972 (Rescorla and Wagner 1972). It attempts to describe in a formalized way the change in associative strength (V) between the conditioned stimulus (CS) and the subsequent unconditioned stimulus (US). In an experiment, when a compound stimulus (AX) is followed by an unconditioned stimulus US_1, the rules for the change in associative strength of stimulus components A and X are as follows:

$$\Delta V_A = [\alpha_A \beta_1] (\lambda_1 - V_{AX}),$$

$$\Delta V_X = [\alpha_X \beta_1] (\lambda_1 - V_{AX}). \tag{7.1}$$

Where in Eq. 7.1: $V_{AX} = V_A + V_X$; λ_1 is the maximum conditioning produced by US_1, indicating the limit of learning; α and β are rate parameters dependent on the conditioned stimulus and the unconditioned stimulus, respectively. These parameters are considered to have fixed values based on the physical characteristics of the specific conditioned and unconditioned stimuli. In any given trial, the current associative strength V_{AX} is compared with λ, and the difference between the two is considered as the error to be corrected, which occurs through the change in the resulting associative strength (ΔV). Therefore, the Rescorla–Wagner model is an error-correction model.

It should be noted that reactive conditioning is a type of induced behavior that comes from the conditioned stimulus and is a passive learning process.

7.2.3 The Law of Effect

Some researchers believe that Pavlov's reactive conditioning experiments are too simple and that other experimental methods should be explored to study the mechanisms of behavior learning in higher animals such as humans.

Edward L. Thorndike was a professor of psychology at Columbia University in the United States. In 1898, he created a box to observe the behavior of cats. The cats could escape from the box through simple actions, such as pulling a rope or pushing a rod. His box was known as Thorndike's puzzle box.[3]

He placed a fish outside the box. When the cat was first put into the puzzle box, its reaction was basically instinctive, blindly performing some actions, and it took a long time to get out of the box to eat the fish. In the following experiments, the frequency of the cat's ineffective actions decreased, while the frequency of successful actions increased. As the number of experiments increased, the speed at which the cat escaped from the puzzle box became faster and faster.

In 1989, the *Psychological Review: Monograph Supplements* published a paper written by Edward Thorndike, titled "Animal Intelligence: An Experimental Study of The Associative Processes in Animals".

In 1905, Thorndike proposed the famous the law of effect as below:

> Responses that produce a satisfying effect in a particular situation become more likely to occur again in that situation, and responses that produce a dissatisfying effect become less likely to occur again in that situation.

That is to say, behavior often depends on previous effects, some effects can enhance behavior, while others can weaken behavior.

The evolutionary theory believes that if a certain characteristic provides an advantage for reproduction, then this characteristic will continue to exist. From this point of view, the law of effect is similar to the evolutionary theory.

The two terms "satisfying" and "dissatisfying" that appear in the law of effect later evolved into "reinforcement" and "punishment", leading to the birth of the theory of operant conditioning.

In 1970, Richard J. Herrnstein, a psychologist at Harvard University in the United States, published an article titled "On the Law of Effect" in the *Journal of the Experimental Analysis of Behavior*, in which he discussed various correlations between the rate of responding and the rate of reinforcement found in single, multiple, and concurrent reinforcement experiments (Herrnstein 1970).

7.2.4 Operant Conditioning

Inspired by the research work of Edward Thorndike, Burrhus F. Skinner began his research on operant conditioning. Burrhus F. Skinner is a famous psychologist and behaviorist in the United States.

Similar to Thorndike's puzzle box, Skinner also made an experimental box in 1930, called operant conditioning chamber, also known as Skinner box,[4] a rat was placed in the box, and a lever controlling food was installed to verify his theory of operant conditioning.

---

[3] https://en.wikipedia.org/wiki/Edward_Thorndike.

[4] https://en.wikipedia.org/wiki/B._F._Skinner.

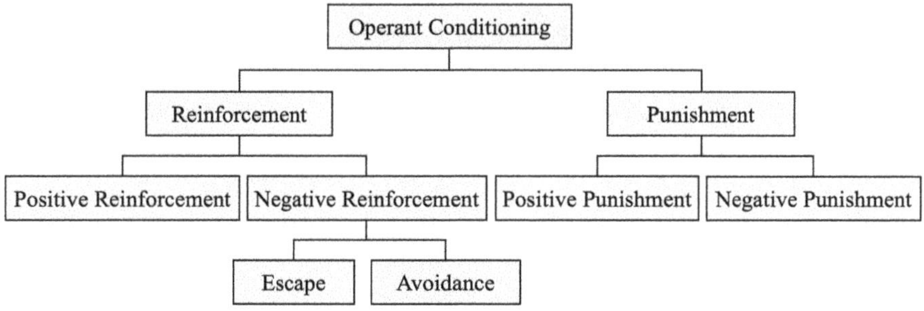

Fig. 7.1 Hierarchical relationship of reinforcement, punishment, and their subtypes

He put a rat into the Skinner box. When the rat was starving and running around, it unintentionally pressed the lever in the box, and a piece of food fell out. The rat got what it wanted. After a while, the rat learned to press the lever when it was hungry.

The results of Skinner's experiment showed that some experiments increase the likelihood of behavior (reinforcement), while others decrease its likelihood of behavior (punishment).

From the aforementioned experimental design of operant conditioning, I think if "reinforcement" and "punishment" are antonyms, then it may be easier to make sense. Therefore, I looked up the behavioral psychology explanation of those two words that are written in *Wikipedia* as follows:

> In operant conditioning, punishment is any change in a human or animal's surroundings which, occurring after a given behavior or response, reduces the likelihood of that behavior occurring again in the future.[5]

Reinforcement can be further divided into positive reinforcement and negative reinforcement, whereas punishment can be further divided into positive punishment and negative punishment. Among them, negative reinforcement is further divided into escape and avoidance.

The hierarchical relationship of reinforcement, punishment, and their subtypes in operant conditioning is shown in Fig. 7.1.

And, the brief descriptions on reinforcement, punishment, and their subtypes in operant conditioning are depicted in Fig. 7.2.

To further understand the various subtypes of operant conditioning, examples are given as follows.

For positive reinforcement, take the rat in the box as an example. When the rat presses a lever, it will get its favorite food, and the repetition rate of this behavior will increase. This is positive reinforcement. For negative reinforcement, there are two types: escape and avoidance. The behavior of covering your eyes with your

---

[5] https://en.wikipedia.org/wiki/Punishment_(psychology).

Reinforcement	*Increase behavior*	Positive Reinforcement		*Add appetitive stimulus following correct behavior*
		Negative Reinforcement	Escape	*Remove noxious stimulus following correct behavior*
			Avoidance	*Behavior avoids noxious stimulus*
Punishment	*Decrease behavior*	Positive Punishment		*Add noxious stimulus following behavior*
		Negative Punishment		*Remove appetitive stimulus following behavior*

Fig. 7.2 Brief descriptions on reinforcement, punishment, and their subtypes

hands to avoid direct sunlight belongs to the former, while the behavior of wearing sunglasses when going out belongs to the latter.

For positive punishment and negative punishment, take the students during class as an example: When the teacher finds a student playing with a mobile phone during class and criticizes him, the repetition rate of this behavior will decrease. This is positive punishment. Because the student plays with the mobile phone during class and does not listen to persuasion, the teacher temporarily confiscates his mobile phone, which decreases similar behavior. This is negative punishment.

It can be said that operant conditioning is a learning process that changes the intensity of behavior through reinforcement or punishment. It is also a process used to achieve this kind of learning.

Burrhus F. Skinner published his monograph in 1938, titled *The Behavior of Organisms: An Experimental Analysis* (Skinner 1938). His book has been cited by many researchers, and Skinner is known as the father of operant conditioning.

Eighty-one years later, the B. F. Skinner Foundation republished this book in 2019 (Skinner 1938).

It is worth emphasizing that operant conditioning causes spontaneous behavior of intelligent agents, is an active learning process, and is also a behavior decision process.

7.2.5 Reinforcement Schedules

In operant conditioning, the timing and frequency of reinforcement for a certain behavior can have a significant impact on the intensity and ratio of its response. Some researchers have conducted in-depth research on this and proposed the theory of reinforcement schedules.

In 1957, Charles B. Ferster and Burrhus F. Skinner coauthored the book "Schedules of Reinforcement" (Ferster and Skinner 1957), which detailed the reinforcement schedules and their mechanisms. Reinforcement schedules are the rules to determine which instances of behavior will be reinforced. These rules are defined according to the time and number of responses required for reinforcement. Different reinforcement schedules have different effects on operant behavior.

The best method to train animal behavior is through reinforcement. As previously mentioned, Skinner used reinforcement to teach rats to press a lever in a Skinner box to get food. Each time the rat pressed the lever, a piece of food would come out, and this training method and process are known as reinforcement schedules.

If a behavior is reinforced every time it occurs, it is called continuous reinforcement. Research shows that continuous reinforcement is the fastest way to establish new behavior or eliminate old behavior. However, many behaviors are not reinforced every time they occur, but are often another type of reinforcement schedules, namely intermittent reinforcement.

Some researchers have designed four different types of intermittent reinforcement schedules, as shown in Table 7.1, namely fixed interval, fixed ratio, variable interval, and variable ratio. Among them, fixed refers to the number of responses or the length of time between reinforcements being constant, variable means the number of responses or the length of time between reinforcements can change, interval indicates that the schedule is based on the time between reinforcements, and ratio indicates that the schedule is based on the number of responses between reinforcements.

Figure 7.3 reveals the four types of intermittent reinforcement schedules and their different response patterns, where the x-axis represents time, and the y-axis is the cumulative number of responses. From this figure, we can see:

(1) Variable ratio schedule:
    be unpredictable and produces a high and stable response rate, with almost no pause after reinforcement.
(2) Fixed ratio schedule:
    be predictable and produces a high response rate, with a pause after reinforcement.

Table 7.1 Four types of intermittent reinforcement schedules

Schedule	Description	Result
Fixed interval	Reinforcement is delivered at predictable time intervals (e.g., after 5, 10, 15, and 20 minutes)	Moderate response rate with significant pause after reinforcement
Fixed ratio	Reinforcement is delivered at predictable response numbers (e.g., after 2, 4, 6, and 8 responses)	High response rate with pause after reinforcement
Variable interval	Reinforcement is delivered at unpredictable time intervals (e.g., after 5, 7, 10, and 20 minutes)	Moderate yet steady response rate
Variable ratio	Reinforcement is delivered at unpredictable response numbers (e.g., after 1, 4, 5, and 9 responses)	High and steady response rate

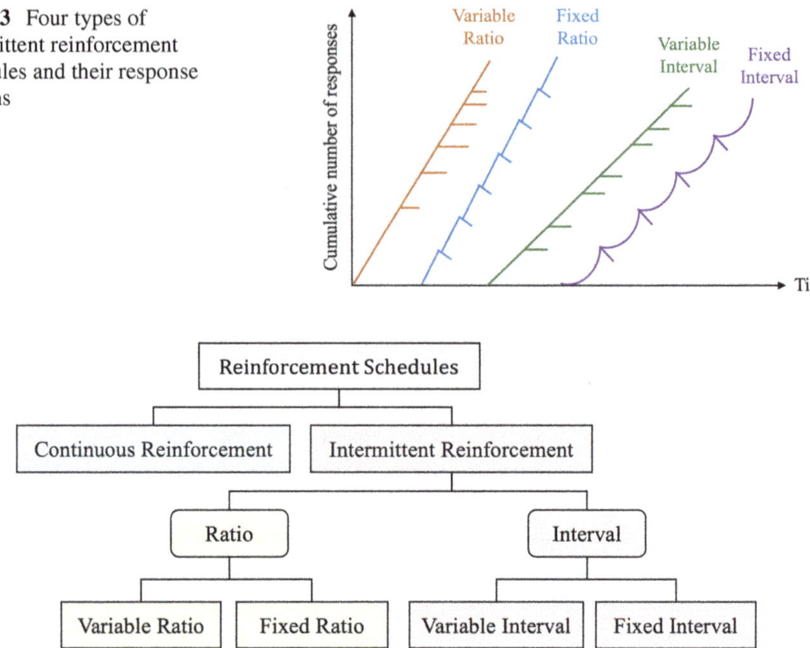

Fig. 7.3 Four types of intermittent reinforcement schedules and their response patterns

Fig. 7.4 Categories of continuous and intermittent reinforcement in reinforcement schedules

(3) Variable interval schedule:
     be unpredictable and produces a moderate, stable response rate.
(4) Fixed interval schedule:
     producing a fan-shaped response pattern, reflecting a significant pause after reinforcement.

In summary, there are two types of reinforcement schedules, namely continuous reinforcement and intermittent reinforcement. Intermittent reinforcement can be further divided into four subtypes, namely variable and fixed ratio, and variable and fixed interval. Here it is represented in the form of Fig. 7.4 for further understanding and memory.

7.3 Behavioral Decision Theory

Behavioral decision theory examines the psychological processes of judgment, decision, and behavior of intelligent agents. The behavioral decision process of an intelligent agent begins with the observation of the environment, makes judgments and decisions based on the feedback from the environment, and then takes appropriate actions.

This section first introduces decision theory and then discusses behavioral decision theory on this basis.

7.3.1   *Decision Theory*

Decision is the action and process of making a choice among several possible options. In the field of psychology, decision is considered a cognitive process.

One of the pioneers of artificial intelligence, Herbert Simon, is a renowned scholar engaged in decision research. He published a book titled *Administrative Behavior: A Study of Decision-making Processes in Administrative Organization* in 1947.

Decision theory, established for decision, studies how to make the best or nearly optimal decisions. Decision theory combines psychology, statistics, philosophy, and mathematics to analyze the decision process.

Decision theory provides a useful theoretical framework for assessing the merits of a specific action plan. There are mainly two types of decision theory frameworks, namely decisions under certainty and decisions under uncertainty, which are aimed at decision under certain and uncertain information respectively.

Decision theory is rooted in utility theory, including linear and nonlinear utility theory, used to analyze the utility of the behavior of the decision agent in a given state and the expected utility. The expected utility is the average value of all possible outcomes of the behavior.

To solve uncertain decision problems, methods such as probability theory are also used to obtain the maximum expected utility of a behavior under uncertain conditions.

There are three subtypes in decision theory:

(1) Normative decision:

It is a theory about how to make rational and optimal decisions under given assumptions and constraints. Normative decision focuses on the identification of optimal decisions. By designing an ideal decision agent, it can perform calculations perfectly and accurately and is completely rational in some sense. The practical application of normative decision is called decision analysis, which aims to design appropriate tools, methods, and software (such as decision support systems) to help people make better decisions.

(2) Descriptive decision:

This is a theory about how intelligent agents actually make decisions. It focuses on describing the behaviors of decision-makers and usually assumes that these behaviors are generated under certain consistency rules. These rules form a procedural or axiomatic framework to coordinate behaviors that do not conform to the expected utility hypothesis, or they can explicitly give the function form of time-inconsistent utility. Relatively speaking, descriptive decisions usually describe observed behaviors under the premise that the decision-maker's behavior conforms to certain consistent rules.

(3) Prescriptive decision:
It provides guidelines for decision, so that makes the best possible decisions under uncertain conditions.

The aforementioned subtypes of decision theory reflect the nature of human actual decision to a certain extent, and there are some inherent connections between them.

7.3.2 Behavioral Decision

In 1954, Ward Edwards, then a psychology professor at Johns Hopkins University, published a paper titled "The Theory of Decision Making" (Edwards 1954) in the *Psychological Bulletin*. This is a review paper that summarizes typical decision problems in psychology, economics, and other fields and also touches on behavior theory. This paper made decision a research topic in the field of psychology and is considered to be the birthmark of behavioral decision theory (Keren 1996).

In 1961, Ward Edwards published a review paper titled "Behavioral Decision Theory" (Edwards 1961) in the *Annual Review of Psychology*.

Behavioral decision theory is built on the foundation of decision theory and behavioral science, used to clarify psychological activities related to decision behavior. Research on behavioral decision theory shows that the decision process of intelligent agents depends on the environment and continuously adjusts its decision behavior according to the feedback of the environment on behavior.

Behavioral decision theory is a descriptive decision theory about the judgment, decision, and behavior of intelligent agents. Behavioral decision theory integrates the theories and methods of normative decision, descriptive decision, and prescriptive decision in decision theory, helping decision-makers make better behavioral decisions.

On February 1, 2005, Ward Edwards passed away. A commemorative article titled "Ward Edwards: Father of Behavioral Decision Theory" (Fryback 2005) was published in *Medical Decision-Making*.

7.4 Behavioral Decision Process

The behavior of intelligent agents is learned through interaction with the environment. Behavioral learning can also be seen as a behavioral decision process adopted by intelligent agents. The conditions in this process are the state of the environment, and the actions taken are the actions under this state. The decision in active learning also includes the elements of reinforcement and punishment of the environment.

This section begins with the St. Petersburg paradox and then provides a formal definition of the decision process of an intelligent agent. The aim is to unify several typical types of decisions in machine learning under this framework, in order to systematically grasp the decision problems in machine learning.

7.4.1 St. Petersburg Paradox

The St. Petersburg paradox,[6] also known as the St. Petersburg lottery, is a classic problem in behavioral decision proposed by Swiss mathematician and physicist Daniel Bernoulli.

Daniel Bernoulli elaborated on this paradox in a paper published in the Commentaries of the Imperial Academy of Science of Saint Petersburg in 1738, when he was living in St. Petersburg, Russia, hence the name of the city. It is said that the paradox was first proposed by Daniel Bernoulli's cousin, Swiss mathematician Nicolas Bernoulli, in 1713. The content is as follows:

> A casino offers a coin-tossing game for a single player. Initially, the casino puts 2 dollars of betting money into the pot. Each time the coin shows heads the casino will double the bet, when it shows tails the game ends and the player wins all the money in the pot. That is, the player wins 2 dollars, if the coin shows tails on the first toss; the player wins 4 dollars, if the coin shows heads on the first toss and tails on the second; the player wins 8 dollars, if the coin shows heads on the first two tosses and tails on the third; and so on.

Question 1:   what is the minimum admission fee that the player needs to pay to ensure that the casino does not lose money?

To answer the aforementioned question, one needs to calculate the expected return of each coin toss: since the coin only has heads and tails, the probability of the player winning 2 dollars is 1/2, the probability of winning 4 dollars is 1/4, the probability of winning 8 dollars is 1/8, and so on. Assuming that the coin always shows heads, and the game continues indefinitely, the expected return is:

$$\mathbb{E}[\cdot] = \sum_{k=1}^{\infty} \frac{1}{2^k} \cdot 2^k = \frac{1}{2} \cdot 2 + \frac{1}{4} \cdot 4 + \frac{1}{8} \cdot 8 + \cdots = 1 + 1 + 1 + \cdots = \sum_{k=1}^{\infty} 1 = \infty.$$

Therefore, it is possible for the player to win an infinite number of dollars.

For the casino, it needs to charge the player an infinite amount of admission fee to avoid losing money.

Question 2:   is any player willing to pay an infinite amount of admission fees to participate in this game?

---

[6] https://en.wikipedia.org/wiki/St._Petersburg_paradox.

A rational player should make such a decision: once the coin shows tails, the game will end, so it is almost impossible to win an infinite amount of money and therefore, should not pay an infinite amount of admission fees to participate in this game. This is the St. Petersburg paradox.

Daniel Bernoulli proposed a solution to the St. Petersburg paradox in this paper, challenging the decision criterion of expected return with the expected utility of the decision process, thus giving birth to the expected utility theory of decision.

7.4.2   Decision Process

Here we first give a formal definition of the decision process, in order to describe its utility and expected utility in the next section.

Definition 7.1 (Decision Process)   The decision process (DP) of an intelligent agent can be formalized as a 3-tuple DP $= \langle S, A, P \rangle$. Where: S and A are both random variables, representing the set of states and the set of actions of the decision process, respectively, and $P(s)$ denotes the probability of the state $s \in S$.

7.4.3   Expected Utility

Before giving the definition of expected utility, it is better to know what utility is.

Definition 7.2 (Utility)   Given a decision process DP $= \langle S, A, P \rangle$, the current state $s \in S$, the action to be taken $a \in A$, then its utility U is a function, denoted as $U(s, a)$, its result is a value in the set of real number $\mathbb{R}$, that is $U : S \times A \rightarrow \mathbb{R}$.

The utility in the decision process refers to the desirability or value of a decision behavior.

Definition 7.3 (Expected Utility)   Given a decision process DP $= \langle S, A, P \rangle$ and utility $U(s, a)$, then its expected utility (EU) is:

$$EU(S, A) = \mathbb{E}[S, A] = \sum_{s \in S, a \in A} P(s)\, U(s, a). \tag{7.2}$$

Where, P denotes the probability of state $s \in S$ occurring, $P(s) \geq 0$ and $\sum P(s) = 1$.

From the aforementioned definition, it can be seen that the expected utility of a decision process is its expected value, which is the sum of the product of the state probability and the utility.

Definition 7.4 (Maximum Expected Utility)   Given the expected utility of a decision process EU (S, A), the maximum expected utility (MEU) of the decision process is:

$$MEU\,(S,\,A) = \arg\max_{a\in A} EU\,(S,\,A)\,. \tag{7.3}$$

Intelligent agents in the decision process will have a certain tendency, known as preferences, but often prefer actions with the maximum expected utility.

Definition 7.5 (Principle of Maximum Expected Utility)   An intelligent agent capable of making rational behavioral decisions, in order to achieve the desired results, will inevitably choose the decision with the maximum expected utility in the decision process (Russell and Norvig 2009).

The aforementioned principle of maximum expected utility is also an important principle in economic activities: a rational investor always hopes that his investment can get the maximum return.

In order to further understand the decision process, expected utility, and other concepts introduced earlier, here is an example.

Exercise 7.1 (Production Quantity and Demand Quantity) There is a bakery where customers buy a certain amount of whole wheat bread every day, referred to as "daily demand". Since whole wheat bread is the staple food of these customers, the daily demand and its probability are relatively stable over a certain period of time and will not change significantly due to seasons or other reasons. According to past experience, the daily demand and its probability are as shown in Table 7.2.

This bakery produces a certain number of whole wheat breads every day, denoted as a (daily production). If the daily production exceeds the daily demand, the remaining whole wheat bread will be sold at a discount the next day. It is also assumed that the cost price of whole wheat bread is 8 dollars each, the retail price on the same day is 10 dollars each, and the discount price the next day is 6 dollars each.

Following is a comparison of the daily demand, daily production, revenue, and expected revenue of whole wheat bread.

Each cell in the middle part of Table 7.3 represents the daily return, which is calculated by the following formula:

$$\text{Profit} = \text{Sales} - \text{TotalCost}.$$

If the daily production exceeds the daily demand, the sales include the retail price of the day and the discount price of the next day. For example, if 100 whole wheat breads are produced in a day, each costing 8 dollars, the total cost is 800 dollars; if the daily demand is 70, each retail price is 10 dollars, the remaining 30 will be sold

Table 7.2 Daily demand for whole wheat bread and its probability

Daily demand	70	80	90	100	110	120	130
Probability	0.10	0.15	0.15	0.20	0.15	0.15	0.10

Table 7.3 Comparison of whole wheat bread revenue

Probability	Daily demand	Daily production						
		70	80	90	100	110	120	130
0.10	70	140	120	100	80	60	40	20
0.15	80	140	160	140	120	100	80	60
0.15	90	140	160	180	160	140	120	100
0.20	100	140	160	180	200	180	160	140
0.15	110	140	160	180	200	220	200	180
0.15	120	140	160	180	200	220	240	220
0.10	130	140	160	180	200	220	240	260
Expected return		140	156	166	170	166	156	140

at a discount price of 6 dollars the next day, making the total sales 880 dollars; the profit is then 80 dollars after subtracting the total cost from the sales.

If the daily production is equal to or less than the daily demand, naturally, the sales only include the retail price of the daily production.

Describing the aforementioned example in terms of decision process, $s \in S$ is the daily demand, $a \in A$ is the daily production, $P(s)$ is the probability of daily demand, and $U(s, a)$ is the daily profit. For example, if the daily demand $s = 70$ and the daily production $a = 100$, then the $U(70, 100) = 80$ dollars.

The daily production of whole wheat bread $a = 90$ has an expected utility equal to the expected return, that is, EU $(s, a = 90) = 166$.

The maximum expected utility MEU $(s, a = 100) = 170$, which is the result of producing 100 whole wheat breads in a day.

7.5 Bayesian Decision Models

A Bayesian decision model is a theoretical model of how to make decisions under uncertainty, based on the Bayesian model introduced in Chap. 3.

This section will mainly introduce the Bayesian decision theory and Bayesian decision networks.

7.5.1 Bayesian Decision Theory

Chapter 3 introduced the Bayesian theorem, that is, if the likelihood can be calculated under the premise of known prior probability and evidence, its posterior probability can be obtained. In the Bayesian theorem, both the likelihood and the posterior probability are expressed in the form of conditional probability.

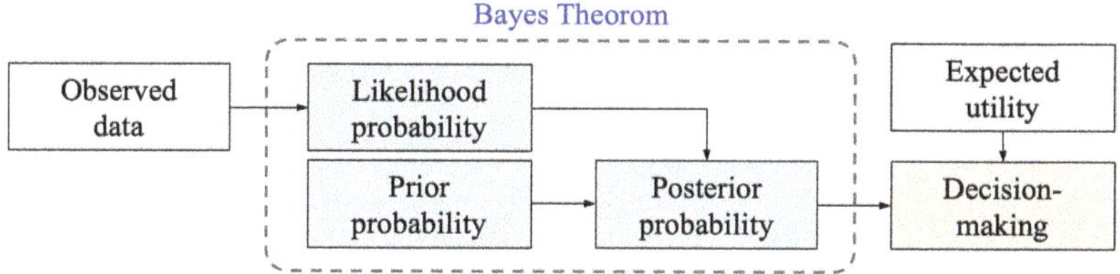

Fig. 7.5 Schematic diagram of Bayesian decision

The essence of the Bayesian theorem is to regard the conditional probability of random variables as the degree of belief in known information, and this degree of belief can be used as the basis for decision.

Bayesian decision theory (BDT) refers to the decision method based on Bayesian theorem and expected utility. Specifically: for uncertain observation data, derive its likelihood, calculate its posterior probability based on the likelihood and prior probability, then calculate its expected utility, and make the best decision according to the principle of maximum expected utility. Figure 7.5 is a schematic diagram of Bayesian decision theory.

Given a decision process $DP = \langle S, A, P \rangle$, where the state set is $S = \{s_1, s_2, \ldots, s_m\}$, the action set is $A = \{a_1, a_2, \ldots, a_n\}$, and probability $P = \{P(s_1), P(s_2), \ldots, P(s_m)\}$, satisfying $P(s_i) \geq 0$, and $\sum_{i=1}^{m} P(s_i) = 1$.

Based on Bayes theorem, we have:

$$P(s_i | a_j) = \frac{P(a_j | s_i) P(s_i)}{P(a_j)}. \tag{7.4}$$

In the aforementioned equation, the evidence $P(a_j)$ is the marginal probability given the action $A = a_j$ and the state $S = s_i$. The marginal probability $P(a_j)$ formula is substituted into the above equation, then the following expression holds:

$$P(s_i | a_j) = \frac{P(a_j | s_i) P(s_i)}{P(a_j)} = \frac{P(a_j | s_i) P(s_i)}{\sum_{s_i} P(a_j, s_i)} = \frac{P(a_j | s_i) P(s_i)}{\sum_{s_i} P(a_j | s_i) P(s_i)}. \tag{7.5}$$

Therefore, we have the following definition on the expected utility of Bayesian decision.

Definition 7.6 (Expected Utility of Bayesian Decision) The expected utility of Bayesian decision (BD) is defined as follows:

$$\mathrm{EU}_{BD}(S, A) = \mathbb{E}[S, A] = \sum_{s_i \in S, a_j \in A} P(s_i) P(s_i | a_j)$$

$$= \sum_{s_i \in S, a_j \in A} P(s_i) \frac{P(a_j | s_i) P(s_i)}{\sum_{s_i} P(a_j | s_i) P(s_i)}. \tag{7.6}$$

The maximum expected utility of Bayesian decision is as follows:

$$\mathrm{MEU}_{BD}(S, A) = \arg\max_{a \in A} \mathrm{EU}_{BD}(S, A). \tag{7.7}$$

Bayesian decision theory can be applied to machine learning such as classification and behavior decision.

7.5.2 Bayesian Decision Networks

Bayesian networks have already been introduced in Chap. 3 of this book, which are a type of directed acyclic graph model.

Bayesian decision networks are a generalized form of Bayesian networks, which can be used to represent and solve uncertain decision problems.

Definition 7.7 (Bayesian Decision Network) A Bayesian decision network (BDN) is an ordered 4-tuple, $\mathrm{BDN} = \langle \mathrm{BDG}, P_{BN}, R, \mathbb{E} \rangle$, where:

- BDG is standing for directed acyclic Bayesian decision graph, $\mathrm{BDG} = \langle \widetilde{X}, \overrightarrow{E} \rangle$, where: $\widetilde{X} = X \cup D \cup U$, X is the set of random variable nodes, D is the set of decision variable nodes, U is the set of utility variable nodes, and $X \cap D \cap U = \emptyset$; $\overrightarrow{E}$ is a directed edge, and it satisfies the condition of a directed acyclic graph, $\forall X_i \in \widetilde{X}, \nexists (X_i \to X_i)$.
- P_{BN} is the set of probability distributions of directed graphs, $P_{BN} = \{P_{X_i} | X_i \in X\}$, where each random variable probability P_{X_i} is defined the same as in Definition 3.17.
- R is the set of reward values.
- $\mathbb{E}$ is the expected utility.

From the aforementioned definition of $\widetilde{X}$, it is known that there are three types of nodes in the Bayesian decision network, namely: random variable nodes, decision variable nodes, and utility variable nodes, which are represented by circles, rectangles, and diamonds, respectively.

Corresponding to the aforementioned three types of nodes, the edges in the Bayesian decision network are also divided into three types, namely: the relationship

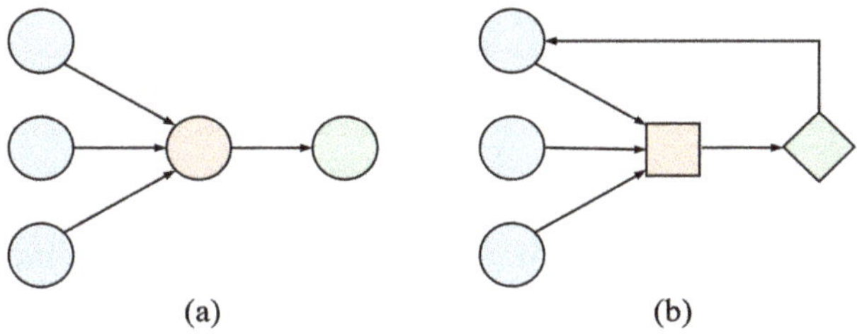

Fig. 7.6 Schematic diagram of Bayesian network and Bayesian decision network

between the random variable node and the input edge is a conditional probability relationship, the relationship between the decision variable node and the input edge is a decision relationship, and the relationship between the utility variable node and the input edge is defined as a utility relationship.

In order to facilitate the comparison between the Bayesian network and the Bayesian decision network, two schematic diagrams are given in Fig. 7.6, where: Fig. 7.6a is the Bayesian network, and Fig. 7.6b is the Bayesian decision network. The similarities and differences between the two can be seen from these two figures.

Bayesian decision networks are a compact graphical and mathematical representation. They are referred to as a generalized form of Bayesian networks because, in addition to the random variable nodes identical to those in Bayesian networks, Bayesian decision networks also add decision variable nodes and utility variable nodes. Decision variable nodes represent decision options, and utility variable nodes represent expected utilities, which are used to calculate the expected utilities of various decision options. Therefore, Bayesian decision networks can not only model probabilistic inference problems but also model and solve decision problems following the principle of maximum expected utility.

Bayesian decision networks have intuitive semantics and are easy to understand. They have become an alternative to decision trees to overcome the problem of the number of branches in decision trees growing exponentially with the increase of variable nodes. Bayesian decision networks can also be combined with combinatorial game theory as a solution for game trees.[7]

Bayesian decision networks can be abbreviated as decision diagrams or decision networks, also known as influence diagrams (Howard and Matheson 2005; Pearl 2005; Kjaerulff and Madsen 2008).

7.6 Markov Decision Models

Chapter 3 of this book has already discussed the Markov model.

---

[7] https://en.wikipedia.org/wiki/Game_tree.

Table 7.4 Two types of Markov processes with decision mechanisms

System state	Fully observable	Partially observable
Decision mechanism	Markov decision process	Partially observable Markov decision process

Markov decision models are Markov models with decision mechanisms. As shown in Table 7.4, according to the state of the stochastic process, it is divided into fully observable and partially observable, and the corresponding decision mechanisms are Markov decision processes and partially observable Markov decision processes.

Further, these two types of Markov processes with decision mechanisms are introduced separately.

7.6.1 Markov Decision Process

Chapter 3 introduced the Markov process and gave a definition (see Definition 3.12). The Markov decision process can be seen as a Markov process with decision mechanism, where adding actions that change the state, and rewards of the environment.

Following is the definition of the Markov decision process.

Definition 7.8 (Markov Decision Process) A Markov decision process (MDP) is a decision process that satisfies the Markov property and can be represented as a 4-tuple, MDP $= \langle S, A, P, R \rangle$. Where S is a state set, A is an action set, P is a transition probability, and R is a reward function.

In the aforementioned definition, the transition probability P is a conditional probability of next state $S(t+1) = s_{t+1}$, given current state $S(t) = s_t$ and current action $A(t) = a_t$, can be expressed as:

$$P(s_{t+1}|s_t, a_t) = \mathbb{P}[S(t+1) = s_{t+1}|S(t) = s_t, A(t) = a_t]. \tag{7.8}$$

Where $\mathbb{P}[\cdot]$ is a transition matrix, the same as in the Definition 3.12.

The reward function $R : S \times A \to R \in \mathbb{R}$, that is, in the state $S(t) = s_t$, action $A(t) = a_t$, and the next state $S(t+1) = s_{t+1}$ under the condition of obtaining rewards $R(t+1) = r_{t+1}$, expressed as:

$$\begin{aligned} r_{t+1} &= R(s_t, a_t) \\ &= \mathbb{E}[R(t+1) = r_{t+1} \mid S(t) = s_t, A(t) = a_t, S(t+1) = s_{t+1}]. \end{aligned} \tag{7.9}$$

The reward r_{t+1} comes from the environment, and the result is a real number, that is, $r_{t+1} \in \mathbb{R}$.

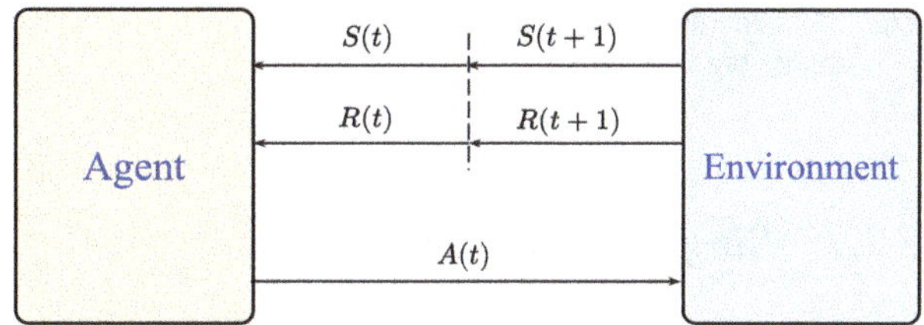

Fig. 7.7 Schematic diagram of the Markov decision process

Figure 7.7 is a schematic diagram of the state, reward, and action between the agent and the environment in the Markov decision process.

The Markov decision process is a decision theory framework for sequential decision in discrete-time stochastic control, and it is also the theoretical framework for the reinforcement learning paradigm in machine learning.

7.6.2 Partially Observable Markov Decision Process

The partially observable Markov decision process is an extension of the Markov decision process in a partially observable environment.

Definition 7.9 (Partially Observable Markov Decision Process) A partially observable Markov decision process (POMDP) can be represented as a 5-tuple, namely: POMDP $= \langle S,\ A,\ P,\ R,\ P_o \rangle$, where: P_o is the conditional observation function, and the other elements in the 5-tuple have the same definitions as in the Markov decision process.

The so-called conditional observation function P_o manifests as conditional observation probability, which is a conditional probability of next observation $O\,(t+1) = o_{t+1}$, given current action $A\,(t) = a_t$ and the next state $S\,(t+1) = s_{t+1}$. It is expressed as:

$$P_o\,(o_{t+1}|a_t,\ s_{t+1}) = \mathbb{P}\,[O\,(t+1) = o_{t+1} \mid A\,(t) = a_t, S\,(t+1) = s_{t+1}].$$
$$(7.10)$$

The conditional observation probability satisfies $\sum_{o \in O} P_o\,(o'|a,\ s') = 1$. At each time period, $P\,(s_{t+1}|s_t, a_t)$ causes changes in the environment state, and also through $P_o\,(o_{t+1}|a_t,\ s_{t+1})$ obtains an observation of the environment o_{t+1}.

7.7 Behavioral Decision Types

The behavioral decisions of intelligent agents can be divided into three types, namely: single-stage decision, multistage decision, and sequential decision.

Fig. 7.8 Schematic diagram
of a single-stage decision

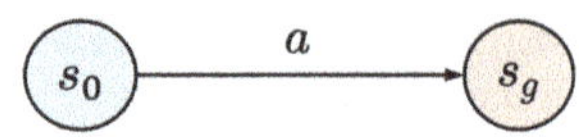

7.7.1 Single-Stage Decision

Single-stage decision is a simplest decision process in behavioral decisions.

Definition 7.10 (Single-Stage Decision) Given a decision process DP $=$ $\langle S, A, P \rangle$, let the initial state be $s_0 \in S$, and the target state be $s_g \in S$. Then the decision process is called single-stage decision, if taking an action $a \in A$, the initial state s_0 is transformed into the target state s_g with only one stage, that is $P\left(s_g | s_0, a\right)$.

A schematic diagram of a single-stage decision is as depicted in Fig. 7.8. The decision process of a single-stage decision has only one step, also known as a one-off decision.

The decision step here is a logical concept.

7.7.2 Multistage Decision

Multistage decision is composed of several interrelated single-stage decision, defined as follows.

Definition 7.11 (Multistage Decision) Given a decision process DP $= \langle S, A, P \rangle$, initial state $s_0 \in S$, the action of the ith step $a_i \in A$, the target state $S_g \subset S$. Then the decision process is called multistage decision, if its action space can be formed from the initial state to the target state, after several state transitions $P\left(s_{i+1} | s_i, a_i\right)$ can reach any target state from the initial state s_0 to $s_{i+1} = s_g \in S_g$.

The decision space in the aforementioned definition can be represented as a tree or a graph. And a decision forest composed of multiple decision trees also belongs to the generalized decision space.

A decision tree, as the name suggests, is a multistage decision method that uses a tree-like structure. The decision tree is also known as decision tree analysis, which is a universal predictive modeling method that covers multiple different fields.

The characteristics of the decision tree are simplicity, intuitiveness, and interpretability.

Generally speaking, a decision tree is built through algorithms and can identify and split datasets based on different conditions. The decision tree is a nonparametric machine learning algorithm and belongs to the supervised learning paradigm. Its goal is to build a model that infers simple decision rules to predict the value of the target state by analyzing data features.

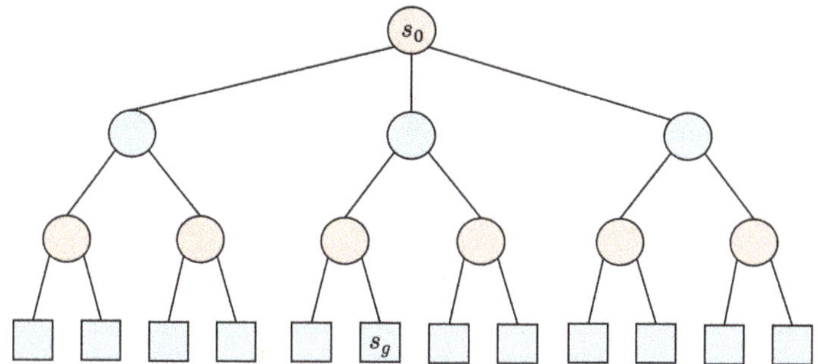

Fig. 7.9 Schematic diagram of the decision tree in multistage decision

A schematic diagram of a decision tree is as depicted in Fig. 7.9. The nodes represent decision states, the branches represent decision options, and the leaves represent the target states of the decision. Decision rules usually take the form of if-then-else statements. The deeper the tree of this type of model, the more complex the decision steps, and the more suitable it is for the problem to be predicted.

There are mainly two types of decision trees: one is called a classification tree, corresponding to the classification task in machine learning; the other is called a regression tree, corresponding to the regression task in machine learning. A classification tree is a decision tree whose target state can take discrete values; a regression tree is a decision tree whose target state can take continuous values (usually real numbers).

In addition to decision trees, decisions in graph space are called decision graphs, and decisions in multiple tree spaces are called decision forests, also known as random decision forests or random forests.

In addition to being used as an independent method, decision trees can be combined with deep neural networks to form deep neural decision trees (Yang et al. 2018). Similarly, decision forests can be combined with deep neural networks to build deep neural decision forests (Kontschieder et al. 2015).

7.7.3 Sequential Decision

Definition 7.12 (Sequential Decision) Given a Markov decision process, MDP $= \langle S, A, P, R \rangle$, initial state $s_0 \in S$, the action of ith step $a_i \in A$ and the reward of the environment $r_i \in R$, then according to each step's state and rewards, and taking its action, that is, $P(s_{i+1}|s_i, a_i, r_i)$, finally reaching the target state $s_{i+1} = s_g$, known as sequential decision or sequential decision-making.

A schematic diagram of sequential decision is as depicted in Fig. 7.10.

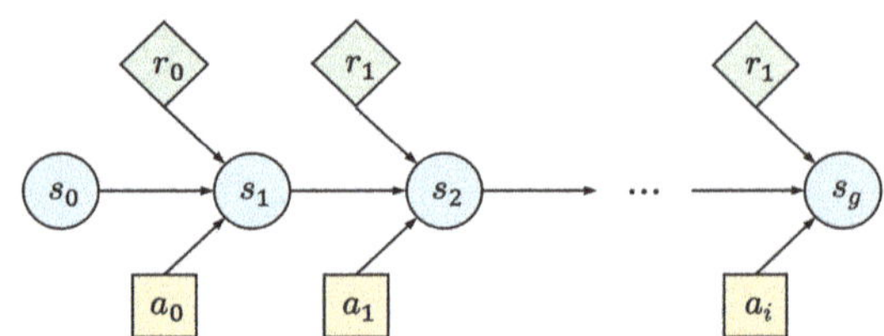

Fig. 7.10 Schematic diagram of sequential decision

Sequential decision and multistage decision are both procedural methods of step-by-step decision, but the differences between sequential decision and multistage decision are:

(1) Sequential decision is dynamic decision based on state and rewards, while multistage decision is static decision based on conditions.
(2) Sequential decision forms a chain-like one-dimensional decision space, while multi-stage decision is a tree-like or graph-like two-dimensional decision space.
(3) Sequential decision does not have a predetermined decision path, while multi-stage decision has a condition-determined decision path.

Reinforcement learning belongs to the machine learning paradigm of sequential decision, which will be discussed in Chap. 10.

Further Reading

1. Richard J. Herrnstein. "On the Law of Effect", *Journal of the Experimental Analysis of Behavior*, 13(2), 1970.
   [Notes] This paper discusses various correlations found between response rates and reinforcement rates in single, multiple, and concurrent reinforcement experiments.
2. Burrhus F. Skinner. The Behavior of Organisms: An Experimental Analysis. *BF Skinner Foundation*, 2019.
   [Notes] This book was originally published in 1938 by Burrhus F. Skinner, who is an American psychologist and behaviorist, known as the "father of operant conditioning". And it is reprinted in 2019 by BF Skinner Foundation.
3. Ward Edwards. "Behavioral decision theory." *Annual Review of Psychology*, 12(1), 1961.
   [Notes] This is a review paper published by Ward Edwards, who was a professor of psychology at Johns Hopkins University, known as the "father of behavioral decision theory".
4. Uffe B. Kjærulff, and Anders L. Madsen. "Bayesian Networks and Influence Diagrams: A Guide to Construction and Analysis." *Springer*, 2008.
   [Notes] This book mainly discusses the construction and analysis of probabilistic graphical networks such as Bayesian networks and influence diagrams.

References

Buckham, J. W. (1924). Behaviorism.

Edwards, W. (1954). The theory of decision making. *Psychological Bulletin* 51(4): 380.

Edwards, W. (1961). Behavioral decision theory. *Annual Review of Psychology* 12(1): 473–498.

Ferster, C. B., and B. F. Skinner. (1957). *Schedules of reinforcement*. Cambridge, MA: B. F. Skinner Foundation.

Fryback, D. (2005). Ward Edwards: father of behavioral decision theory. *Medical Decision Making* 25(4): 468–471.

Geman, S., E. Bienenstock, and R. Doursat. (1992). Neural networks and the bias/variance dilemma. *Neural Computation* 4(1): 1–58.

Gormezano, I., N. Schneiderman, E. Deaux, and I. Fuentes. (1962). Nictitating membrane: Classical conditioning and extinction in the Albino Rabbit. *Science* 138(3536): 33–34.

Herrnstein, R. J. (1970). On the law of effect. *Journal of the Experimental Analysis of Behavior* 13(2): 243–266.

Howard, R. A., and J. E. Matheson. (2005). Influence diagrams. *Decision Analysis* 2(3): 127–143.

Keren, G. (1996). Perspectives of behavioural descion making: some critical notes. *Organizational behavior and human decision processes* 65(3): 169–178.

Kjaerulff, U. B., and A. L. Madsen. (2008). *Bayesian networks and influence diagrams: A guide to construction and analysis*. New York: Springer.

Kontschieder, P., M. Fiterau, A. Criminisi, and S. R. Bulo. (2015). Deep neural decision forests. In *IEEE International Conference on Computer Vision (ICCV)*.

Pearl, J. (2005). Influence diagrams–historical and personal perspectives. *Decision Analysis* 2(4): 232–234.

Rescorla, R. A., and A. R. Wagner. (1972). A theory of Pavlovian conditioning: variations in the effectiveness of reinforcement and nonreinforcement. *Classical Conditioning II: Current Research and Theory* 2: 64–99.

Russell, S., and P. Norvig. (2009). *Artificial intelligence: A modern approach*, 3rd ed. Pearson Education.

Skinner, B. F. (1938). *The behavior of organisms: An experimental analysis*. Cambridge, MA: BF Skinner Foundation.

Thorndike, E. L. (1898). Animal intelligence: an experimental study of the associative processes in animals. *The Psychological Review: Monograph Supplements* 2(4).

Watson, J. B. (1913). Psychology as the behaviorist views it. *Psychological Review* 20(2): 158.

Yang, Y., I. G. Morillo, and T. M. Hospedales. (2018). Deep Neural Decision Trees. arXiv preprint arXiv:1806.06988.

Part III
Paradigms

In science and philosophy, a paradigm refers to a distinct set of unique concepts or thought patterns, including theories, methods, axioms, and norms, which are widely recognized and accepted to a field.

The historian and philosopher of science, Thomas S. Kuhn (1922–1996), elaborated on the paradigm of science in his book *The Structure of Scientific Revolutions*. The book was first published in 1962, and a 50th anniversary edition was released in 2012 (Kuhn 2012).

In the field of machine learning, several paradigms are widely recognized and accepted, which can be referred to as machine learning paradigms, or simply learning paradigms.

Regarding the term "paradigms" of machine learning, I have had conducted some investigation. Most literatures use the same term (Wiering and Otterlo 2012), but some ones refer to it as the types of machine learning (Lantz 2015) or the scenarios of machine learning (Mohri et al. 2012).

No matter what they are called types or scenarios of machine learning, their connotations are largely similar to the paradigms of machine learning.

There are three representative paradigms in machine learning, namely: supervised learning, unsupervised learning, and reinforcement learning; their schematic diagrams are depicted in following figure.

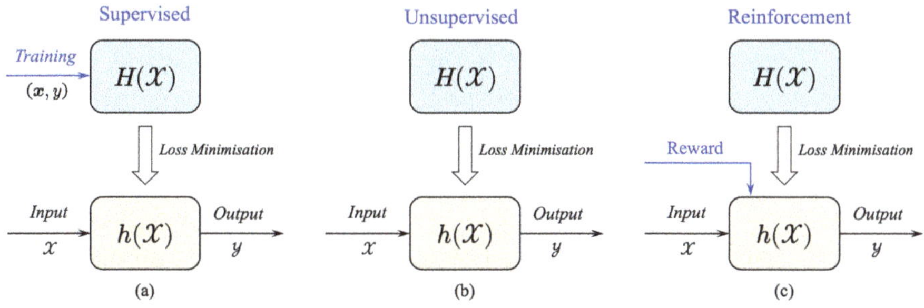

Moreover, some learning quasi-paradigms emerged, such as meta-learning, transfer learning, self-supervised learning, and n-shot learning. The reason why it is so called "quasi-paradigms" is because each of those quasi-paradigms has certain characteristics of paradigm but not yet formed an independent paradigm completely.

The importance of studying learning paradigms lies in the following two considerations: first, it can understand various machine learning algorithms from the viewpoint of paradigms, thereby deepening understanding at the level of machine learning methodology, and second, it can analyze the machine learning tasks that need to be solved from the viewpoint of paradigms, to better design machine learning algorithms and achieve the best machine learning results.

This part first discusses the three machine learning paradigms of supervised learning, unsupervised learning, and reinforcement learning in three chapters and then discusses several quasi-paradigms of machine learning in one chapter.

References

Kuhn, T. S. (2012). *The structure of scientific revolutions* (4th ed.). Chicago: University of Chicago Press.

Lantz, B. (2015). *Machine learning with R.* Birmingham: Packt Publishing.

Mohri, M., A. Rostamizadeh, and T. Ameet. (2012). *Foundations of machine learning.* Cambridge: MIT Press.

Wiering, M., and M. V. Otterlo. (2012). *Reinforcement learning: State-of-the-art.* Berlin: Springer Publishing Company.

Chapter 8
Supervised Learning Paradigm

Abstract Supervised learning is one of the three major paradigms of machine learning. This chapter begins from the definition of supervised learning and explains its working principle using formal and illustrated descriptions. Second, we introduce the classic tasks of supervised learning. Next, we respectively discuss the two important factors to guarantee the performance for a supervised learning algorithm: one is the bias-variance problems including bias-variance trade-off and bias-variance decomposition, and another one is the risk minimization principles, namely expected risk minimization, empirical risk minimization, and structural risk minimization. We then introduce several variants of supervised learning, such as weakly supervised learning, semi-supervised learning, and label-free supervision, where the label-free supervision is a supervised learning approach based on domain knowledge without manual labeling. Finally, we introduce the no free lunch theorems, which are related to machine learning.

8.1 Definition

Supervised learning is such a machine learning paradigm that is the longest-standing and most content-rich in the field of machine learning. Its application field is extremely wide, for example: image classification, optical character recognition (OCR), handwriting recognition, information retrieval, recommendation systems, spam mail detection, speech recognition, bioinformatics, and chemical biology.

The main feature of the supervised learning paradigm is to learn from a large amount of labeled data.

First of all, let us take a look at what supervised learning is.

Definition 8.1 (Supervised Learning) Supervised learning (SL) is a learning paradigm with labeled data that is used to train the learning algorithm to obtain the best learner called the hypothesis. This hypothesis is then used to predict unknown input data to obtain corresponding output results.

The hypothesis is also known as the model of supervised learning.

W. Wang, *Principles of Machine Learning*,
https://doi.org/10.1007/978-981-97-5333-8_8

As the name suggests, supervised learning is like having a teacher supervise the entire learning process.

The training data of the supervised learning paradigm are labeled. The labeling refers to marking the expected output value for each input data, thereby forming a data pair. For example: labeling an email as spam or non-spam, and labeling a handwritten Arabic number as the corresponding number.

8.2   Working Principle

This section first gives a formal description of supervised learning and then further explains it with an illustration.

8.2.1   Formal Description

Without loss of generality, we use a formalized method here to describe the working principle of supervised learning, omitting issues such as the vector dimension of the input space and parameterization.

Supervised learning (SL) can be represented as a 5-tuple, $SL = \langle \mathcal{X}, \mathcal{Y}, S, H, \mathcal{L} \rangle$. Where: $\mathcal{X}$ denotes the input space, $\mathcal{Y}$ denotes the output space, S is the labeled training sample set, H is the hypothesis set, and $\mathcal{L}$ is the loss function.

The formation process of the training sample set is as follows. Through a probability distribution $P(x)$ in the input space $\mathcal{X}$ to obtain n independent and identically distributed (i.i.d.) observation data, $D = \{x_i \mid x_i \in \mathcal{X} \text{ and } i = 1, \ldots, n\}$.

Let the following target function of supervised learning

$$f : \mathcal{X} \rightarrow \mathcal{Y} \tag{8.1}$$

be a conditional probability distribution $P(y|x)$, using it to label the corresponding result $y_i \in \mathcal{Y}$ for each observed data $x_i \in \mathcal{X}$.

Based on the joint probability distribution $P(x, y) = P(y|x) P(x)$, we obtain the data pair (x, y), forming the following labeled training sample set:

$$S = \{(x_i, y_i) \mid i = 1, \ldots, n\} \subseteq \mathcal{X} \times \mathcal{Y}. \tag{8.2}$$

The hypothesis function set H is a supervised learning algorithm designed for a specific task of supervised learning, and satisfies:

$$H : \mathcal{X} \rightarrow \mathcal{Y}. \tag{8.3}$$

Supervised learning is to use the labeled training sample set S to train the hypothesis function set H and obtain a hypothesis $h \in H$ with the smallest expected error, so that the predicted value $h(x) = \hat{y}$ closest to the target output in the training samples y.

The role of the loss function is to measure the error between the predicted value $\hat{y}$ of the hypothesis function $h(x)$ and the target output (i.e. the labeled result) y and to minimize its expected error, denoted as:

$$\mathcal{L}(h(x), y) = \mathcal{L}(\hat{y}, y) = \arg\min_{h \in H} \mathbb{E}[h(x) - y]. \tag{8.4}$$

The result is a real number. That is to say, the loss function $\mathcal{L}(\hat{y}, y)$ is a mapping from the Cartesian product of the output space to the set of real numbers $\mathbb{R}$, that is:

$$\mathcal{L}: \mathcal{Y} \times \mathcal{Y} \to \mathbb{R}. \tag{8.5}$$

The loss function depends on the task and algorithm involved in supervised learning, because different task and algorithm have different loss function. Here, the $\mathcal{L}(\hat{y}, y)$ is just an abstract representation of the loss function. The choice of the loss function is a determining factor in supervised learning for selecting the hypothesis function $h(x) = \hat{y}$. The loss function also affects the convergence speed of supervised learning.

8.2.2   Illustrated Description

A schematic diagram of the supervised learning process is dedicated in Fig. 8.1, which includes the training process of the hypothesis $h(x) \in H(\mathcal{X})$, and the process of using the trained hypothesis $h(\mathcal{X})$ to predict unknown data.

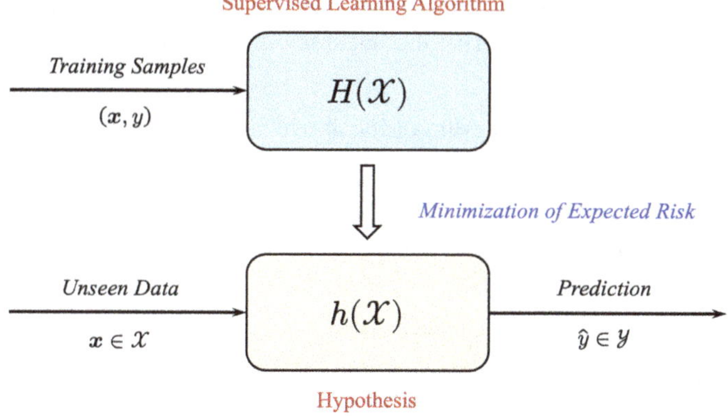

Fig. 8.1 Schematic diagram of supervised learning

8.2.3 Operational Steps

Supervised learning generally adopts the following steps:

(1) Collecting the datasets
     That is, collecting data and constructing corresponding datasets, including training dataset, validation dataset, and test dataset. For example, for handwriting recognition, the data are a set of images of handwritten characters or words. We may also choose a suitable public dataset.
(2) Labeling the training data
     Each piece of training data needs to be labeled with its output value, forming data pairs and thereby constructing a training dataset. The labeling of some training data (such as medical images) needs to be completed by professionals in the field.
(3) Determining the feature extraction approach
     There are usually two approaches to extracting features from input data, namely: hand-crafted features and learned features. The former uses a certain feature descriptor and its feature extraction algorithm to extract features manually, and the latter through deep neural network to learn features or representations automatically.
(4) Designing the supervised learning algorithm
     Design or choose a supervised learning algorithm to solve a given field-specific task. For example, for an image classification task, we can choose a support vector machine (SVM) classification algorithm, or a convolutional neural network.
(5) Training the algorithm
     Use the labeled training dataset to train the algorithm. We may set some specific parameters for the algorithm and then adjust and optimize these parameter values one by one with the labeled training data until its performance is optimized, thereby obtaining the optimal hypothesis.
(6) Evaluating the hypothesis
     To obtain an optimal hypothesis, we need to evaluate this hypothesis with the validation dataset to verify the accuracy and generalization ability of this hypothesis.

Figure 8.2 is a schematic diagram of the above six steps for supervised learning.

8.3 Classic Tasks

The classic tasks covered by the supervised learning paradigm mainly include three types, namely: classification, regression, and ranking.

Classification is the most common task in supervised learning, that is, for known categories, train the classification algorithm with a set of labeled data, obtain its

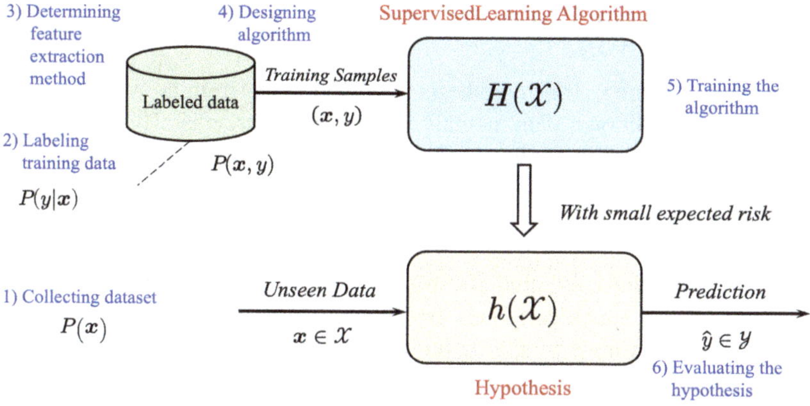

Fig. 8.2 Six steps for supervised learning

classification hypothesis, and then use it to predict the category of unknown data. The output result of classification is discrete categories.

Regression, also known as regression analysis, trains to obtain the linear or nonlinear relationship between input and output variables through a set of labeled data and then uses it to predict the output value of unknown data. The output result of regression is continuous values.

Ranking is to train through a set of labeled data to obtain the ordered relationship between input data and then use it to predict the relative relationship between unknown data.

The difference between the aforementioned three tasks is mainly reflected in the output space $\mathcal{Y}$. Therefore, based on the formalized description of supervised learning given in the previous section, we further formalize its output description:

- Classification:
  For this task, its output space $\mathcal{Y}$ is a set of known discrete categories, i.e., $\mathcal{Y} = \{c_1, c_2, \ldots, c_k\}$.
- Regression:
  For this task, the output space $\mathcal{Y}$ is continuous real numbers, i.e., $\mathcal{Y} \subseteq \mathbb{R}$.
- Ranking:
  For this task, the output space $\mathcal{Y}$ is a set with relative order after ranking, i.e., $\mathcal{Y} = \langle y, \preceq \rangle$, where: $y \in \mathcal{Y}$, and the symbol $\preceq$ is used to represent an ordered relationship.

Detailed introductions to classification and regression tasks and their representative algorithms will be given in Chaps. 12 and 13 of this book, respectively.

8.4   Bias-Variance Problem

In supervised learning, the hypothesis obtained after training will have errors when predicting unknown data, namely the bias and variance. These are a pair of parameters that can easily conflict, and these parameters have such a characteristic, that is: reducing bias will lead to an increase in variance, and vice versa.

This phenomenon is called the "bias-variance problem" or the "bias-variance dilemma".

It is worth emphasizing that the bias-variance problem is the core issue of supervised learning, specifically the core issue of classification and regression tasks in the supervised learning paradigm, but there is no such problem in the paradigms of unsupervised learning and reinforcement learning.

This section first gives the definitions of bias and variance, then discusses the bias-variance tradeoff and bias-variance decomposition. The joint probability distribution of generating training samples $P(x, y)$ is denoted as the target function $f(x, y)$ form, that is, $y = f(x)$.

8.4.1   Bias and Variance

8.4.1.1   Bias

The bias is an error caused by the erroneous hypothesis in supervised learning.

Definition 8.2 (Bias)   Bias, mathematically, is the difference between the expected value of prediction and the target value of the label, that is:

$$\text{Bias}\,[h\,(x)] = \mathbb{E}\,[h\,(x)] - y. \tag{8.6}$$

Bias is a measure of the accuracy of prediction derived from the hypothesis in supervised learning.

Since the average predicted value of the hypothesis converges to the expected value, it can be considered that the bias is approximately equal to the difference between the average predicted value and the expected value.

8.4.1.2   Variance

The variance is an error from sensitivity to small fluctuations in the training set.

Definition 8.3 (Variance)   Variance, mathematically, is the expectation of the square of the difference between the predicted value and the expected value of prediction, that is:

$$\text{Var}\,[h\,(x)] = \mathbb{E}\left[(h\,(x) - \mathbb{E}\,[h\,(x)])^2\right]. \tag{8.7}$$

Variance is a measure of the precision of prediction derived from the hypothesis in supervised learning.

Similarly, since the average predicted value of the hypothesis converges to the expected value, the variance can be seen as the degree to which the predicted value of the hypothesis deviates from its average predicted value.

8.4.2   Bias-Variance Tradeoff

The bias-variance trade-off is a comprehensive consideration of bias and variance, two parameters that are prone to conflict.

Analyzing the definitions of bias and variance given in the previous section, it can be seen that these two easily conflicting parameters have the following properties:

"Bias" is the error caused by the deviation of the predicted result of the hypothesis from the target. Algorithms with high bias are usually not sufficiently fitted to their training data, thus failing to accurately capture the regularity in the data, resulting in an "underfitting" problem. However, from another perspective, algorithms with moderately high bias usually generate simpler models.

"Variance" is the error caused by the sensitivity of the hypothesis to the noise in the training set. High variance can cause the hypothesis to pay too much attention to the fluctuations in the training data, resulting in an "overfitting" situation. That is, the hypothesis with high variance performs well on training data, but predictions on unknown data often have a higher error rate.

A bullseye diagram of bias-variance trade-off is depicted in Fig. 8.3.

Let's discuss the physical meaning of bias and variance first.

The y obtained by the target function $f(x) = y$ is the actual value labeled on the training data, like the bullseye; each predicted value $\hat{y}$ obtained by $h(x) = \hat{y}$, is like a predicted point on the target. Therefore, the bias is the distance that the predicted

Fig. 8.3 Bullseye of bias-variance trade-off

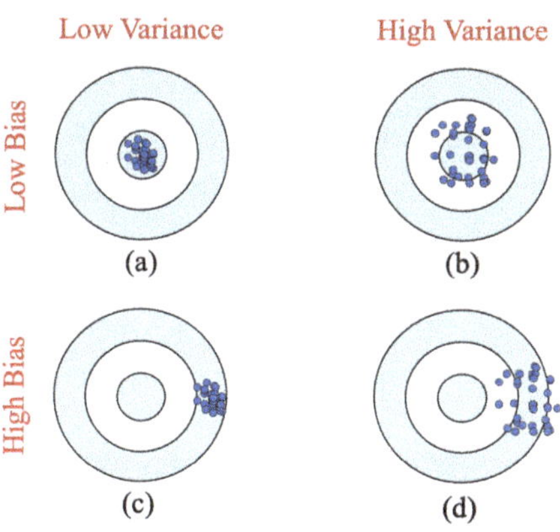

point deviates from the bullseye, and the variance is the degree of dispersion of the predicted point.

And, from Fig. 8.3, we can be seen:

- When the bias and variance are low, as shown in Fig. 8.3a, the prediction of unknown data is most accurate.
- When the bias is low and the variance is high, as shown in Fig. 8.3b, the predicted points of unknown data are scattered around the bullseye.
- When the bias is high and the variance is low, as shown in Fig. 8.3c, the predicted points of unknown data all deviate from the bullseye.
- When the bias and variance are high, as shown in Fig. 8.3d, the predicted points of unknown data are scattered away from the bullseye.

The bias-variance trade-off is one of the core issues in supervised learning. Ideally, we hope to train a hypothesis similar to Fig. 8.3a, which can accurately capture the regularity in its training data and generalize well to predict unknown data.

8.4.3 Bias-Variance Decomposition

Bias-variance decomposition was originally proposed for squared loss (Geman et al. 1992) and later used for zero-one loss (Friedman 1997). Some researchers have also proposed methods for bias and variance decomposition for any loss function, called unified bias-variance decomposition (Domingos 2000).

Bias-variance decomposition is a very useful and widely used tool for analyzing the expected generalization error of supervised learning models. Therefore, this book discusses it in this chapter on the supervised learning paradigm.

Mean squared error (MSE) is also called mean square deviation (MSD). Taking MSE as an example to introduce the problem of bias-variance decomposition of the hypothesis, there are following two cases.

(1) First, assume that there is no noise in the labeled training samples, that is, $y = f(x)$. Therefore:

$$\text{MSE}(x) = \mathbb{E}\left[(y - h(x))^2\right] = \mathbb{E}\left[(f(x) - h(x))^2\right]. \tag{8.8}$$

Decompose the aforementioned equation.

On the right side of the equation $\mathbb{E}\left[(f(x) - h(x))^2\right]$ subtract and add $\mathbb{E}[h(x)]$, the equation remains unchanged:

$$\mathbb{E}\left[(f(x) - h(x))^2\right] = \mathbb{E}\left[(f(x) - \mathbb{E}[h(x)] + \mathbb{E}[h(x)] - h(x))^2\right]$$

$$= \mathbb{E}\left[((f(x) - \mathbb{E}[h(x)]) + (\mathbb{E}[h(x)] - h(x)))^2\right].$$

At this point, the expectation $\mathbb{E}[\cdot]$ has changed from the square of the difference to the square of the sum. By applying the equation of the square of the sum and further expanding it, we have:

$$\mathbb{E}\left[(f(x) - h(x))^2\right] = \mathbb{E}\left[(f(x) - \mathbb{E}[h(x)])^2\right] + \mathbb{E}\left[(\mathbb{E}[h(x)] - h(x))^2\right] +$$
$$2\mathbb{E}\left[(f(x) - \mathbb{E}[h(x)])(\mathbb{E}[h(x)] - h(x))\right]$$
$$= \mathbb{E}\left[(f(x) - \mathbb{E}[h(x)])^2\right] + \mathbb{E}\left[(\mathbb{E}[h(x)] - h(x))^2\right] +$$
$$2\mathbb{E}\left[f(x)\mathbb{E}[h(x)]\right] - 2\mathbb{E}\left[\mathbb{E}[h(x)]^2\right] -$$
$$2\mathbb{E}\left[f(x)h(x)\right] + 2\mathbb{E}\left[h(x)\mathbb{E}[h(x)]\right]$$
$$= \mathbb{E}\left[(f(x) - \mathbb{E}[h(x)])^2\right] + \mathbb{E}\left[(\mathbb{E}[h(x)] - h(x))^2\right] +$$
$$2\Big(f(x)\mathbb{E}[h(x)] - \mathbb{E}[h(x)]^2 - f(x)\mathbb{E}[h(x)]$$
$$+ \mathbb{E}[h(x)]^2\Big).$$

The third term on the right side of the above expression is zero. Applying the bias Eq. 8.6 and the variance Eq. 8.7, we get:

$$\mathbb{E}\left[(y - h(x))^2\right] = E\left[(f(x) - E[h(x)])^2\right] + E\left[(E[h(x)] - h(x))^2\right]$$
$$= (\text{Bias}[h(x)])^2 + \text{Var}[h(x)].$$

(2) Secondly, let the training sample set S contain noise ϵ, therefore $y = f(x) + \epsilon$. Omitting the intermediate derivation steps, the following equation holds:

$$\mathbb{E}\left[(y - h(x))^2\right] = \mathbb{E}\left[(f(x) + \epsilon - h(x))^2\right]$$
$$= \mathbb{E}\left[(f(x) - \mathbb{E}[h(x)])^2\right] + \mathbb{E}\left[(\mathbb{E}[h(x)] - h(x))^2\right]$$
$$+ \mathbb{E}\left[\epsilon^2\right]$$
$$= (\text{Bias}[h(x)])^2 + \text{Var}[h(x)] + \text{Var}(\epsilon). \tag{8.9}$$

Where, $\text{Var}(\epsilon) = \sigma^2$, is known as the Irreducible Error.

Combining the aforementioned equations, the mean squared error (MSE) can be expressed in the following form:

$$\text{MSE}(x) = (\text{Bias}[h(x)])^2 + \text{Var}[h(x)] + \sigma^2. \tag{8.10}$$

Fig. 8.4 Bias-variance decomposition

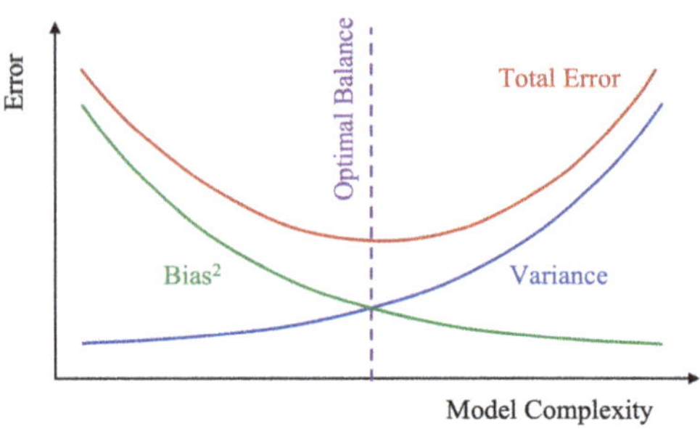

The mean squared error (MSE) can also be expressed as:

$$\text{Total Error} = \text{Bias}^2 + \text{Variance} + \text{Irreducible Error}. \tag{8.11}$$

The meaning of the aforementioned expression is clear and easy to understand, that is: the total error equals the sum of the three parameters, i.e., square of the bias, variance, and the irreducible error.

A schematic diagram of Eq. 8.11 is shown in Fig. 8.4, revealing the relationship between the complexity of the supervised learning model and the total error, square of bias, and variance.

From this, it can be seen that to build a better model of supervised learning, a suitable balance point between bias and variance must be found to minimize the total error. The intersection point between the dotted line of optimal balance and the solid line of total error in Fig. 8.4 is the optimal result. A model that satisfies the optimal balance between bias and variance will not have the problems of underfitting or overfitting.

In conclusion, the issue of bias and variance is one of the important factors in understanding the supervised learning paradigm.

8.5   Risk Minimization Principles

The risk minimization principles are also the important factors to consider in the supervised learning paradigm, which provide the theoretical performance boundaries for a supervised learning algorithm.

The risk minimization principles discussed in this section include expected risk minimization, empirical risk minimization, and structural risk minimization.

8.5.1 Expected Risk Minimization

Definition 8.4 (Expected Risk)   The expected risk of a hypothesis $h \in H$ is represented by risk function $R(h)$, which means the expected risk of loss function $\mathcal{L}(h(\mathbf{x}), y)$, that is:

$$R(h) = \mathbb{E}_{X,Y}[\mathcal{L}(h(\mathbf{x}), y)] = \int_{X,Y} \mathcal{L}(h(\mathbf{x}), y)\, p(\mathbf{x}, y)\, d\mathbf{x}dy. \tag{8.12}$$

Expected risk is also known as *expected error* or *generalized error*.

The expected risk is used to determine whether the predicted value of the hypothesis $h(\mathbf{x})$ is closest to the target value y.

For instance, if we consider the classification tasks in supervised learning, its expected risk is used to calculate the expectation that $h(\mathbf{x})$ is not equal to y, i.e.,

$$R(h) = \mathbb{E}_{X,Y}[\mathbb{I}(h(\mathbf{x}) \neq y)] = \int_{X,Y} \mathbb{I}(h(\mathbf{x}) \neq y)\, p(\mathbf{x}, y)\, d\mathbf{x}dy.$$

Where, $\mathbb{I}(\omega)$ is the indicator function for the expression ω, which is defined as follows:

$$\mathbb{I}(\omega) = \begin{cases} 1 & \text{if } \omega = \text{true} \\ 0 & \text{otherwise} \end{cases}$$

For the classification tasks, $\mathbb{I}(h(\mathbf{x}) \neq y) = 1$ if $h(\mathbf{x}) \neq y$ is true, else 0.

Definition 8.5 (Expected Risk Minimization)   Expected risk minimization refers to finding an optimal hypothesis h^* in the hypothesis set H through training, so that its expected risk $R(h)$ is minimized, that is:

$$h^* = \arg\min_{h \in H} R(h). \tag{8.13}$$

Typically, for a supervised learning algorithm, the loss function $\mathcal{L}(h(\mathbf{x}), y)$ is known, however, in the joint probability distribution $p(\mathbf{x}, y) = p(y|\mathbf{x})\, p(\mathbf{x})$, both of $p(\mathbf{x})$ and $p(y|\mathbf{x})$ are true but unknown, therefore it is impossible to calculate $p(\mathbf{x}, y)$, and thus it is also impossible to calculate its expected risk:

$$R(h) = \int_{X,Y} \mathcal{L}(h(\mathbf{x}), y)\, p(\mathbf{x}, y)\, d\mathbf{x}dy.$$

From this, it can be inferred that the minimized expected risk h^* is just a theoretical value.

However, it is possible to calculate its empirical risk and structural risk through the loss function $\mathcal{L}(h(x), y)$, which involves the problem of minimizing these two types of risks.

Empirical risk minimization and structural risk minimization have become two important principles of risk minimization (Vapnik 1991), used to evaluate two different aspects of the performance of supervised learning algorithms. This will be elaborated in the following two subsections.

8.5.2 Empirical Risk Minimization

The core idea of empirical risk minimization is that we cannot precisely know the actual working condition of a supervised learning algorithm (expected risk) because we do not know the true probability distribution of the data processed by the algorithm, but we can compute its approximation, called empirical risk, on the set of labeled training data.

Definition 8.6 (Empirical Risk) For a given training sample set, $S = \{(x_i, y_i) \mid i = 1, \ldots, n\}$, its empirical risk is denoted as $R_{emp}(h)$, then its computing equation is as follows:

$$R_{emp}(h) = \frac{1}{n} \sum_{i=1}^{n} \mathcal{L}(h(x_i), y_i). \tag{8.14}$$

Empirical risk is also known as *empirical error*.

How to reduce empirical risk is one of the important issues in supervised learning.

Definition 8.7 (Empirical Risk Minimization) Empirical risk minimization (ERM) is also known as the principle of empirical risk minimization, which means that the learning algorithm should choose a hypothesis that minimizes empirical risk h_{ERM}, that is:

$$h_{ERM} = \arg \min_{h \in H} R_{emp}(h). \tag{8.15}$$

The principle of empirical risk minimization is abbreviated as the ERM principle (Vapnik 1991).

Therefore, the supervised learning algorithm based on the ERM principle is to solve the above problem of empirical risk minimization (Fig. 8.5).

Fig. 8.5 Schematic diagram
of empirical risk
minimization

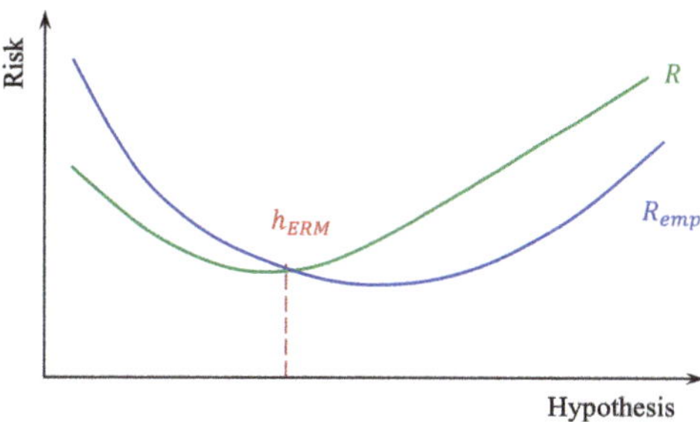

The performance of a supervised learning algorithm that satisfies the ERM
principle can be guaranteed to meet the following uniform deviation bounds:

$$P\left(\sup_{h\in H}\left|R_{emp}\left(h\right)-R\left(h\right)\right|\geq\epsilon\right)\leq\delta.\tag{8.16}$$

Where, the mathematical symbol sup $(\cdot)$ denotes the supremum.

For all $n\geq 1$ and $\epsilon>0$, the relationship between the uniform deviation bounds
δ and the growth function $G_H\left(n\right)$ discussed in Chap. 4 is as follows:

$$P\left(\sup_{h\in H}\left|R_{emp}\left(h\right)-R\left(h\right)\right|\geq\epsilon\right)\leq\delta=8G_H\left(n\right)e^{\frac{-n\epsilon^2}{32}}.\tag{8.17}$$

This means the generalization bounds.

8.5.3   *Structural Risk Minimization*

Definition 8.8 (Structural Risk)   In supervised learning, it is necessary to train
a generalized model based on a finite training dataset. The problem that arises is
overfitting, that is, the model is too suitable for the specificity of the training dataset,
and the generalization effect on new unknown data is poor. This is the so-called
structural risk.

Definition 8.9 (Structural Risk Minimization)   The principle of structural risk
minimization (SRM) is to solve the problem of structural risk by balancing the
complexity of the model and the fit of the training data. This is an inductive principle
in supervised learning.

Structural risk introduces a sequence of hypothesis sets with a nested structure:

$$H_1\subset H_2\subset\cdots\subset H_m\subset\cdots.$$

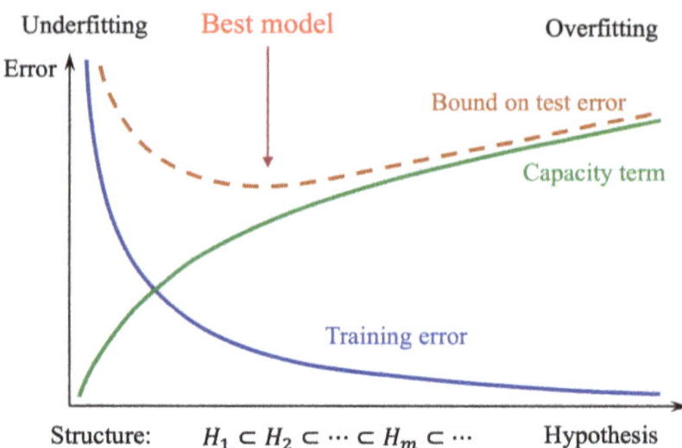

Fig. 8.6 Schematic diagram of structural risk minimization

Given a training sample set $S = \{(\boldsymbol{x}_i, y_i)\ i = 1, \ldots, n\}$, for each $h \in H$, and $m \in \mathbb{N}$, there exists the structural risk $R_{str}(h)$ as follows:

$$R_{str}(h) = \frac{1}{n} \sum_{i=1}^{n} \mathcal{L}(h(\boldsymbol{x}_i), y_i) + \text{pen}(m, n) = R_{emp}(h) + \text{pen}(m, n). \qquad (8.18)$$

Where, $\text{pen}(\cdot, \cdot)$ denotes the penalty function of complexity, which is used to preferentially select models with smaller estimation errors and measure the capacity or size of the model.

The usual practice is to choose a large model H and define the regularization matrix of H, which is usually denoted as a norm, hence:

$$R_{str}(h) = R_{emp}(h) + \lambda \|h\|^2. \qquad (8.19)$$

The λ in the aforementioned equation is a regularization parameter, used to make an appropriate trade-off between the model's generalization and complexity (Fig. 8.6).

Structural risk minimization is for each H_m to find the smallest solution $h_{ERM}(\boldsymbol{x})$ in structural risk $R_{str}(h)$, that is:

$$h_{ERM}(\boldsymbol{x}) = \arg \min_{h \in H} [R_{str}(h)]. \qquad (8.20)$$

8.6 Variants of Supervised Learning

The supervised learning paradigm is the oldest and most widely used paradigm in the machine learning field.

However, supervised learning requires a large amount of labeled training data for the training of supervised learning models, and the accuracy and precision of a model depend on the quantity and quality of the training samples.

The labeling of training samples has the following problems:

- The labeling is basically done manually, usually referred to as manual labeling or hand labeling. For thousands or tens of thousands of training samples, the time-consuming, labor-intensive, and large workload of manual labeling is self-evident.
- The labeling work is often done by domain experts, for example, the training samples for medical image segmentation require medical imaging experts to label.

Researchers in the field of machine learning are increasingly realizing that humans can often gain some universal knowledge from a small amount of data, so machine learning not only needs to solve the problem of learning from big data but also needs to solve the problem of learning from limited labeled data.

Over the years, some variants have gradually evolved based on the supervised learning paradigm, including weakly supervised learning, semi-supervised learning, and even label-free supervised learning. The differences between these variants mainly lie in the quality or quantity of training samples, or else the training with domain knowledge.

The following will be introduced separately.

8.6.1 Weakly Supervised Learning

Weakly supervised learning, also known as weak supervision, uses low-quality, small-quantity, so-called "weak" labeled training samples to train machine learning models.

The main reasons for the appearance of weakly labeled training samples are as follows:

- The quantity of labeled training data is limited.
- The accuracy of labeled training data is not good.
- The technicians who do the labeling work lack specialist experience.

Therefore, weakly supervised learning can be divided into the following three types (Zhou 2018):

(1) Incomplete supervision:
    Only a part of the training data is labeled, which is not enough to carry out complete supervised learning training for the machine learning model.
(2) Inexact supervision:
    The labeling of training data is rough and lacks accuracy, which is called coarse-grained labels.

(3) Inaccurate supervision:

The technicians who do the labeling of training data have limited experience, causing some labels to be wrong and cannot be used as the real basis for training, making it difficult to become the ground truth.

Despite this, without the premise of having "strong" labeled training samples like supervised learning, weakly supervised learning is a worthwhile path to explore. From another viewpoint, weakly supervised learning does not strictly require labeled training samples, which can objectively greatly reduce the cost of manual labeling data.

8.6.2   Semi-Supervised Learning

The so-called semi-supervised learning refers to training using a large amount of unlabeled data in addition to a small amount of labeled data. That is to say, semi-supervised learning is between supervised learning (using labeled training data) and unsupervised learning (using unlabeled training data); hence, it is called semi-supervised learning.

In fact, semi-supervised learning was originally a human learning method, and its use in machine learning naturally has its theoretical and practical value.

Due to the related costs of labeling data, it is very difficult to obtain a sufficient number of accurately labeled training datasets, while obtaining unlabeled data does not involve the problem of labeling costs.

Many experiments have shown that combining a large amount of unlabeled data with a small amount of labeled data to train machine learning models can significantly reduce the cost of labeled data and significantly improve the accuracy of the model. In this case, semi-supervised learning has great practical value.

Semi-supervised learning can be seen as a special case of weakly supervised learning. Conversely, weakly supervised learning can be seen as a generalization of semi-supervised learning.

8.6.3   Label-Free Supervised Learning

At the *Thirty-First AAAI Conference on Artificial Intelligence* (AAAI) held in San Francisco, USA in February 2017, a paper titled "Label-Free Supervision of Neural Networks with Physics and Domain Knowledge" (Stewart and Ermon 2017) won the Best Paper Award. This academic conference is one of the top conferences in the field of artificial intelligence research.

The novelty of this paper lies in supervising the training of neural networks with specific constraint data in the output space, instead of using labeled data; these constraint data are derived from prior domain knowledge, such as known physical

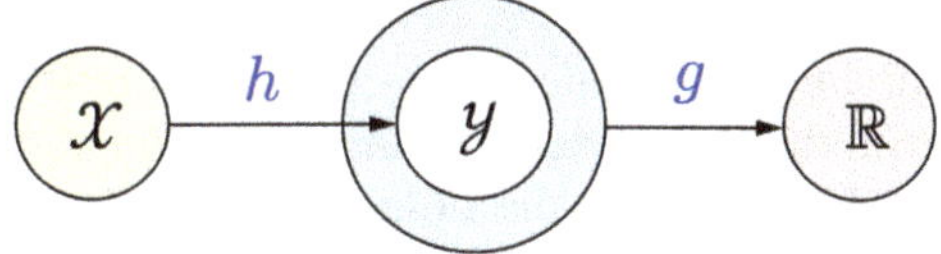

Fig. 8.7 Label-free supervised learning through output space constraints

laws. By finding a mapping h to capture the structure required by g, it is possible to obtain the mapping function h without providing the label $\mathcal{Y}$. As shown in Fig. 8.7.

In traditional supervised learning, a hypothesis function $h : \mathcal{X} \rightarrow \mathcal{Y}$. To quantify performance, a loss function needs to be provided $\mathcal{L} : \mathcal{X} \times \mathcal{Y} \rightarrow \mathbb{R}$.

A hypothesis function is calculated through the following loss function h^*:

$$h^* = \arg\min_{h \in H} \sum_{i=1}^{n} \mathcal{L}\left(h\left(\boldsymbol{x}_i\right), y_i\right). \tag{8.21}$$

Where the optimization is within the predefined set of hypothesis functions H is performed on. For the example given in this paper, H is a neural network set by its weights.

It should be pointed out that, in supervised learning, there is a type of problem called structured prediction, that is, predicting structured objects, not predicting scalar discrete or real values. In other words, the output space of structured prediction is complex object space, not simple scalar, or real spaces. For example, for image classification problems, the input space $\mathcal{X}$ is complex images, and the output space $\mathcal{Y}$ is simple binary or multiclass discrete values. However, for structured prediction problems, both the input space $\mathcal{X}$ and the output space $\mathcal{Y}$ are complex. Structured prediction is widely used in bioinformatics, natural language processing, speech recognition, and computer vision.

This paper proposes a solution to the structured prediction problem in computer vision.

If prior knowledge has shown that the output of h^* has unique properties in the function set in H, these properties can be used to train the machine learning algorithm without directly providing labeled samples y. Specifically, consider a supervised method that does not provide labels y, but optimizes the necessary properties of the output g. Hence, the following equation:

$$\hat{h}^* = \arg\min_{h \in H} \sum_{i=1}^{n} g\left(\boldsymbol{x}_i, h\left(\boldsymbol{x}_i\right)\right) + R\left(h\right). \tag{8.22}$$

Where, $g : \mathcal{X} \times \mathcal{Y} \rightarrow \mathbb{R}$ is the constraint function, and $R : H \rightarrow \mathbb{R}$ is the regularization term.

By using stochastic gradient descent to optimize Eq. 8.22, it can eliminate the need for labeled samples.

In this regard, the paper describes an experiment tracking a free-falling object: the input is a video of an object being thrown into the air and then falling freely, with the aim of determining the height of the object in each video frame.

According to the domain knowledge of physics, the height of the object is a parabola that changes with time, that is:

$$y_i = y_0 + v_0\,(i\,\Delta t) + a\,(i\,\Delta t)^2 . \tag{8.23}$$

Subtract the fixed acceleration term, fit a curve corresponding to the initial height and speed parameters, and then add the acceleration component. The constraint loss is:

$$g\,(x, h\,(x)) = g\,(h\,(x)) = \sum_{i=1}^{N} \left| \hat{y}_i - h\,(x)_i \right| . \tag{8.24}$$

The paper also presents two other experiments: tracking the position of a walking man, and detecting objects with causal relationships. Through these experiments, it is verified the effectiveness of this method in simulated computer vision tasks.

In the absence of any labeled samples, this paper proposes a new approach to structured prediction in computer vision, i.e., successfully training convolutional neural networks based on physics and domain knowledge.

8.7 No Free Lunch Theorems

There are such theorems in machine learning that are no free lunch (NFL) theorems. The theorems attempt to reveal the fundamental relationships between the functionalities and the performances of machine learning algorithms, as implied by the popular adage

"No such thing as a free lunch."

8.7.1 Origin of the Theorems

The no free lunch theorems were first proposed by David H. Wolpert and William G. Macready in a technical report (SFI-TR-95-02-010) at the Santa Fe Institute in the United States in 1995 for search problems, and its name is "No Free Lunch Theorems for Search" (Wolpert and Macready 1995).

In 1996, David Wolpert published a paper titled "The Lack of A Priori Distinctions between Learning Algorithms", in which he discussed the No Free Lunch Theorems for learning algorithms (Wolpert 1996).

In 1997, David Wolpert and William Macready further elaborated on these theorems for optimization algorithms and published a paper titled "No Free Lunch Theorems for Optimization" (Wolpert and Macready 1997).

In 2001, David Wolpert published a paper titled "The Supervised Learning No-Free-Lunch Theorems" specifically for supervised learning (Wolpert 2001).

In 2005, David Wolpert and William Macready published a paper titled "Coevolutionary Free Lunches" (Wolpert and Macready 2005).

In 2013, David Wolpert published a paper titled "What the No Free Lunch Theorems Really Mean" (Wolpert 2013).

8.7.2   Meaning of the Theorems

In the metaphor of "no free lunch", each "restaurant" (problem-solving process) has a menu listing each "dish" (problem) and its "price" (performance of the problem-solving process). Suppose each restaurant's menu is exactly the same, only the prices are different. For a customer who likes to order every dish, the average cost of dining does not depend on the choice of restaurant. However, if a vegetarian and a meat-eater dine on a Dutch treat, the former's cost will increase. To reduce the average cost, it is necessary to know in advance: (i) what dishes will be ordered, (ii) in which restaurant these dishes will be ordered. That is to say, the performance of problem-solving depends on using prior information to match the process with the problem.

Formally speaking, when the probability distribution of problem instances makes all problem-solving results the same, there is no free lunch. For search problems, the problem instance is a target function, and the result is a series of values obtained when evaluating candidate solutions within the function domain. The standard interpretation of the result is that the search is an optimization process. There is no free lunch in the search only when the distribution of the target function remains unchanged under the space permutation of candidate solutions. This condition is not entirely true in practice, but the no free lunch theorems show that they are approximately true.

8.7.3   Two Sub-Theorems

Let R denote the generalization error, S denote the training set, m denote the number of elements (i.e., labeled input–output pairs) in the training set, f denote the target function that maps inputs to outputs, and h denote the hypothesis (i.e., the function approximating f obtained after training the algorithm).

The no free lunch theorems are composed of two sub-theorems as follows.

Fig. 8.8 The error between hypothesis h and target function f

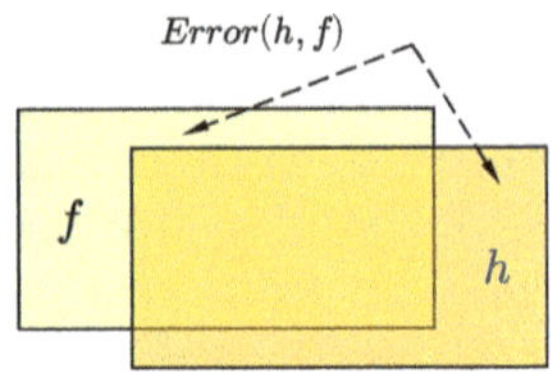

Theorem 8.1 (Expectation of Algorithm) *The expectation* $\mathbb{E}[R|S]$ *can be expressed as a (non-Euclidean) inner product between the distributions* $P(h|S)$ *and* $P(f|S)$:

$$\mathbb{E}[R|S] = \sum_{h,f} \text{Error}(h, f)\, P(h|S) P(f|S). \tag{8.25}$$

Where: $\mathbb{E}[\cdot]$ denotes the expectation, $\text{Error}(h, f) = (h - f) \cup (f - h)$ is the error between the hypothesis h and the target function f, see Fig. 8.8.

The result of Eq. 8.25 also applies to $\mathbb{E}[R|m]$, $\mathbb{E}[R|f, S]$, and $\mathbb{E}[R|f, m]$.

Regarding the aforementioned theorem, David Wolpert pointed out that the result depends on how well the alignment between the learning algorithm $P(h|S)$ and the actual posterior probability $P(f|S)$. This leads people to ask questions like: Which set of posteriors makes algorithm G_1 better than algorithm G_2?" This also means that unless $P(f|S)$ can be proven to have a certain form, it is impossible to prove that a specific $P(h|S)$ and $P(f|S)$ are completely aligned, so it is impossible to prove anything about how good the generalization of the learning algorithm is.

Theorem 8.2 (Expectation of Two Algorithms Are Equivalent) *Consider the error function off-training-set. Let* $\mathbb{E}_i[\cdot]$ *denote the expectation obtained by the ith learning algorithm. For any two learning algorithms* $P_1(h|S)$ *and* $P_2(h|S)$, *independent of the sampling distribution:*

(a) Uniformly averaged over all f, and for a training set of m elements, $\mathbb{E}_1[R|f, m] - \mathbb{E}_2[R|f, m] = 0$;
(b) Uniformly averaged over all f, and for any training set S, $\mathbb{E}_1[R|f, S] - \mathbb{E}_2[R|f, m] = 0$;
(c) Uniformly averaged over all $P(f)$, and for the training set with m elements, $\mathbb{E}_1[R|m] - \mathbb{E}_2[R|m] = 0$;
(d) Uniformly averaged over all $P(f)$, and for any training set S, $E_1[R|S] - E_2[R|S] = 0$.

In other words, for what is usually referred to as risk $E[R|S]$, $\mathbb{E}[R|m]$, $\mathbb{E}[R|f, S]$, or $\mathbb{E}[R|f, m]$ by any measure, the average performance of all algorithms is equal. In other words, for any two learning algorithms, in some cases algorithm A is superior to algorithm B, and vice versa in other cases; overall, they are equal.

By the way, there is a different view on the no free lunch theorems in the machine learning community. The reference (Giraud-Carrier and Provost 2005) points out that under some very simple and basic assumptions, the no free lunch theorems have almost nothing to do with the research of machine learning.

Table 8.1 Applied scope targeted by the no free lunch theorems

Category	Applied scope	When
Search	Search problems	1995
Optimization	Optimization problems	1997
Evolution	Coevolution problems	2005
Learning	Learning algorithms	1996
	Supervised learning	2001

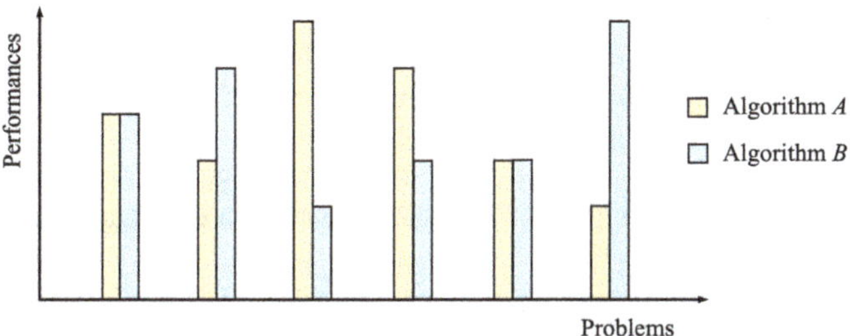

Fig. 8.9 Schematic diagram of no free lunch theorems

8.7.4 Applied Scopes

In summary, the applied scopes targeted by the no free lunch theorems, as shown in Table 8.1, can be divided into two parts: one is for search, optimization, and evolution problems, and the other is for machine learning.

The no free lunch theorems for search and optimization problems applies to finite spaces and non-resampling algorithms. The conclusion of David Wolpert and William Macready is that when any two search or optimization algorithms are averaged over all possible problems, their results are equivalent. That is to say, if the performance of algorithm A is better than algorithm B in one type of problem, it will be the opposite in another type of problem, and their average performance is equivalent, as shown in Fig. 8.9. There is no algorithm that performs best on all possible problems, there is no free lunch.

The no free lunch theorems for machine learning, specifically for supervised learning, refers to the case where the loss function is the misclassification rate. If the focus is on off-training-set error, there is no a priori difference between learning algorithms.

Further Reading

1. Vladimir N. Vapnik. "Principles of Risk Minimization for Learning Theory." *Conference on Neural Information Processing Systems* (NIPS), 1991.
   [Notes] This paper proposed the principles of empirical risk minimization and structural risk minimization, as well as illustrated the systematic improvements in prediction power in the application to zip-code recognition.

2. Zhihua Zhou. "A Brief Introduction to Weakly Supervised Learning." *National Science Review*, 5(1) 2017.

   [Notes] This article provided a detailed survey of weakly supervised learning, in which, weakly supervised learning is divided into three types, namely incomplete supervision, inexact supervision, and inaccurate supervision.

3. Jesper van Engelen, and Holger Hoos. "A Survey on Semi-Supervised Learning." *Machine Learning*, 109(2), 2020.

   [Notes] This article provided a detailed survey of semi-supervised learning. It proposed a taxonomy of semi-supervised classification algorithms, which sheds light on the different conceptual and methodological approaches for incorporating unlabeled data into the training process.

4. Russell Stewart, and Stefano Ermon. "Label-Free Supervision of Neural Networks with Physics and Domain Knowledge." *AAAI Conference on Artificial Intelligence* (AAAI). 2017.

   [Notes] This is the paper presented at AAAI 2017 and won the Best Paper Award. It has been introduced briefly in Sect. 8.6.3 of this chapter.

5. David H. Wolpert. "The Supervised Learning No-Free-Lunch Theorems." *Soft Computing and Industry*, 2002.

   [Notes] This paper discussed the no free lunch theorems in the context of supervised learning in machine learning.

References

Domingos, P. (2000). A unifeid bias-variance decomposition and its applications. In *International Conference on Machine Learning (ICML)*.

Friedman, J. H. (1997). On bias, variance, 0/1-loss, and the curse-of-dimensionality. *Data Mining and Knowledge Discovery* 1(1): 55–77.

Geman, S., E. Bienenstock, and R. Doursat. (1992). Neural networks and the bias/variance dilemma. *Neural Computation* 4(1): 1–58.

Giraud-Carrier, C., and F. Provost. (2005). Toward a justification of meta-learning: Is the no free lunch theorem a show-stopper. In *ICML-2005 Workshop on Meta-learning*.

Stewart, R., and S. Ermon. (2017). Label-free supervision of neural networks with physics and domain knowledge. In *Thirty-First AAAI Conference on Artificial Intelligence (AAAI-17)*.

Vapnik, V. N. (1991). Principles of risk minimization for learning theory. In *Conference on Neural Information Processing Systems (NIPS)*.

Wolpert, D. H. (1996). The lack of a priori distinctions between learning algorithms. *Neural Computation* 8(7): 1341–1390.

Wolpert, D. H. (2001). The supervised learning no-free-lunch theorems. In *World Conference on Soft Computing*.

Wolpert, D. H. (2013). *What the no free lunch theorems really mean: how to improve search algorithms*. Ubiquity, an ACM publication.

Wolpert, D. H., and W. G. Macready. (1995). *No free lunch theorems for search*. Citeseer.

Wolpert, D. H., and W. G. Macready. (1997). No free lunch theorems for optimization. *IEEE Trans on Evolutionary Computation* 1(1): 67–82.

Wolpert, D. H., and W. G. Macready. (2005). Coevolutionary free lunches. *IEEE Transactions on Evolutionary Computation* 9(6): 721–735.

Zhou, Z. H. (2018). A brief introduction to weakly supervised learning. *National Science Review* 5(1): 44–53.

Chapter 9
Unsupervised Learning Paradigm

Abstract Unsupervised learning is second one of the three major paradigms of machine learning. We first give the definition of unsupervised learning, explain its working principle using formal and illustrated descriptions, and introduce the classic tasks of unsupervised learning. Due to the outstanding ability of humans in unsupervised learning, generative models beyond the classic unsupervised learning have become the important topics. In this regard, we focus on the introduction to the generative models proposed in recent years, namely energy models, autoencoding models, generative adversarial models, normalizing flow models, autoregressive models, and diffusion models. It can be thought that the generative models have given birth to large language models.

9.1 Definition

Unsupervised learning is a learning paradigm that is fundamentally different from the supervised learning paradigm.

So far, many researchers have proposed several unsupervised learning tasks and constructed various unsupervised learning models. However, due to the relatively independent nature of these tasks and models, it is difficult to abstract common problems and principles to follow, as in supervised learning.

However, precisely because human learning is mainly unsupervised, unsupervised machine learning becomes even more important.

At the beginning, let us take a look at what the unsupervised learning is.

Definition 9.1 (Unsupervised Learning) Unsupervised learning (UL) is a learning paradigm to machine learning with unlabeled data, which analyzes or processes the data based on a certain established criterion to obtain the desired results.

Since unsupervised learning does not have labeled data, it can be seen as a self-teaching method.

Here, a simple comparison is made between the two learning paradigms of supervised learning and unsupervised learning. In the former, the intelligent agent can obtain labeled data as a training dataset and train the machine learning algorithm

W. Wang, *Principles of Machine Learning*,
https://doi.org/10.1007/978-981-97-5333-8_9

Table 9.1 Comparison between supervised learning and unsupervised learning

Supervised Learning	Unsupervised Learning
The training data for supervised learning algorithm are labeled	All data for unsupervised learning algorithm is unlabeled
The algorithm is trained based on labeled data	Analysis or processing is carried out based on unlabeled data

based on these training data; in the latter, the data are unlabeled, and analysis or corresponding processing is carried out based on unlabeled data. The differences between the two are shown in Table 9.1.

9.2   Working Principle

This section first gives a formal description of unsupervised learning and then further explains it with a diagram.

9.2.1   Formal Description

Without loss of generality, this section uses a formal method to describe the working principle of unsupervised learning, omitting the vector dimensions of the input and output spaces and parameterization issues.

Unsupervised learning (UL) can be represented as a 4-tuple, $UL = \langle X, Y, H, \mathcal{L} \rangle$. Where: X denotes the input space, Y denotes the output space, H is the hypothesis set, $\mathcal{L}$ is the loss function.

The input data are obtained through a probability distribution $P(x)$ in the input space X to get n independent and identically distributed (i.i.d.) observation data, $D = \{x_i \mid x_i \in X \text{ and } i = 1, \ldots, n\}$.

Let the following target function of unsupervised learning

$$f : X \rightarrow Y \tag{9.1}$$

be a mapping function from the input space to the output space $f(x) = y$, that can also be expressed as a conditional probability distribution $P(y|x)$, where $x \in X, y \in Y$.

Design a machine learning algorithm for the aforementioned target function, that is, the hypothesis set H, which satisfies:

$$H : X \rightarrow Y . \tag{9.2}$$

By directly analyzing the input dataset D, find the best hypothesis function $h \in H$, $h(\boldsymbol{x}) = \hat{y}$, so that the loss between the hypothesis $h(\boldsymbol{x})$ and the target function $f(\boldsymbol{x})$ is minimized, that is:

$$\mathcal{L}(h(\boldsymbol{x}), f(\boldsymbol{x})) = \mathcal{L}(\hat{y}, y) = \underset{h \in H}{\arg\min}\, \mathbb{E}[h(\boldsymbol{x}) - f(\boldsymbol{x})]. \tag{9.3}$$

The result is a real number. That is, the loss function $\mathcal{L}(\hat{y}, y)$ maps the Cartesian product of the output space to the set of real numbers $\mathbb{R}$, that is:

$$\mathcal{L} : \mathcal{Y} \times \mathcal{Y} \to \mathbb{R}. \tag{9.4}$$

The loss function of unsupervised learning depends on the specific task and its model, because the loss function varies with the different tasks and models of unsupervised learning. Here, $\mathcal{L}(\cdot, \cdot)$ is an abstract representation of the loss function, the purpose of which is to minimize the error between the hypothesis function and the target function, in other words, to make the hypothesis function as close as possible to its target function.

9.2.2 Illustrated Description

Figure 9.1 is a schematic diagram of unsupervised learning, from which it can be seen that unsupervised learning does not have a training process with labeled data.

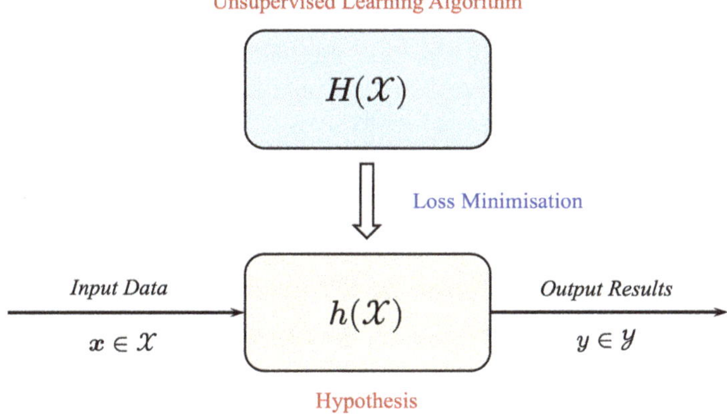

Fig. 9.1 Schematic diagram of unsupervised learning

9.3 Classic Tasks

The classic tasks covered by the unsupervised learning paradigm mainly include three types: clustering, dimensionality reduction, and association.

Clustering is the process of dividing data into several clusters based on the analysis of input data. It can be divided into classical clustering methods and neo-classical clustering methods, in which the former is also known as traditional clustering methods, and the latter is also known as modern clustering methods including deep learning-based clustering.

Dimensionality reduction is the mapping of high-dimensional data to low-dimensional space. The main approaches to dimensionality reduction can be divided into three categories, namely linear dimensionality reduction, nonlinear dimensionality reduction, and deep dimensionality reduction.

Association refers to the discovery of associative relationships between data objects in large databases and the establishment of corresponding association rules. The classic *shopping basket analysis* in large supermarket is known as one of the examples.

The main differences between these three tasks are reflected in the relationship between the input and output spaces. Therefore, based on the formal description of unsupervised learning given in the previous section, a formal description of the relationship between the input and output of these three tasks is also given.

- Clustering:
  For this task, its output is the k clusters $\{G_j \mid j = 1, \ldots, k\} \subseteq \mathcal{Y}$, obtained by grouping the input data. Where each cluster G_j is also a subset of the input data, that is, $G_j \subseteq \mathcal{X}$; and the number of clusters k is unknown in advance and is obtained after clustering analysis.
- Dimensionality Reduction:
  For this task, its output is the new data after reducing the dimensionality of the input data, $h : \mathcal{X} \to \mathcal{Y}$. Where, h is the optimal dimension reduction function, $\mathcal{X} \subseteq \mathbb{R}^h$ is the high-dimensional input space, $\mathcal{Y} \subseteq \mathbb{R}^l$ is the low-dimensional output space, $l \ll h$ and $1 \leq l \leq 3$.
- Association:
  For this task, its output is a set of association rules. Let $\mathcal{X} = \{x_1, x_2, \ldots, x_n\}$ be a set of data items with binary attributes, if there are a subset of data items $X \subseteq \mathcal{X}$, and a single data item $Y \in \mathcal{X}$, then the association rule $X \Rightarrow Y \in \mathcal{Y}$.

The clustering task and dimensionality reduction task will be detailed in Chaps. 14 and 15 of this book, respectively. The association has been explained in Sect. 6.4 of Chap. 6.

9.4 Beyond the Classic

It's worth noting that the paradigm of unsupervised learning should not be limited to classic tasks such as clustering, dimensionality reduction, and association. This is because humans have a wide range of unsupervised learning abilities, and machines should also have relatively broad unsupervised learning functions.

With the continuous development of deep learning, some researchers combine deep neural networks with some other approaches in artificial intelligence, machine learning, probability theory, statistics, physics, etc., and propose some emerging unsupervised models, namely generative models.

Further, we first discuss the importance of unsupervised learning and then introduce some typical generative models.

9.4.1 Importance

In the process of the vigorous development of machine learning and deep learning, unsupervised learning is increasingly attracting people's attention, and some famous researchers have also expressed some views on the importance of unsupervised learning. The three giants of deep learning, namely Yann LeCun, Yoshua Bengio, and Geoffrey Hinton, published a review article titled "Deep Learning" (LeCun et al. 2015) in the journal *Nature* in May 2015, in which they discussed the importance of unsupervised learning and pointed out:

> Unsupervised learning had a catalytic effect in reviving interest in deep learning, but has since been overshadowed by the successes of purely supervised learning. Although we have not focused on it in this Review, we expect unsupervised learning to become far more important in the longer term.

Yann LeCun is a professor at New York University and was the director of AI research at Facebook. In his invited talk at the 2016 *Conference on Neural Information Processing Systems* (NIPS), he made the following analogy when talking about unsupervised learning, supervised learning, and reinforcement learning:

> If intelligence was a cake, unsupervised learning would be the cake, supervised learning would be the icing on the cake, and reinforcement learning would be the cherry on the cake. ...... We know how to make the icing and the cherry, but we don't know how to make the cake. We need to solve the unsupervised learning problem before we can even think of getting to true AI.[1]

In the aforementioned comments, Yann LeCun particularly emphasized the importance of unsupervised learning, especially since he believes that an effective way to solve unsupervised learning has not been found. As for whether his use of cake, icing, and cherry to describe the relationship between unsupervised learning, supervised learning, and reinforcement learning is accurate, that's another story.

---

[1] https://www.youtube.com/watch?v=Ount2Y4qxQo.

Yoshua Bengio is a professor at the University of Montreal, Canada, and the Canada Research Chair in Statistical Learning Algorithms. During his visit to Microsoft's headquarters in Redmond, Washington, USA, on April 6, 2017, he expressed the following views on unsupervised learning:[2]

> So this is one area where the current state of the art in machine learning and deep learning is way below what humans can do. A two-year-old child can learn by simply observing and interacting with the world. For example, she can understand physics without having to take classes and understand gravity and pressure and so on by playing and observing. This is unsupervised learning. ...... It's very important, because in order for machines to go beyond these very limited tasks they currently are good at we'll need unsupervised learning.

9.4.2 Generative Models

As is well known, supervised learning requires a large number of labeled training samples. The labeling usually needs to be done manually, but often requires professionals to complete. Therefore, the large amount of work, high cost, and even the difficulty of labeling brought about by manual annotation become a bottleneck for supervised learning.

However, for unsupervised learning, although there is no need for manual annotation of data, it should not be limited to classic unsupervised learning tasks such as clustering, dimensionality reduction, and association.

Humans mainly acquire new knowledge and skills through unsupervised learning, and various unsupervised modes in natural intelligence have become new paths that machine learning is exploring. Therefore, generative models, as modern models of unsupervised learning, have emerged.

Generative models are a type of multimedia content generation model based on new deep neural network architectures. Specifically, given the observed data of a certain domain distribution x, its goal is to learn to model its real data distribution $p(x)$. Through pre-training, this model can generate any new sample x', and it can also be used to evaluate the likelihood of observed or sampled data.

Generative models belong to the unsupervised learning paradigm, using pre-training methods, but their training is different from supervised learning: the training of generative models often uses unlabeled training data, while the training of supervised learning requires labeled training samples.

Some researchers have combined deep neural networks with other classical theories in artificial intelligence, machine learning, probability theory, statistics, and even physics, and proposed some novel generative models. Those models are characterized by:

- based on classical theories;
- using deep neural networks;

---

[2] https://blogs.microsoft.com/ai/a-conversation-ai-pioneer-yoshua-bengio/.

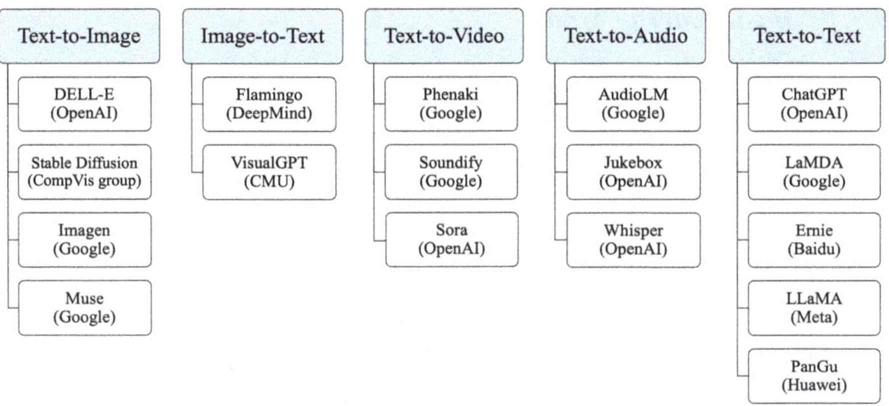

Fig. 9.2 Representative multimedia generative models and their research institutions

- used for multimedia content generation;
- unlabeled training data; and
- pre-training.

The continuous development of generative models has made generative AI one of the hot research fields in recent years. Its main feature is to generate one type of media content from another, including: text-to-image, image-to-text, text-to-video, text-to-audio, and text-to-text.

Some representative multimedia generative models and their research organizations are depicted in Fig. 9.2.

In the following sections, we will separately introduce some typical generative models, namely: energy-based models, autoencoder models, generative adversarial models, normalizing flow models, autoregressive models, and diffusion models.

9.5   Energy Models

Energy models, also referred to as energy-based models (EBMs), are a type of generative model that introduces statistical mechanics into machine learning.

This section first introduces the working principle and optimization methods of basic energy model and then introduces Boltzmann model and related research approaches of the energy models.

9.5.1 Basic Energy Model

There are two important properties in the axioms of probability theory, which were introduced in Chap. 3 of this book, namely non-negativity, and normalization. The following uses a parameterized probability density function to further describe these two properties.

Let $p_\theta(x)$ be the probability density function where x is the input data and $x \in \mathcal{X} \subseteq \mathbb{R}^n$, and θ is a parameter. Then, the nonnegativity of probability is expressed as $p_\theta(x) \geq 0$, and normalization is expressed as $\int p_\theta(x)dx = 1$.

The energy model can avoid the normalization problem of the aforementioned probability distribution, making the design of machine learning algorithms more flexible, and can provide a unified framework for many probabilistic and non-probabilistic learning (Ranzato et al. 2007).

9.5.1.1 Working Principle

The energy model refers to the non-normalized probability scalar as energy and characterizes its dependence by the association of energy with the combination configuration of input variables and latent variables. Inference in the energy model includes finding the latent variables (values) that minimize energy given a set of input variables (values). The energy model learns a function that associates low energy with the correct values of latent variables and high energy with incorrect values.

Without loss of generality, here we only introduce the single-variable energy model:

$$p_\theta(x) = \frac{e^{-E_\theta(x)}}{Z_\theta(x)}. \tag{9.5}$$

Where: $E_\theta(x) : \mathbb{R}^D \to \mathbb{R}$ is an energy function with parameter θ, which can be represented by a nonlinear regression function (Song and Kingma 2021); and $Z_\theta(x)$ denotes the normalizing constant, and:

$$Z_\theta(x) = \int e^{-E_\theta(x)}dx. \tag{9.6}$$

$Z_\theta(x)$ is also known as the partition function.

9.5.1.2 Maximum Likelihood Training

A key issue with energy models is how to train them. Since $Z_\theta(\cdot)$ is the function of parameter θ, therefore, the evaluation and differentiation to its parameters involve an integral that is usually difficult to solve. A common method to obtain the probability model from independent and identically distributed data is maximum likelihood estimation (MLE).

Let $p_\theta(x)$ be a probability model with parameter θ, and $p_{\text{data}}(x)$ denote the actual data distribution of the dataset. By maximizing the expected log-likelihood function of this data distribution, $p_\theta(x)$ can be made to fit $p_{\text{data}}(x)$:

$$\mathbb{E}_{x \sim p_{\text{data}}(x)}\left[\log p_\theta(x)\right].$$

The above expectation can be easily estimated with samples from this dataset. The maximization of likelihood is equivalent to the minimization of the KL divergence between $p_{\text{data}}(x)$ and $p_\theta(x)$ because:

$$-\mathbb{E}_{x \sim p_{\text{data}}(x)}\left[\log p_\theta(x)\right] = D_{KL}\left(p_{\text{data}}(x) \parallel p_\theta(x)\right) - \mathbb{E}_{x \sim p_{\text{data}}(x)}\left[\log p_{\text{data}}(x)\right].$$

Where, the second term on the right side of the equation $\mathbb{E}_{x \sim p_{\text{data}}(x)}\left[\log p_{\text{data}}(x)\right]$ does not depend on θ, and can be considered as a constant. Therefore:

$$-\mathbb{E}_{x \sim p_{\text{data}}(x)}\left[\log p_\theta(x)\right] = D_{KL}\left(p_{\text{data}}(x) \parallel p_\theta(x)\right) - \text{constant}.$$

Due to the normalization constant $Z_\theta(\cdot)$ being difficult to solve, the likelihood of the energy model cannot be directly calculated like the maximum likelihood method. However, the gradient of the log-likelihood can be estimated using the Markov chain Monte Carlo (MCMC) method, allowing the likelihood to be maximized using gradient ascent. Specifically, the gradient of the log-probability based on the energy model is decomposed into the sum of two terms:

$$\nabla_\theta \log p_\theta(x) = -\nabla_\theta E_\theta(x) - \nabla_\theta \log Z_\theta. \tag{9.7}$$

The first gradient term on the right side of the equation, $-\nabla_\theta E_\theta(x)$, can be directly calculated using automatic differentiation. The difficulty lies in the second gradient term, $\nabla_\theta \log Z_\theta$, which is difficult to calculate accurately.

This gradient term can be rewritten as the following mathematical expectation:

$$\nabla_\theta \log p_\theta(x) = -\nabla_\theta E_\theta(x) - \nabla_\theta \log Z_\theta$$

$$= \mathbb{E}_{x \sim p_\theta(x)}\left[-\nabla_\theta E_\theta(x)\right].$$

Therefore, the unbiased one-sample Monte Carlo estimate of the log-likelihood gradient can be obtained from the following equation:

$$\nabla_\theta \log p_\theta(x) \simeq -\nabla_\theta E_\theta(\tilde{x}). \tag{9.8}$$

Where $\tilde{x} \sim p_\theta(x)$ is a random sample from the distribution of x given by the energy-based model.

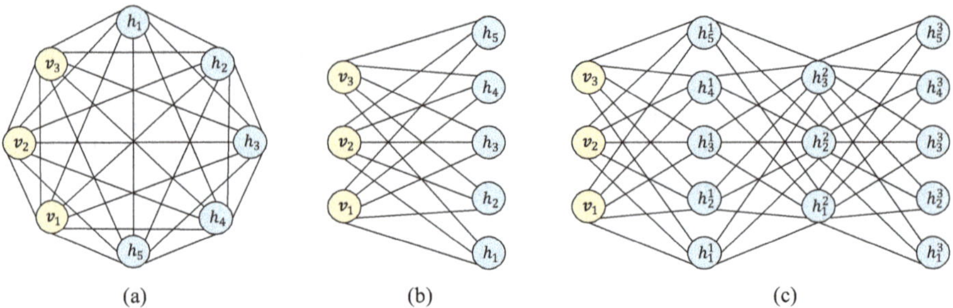

Fig. 9.3 Energy-based Boltzmann models

9.5.2 Boltzmann Models

Here we introduce three types of energy-based Boltzmann models, namely: Boltzmann machines, restricted Boltzmann machines, and deep Boltzmann machines, as shown in Fig. 9.3.

9.5.2.1 Boltzmann Machines

Boltzmann machines (BMs) are undirected networks composed of symmetrically coupled random binary units (Salakhutdinov and Hinton 2009).

As depicted in Fig. 9.3a, a Boltzmann machine is a network of symmetrically connected, includes a set of visible units $v \in \{0, 1\}^D$, and a set of hidden units $h \in \{0, 1\}^P$, where D and P denote the numbers of visible and hidden units, respectively. The energy of the state $\{v, h\}$ is defined as follows:

$$E_\theta(v, h) = -\frac{1}{2}v^T \mathbf{L}v - \frac{1}{2}h^T \mathbf{J}h - v^T \mathbf{W}h. \tag{9.9}$$

Where $\theta = \{\mathbf{W}, \mathbf{L}, \mathbf{J}\}$ is the parameter of this model, and $\{\mathbf{W}, \mathbf{L}, \mathbf{J}\}$ are three symmetric interaction terms: $\mathbf{W}$ denotes visible-to-hidden, $\mathbf{L}$ denotes visible-to-visible, and $\mathbf{J}$ denotes hidden-to-hidden.

The Boltzmann machine is trained by contrastive divergence and finds its equilibrium state through Gibbs sampling. However, this requires an exponential amount of time in terms of the number of units, making the construction of larger models impractical.

9.5.2.2 Restricted Boltzmann Machines

A restricted Boltzmann machine (RBM) is a special type of Boltzmann machine.

The so-called restriction refers to limiting the connections between visible units and hidden units to solve the exponential time problem existing in the Boltzmann machines. Therefore, the restricted Boltzmann machines become a two-layer undirected network: one layer is the visible layer composed of visible units, and the other layer is the hidden layer composed of hidden units, as shown in Fig. 9.3b. Its state $\{v, h\}$ is defined by the following energy:

$$E_\theta(v, h) = -v^\mathsf{T} \mathbf{W} h. \tag{9.10}$$

The parameter of this model is $\theta = \{\mathbf{W}\}$, that is, there is only visible to hidden and no other two symmetric interaction items in the Boltzmann machine.

The restricted Boltzmann machines can be trained using the maximum likelihood method. Sampling of the restricted Boltzmann machine can be done using Gibbs sampling or Markov chain Monte Carlo (MCMC) methods.

The restricted Boltzmann machines laid the foundation for deep Boltzmann machines.

9.5.2.3 Deep Boltzmann Machines

A deep Boltzmann machine (DBM) is an undirected deep network with one visible layer and several hidden layers, equivalent to adding several hidden layers to the restricted Boltzmann machines.

For the deep Boltzmann machines shown in Fig. 9.3c, which has one visible layer and three hidden layers, its state $\{v, h^1, h^2, h^3\}$ is defined by the following energy:

$$E_\theta(v, h^1, h^2, h^3) = -v^\mathsf{T} \mathbf{W}^1 h^1 - h^{1\mathsf{T}} \mathbf{W}^2 h^2 - h^{2\mathsf{T}} \mathbf{W}^3 h^3. \tag{9.11}$$

Where the parameter of this model $\theta = \{\mathbf{W}^1, \mathbf{W}^2, \mathbf{W}^3\}$ denotes the three symmetric interaction terms from visible to hidden, hidden to hidden, and hidden to hidden, respectively.

9.5.3 Related Works

Yann LeCun has always been a researcher and advocate of energy-based models. As early as 2004, he and two researchers from the NEC Labs in the United States published a paper titled "Synergistic Face Detection and Pose Estimation with Energy-Based Models" (Osadchy et al. 2004) at the *Conference on Neural Information Processing Systems* (NIPS). In 2007, he and several researchers from New York University published a paper titled "A Unified Energy-based Framework for Unsupervised Learning" at the *International Conference on Artificial Intelligence and Statistics* (Ranzato et al. 2007).

In 2011, several researchers from Stanford University published a paper titled "Learning Deep Energy Models" (Ngiam et al. 2011) at the "International Conference on Machine Learning" (ICML), proposing a research approach to implement energy models using deep feedforward networks.

In recent years, some researchers have been engaged in research and applications based on energy models and have published some distinctive articles, including the paper titled "Concept Learning with Energy-based Models" (Mordatch 2018) submitted to the electronic preprint repository *arXiv* in 2018, and the paper titled "Implicit Generation and Modeling with Energy Based Models" (Du and Mordatch 2019) published at the *Conference on Neural Information Processing Systems* (NeurIPS) in 2019.

9.6 Autoencoding Models

Autoencoder models originated from the Autoencoder (AE) proposed in the 1980s.

An autoencoder is an unsupervised neural network model composed of an encoder and a decoder, used to select and extract effective features from input data. Its basic idea has become a learning generative model based on data encoding.

By the way, the prefix "auto" of "autoencoder" is said to be derived from Greek, meaning "self". Similarly, the prefix "auto" of "autoregressive" in Sect. 9.9 of this chapter has the same meaning.

With the continuous development of autoencoder models, several variants have appeared, including undercomplete autoencoder (UAE), sparse autoencoder (SAE), denoising autoencoder (DAE), and variational autoencoder (VAE).

This section first discusses the basic architecture and working principle of the autoencoder and then introduces the aforementioned types of autoencoders in several subsections.

9.6.1 Autoencoders

The working principle of the autoencoder is as follows: the encoder maps the input data to an intermediate code, and the decoder decodes this intermediate code and outputs it, which is equivalent to reconstructing the input data. There is a hidden latent space layer inside the autoencoder, where the intermediate code is formed. Figure 9.4 is a schematic diagram of the basic model of the autoencoder.

Fig. 9.4 Schematic diagram of the autoencoder

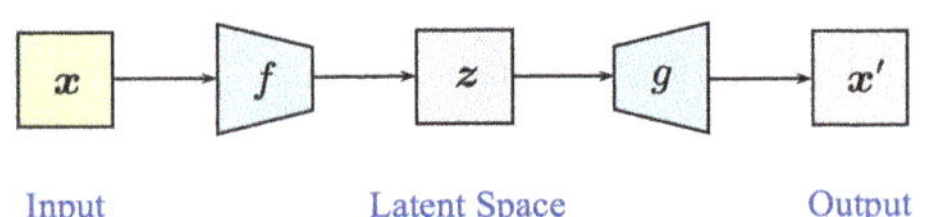

For ease of narration, let's first give a formal definition of the autoencoder.

Definition 9.2 (Autoencoders) Let the data space $\mathcal{X} \subseteq \mathbb{R}^u$ be a u-dimensional real vector space, $\mathcal{Z} \subseteq \mathbb{R}^v$ be a v-dimensional real vector space referred to as latent space, $x, x' \in \mathcal{X}$ are the input and output data respectively, and $z \in \mathcal{Z}$ is the intermediate code in the latent space. Let f denote the encoder, and g denote the decoder. The encoder f maps the input data x to the intermediate code in the latent space z, i.e., $z = f(x)$; the decoder then reconstructs this intermediate code z into output data x', that is, $x' = g(z)$. Therefore, a neural network is known as an autoencoder (AE), if it satisfies the following expression

$$x' = g(z) = g(f(x)), \tag{9.12}$$

and $x' \not\equiv x$.

In the aforementioned definition, $x' \not\equiv x$ indicates that the output data x' is not a copy of the input data x but the result of its reconstruction.

Based on the linear or nonlinear characteristics of the autoencoder, it can be further divided into linear autoencoder and nonlinear autoencoder.

9.6.2 Undercomplete Autoencoders

Definition 9.3 (Undercomplete Autoencoders) For an autoencoder that satisfies Definition 9.2, it is called an undercomplete autoencoder (UAE), if and only if the dimension of its latent space $\mathcal{Z} \subseteq \mathbb{R}^v$ is less than the dimension of the data space $\mathcal{X} \subseteq \mathbb{R}^u$, i.e., $v < u$.

Here, let's first understand what is the mathematical definition on "undercomplete". In the *Wiktionary*, the term is defined as:
Describing a frame (in linear algebra) having a set of functions less than a basis.[3].

According to the definition of undercomplete, it is known that an undercomplete autoencoder is one whose latent space dimension is less than the dimension of the input and output space.

Due to the bottleneck effect of its latent space dimension, the undercomplete autoencoder can be used to learn the significant features of the input data, which is called undercomplete representation. Figure 9.5 is its schematic diagram.

The loss function of the undercomplete autoencoder is shown further.

$$\mathcal{L}(x, x') = \mathcal{L}(x, g(f(x))). \tag{9.13}$$

---

[3] https://en.wiktionary.org/wiki/undercomplete

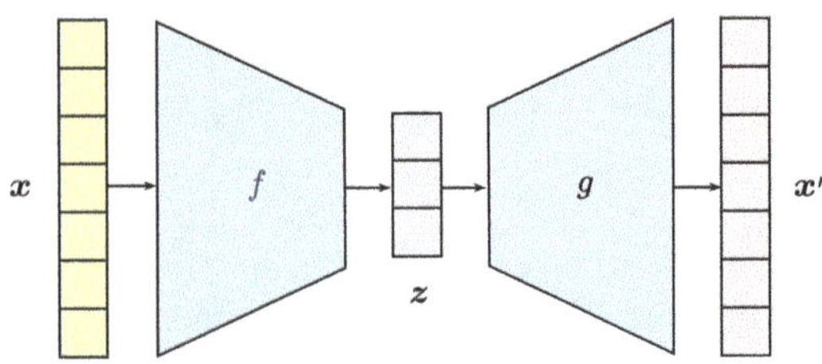

Fig. 9.5 Schematic diagram of an undercomplete autoencoder

The undercomplete autoencoder is a typical autoencoder that can be used for dimensionality reduction.

By the way, in Definition 9.2, it is not specified the relationship between the dimension of the latent space $\mathcal{Z} \subseteq \mathbb{R}^{v}$ and the dimension of data space $\mathcal{X} \subseteq \mathbb{R}^{u}$. The function of autoencoders will change with the relationship of dimensions. There are three situations:

- If $v < u$, it is the "undercomplete autoencoder" introduced in this section.
- If $v > u$, it is called an "overcomplete autoencoder".
- If $v = u$, it becomes a "fully-connected autoencoder".

This book does not elaborate on the "overcomplete autoencoder" and "fully connected autoencoder", if you are interested on them, you can consult the related materials by yourself.

9.6.3   Sparse Autoencoders

In addition to reducing dimensions in the latent space like the undercomplete autoencoders, constraints can also be added to the loss function to give it good representation learning capabilities. The sparse autoencoder is one of the instances.

Definition 9.4 (Sparse Autoencoders)   A sparse autoencoder (SAE) is an autoencoder with a sparsity constraint, but the dimension of its latent space is not limited and can be equal to or even greater than the dimension of the input space.

The so-called sparsity constraint is to control the number of active neurons in the latent space of the autoencoder at the same time. For example, if it is controlled so that only 30% of the neurons are active, its sparsity is 30%. Figure 9.6 is a schematic diagram of a sparse autoencoder.

The sparse autoencoder adds a regularization constraint term to the loss function, that is, the sparsity penalty term $\Omega(z)$, therefore its loss function is represented in the following form:

$$\mathcal{L}\left(x, x'\right) + \Omega(z) = \mathcal{L}(x, g(f(x))) + \Omega(z). \tag{9.14}$$

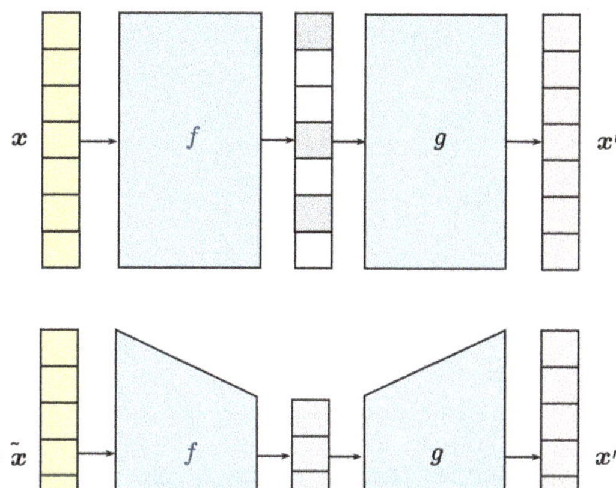

Fig. 9.6 Schematic diagram of a sparse autoencoder

Fig. 9.7 Schematic diagram of the denoising autoencoder

This sparsity penalty term $\Omega(z)$ can be adopted by different approaches, such as: KL divergence, L^1 or L^2 regularization.

In addition, some researchers have proposed a method, called "k-sparse autoencoders" (Makhzani and Frey 2014), to limit the number of active neurons manually. It was published at the *International Conference on Learning Representations* (ICLR) in 2014.

9.6.4   Denoising Autoencoders

Definition 9.5 (Denoising Autoencoders)   A denoising autoencoder (DAE) learns useful features in the input data by changing the reconstruction error term in the loss function. Its loss function is shown in the following.

$$\mathcal{L}\left(x, x'\right) = \mathcal{L}\left(x, g\left(f(\widetilde{x})\right)\right). \tag{9.15}$$

Where $\widetilde{x} = x \oplus \text{noise}$.

The schematic diagram of the denoising autoencoder is shown in Fig. 9.7. Unlike other autoencoders, the input of the denoising autoencoder $\widetilde{x}$ is the original data x with some additive noise. The so-called additive white Gaussian noise is a common noise that can be added.

9.6.5 Variational Autoencoders

Definition 9.6 (Variational Autoencoders) A variational autoencoder (VAE) is an artificial neural network that uses probabilistic graphical models and variational Bayes method.[4]

The variational autoencoder was proposed in 2013 by Diederik P. Kingma and Max Welling from the University of Amsterdam and then published at the *International Conference on Learning Representations* (ICLR) in 2014 (Kingma and Welling 2014).

Although the architecture of the variational autoencoder is similar to that of the basic autoencoders, the variational autoencoder is a deep generative model based on probabilistic methods, with a completely different mathematical formulation. The latent space is in this case composed by a mixture of distributions instead of a fixed vector.

9.6.5.1 Working Principle

Let's consider a variational autoencoder that acquires input data satisfying the independent and identically distributed (i.i.d.) condition via a marginal probability distribution, denoted as $p_\theta(x)$, where θ denotes a parameter.

As a deep generative model with latent variable z, variational autoencoder should be modeled as a joint probability of input variable x and latent variable z, written as:

$$p_\theta(x, z) = p_\theta(z|x)\, p_\theta(x), \tag{9.16}$$

$$p_\theta(x, z) = p_\theta(x|z)\, p_\theta(z). \tag{9.17}$$

In Eq. 9.16, $p_\theta(z|x)$ is the conditional probability of mapping the input data x to the latent space code z, while in Eq. 9.17, $p_\theta(x|z)$ is the conditional probability of reconstructing the latent space code z into the output data x'. From the perspective of the autoencoder, $p_\theta(z|x)$ is the encoder, and $p_\theta(x|z)$ is the decoder.

According to Bayes' theorem, the encoder should be solved by the following expression:

$$p_\theta(z|x) = \frac{p_\theta(x|z)\, p_\theta(z)}{p_\theta(x)}. \tag{9.18}$$

Where, $p_\theta(z|x)$ is the posterior probability, $p_\theta(x|z)$ is the likelihood, $p_\theta(z)$ is the prior, $p_\theta(x)$ is the evidence.

---

[4] https://en.wikipedia.org/wiki/Variational_Bayesian_methods.

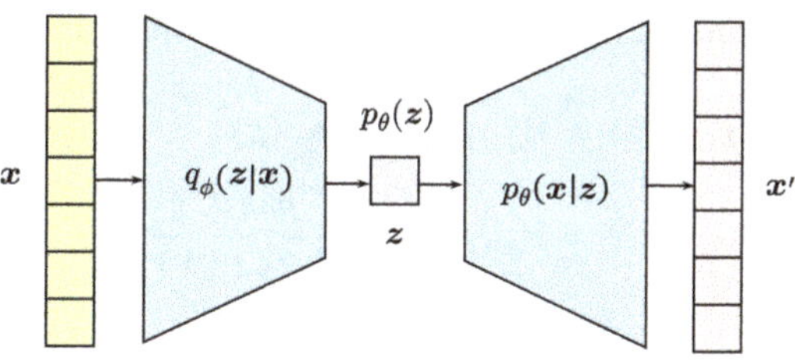

Fig. 9.8 Schematic diagram of the variational autoencoder

The evidence $p_\theta(x)$ can be calculated through marginal probability. With the latent space intermediate code z as the margin, we have:

$$p_\theta(x) = \int_z p_\theta(x, z)dz = \int_z p_\theta(x|z) \, p_\theta(z)dz. \tag{9.19}$$

As can be seen from the aforementioned equation, $p_\theta(x)$ is calculated by integrating the latent space's intermediate code z, which requires exponential computation time. Therefore, the posterior probability $p_\theta(z|x)$ is a computational intractability problem.

The basic idea of variational inference is to treat inference as a means of obtaining approximate results. Therefore, the variational autoencoder uses a conditional probability $q_\phi(z|x)$ to obtain the approximate result of posterior probability $p_\theta(z|x)$. That is:

$$q_\phi(z|x) \approx p_\theta(z|x). \tag{9.20}$$

Where ϕ is referred to as the variational parameter.

Therefore, in the variational autoencoder, the approximate posterior distribution $q_\phi(z|x)$ is treated as its probabilistic encoder, $z \sim q_\phi(z|x)$; the conditional likelihood distribution $p_\theta(x|z)$ is treated as its probabilistic decoder, $x' \sim p_\theta(x|z)$.

The schematic diagram of the variational autoencoder is shown in Fig. 9.8.

9.6.5.2 Evidence Lower Bound

The goal of the variational autoencoder, like other variational methods, is to optimize the evidence lower bound (ELBO).

The evidence lower bound is usually derived from Jensen's inequality,[5] but the variational autoencoder proposes an alternative method. Here are its derivation process and results. For any choice of variational inference ϕ with variational

---

[5] https://en.wikipedia.org/wiki/Jensen's_inequality.

parameters $q_\phi(z|x)$, the following expression and its derivation process hold:

$$
\begin{aligned}
\log p_\theta(x) &= \mathbb{E}_{q_\phi(z|x)}\left[\log p_\theta(x)\right]\\
&= \mathbb{E}_{q_\phi(z|x)}\left[\log\left[\frac{p_\theta(x, z)}{p_\theta(z|x)}\right]\right]\\
&= \mathbb{E}_{q_\phi(z|x)}\left[\log\left[\frac{p_\theta(x, z)}{q_\phi(z|x)}\frac{q_\phi(z|x)}{p_\theta(z|x)}\right]\right]\\
&= \mathbb{E}_{q_\phi(z|x)}\left[\log\left[\frac{p_\theta(x, z)}{q_\phi(z|x)}\right]\right] + \mathbb{E}_{q_\phi(z|x)}\left[\log\left[\frac{q_\phi(z|x)}{p_\theta(z|x)}\right]\right].
\end{aligned}
\tag{9.21}
$$

In the aforementioned expression, the second term on the right side of the equation is the non-negative KL divergence (Kullback-Leibler divergence) between $q_\phi(z|x)$ and $p_\theta(z|x)$, which is:

$$
D_{KL}\left(q_\phi(z|x) \,\|\, p_\theta(z|x)\right) = \mathbb{E}_{q_\phi(z|x)}\left[\log\left[\frac{q_\phi(z|x)}{p_\theta(z|x)}\right]\right] \geq 0.
$$

The KL divergence is 0, if and only if $q_\phi(z|x)$ is the true posterior distribution. The first term on the right side of Eq. 9.21 is the evidence lower bound (ELBO), also known as the variational lower bound:

$$
\begin{aligned}
\mathcal{L}_{\theta,\phi}(x) &= \mathbb{E}_{q_\phi(z|x)}\left[\log\left[\frac{p_\theta(x, z)}{q_\phi(z|x)}\right]\right]\\
&= \mathbb{E}_{q_\phi(z|x)}\left[\log p_\theta(x, z) - \log q_\phi(z|x)\right].
\end{aligned}
\tag{9.22}
$$

Therefore, Eq. 9.21 can be written in the following form:

$$
\log p_\theta(x) = \mathcal{L}_{\theta,\phi}(x) + D_{KL}\left(q_\phi(z|x) \,\|\, p_\theta(z|x)\right).
\tag{9.23}
$$

Due to the non-negativity of the KL divergence, this ELBO is the lower bound of the log-likelihood of the data, that is:

$$
\mathcal{L}_{\theta,\phi}(x) = \log p_\theta(x) - D_{KL}\left(q_\phi(z|x) \,\|\, p_\theta(z|x)\right) \leq \log p_\theta(x).
\tag{9.24}
$$

As can be seen from the aforementioned equation, $\mathcal{L}_{\theta,\phi}(x) \leq \log p_\theta(x)$. Therefore, the evidence is lower bound $\mathcal{L}_{\theta,\phi}(x)$ and the gap between the marginal likelihood $\log p_\theta(x)$ is referred to as the tightness of the boundary. In terms of KL divergence, $q_\phi(z|x)$ the closer it is to the true posterior distribution $p_\theta(z|x)$, the smaller the gap.

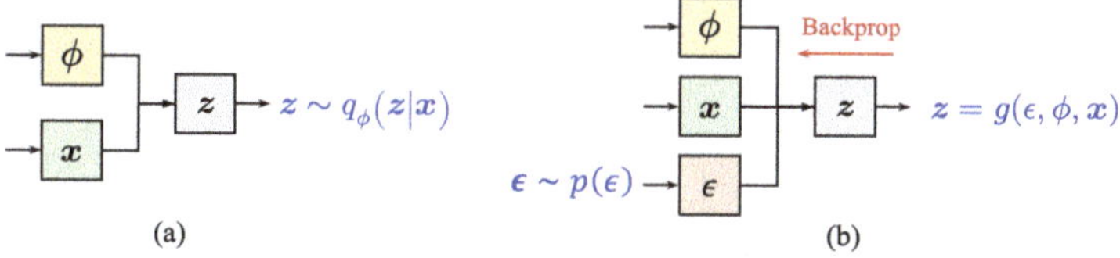

Fig. 9.9 Schematic diagram of the reparameterization trick

9.6.5.3 Re-Parameterization Trick

Since the variational autoencoder uses the probability distribution $q_\phi(z|x)$ to obtain the latent variable z from input variable x, i.e., $z \sim q_\phi(z|x)$, so that it needs to be differentiable by a change of variables, this is called re-parameterization trick.

The original formulation, $z \sim q_\phi(x|x)$, is related to parameter ϕ and input variable x, as shown in Fig. 9.9a. The re-parameterization is a differentiable transformation of z into ϕ, x, as well as another random variable ϵ.

The latent variable z can be expressed as the differentiable transformation of another random variable ϵ, that is:

$$z = f(\phi, x, \epsilon). \tag{9.25}$$

Where the distribution of random variable $\epsilon \sim p(\epsilon)$ is independent of x and ϕ. In order to facilitate comparison with $z \sim q_\phi(z|x)$, the aforementioned equation can be denoted as:

$$z \sim g_\phi(z|x, \epsilon). \tag{9.26}$$

The re-parameterized formulation is depicted in Fig. 9.9b.

Suppose the latent variable z be multivariate Gaussian distributions, so it can be described as:

$$z \sim q_\phi(z|x) = \mathcal{N}(\mu, \sigma^2).$$

Given $\epsilon \sim \mathcal{N}(0, 1)$, and the symbol $\odot$ denotes the element-wise product, then by reparameterization, we have:

$$z = \mu + \sigma \odot \epsilon.$$

Achieved by the above trick, the re-parameterized variational autoencoder is shown in Fig. 9.10.

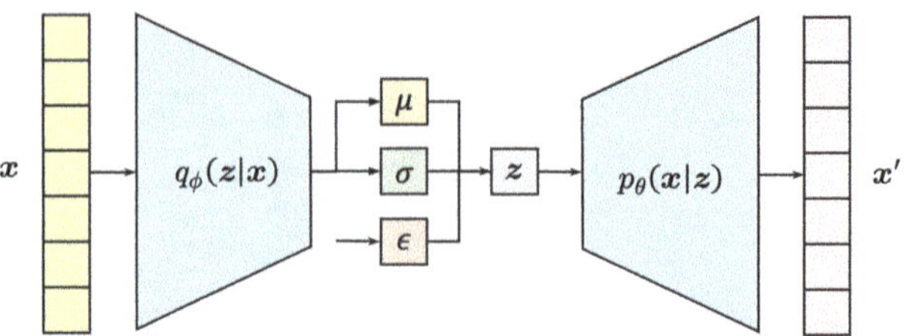

Fig. 9.10 Schematic diagram of re-parameterization of the variational autoencoder

9.7 Generative Adversarial Models

The so-called generative adversarial models refer to deep neural network-based generative adversarial models, which originated from generative adversarial networks.

9.7.1 Generative Adversarial Networks

The generative adversarial networks (GANs) were proposed by Ian Goodfellow et al., in a paper titled "Generative Adversarial Nets," which was published at the *Conference on Neural Information Processing Systems* (NIPS) in 2014 (Goodfellow et al. 2014).

When the aforementioned paper was published, Ian Goodfellow was pursuing his Ph.D. degree at the University of Montreal in Canada, and his advisors were Yoshua Bengio and Aaron Courville.

In 2016, during the surge of interest in generative adversarial networks in the field of artificial intelligence and machine learning, Yann LeCun, one of the giants of deep learning, commented on it, calling it "the most interesting idea in the last 10 years in machine learning".

9.7.1.1 Working Principle

A generative adversarial network (GAN) belongs to unsupervised learning. The core idea of the GAN model is based on the minimax theorem in zero-sum games, of which a generator and a discriminator operate in a manner of two player game.

As shown in Fig. 9.11, the basic structure of a GAN model mainly consists of a generator (G) and a discriminator (D), both of which use multilayer neural networks.

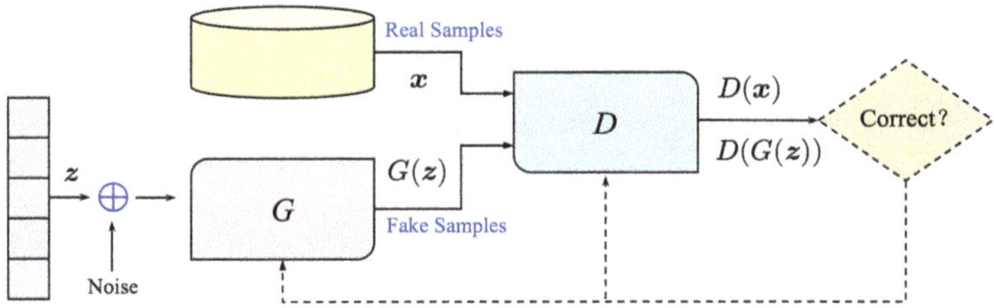

Fig. 9.11 Basic structure of a generative adversarial network (GAN)

9.7.1.2 Training Process

It first adds random noise to the input latent space data z in the generative adversarial network, and generates a fake sample $G(z)$ through the generator. The discriminator then takes the real sample x and the fake sample $G(z)$ as inputs, and obtains $D(x)$ and $D(G(z))$ respectively. For the real sample x, the discriminator tries to make $D(x)$ true. For the fake samples $G(z)$, the generator tries to make it true, while the discriminator tries to make $D(G(z))$ false.

Generative adversarial networks are based on the minimax theorem of zero-sum games, that is, calculating $V(G, D)$ in which the minimum value of the generator G and the maximum value of the discriminator D are calculated. The expression is as follows:

$$\min_{G} \max_{D} V(G, D) = E_{x \sim P_{\text{data}}(x)} \left[\log D(x)\right] + E_{z \sim P_z(z)} \left[\log(1 - D(G(z)))\right].$$

$$(9.27)$$

The discriminator strives to distinguish the difference between real samples and fake samples, trying to increase $D(x)$ and $D(G(z))$ differences, while the generator tries to make the fake samples $G(z)$ consistent with the real samples x, making it difficult for the discriminator to distinguish. Therefore, the optimization process of these two modules is like a game between two players. After multiple rounds of adversarial training, their performance is continuously optimized during the iteration process. When the discriminator finally cannot distinguish the difference between $D(x)$ and $D(G(z))$, then reach a zero-sum state, it is considered that the GAN has reached its optimum.

So that, for easy memory, we give the following definition on generative adversarial models.

Definition 9.7 (Generative Adversarial Models) A generative adversarial model (GAM) is one of the deep generative models composed of generator and discriminator modules, based on the minimax theorem of two-player games.

9.7.2   Related Works

As one of the novel deep generative models, a generative adversarial network has attracted the attention of many researchers, and many GAN papers have appeared after 2014, far exceeding the number of papers on convolutional neural networks (CNN) after 2012. For this reason, someone has specifically set up a "The GAN Zoo[6]" on Github, and provided a graph showing the monthly cumulative growth trend of GAN papers.

The Readme in "The GAN Zoo" also writes:

"Every week, new GAN papers are coming out and it's hard to keep track of them all, not to mention the incredibly creative ways in which researchers are naming these GANs! So, here's a list of what started as a fun activity compiling all named GANs!"

Readers can find all the information about GAN papers from "The GAN Zoo".

On December 12, 2018, Tero Karras et al. from Nvidia Corporation in the United States submitted a paper to the electronic preprint repository *arXiv*, titled "A Style-based Generator Architecture for Generative Adversarial Networks" (Karras et al. 2018), also known as StyleGAN.

This network uses a style transfer method to separate facial details for individual adjustments, thereby significantly surpassing existing GAN models in generating facial images.

9.8   Normalizing Flow Models

The normalizing flow model (Rezende and Mohamed 2015), also known as the flow-based generative model (Ho et al. 2019), is a type of generative model.

Definition 9.8 (Normalizing Flow Models)   The normalized flow model is a model for modeling probability distributions, which uses the change of variables in probability to transform simple distributions into complex distributions, and then performs inverse transformations to restore them to simple distributions.

This section first introduces the change of variables used in normalizing flow and then introduces typical normalizing flow methods.

---

[6] https://github.com/hindupuravinash/the-gan-zoo.

9.8.1   Change of Variables

The change of variables refers to a basic technique in mathematics to simplify problems, in which the original variables are replaced with functions of other variables,[7] also known as the *change-of-variables theorem*. In normalizing flows, it specifically refers to the change of variables in probability distribution transformation.

Let the random variable $z \in \mathbb{R}^D$ be a latent variable, its probability density function $p_\theta : \mathbb{R}^D \to \mathbb{R}^D$, that is:

$$z \sim p_\theta(z). \tag{9.28}$$

Where $p_\theta(z)$ is a simple, easy-to-handle probability distribution, such as the Gaussian distribution.

The Gaussian distribution is also known as the normal distribution, which is why this model is called the "normalizing flow" model.

A function that has its inverse function is called an invertible function (also known as a bijective function). Using the invertible function $g_\theta(z)$ to generate random variables of complex distribution $x \in \mathbb{R}^D$, can be obtained through its inverse function $g_\theta^{-1}(x)$, and then transform x back to the original random variable z. Therefore, we have:

$$x = g_\theta(z), \tag{9.29}$$

$$z = g_\theta^{-1}(x). \tag{9.30}$$

Where: the function of Eq. 9.29 is to obtain the random variable x from the random variable z, which is a transformation from a normal distribution to a complex distribution, and the probability density function $g_\theta(z)$ completes its generation process; while the function of Eq. 9.30 serves to transform the random variable x back into the random variable z, which is a transformation from a complex distribution to a normal distribution. The inverse probability density function $g_\theta^{-1}(x)$ completes its normalization process.

Next, consider how to infer the unknown probability density function $p_\theta(x)$ of the new random variable x.

Due to the correspondence between the random variables x and z, in order to maintain their probabilities equal, the change of $p_\theta(x)$ with dx must be equal to the change of $p_\theta(z)$ with dz, therefore we have:

$$p_\theta(x)\,dx = p_\theta(z)\,dz. \tag{9.31}$$

Since x and z is a vector of two different coordinate systems, the interval between dz and dx is a determinant, then use $z = g_\theta^{-1}(x)$ to perform change of variables,

---

[7] https://en.wikipedia.org/wiki/Change_of_variables.

then we have:

$$p_\theta\left(\boldsymbol{x}\right) = p_\theta\left(\boldsymbol{z}\right)\left|\det\left(\frac{d\boldsymbol{z}}{d\boldsymbol{x}}\right)\right|$$

$$= p_\theta\left(\boldsymbol{z}\right)\left|\det\left(\frac{d{g_\theta}^{-1}\left(\boldsymbol{x}\right)}{d\boldsymbol{x}}\right)\right|. \tag{9.32}$$

Where, $\det\left(\frac{d{g_\theta}^{-1}\left(\boldsymbol{x}\right)}{d\boldsymbol{x}}\right)$ is the Jacobian determinant of function ${g_\theta}^{-1}\left(\boldsymbol{x}\right)$ (Rezende and Mohamed 2015). Since $\det\left(A^{-1}\right) = \left(\det\left(A\right)\right)^{-1}$, we have:

$$p_\theta\left(\boldsymbol{x}\right) = p_\theta\left(\boldsymbol{z}\right)\left|\det\left(\frac{dg_\theta\left(\boldsymbol{x}\right)}{d\boldsymbol{x}}\right)\right|^{-1}. \tag{9.33}$$

Taking the logarithm of the aforementioned equation, we have:

$$\log\,p_\theta\left(\boldsymbol{x}\right) = \log\,p_\theta\left(\boldsymbol{z}\right) - \sum_{i=1}^{n}\log\left|\det\left(\frac{dg_\theta\left(\boldsymbol{x}\right)}{d\boldsymbol{x}}\right)\right|. \tag{9.34}$$

9.8.2   Normalizing Flow

Let $\{\boldsymbol{h}_i\}_{i=0}^{n}$ denote the intermediate random variables, and $\boldsymbol{h}_n = \boldsymbol{x}, \boldsymbol{h}_0 = \boldsymbol{z}$. Let the probability density function $g_\theta = g_\theta^n \circ \cdots \circ g_\theta^2 \circ g_\theta^1$, then $\boldsymbol{x} = g_\theta\left(\boldsymbol{z}\right)$ can be written as:

$$\boldsymbol{h}_n = g_\theta^n \circ \cdots \circ g_\theta^2 \circ g_\theta^1\left(\boldsymbol{h}_0\right). \tag{9.35}$$

Therefore, Eq. 9.34 can be written as:

$$\log p_\theta\left(\boldsymbol{h}_n\right) = \log p_\theta\left(\boldsymbol{h}_0\right) - \sum_{i=1}^{n}\log\left|\det\left(\frac{dg_\theta^i\left(\boldsymbol{h}_{i-1}\right)}{d\boldsymbol{h}_{i-1}}\right)\right|. \tag{9.36}$$

It can be seen that through a series of probability density functions, $\{g_\theta^i\}_{i=1}^{n}$ and $\boldsymbol{h}_i = g_\theta^i\left(\boldsymbol{h}_{i-1}\right)$, transforming the normal distribution into a complex distribution. Similarly, let the inverse probability density function

$$g_\theta^{-1} = \left(g_\theta^1\right)^{-1} \circ \left(g_\theta^2\right)^{-1} \circ \cdots \circ \left(g_\theta^n\right)^{-1},$$

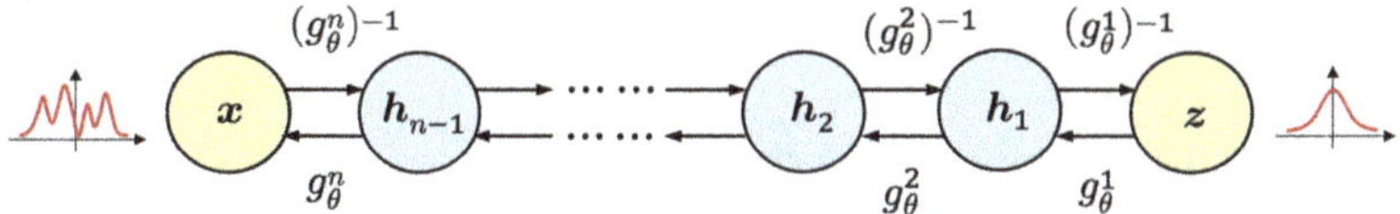

Fig. 9.12 Schematic diagram of the normalization flow distribution transformation

the following expression holds:

$$\boldsymbol{h}_0 = \left(g_\theta^1\right)^{-1} \circ \left(g_\theta^2\right)^{-1} \circ \cdots \circ \left(g_\theta^n\right)^{-1}(\boldsymbol{h}_n). \tag{9.37}$$

From this, it can be seen that through a series of inverse probability density functions, $\{(g_\theta^i)^{-1}\}_{i=n}^1$ and $\boldsymbol{h}_{i-1} = \left(g_\theta^i\right)^{-1}(\boldsymbol{h}_i)$, also transforms the complex distribution into a normal distribution.

A schematic diagram of the normalization flow distribution transformation is shown in Fig. 9.12.

It is worth emphasizing that, $\left(g_\theta^i\right)^{-1}$ is the inverse probability density function of g_θ^i. Thus, normalizing flow is a method of transforming probability density through a series of reversible functions. By repeatedly applying reversible functions, a bidirectional mapping (bijection) sequence of flows is formed. The normal distribution flows through the positive mapping flow to obtain a complex distribution, and the complex distribution flows through the inverse mapping flow to obtain a normal distribution.

9.8.3 Related Works

Many researchers are constantly engaged in academic and applied research on normalizing flows. For example, normalizing flow models are used for image generation (Kingma and Dhariwal 2018), audio generation (Ping et al. 2020), video generation (Kumar et al. 2019), molecular graph generation (Shi et al. 2019), and 3D point cloud generation (Yang et al. 2019).

9.9 Autoregressive Models

Autoregressive models (ARs) are a type of generative models for dealing with time series problems.

This section first introduces the concept of autoregressive models, then introduces deep autoregressive models, and on this basis introduces several representative deep autoregressive generative models.

9.9.1 k-Order Autoregressive Models

Before giving the definition of k-order autoregressive models, we define the autoregressive property first.

Definition 9.9 (Autoregressive Property) Given a set of variables $\{x_1, x_2, \ldots, x_t\}$ over time t, it is called autoregressive property if where the variable x_t depends linearly on some or all its previous variables $\{x_1, x_2, \ldots, x_{t-1}\}$.

Definition 9.10 (k-Order Autoregressive Models) A k-order autoregressive model $AR(k)$ is a statistical model with the autoregressive property, which is defined as follows:

$$x_t = \sum_{i=1}^{k} \varphi_i x_{t-i} + \varepsilon_t. \tag{9.38}$$

Where, $\varphi_1, \ldots, \varphi_k$ are the parameters of the model, and ε_t is white noise.

From the aforementioned definition, it can be seen that the k-order determines the result of x_t:

if $k = 0$, then x_t equals white noise ε_t;
if $k = 1$, then x_t is only related to the variable at the previous moment x_{t-1}; and
if $k = t$, then x_t is related to all its previous variables $\{x_1, x_2, \ldots, x_{t-1}\}$.

Both autoregression and linear regression belong to regression models. However, linear regression is to predict y_t based on x_t, while autoregression is to predict x_t itself based on its previous variables, which is the origin of the compound word "autoregressive" formed by the prefix "auto-" and "regressive".

Autoregressive models can be widely used for time series problems that satisfy the autoregressive property and change over time.

9.9.2 Deep Autoregressive Models

Definition 9.11 (Deep Autoregressive Models) A deep autoregressive model (DAM) is a deep network model with autoregressive property, also known as an autoregressive generative model.

The deep autoregressive models can be modeled based on the chain rule of joint probability (Gregor et al. 2014). Given a set of random variable $\{x_1, x_2, \ldots, x_t\}$ over time t, and $p_\theta(\cdot)$ denotes a probability density function, then its joint probability can be expressed as:

$$p_\theta(x_1, x_2, \ldots, x_t) = \prod_{i=1}^{t} p_\theta(x_i | x_1, x_2, \ldots, x_{i-1}). \tag{9.39}$$

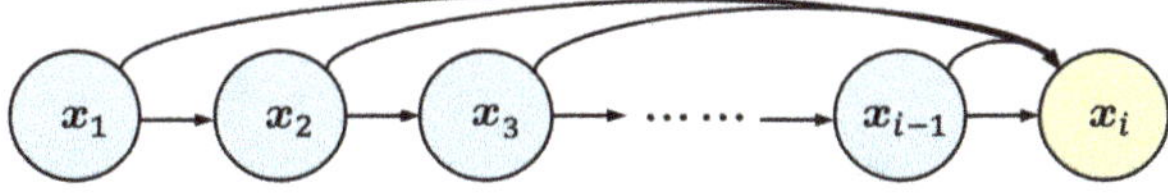

Fig. 9.13 Directed graph of the deep autoregressive model

The directed graph of deep autoregressive model $p_\theta(x_i|x_1, x_2, \ldots, x_{i-1})$ is shown in Fig. 9.13.

Taking the logarithm on Eq. 9.39, we get:

$$\log p_\theta(x_1, x_2, \ldots, x_t) = \sum_{i=1}^{n} \log p_\theta(x_i|x_1, x_2, \ldots, x_{i-1}). \tag{9.40}$$

In this way, the multiplication on the right side of Eq. 9.39 is transformed into the addition, so that the expression form of Eq. 9.40 becomes similar to Eq. 9.38. Since the deep autoregressive model is based on tractable likelihood estimation, it maximizes the likelihood based on the given data during the training process, which can be expressed as:

$$\max_{\theta} \sum_{x_i} \log p_\theta(x_1, x_2, \ldots, x_i) = \max_{\theta} \sum_{x_i} \sum_{i=1}^{n} \log p_\theta(x_i|x_1, x_2, \ldots, x_{i-1}). \tag{9.41}$$

Neural networks can be used to predict each value in a sequence of random variables, and the Softmax function is used to output its probability distribution.

The deep autoregressive model is a generative model that needs to optimize the closeness between the training data and the model distribution, so it usually minimizes the KL divergence between the data $p_{\text{data}}(\cdot)$ and model $p_\theta(\cdot)$ probability distributions:

$$\min_{\theta} D_{KL}(p_{\text{data}}(\cdot) \| p_\theta(\cdot)) = \log p_{\text{data}}(\cdot) - \log p_\theta(\cdot) \tag{9.42}$$

9.9.3 Related Works

Some researchers have conducted in-depth research on deep autoregressive models, including representative deep autoregressive models such as PixelCNN (van Oord et al. 2016) for image pixel prediction, WaveNet (Oord et al. 2016) for original audio generation, and the decoder in Transformer (Vaswani et al. 2017) for natural language processing and computer vision.

9.10 Diffusion Models

Diffusion models (Dhariwal and Nichol 2021), which have emerged in recent years, also belong to unsupervised generative models.

The diffusion model originated from a paper titled "Deep Unsupervised Learning using Nonequilibrium Thermodynamics" (Sohl-Dickstein et al. 2015) published at the *International Conference on Machine Learning* (ICML) in 2015.

Nonequilibrium thermodynamics is a branch of thermodynamics that studies physical systems in non-thermodynamic equilibrium states. Among them, extended irreversible thermodynamics is an important part of nonequilibrium thermodynamics research. A typical example is a drop of ink gradually diffusing in water, where each step of the diffusion process can be described by a probability distribution.

Inspired by nonequilibrium thermodynamics, the paper constructs a diffusion probability model by gradually perturbing the data on multiple scales through the forward diffusion process and then learning to gradually restore the data through the reverse diffusion process.

The diffusion model is a generation model. However, its data generation quality and speed are not as good as the generative adversarial model (GAN).

At the 2020 *Conference on Neural Information Processing Systems* (NeurIPS), a paper titled "Denoising Diffusion Probabilistic Models" (Ho et al. 2020) improved the performance of the original diffusion probability model and demonstrated its powerful ability in high-quality image synthesis.

Here, we first give the definitions of diffusion and diffusion probability models, then introduce the basic principles of diffusion models taken on the denoising diffusion probability model.

9.10.1 *Diffusion Probabilistic Models*

Definition 9.12 (Diffusion)   Diffusion is the net process of movement of molecules or particles under a concentration gradient, generally from a region of higher concentration to a region of lower concentration.

For instance, a tea bag immersed in a cup of hot water will diffuse into the water and change its color.

The concept of diffusion is widely used in many fields, including physics, chemistry, biology, sociology, economics, statistics, data science, and finance.

Definition 9.13 (Diffusion Probabilistic Models)   A diffusion probabilistic model, also known as diffusion model in short, is a type of latent variable generative model inspired by diffusion principle, which consists of the forward diffusion process to slowly add random noise to data and the reverse diffusion process to reconstruct desired data from the noise.

The diffusion model, as a generative model, is used to generate data similar to the training data.

9.10.2 Denoising Diffusion Processes

The working principle of denoising diffusion probability model, fundamentally, its forward diffusion process is to continuously add Gaussian noise to the training data, and then its reverse diffusion process is to learn the denoising to recover the data. After training, this model can be used to generate data by randomly sampling noise and its learned denoising process.

Specifically, the diffusion model is a latent variable model that uses a Markov chain for latent space mapping. This Markov chain gradually adds Gaussian noise to the original image to obtain an approximate posterior probability $q(x_{1:T}|x_0)$, where $x_1, x_2, \ldots, x_T$ are latent variables with the same dimensions. The joint probability distribution $p_\theta(x_{0:T})$ is the reverse process of the model, learning a Gaussian transition, starting from $p(x_T) = \mathcal{N}(x_T; 0, \mathbf{I})$.

A schematic diagram of the denoising diffusion probability model is as dictated in Fig. 9.14. It shows the Markov chain of the model, the process of gradually adding noise to the original image, and its reverse process.

Therefore, the diffusion model includes two steps, namely the forward diffusion process and the reverse diffusion process.

9.10.2.1 Forward Diffusion Process

The forward diffusion process is based on a Markov chain, which gradually adds Gaussian noise to the data x_0 according to the variance schedule $\beta_1, \beta_2, \ldots, \beta_T$. The x_t is obtained through the following conditional probability:

$$q(x_t|x_{t-1}) = \mathcal{N}\left(x_t; \sqrt{1 - \beta_t} x_{t-1}, \beta_t \mathbf{I}\right). \tag{9.43}$$

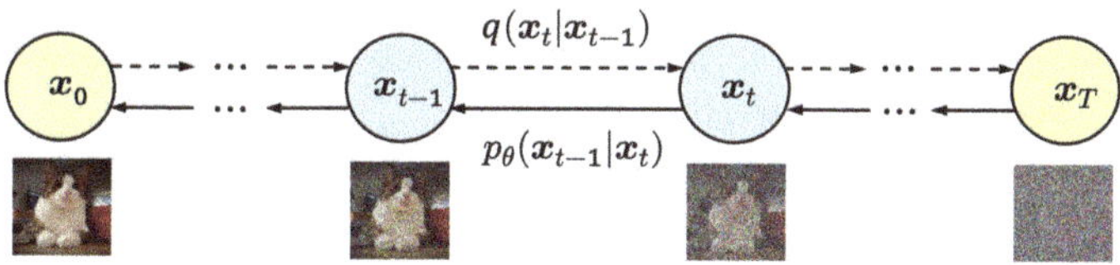

Fig. 9.14 Schematic diagram of the denoising diffusion probability model

Therefore, the Markov chain expression of the forward diffusion process can be obtained:

$$q\left(\boldsymbol{x}_{1:T}|\boldsymbol{x}_0\right) = \prod_{t=1}^{T} q\left(\boldsymbol{x}_t|\boldsymbol{x}_{t-1}\right). \tag{9.44}$$

9.10.2.2 Reverse Diffusion Process

The conditional probability of the reverse diffusion process is as follows.

$$p_\theta(\boldsymbol{x}_{t-1}|\boldsymbol{x}_t) = \mathcal{N}\left(\boldsymbol{x}_{t-1}; \mu_\theta(\boldsymbol{x}_t, t), \sum_\theta (\boldsymbol{x}_t, t)\right). \tag{9.45}$$

Based on this, we can obtain the expression of the Markov chain for the reverse diffusion process:

$$p_\theta\left(\boldsymbol{x}_{0:T}\right) = p\left(\boldsymbol{x}_T\right) \prod_{t=T}^{1} p_\theta\left(\boldsymbol{x}_{t-1}|\boldsymbol{x}_t\right). \tag{9.46}$$

9.10.3 Training of Diffusion Models

The training of diffusion models is carried out by optimizing the usual variational bound of the negative log-likelihood:

$$\mathbb{E}_q\left[-\log \frac{p_\theta\left(\boldsymbol{x}_{0:T}\right)}{q\left(\boldsymbol{x}_{1:T}|\boldsymbol{x}_0\right)}\right] \geq \mathbb{E}[-\log p_\theta\left(\boldsymbol{x}_0\right)]. \tag{9.47}$$

The variance in the forward diffusion process β_t can be learned through reparameterization or kept constant as a hyperparameter, and the expressiveness of the reverse denoising process can be partially ensured by choosing the Gaussian condition in $p_\theta\left(\boldsymbol{x}_{t-1}|\boldsymbol{x}_t\right)$, because when β_t is small, the forward and reverse processes have the same function form. A notable feature of the forward process is that it allows for the closed form on any time step t for $\boldsymbol{x}_t$ sample as follows: let $\alpha_t = 1 - \beta_t$, and $\overline{\alpha}_t = \prod_{s=1}^{t} \alpha_s$, then we have:

$$q(\boldsymbol{x}_t|\boldsymbol{x}_0) = \mathcal{N}\left(\boldsymbol{x}_t; \sqrt{\overline{\alpha}_t}\boldsymbol{x}_0, (1 - \overline{\alpha}_t)\mathbf{I}\right). \tag{9.48}$$

Let the random term in Eq. 9.47 be $\mathbb{E}_q\left[-\log\frac{p_\theta(\boldsymbol{x}_{0:T})}{q(\boldsymbol{x}_{1:T}|\boldsymbol{x}_0)}\right] = L$. Therefore, effective training can be carried out on L using stochastic gradient descent.

$$L = \mathbb{E}_q\left[-\log\frac{p_\theta(\boldsymbol{x}_{0:T})}{q(\boldsymbol{x}_{1:T}|\boldsymbol{x}_0)}\right] = \mathbb{E}_q\left[-\log p(\boldsymbol{x}_T) - \sum_{t\geq 1}\log\frac{p_\theta(\boldsymbol{x}_{t-1}|\boldsymbol{x}_t)}{q(\boldsymbol{x}_t|\boldsymbol{x}_{t-1})}\right].$$

$$(9.49)$$

It can also be improved by reducing the variance, so L is rewritten as the following Equation:

$$L = \mathbb{E}_q\left[D_{KL}\left(q\left(\boldsymbol{x}_t|\boldsymbol{x}_0\right) \| p\left(\boldsymbol{x}_T\right)\right)\right]$$
$$+ \sum_{t>1} D_{KL}\left(q\left(\boldsymbol{x}_{t-1}|\boldsymbol{x}_t, \boldsymbol{x}_0\right) \| p_\theta\left(\boldsymbol{x}_{t-1}|\boldsymbol{x}_t\right)\right) - \log p_\theta\left(\boldsymbol{x}_0|\boldsymbol{x}_1\right). \quad (9.50)$$

The KL divergence is directly applied to the aforementioned expression to compare $p_\theta\left(\boldsymbol{x}_{t-1}|\boldsymbol{x}_t\right)$ with the posterior probability of the forward diffusion process, when the condition is $\boldsymbol{x}_0$ time, its posterior is manageable:

$$q\left(\boldsymbol{x}_{t-1}|\boldsymbol{x}_t, \boldsymbol{x}_0\right) = \mathcal{N}\left(\boldsymbol{x}_{t-1}; \tilde{\mu}_t(\boldsymbol{x}_t, \boldsymbol{x}_0), \tilde{\beta}_t\mathbf{I}\right). \quad (9.51)$$

Where

$$\tilde{\mu}_t(\boldsymbol{x}_t, \boldsymbol{x}_0) = \frac{\sqrt{\bar{\alpha}_{t-1}}\beta_t}{(1-\bar{\alpha}_t)}\boldsymbol{x}_0 + \frac{\sqrt{\bar{\alpha}_t}(1-\bar{\alpha}_{t-1})}{(1-\bar{\alpha}_t)}x_t \quad \text{and}$$

$$\tilde{\beta}_t = \frac{(1-\bar{\alpha}_{t-1})}{(1-\bar{\alpha}_t)}\beta_t. \quad (9.52)$$

Therefore, the KL divergence in Eq. 9.50 is a comparison between Gaussian distributions, so they can be calculated in a closed-form expression in a Rao–Blackwellized manner, rather than high-variance Monte Carlo estimation.

9.10.4 Related Works

The success of the denoising diffusion probabilistic model has attracted significant attention in the field of deep generative models, followed by a series of text-image generation systems based on diffusion models, including DALL-E published by OpenAI, Imagen launched by Google, and Stable Diffusion released by Stability AI.

The applications of diffusion models are mainly in the field of computer vision, including image denoising, image inpainting, super resolution, and image generation. In addition, they are applied in natural language processing, multimodal modeling, time series modeling, etc.

However, due to the bidirectional iterative process of forward diffusion and reverse diffusion in diffusion models, they are considered to have a slow sampling speed, which limits their potential for real-time applications. Therefore, in March 2023, OpenAI submitted a paper titled "Consistency Models" on *arXiv*. This model is inspired by diffusion models, but it can generate real samples in a single unidirectional iteration (Song et al. 2023).

Further Reading

1. Sam Bond-Taylor, Adam Leach, Yang Long, and Chris G. Willcocks. "Deep Generative Modelling: A Comparative Review of VAEs, GANs, Normalizing Flows, Energy-Based and Autoregressive Models." *IEEE Transactions on Pattern Analysis and Machine Intelligence* (TPAMI), 2021.
   [Notes] As can be seen from the title of the paper, it compares and reviews deep generative models such as variational autoencoders (VAEs), generative adversarial networks (GANs), normalizing flow models, energy-based models, and autoregressive models.
2. Margarita Osadchy, Matthew Miller, and Yann Cun. "Synergistic face detection and pose estimation with energy-based models." *Conference on Neural Information Processing Systems* (NIPS), 2004.
   [Notes] This is one of the earlier papers on machine learning based on energy models.
3. Diederik P. Kingma, and Max Welling. "Auto-encoding Variational Bayes." *arXiv preprint* arXiv:1312.6114, 2013.
   [Notes] This paper is considered the seminal work on variational autoencoders (VAE) and was then officially published at the *International Conference on Learning Representations* (ICLR) in 2014.
4. Ian Goodfellow, Jean Pouget-Abadie, Mehdi Mirza, Bing Xu, David Warde-Farley, Sherjil Ozair, Aaron Courville, and Yoshua Bengio. "Generative Adversarial Nets." *Conference on Neural Information Processing Systems* (NIPS), 2014.
   [Notes] This paper proposed one of the prominent frameworks for approaching generative AI. It has since become one of the most influential generative models.
5. Danilo Rezende, and Shakir Mohamed. "Variational Inference with Normalizing Flows." *International Conference on Machine Learning* (ICML), 2015.
   [Notes] This paper is considered an important literature in enabling the normalizing flows to be popularized.
6. Karol Gregor, Ivo Danihelka, Andriy Mnih, Charles Blundell, and Daan Wierstra. "Deep AutoRegressive Networks." *International Conference on Machine Learning* (ICML), 2014.

[Notes] The proposed deep autoregressive network is a new deep generative architecture with autoregressive stochastic hidden units, capable of capturing high-level structures in data to generate high-quality samples.

7. Jascha Sohl-Dickstein, Eric Weiss, Niru Maheswaranathan, and Surya Ganguli. "Deep unsupervised learning using nonequilibrium thermodynamics." *International Conference on Machine Learning* (ICML), 2015.
[Notes] This paper is considered the seminal work on diffusion model inspired by nonequilibrium thermodynamics.

8. Jonathan Ho, Ajay Jain, and Pieter Abbeel. "Denoising diffusion probabilistic models." *Conference on Neural Information Processing Systems* (NeurIPS), 2020. [Notes] The research of this paper improved the performance of the original diffusion probabilistic model, demonstrating its powerful ability in high-quality image synthesis, and further promoted the application of diffusion models.

9. Yang Song, Prafulla Dhariwal, Mark Chen, and Ilya Sutskever. "Consistency Models." *arXiv preprint* arXiv:2303.01469, 2023.
[Notes] This paper proposed the one inspired by diffusion models, but it can generate real samples in a single unidirectional iteration.

References

Dhariwal, P., and A. Nichol (2021). Diffusion models beat GANs on image synthesis. In *Conference on Neural Information Processing Systems (NeurIPS)*.

Du, Y., and I. Mordatch. (2019). Implicit Generation and Modeling with Energy Based Models. In *Conference on Neural Information Processing Systems (NeurIPS)*.

Goodfellow, I. J., J. Pouget-Abadie, M. Mirza, B. Xu, D. Warde-Farley, S. Ozair, A. Courville, and Y. Bengio. (2014). Generative adversarial nets. *Conference on Neural Information Processing Systems (NIPS)* 3: 2672–2680.

Gregor, K., I. Danihelka, A. Mnih, C. Blundell, and D. Wierstra. (2014). In *Deep AutoRegressive Networks. International Conference on Machine Learning (ICML)*, PMLR.

Ho, J., X. Chen, A. Srinivas, Y. Duan and P. Abbeel. (2019). Flow++: Improving Flow-Based Generative Models with Variational Dequantization and Architecture Design. In *International Conference on Machine Learning (ICML)*.

Ho, J., A. Jain, and P. Abbeel. (2020). Denoising diffusion probabilistic models. In *Conference on Neural Information Processing Systems (NIPS)* 6840–6851.

Karras, T., S. Laine, and T. Aila. (2018). A Style-Based Generator Architecture for Generative Adversarial Networks. arXiv Preprint arXiv:1812.04948.

Kingma, D. P., and P. Dhariwal. (2018). Glow: Generative flow with invertible 1x1 convolutions. In *Conference on Neural Information Processing Systems (NeurIPS)*.

Kingma, D. P., and M. Welling. (2014). Auto-encoding Variational Bayes. In *International Conference on Learning Representations (ICLR)*.

Kumar, M., M. Babaeizadeh, D. Erhan, C. Finn, S. Levine, L. Dinh, and D. Kingma. (2019). VideoFlow: A conditional flow-based model for stochastic video generation. In *International Conference on Learning Representations (ICLR)*.

LeCun, Y., Y. Bengio, and G. Hinton. (2015). Deep learning. *Nature* 521(7553): 436.

Makhzani, A., and B. Frey. (2014). k-Sparse autoencoders. In *International Conference on Learning Representations (ICLR)*.

Mordatch, I. (2018). Concept Learning with Energy-based Models. arXiv preprint arXiv:1811.02486.

Ngiam, J., Z. Chen, P. W. Koh, and A. Y. Ng. (2011). Learning deep energy models. In *Proceedings of the 28th International Conference on Machine Learning (ICML)*.

Oord, A. v. d., S. Dieleman, H. Zen, K. Simonyan, O. Vinyals, A. Graves, N. Kalchbrenner, A. Senior, and K. Kavukcuoglu. (2016). WaveNet: A Generative Model for Raw Audio. arXiv preprint arXiv:1609.03499.

Osadchy, M., M. Miller, and Y. Cun. (2004). Synergistic face detection and pose estimation with energy-based models. In *Conference on Neural Information Processing Systems (NIPS)*.

Ping, W., K. Peng, K. Zhao, and Z. Song. (2020). WaveFlow: A compact flow-based model for raw audio. In *International Conference on Machine Learning (ICML)*, PMLR.

Ranzato, M. A., Y.-L. Boureau, S. Chopra, and Y. LeCun. (2007). A unified energy-based framework for unsupervised learning. In *International Conference on Artificial Intelligence and Statistics*, PMLR.

Ranzato, M. A., Y.-L. Boureau, S. Chopra, and Y. LeCun. (2007). A unified energy-based framework for unsupervised learning. In *Artificial Intelligence and Statistics*, PMLR.

Rezende, D., and S. Mohamed. (2015). Variational inference with normalizing flows. In *International Conference on Machine Learning (ICLR)*, PMLR.

Salakhutdinov, R., and G. Hinton. (2009). Deep Boltzmann machines. In *Proceedings of the Twelfth International Conference on Artificial Intelligence and Statistics*, PMLR.

Shi, C., M. Xu, Z. Zhu, W. Zhang, M. Zhang, and J. Tang. (2019). GraphAF: A flow-based autoregressive model for molecular graph generation. In *International Conference on Learning Representations (ICLR)*.

Sohl-Dickstein, J., E. Weiss, N. Maheswaranathan, and S. Ganguli. (2015). Deep unsupervised learning using nonequilibrium thermodynamics. In *International Conference on Machine Learning*, PMLR.

Song, Y., and D. P. Kingma. (2021). How to Train Your Energy-Based Models. arXiv preprint: arXiv: 2101.03288.

Song, Y., P. Dhariwal, M. Chen, and I. Sutskever. (2023). Consistency Models. arXiv preprint arXiv:2303.01469.

van Oord, A., N. Kalchbrenner, and K. Kavukcuoglu. (2016). Pixel recurrent neural networks. In *International Conference on Machine Learning (ICML)*, PMLR.

Vaswani, A., N. Shazeer, N. Parmar, J. Uszkoreit, L. Jones, A. N. Gomez, L. Kaiser, and I. Polosukhin. (2017). Attention is all you need. In *Conference on Neural Information Processing Systems (NIPS)*.

Yang, G., X. Huang, Z. Hao, M.-Y. Liu, S. Belongie, and B. Hariharan. (2019). Pointflow: 3D point cloud generation with continuous normalizing flows. In *Proceedings of the IEEE/CVF International Conference on Computer Vision (ICCV)*.

Chapter 10
Reinforcement Learning Paradigm

Abstract Reinforcement learning is the paradigm for intelligent agents to make sequential decision-making through interaction with the environment, based on the model of Markov decision process. There are two types of reinforcement learning, i.e., model-based and model-free. The former means that all elements in the model are known so that it can employ planning method, while the latter refers that there are uncertain elements in the model; therefore, it uses the learning approach. This chapter starts with an overview of the development stages of reinforcement learning. Second, we give the definition and discuss the related elements of reinforcement learning. We next focus on the one of model-based method called dynamic programming and the two of model-free methods, namely Monte Carlo learning and temporal difference (TD) learning. Then we study the eligibility trace mechanisms in reinforcement learning. After that, we introduce deep reinforcement learning. The comparison of the aforementioned reinforcement learning methods is made finally.

10.1 Overview

Reinforcement learning, as one of the three paradigms in machine learning, has solid theoretical foundations and broad application fields. Its theoretical foundations involve behavioral psychology, Markov decision process, and Monte Carlo learning. And its application fields cover artificial intelligence, operations research, control theory, game theory, and information theory.

The development history of reinforcement learning can be divided into three stages, named as ancient reinforcement learning, neoteric reinforcement learning, and modern reinforcement learning.

10.1.1   Ancient Reinforcement Learning

Ancient reinforcement learning is the stage of reinforcement learning based on the study of animal behaviors.

The first representative achievement of this stage is the theory of respondent conditioning, proposed by Ivan Pavlov in 1897. And it also includes the Rescorla–Wagner model proposed by Robert Rescorla and Allan Wagner in 1972. Their research on animal conditioned and unconditioned responses can be regarded as passive behavioral learning theory.

Another representative achievement of this stage is the theory of operant conditioning. This includes the law of effect discovered by Edward Thorndike in 1898 using the "Thorndike puzzle box", the theory of operant conditioning derived by Burrhus Skinner in 1938 through the "Skinner box", and the theory of reinforcement schedules proposed by Charles Ferster and Burrhus Skinner in 1957. Their research on animal behavior in specific environments can be considered as active behavioral learning theory.

It is worth emphasizing that the concept of "reinforcement" learning was first proposed in the theory of operant conditioning.

Chapter 7 of this book has already provided a detailed introduction to the aforementioned behavioral learning theory.

10.1.2   Neoteric Reinforcement Learning

Neoteric reinforcement learning is a stage of studying reinforcement learning based on automatic control theory.

The representative achievements of this stage are as follows.

In 1954, Marvin Minsky proposed the Theory of Neural-Analog Reinforcement Systems (Minsky 1954).

In 1965, M. D. Waltz and King-sun Fu published "Reinforcement Learning Control Systems" (Waltz and Fu 1965).

In 1970, Jerry M. Mendel and Robert W. McLaren published "Reinforcement Learning Control and Pattern Recognition Systems" (Mendel and McLaren 1970).

In summary, the research of neoteric reinforcement learning involves various technical routes and implementation methods, rather than relying on a single theoretical framework to study reinforcement learning.

10.1.3   Modern Reinforcement Learning

Modern reinforcement learning, also known as modern computational reinforcement learning, can be traced back to the 1980s.

A significant milestone is the doctoral thesis titled *Temporal Credit Assignment in Reinforcement Learning* published by Richard S. Sutton in 1984, under the supervision of professor Andrew G. Barto.

Richard S. Sutton is considered one of the founders of modern computational reinforcement learning.

In the field of machine learning, modern computational reinforcement learning is often simply referred to as reinforcement learning.

Reinforcement learning uses the Markov decision process as a formal theoretical framework to study how an intelligent agent interacts with the external environment and takes appropriate action based on the current state and reward from the environment.

The intelligent agent in reinforcement learning is also a decision-maker, forming a sequential and coherent decision-making process in the interaction with the environment; thus, it is also an effective method of sequential decision-making.

In 2013, the London-based company DeepMind combined reinforcement learning with deep learning, bringing deep reinforcement learning to the stage and demonstrating its strong vitality.

Reinforcement learning can be divided into the following two types: (Sutton and Barto 2018):

(1) The state space is finite or even small enough.
     In this case, reinforcement learning can often find an exact solution, and its sequential decision-making process can be represented in the form of tables, so it is called *tabular solution methods*.
(2) The state space is infinite.
     In this case, even with infinite time, the optimal solution cannot be found, so only approximate calculations can be made with limited resources, which are called *approximate solution methods*.

In this chapter, we will focus on the tabular solution methods that are commonly used in reinforcement learning.

10.2 Definition

Definition 10.1 (Reinforcement Learning) Reinforcement learning (RL) is a sequential decision method in machine learning, which is modeled as Markov decision process. The intelligent agent in RL interacts with the environment, taking corresponding actions based on the current state and feedback from the environment, and learns to maximize its cumulative rewards.

Chapter 7 of this book has already given the following definition. A Markov decision process (MDP) is formalized as a 4-tuple MDP $= \langle S, A, P, R \rangle$, where: S is the state set, A is the action set, P is the state transition probability, and R is the reward function.

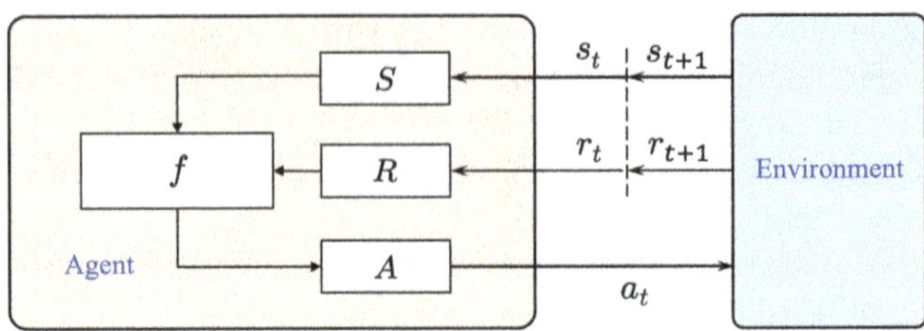

Fig. 10.1 Conceptual model of reinforcement learning

The conceptual model of reinforcement learning is shown in Fig. 10.1. Where, f is the transformation function from state and reward to action, i.e., $f : S \times R \rightarrow A$.

The theoretical framework of reinforcement learning is the Markov decision process.

10.3   Related Elements

For reinforcement learning based on the Markov decision process, what needs to be solved is how to optimally control the Markov decision process, which contains several related elements. The following will introduce these elements separately.

10.3.1   Policy

For the intelligent agent, i.e., the decision maker, in reinforcement learning, the core issue for optimal control of the Markov decision process (MDP) is to find a policy that determines the actions of the intelligent agent.

Definition 10.2 (Policy) Given a Markov decision process MDP $= \langle S, A, P, R \rangle$, its control policy π at time point t is the action a_t to be taken under the state s_t, represented as:

$$\pi \left(a_t | s_t \right) = \mathbb{P}\left[A\left(t \right) = a_t | S\left(t \right) = s_t \right]. \tag{10.1}$$

Where, $\mathbb{P}[\cdot]$ is a transition matrix, which has been introduced in Definition 3.11 in Chap. 3.

The control policy is applied to the Markov decision process (MDP), so that reinforcement learning forms the following operation procedure:

Start from the initial state s_0; according to the policy $\pi \left(a_0 | s_0 \right)$ to get the recommended action a_0, and execute this action; through the reward function $R\left(s_0, a_0 \right)$

to get the reward r_1; then through the transition probability $P\left(s_1|s_0, a_0\right)$ to generate the state s_1; and so forth.

For clarity, we write the aforementioned operation procedure as the following pseudo-code:

$$(0)\ \ t \leftarrow 0$$

$$(1)\ \ a_t \leftarrow \pi\left(a_t|s_t\right)$$

$$(2)\ \ r_{t+1} \leftarrow R\left(s_t, a_t\right)$$

$$(3)\ \ s_{t+1} \leftarrow P\left(s_{t+1}|s_t, a_t\right)$$

$$(4)\ \ t \leftarrow t + 1$$

$$(5)\ \ \text{goto } 1$$

So that it generates the following state-action sequence:

$$\{s_0, a_0, r_1, s_1, a_1, r_2, s_2, a_2, r_3, \ldots\ldots\}.$$

The aforementioned state-action sequence ends with the state s_{goal}, if it is an episodic task. However, if the task is continuous, the state-action sequence will be infinitely extended.

The policy is implemented by the intelligent agent, with the aim of taking the best action in an environment modeled as a Markov decision process.

It is worth noting that in the elements of reinforcement learning, the policy plays a very important role. The policy defines the way the intelligent agent acts in a given state. In general, the policy is a mapping from the perceived state to the action taken based on that state. It corresponds to the stimulus-response rule in behavioral learning. The policy is at the core of optimizing control in the Markov decision process, because only the policy can determine the actions taken by the intelligent agent in a given environment. In general, the policy may be random and can also be interpreted as the conditional probability of which action to take on which state. The goal of the intelligent agent in optimizing control of the Markov decision process is to select an optimal policy π, to maximize cumulative rewards.

10.3.2 Return

Definition 10.3 (Return) The *return* in reinforcement learning refers to the weighted cumulative rewards.

It needs to be emphasized that the return and rewards are related but not equivalent: rewards are part of the return, and the return is derived from the weighted and cumulative rewards.

As defined in Definition 7.8 of Chap. 7, at time t, the reward $r_{t+1} = R(s_t, a_t)$ where the r_{t+1} is called immediate reward at time t, and $r_{t+2}, r_{t+3}, \ldots$ is referred to as the expected rewards at time t, and the return at time t is denoted as G_t. There are two formal definitions of return.

10.3.2.1   Finite Horizon Without Discounts

In this formal definition, the return G_t is the sum of the immediate reward and part of the expected rewards, that is:

$$G_t = r_{t+1} + r_{t+2} + \ldots + r_{t+h} = \sum_{k=0}^{h} r_{t+k+1}. \tag{10.2}$$

Where each term has a weight of 1, and h is referred to as the finite horizon, that is, its time step is finite.

In the process of repeated interaction between the intelligent agent and the environment, most sequential decision tasks naturally form some sub-sequences, referred to as episodes. For example, the episodes exist in such repeated interactive games that include board games, card games, tile games, and video games. An episode ends under a special state called the terminal state. It then resets to the standard initial state or a sample from the standard distribution of the initial state. Sequential decision tasks with these episodes are called episodic tasks. For these episodic tasks, it is necessary to distinguish between non-terminal states and terminal states.

In addition, some sequential decision-making tasks cannot naturally be divided into identifiable episodes, but are continuous actions, without a so-called horizon. For example, a continuous process control or a robotic arm that continuously acts on the assembly line. These tasks are called continuing tasks. Equation 10.2 cannot be directly applied to continuing tasks, because in these tasks, $h \rightarrow \infty$, its expected reward itself belongs to the infinite horizon.

In this case, the weights involved in the return are called discounts. Hence, the following formal definition is given.

10.3.2.2   Infinite Horizon with Discounts

In this formal definition, the return G_t is the sum of the immediate reward and the expected rewards with discounts, that is:

$$G_t = r_{t+1} + \gamma r_{t+2} + \gamma^2 r_{t+3} + \ldots = \sum_{k=0}^{\infty} \gamma^k r_{t+k+1}. \tag{10.3}$$

Where, γ is a parameter, known as the discount factor, which discounts the expected rewards.

Theoretically, $0 \le \gamma \le 1$, known as the discount rate, represents the degree of association between immediate rewards and expected rewards in the return G_t. It can be divided into the following three cases:

(1) If $\gamma = 0$, then $G_t = r_{t+1}$, that is, the return only focuses on the immediate reward r_{t+1}.
(2) If $0 < \gamma < 1$, the value of G_t is convergent, equivalent to the expected reward $r_{t+2}, r_{t+3}, \ldots$ has a boundary, so the infinite horizon problem is transformed into a finite horizon problem, and its return is equal to the immediate reward plus a part of the future discounted reward.
(3) If $\gamma = 1$, then G_t becomes:

$$G_t = r_{t+1} + r_{t+2} + r_{t+3} + \ldots = \sum_{k=0}^{\infty} r_{t+k+1}.$$

In this case, G_t does not converge.

In conclusion, to ensure convergence, the discount rate commonly chosen is $0 \le \gamma < 1$.

Equation 10.2 is referred to as the finite horizon model, while Eq. 10.3 is referred to as the infinite horizon, discounted model. So far, most reinforcement learning algorithms have adopted the latter.

The goal of the intelligent agent is to maximize the return, that is, to maximize the expected return.

10.3.3   Value Functions

The value functions are the way to link returns with policies. The optimal control algorithms of Markov decision processes mainly calculate the optimal policy through the value function.

A value function represents an evaluation, that is, it estimates the quality of a particular state in which the intelligent agent is located, or the quality of a specific action performed by the intelligent agent in that state. Here, the quality is measured by the expected return. The value function is defined for a specific policy.

There are following two types of value functions for reinforcement learning.

10.3.3.1   State-Value Function

The value of a state s under policy π, denoted as $V^{\pi}(s)$, is the expected return starting from s and following π. Here, the infinite horizon discounted model, i.e.,

Eq. 10.3, is used, so $V^\pi (s)$ can be expressed as:

$$V^\pi (s) = \mathbb{E}^\pi [G_t \mid S (t) = s]$$

$$= \mathbb{E}^\pi \left[\sum_{k=0}^{\infty} \gamma^k r_{t+k+1} \mid S (t) = s \right]. \tag{10.4}$$

The aforementioned equation is referred to as the state-value function under policy π.

10.3.3.2   State-Action-Value Function

Similarly, the state-action-value function under policy π, denoted as $Q^\pi (s, a)$, is defined as the expected return starting from state s, taking action a, and following policy π, expressed as:

$$Q^\pi (s, a) = \mathbb{E}^\pi [G_t \mid S (t) = s, A (t) = a]$$

$$= \mathbb{E}^\pi \left[\sum_{k=0}^{\infty} \gamma^k r_{t+k+1} \mid S (t) = s, A (t) = a \right]. \tag{10.5}$$

10.3.4   Bellman Equations

The fundamental property of value functions is that they satisfy certain recursive relationships, manifested as mathematical expectations. The optimization calculation methods of the value functions are called the Bellman equations, which correspond to two types of value functions.

10.3.4.1   Bellman Equation for State-Value Function

For the state-value function, i.e., Eq. 10.4, for any state s under policy π, there exists the following consistency condition between its value and the subsequent states:

$$V^\pi (s) = \mathbb{E}^\pi \left[\sum_{k=0}^{\infty} \gamma^k r_{t+k+1} \mid S (t) = s \right]$$

$$= \mathbb{E}^\pi \left[r_{t+1} + \gamma \sum_{k=1}^{\infty} \gamma^k r_{t+k+1} \mid S (t) = s \right]$$

$$= \mathbb{E}^{\pi} \left[r_{t+1} + \gamma V^{\pi} \left(s_{t+1} \right) \mid S\left(t \right) = s \right]$$

$$= \sum_{s'} P\left(s' \mid s, \pi\left(s \right) \right) \left(R\left(s, a \right) + \gamma V^{\pi}\left(s' \right) \right). \tag{10.6}$$

The aforementioned equation is the Bellman equation for the state-value function V^{π}. It states that the expected value of a state is recursively defined based on immediate rewards and the value of subsequent states, these state-values are discounted and weighted by their transition probabilities. For this set of equations, V^{π} is the unique solution. That is to say, for a given policy π, V^{π} is unique.

For any given Markov decision process, the goal is to find an optimal policy, that is, a policy that yields the maximum reward. This means that for all states $s \in S$, the state-value function in Eq. 10.6 should be maximized.

An optimal policy, denoted as π^{*}, for all $s \in S$ and all policies π, satisfies $V^{\pi^{*}}\left(s \right) \geq V^{\pi}\left(s \right)$. It has been proven that the optimal solution $V^{*} = V^{\pi^{*}}$ satisfies the following equation:

$$V^{*}\left(s \right) = \max_{a \in A} \sum_{s' \in S} P\left(s' \mid s, a \right) \left(R\left(s, a \right) + \gamma V^{*}\left(s' \right) \right). \tag{10.7}$$

Equation 10.7 is known as the Bellman optimality equation. It indicates that the state-value under the optimal strategy must be equal to the expected return of the optimal action in that state. Given this optimal state-value function V^{*}, to select an optimal action, the following equation can be applied:

$$\pi^{*}\left(s \right) = \arg\max_{a} \sum_{s' \in S} P\left(s' \mid s, a \right) \left(R\left(s, a \right) + \gamma V^{*}\left(s' \right) \right). \tag{10.8}$$

Equation 10.8 is called the greedy policy because it uses the value function V^{*} and greedily selects its optimal action.

10.3.4.2   Bellman Equation for State-Action-Value Function

Similarly, we can obtain the optimal state-action-value function of Eq. 10.5 is as follows:

$$Q^{*}\left(s, a \right) = \mathbb{E}^{\pi} \left[\sum_{k=0}^{\infty} \gamma^{k} r_{t+k} \mid S\left(t \right) = s, A\left(t \right) = a \right]$$

$$= \mathbb{E}^{\pi} \left[R\left(s, a, s' \right) + \gamma \max_{a'} Q^{*}\left(s', a' \right) \mid S\left(t \right) = s, A\left(t \right) = a \right]$$

$$= \sum_{s' \in S} P\left(s' \mid s, a \right) \left(R\left(s, a \right) + \gamma \max_{a'} Q^{*}\left(s', a' \right) \right). \tag{10.9}$$

The Q function is useful because it uses transition probabilities to perform the weighted sum of different choices. The relationship between Q^* and V^* is given by the following equations:

$$Q^*(s, a) = \sum_{s' \in S} P\left(s'|s, a\right) \left(R(s, a) + \gamma V^*(s')\right). \tag{10.10}$$

$$V^*(s) = \max_{a \in A} Q^*(s, a). \tag{10.11}$$

Thus, the optimal policy selection can be represented as:

$$\pi^*(s) = \arg\max_{a \in A} Q^*(s, a). \tag{10.12}$$

Equation 10.12 shows that the optimal policy refers to the action that leads to the highest expected utility among a number of possible states.

10.3.4.3   Comparison on Bellman Equations

We can take a comparison on the two Bellman equations used to optimize the state value function and the state-action value function, which are shown as following:

$$V^*(s) = \max_{a \in A} \sum_{s' \in S} P\left(s'|s, a\right) \left(R(s, a) + \gamma V^*(s')\right).$$

$$Q^*(s, a) = \sum_{s' \in S} P\left(s'|s, a\right) \left(R(s, a) + \gamma \max_{a'} Q^*(s', a')\right).$$

To facilitate the understanding, their differences are depicted in Fig. 10.2.

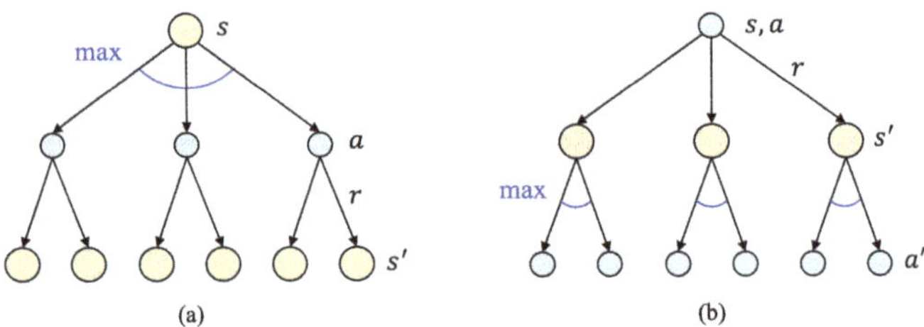

Fig. 10.2   Bellman equations for state value function vs. state-action value function

10.3.5   Model-Based and Model-Free

10.3.5.1   Model-Based Methods

A model-based method refers to that for a Markov decision process, MDP $=$ $\langle S, A, P, R \rangle$, where all the elements of the model are known; therefore, we can directly use the Bellman equation to calculate the value functions for the optimal action selection that maximizes the reward for the environment.

Knowing all the elements of the model, so that we can do planning based on the model, and dynamic programming (DP) is used for the planning.

Why is the term programming related to planning? We can find the explanation in Dasgupta et al. (2006), what they wrote is as follows.

> The origin of the term dynamic programming has very little to do with writing code. It was first coined by Richard Bellman in the 1950s, a time when computer programming was an esoteric activity practiced by so few people as to not even merit a name. Back then programming meant "planning", and "dynamic programming" was conceived to optimally plan multistage processes.

Back to the method of dynamic programming, it is an effective solution to multistage process optimization problems and works by computing value functions that can be used to optimize action selection based on the model.

Using dynamic programming to do the optimal control of the Markov decision process is also known as decision-theoretic planning (Blythe 1999). Jerome A. Feldman and Robert F. Sproull were among the first scholars to work on planning in decision theory (Feldman and Sproull 1977).

Decision-theoretic planning is also known as planning under uncertainty (Blythe 1999; Lavalle 2006). That is, the information received by the agent from the environment is uncertain, the same action does not always lead to the same result, and there may be a trade-off between different outcomes of planning.

10.3.5.2   Model-Free Methods

In most sequential decision-making tasks, it is often difficult to obtain a perfect model of the Markov decision process, that is, there are uncertain elements in the model.

That is, we do not know some environment-dependent elements of the Markov decision process MDP $= \langle S, A, P, R \rangle$, i.e., the transition probability P and the reward function R, so we cannot directly use the Bellman equation to solve for the value function and policy. This situation without a perfect model can be relied on is called model-free.

However, when an intelligent agent takes an action under a state, it will receive rewards from the environment, meaning that the intelligent agent can indirectly perceive the objective existence of the transition probability and reward function under the state and action. Therefore, the intelligent agent forms a simulation strategy through interaction with the environment, generates state and reward samples, and then uses these samples to estimate the value function and strategy.

In summary, the model-free method is a learning method without a perfect model. That is to say, when there is no perfect model of the Markov decision process, the intelligent agent can learn the objectively existing samples of states and rewards in the trial-and-error process of interacting with the environment and then use them to optimize the strategy and value functions.

Therefore, the combination of the Markov decision process and learning can be called decision-theoretic learning.

In the model-free method, there is a continuous process of exploration and exploitation. That is to say, the next action is determined based on the current state and the rewards of the environment, and through multiple iterations and learning of exploration and exploitation, its return is maximized.

The exploration–exploitation dilemma refers to the difficult problem encountered when balancing exploration and exploitation: in order to maximize returns, the model-free method tends to exploit actions that have been taken effectively; but to explore more effective actions, actions that have not been chosen before must be tried. The dilemma lies in that, whether it is exploration or exploitation, it is impossible to proceed without failure. The model-free method needs to strike a balance between exploration and exploitation in order to maximize its return. The model-free method can only get a reliable estimate of the expected rewards by trying various actions; the model-free method tends to choose better actions by trying various actions.

The exploration–exploitation dilemma is a unique problem of the reinforcement learning paradigm, and there is no such phenomenon in the supervised learning and unsupervised learning paradigms.

10.3.6 *On-Policy and Off-Policy*

On-policy learning refers to the situation where the policy for evaluation and improvement is the same as the policy used for decision-making.

Off-policy learning, on the other hand, refers to the situation where the policy for evaluation and improvement is different from the policy used for decision-making. In a sense, off-policy is like an extrapolation problem, trying to make informed decisions based on available data and predict situations that have not been experienced before.

The comparison between on-policy and off-policy is as depicted in Table 10.1.

Table 10.1 On-policy vs. off-policy

On-policy	Off-policy
Using the deterministic outcomes or samples from the target policy to train the algorithm	Training on the distribution of transitions or episodes produced by a different behavior policy rather than that produced by the target policy
Learning policy π from the episodes that generated using the π	Learning policy π from the episodes generated using another policy ϕ
Learning while doing the job	Learning while watching other one doing the job

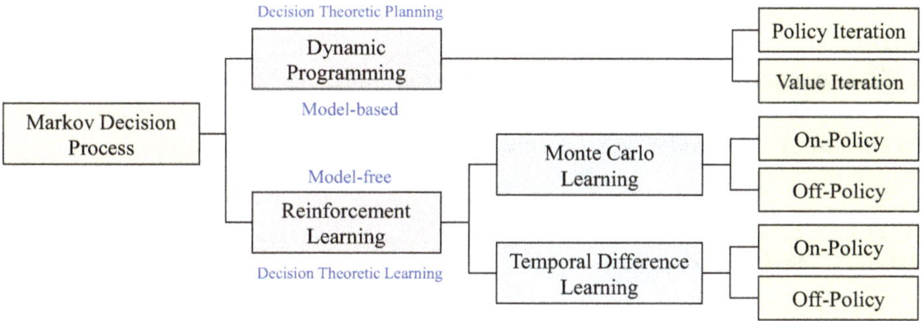

Fig. 10.3 Hierarchical relationships of dynamic programming and reinforcement learning

10.3.7   Hierarchical Relationships

There are commonalities and differences between model-based and model-free: both are modeling based on the Markov decision process, but all the elements in the former model are known and can be planned; while the latter model has uncertain elements and needs to learn from the environment.

The method of model-based is dynamic programming, includes policy iteration and value iteration.

The method of model-free is reinforcement learning, includes Monte Carlo learning and temporal difference learning, and both with on-policy and off-policy.

Dynamic programming is also known as decision-theoretic planning since its theoretical basis is Markov decision process. Similarly, reinforcement learning can be thought as decision-theoretic learning.

Reinforcement learning is also known as approximate dynamic programming (Sutton and Barto 2018; Bertsekas, 2012). Therefore, the similarities and differences between model-based and model-free can also be judged from the viewpoint of dynamic programming and approximate dynamic programming.

For ease to compare, the hierarchical relationships of dynamic programming and reinforcement learning is depicted in Fig. 10.3.

10.4   Dynamic Programming

In the early 1950s, American mathematician Richard Bellman, while studying the optimization problem of multistep decision processes, transformed the multistep process into a series of single-step problems, using the relationships between stages to solve them one by one, thus creating a new method for solving multistep process optimization problems, namely the theory of dynamic programming (Bellman 1952).

In decision theory planning, dynamic programming is used to optimize control of Markov decision processes and calculate the optimal policy of Markov decision processes.

The two core methods of dynamic programming are value iteration (Bellman 1957) and policy iteration (Howard 1960) proposed by Richard Bellman and Ronald A. Howard, respectively, based on the Bellman equations.

10.4.1   Value Iteration

The value iteration focuses entirely on directly evaluating its value function. Its algorithm calculates the necessary updates in real time, combining the process of policy evaluation and policy improvement. Therefore, the Bellman optimization equation of Eq. 10.10 has become the following update rule:

$$Q_{t+1}(s, a) = \sum_{s' \in S} P(s' \mid s, a)(R(s, a) + \gamma V_t(s')).$$

$$V_{t+1}(s) = \max_a Q_{t+1}(s, a). \tag{10.13}$$

The value iteration algorithm starts from the value function of all states. V_0 to start and iteratively update the value of each state according to Eq. 10.13 to obtain the next value function V_t, where $t = 1, 2, 3, \cdots$. The sequence of value functions formed by the value iteration algorithm is shown in Fig. 10.4.

In fact, in the value iteration algorithm, the intermediate Q value function is also generated, forming the sequence shown in Fig. 10.5.

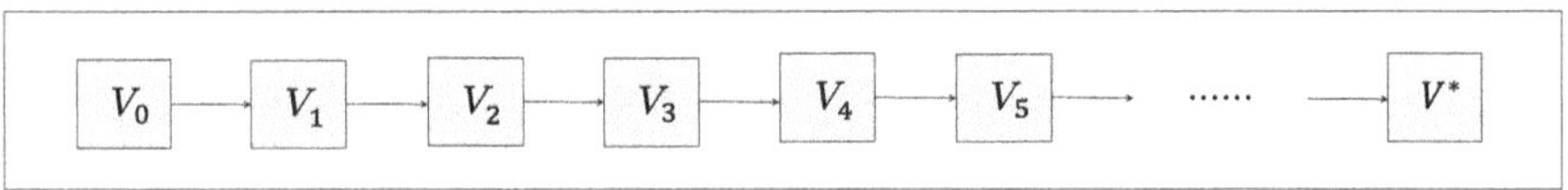

Fig. 10.4 Sequence of value functions generated by value iteration

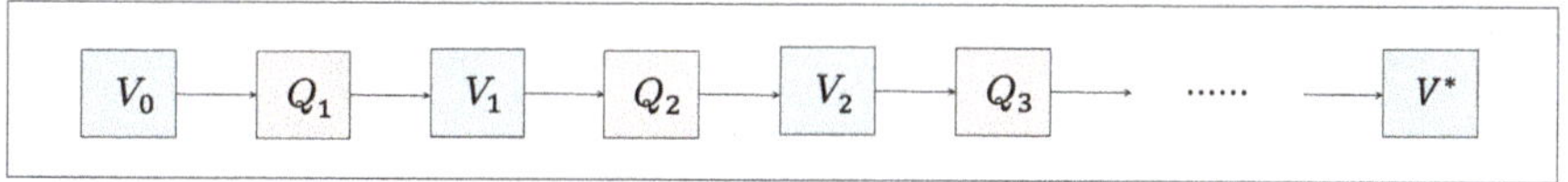

Fig. 10.5 Sequence of value functions and Q value functions generated by value iteration

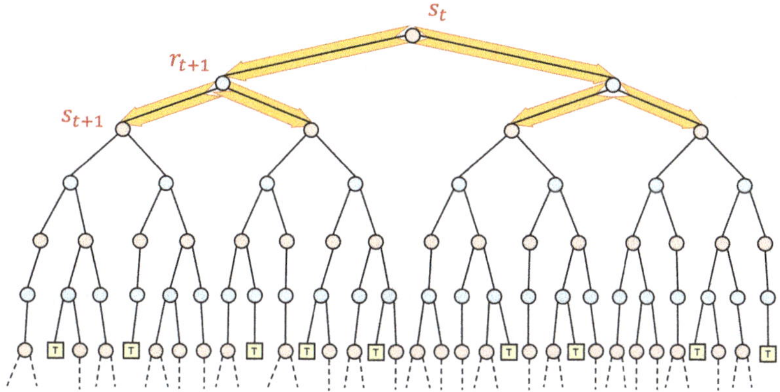

Fig. 10.6 Backup diagram for dynamic programming to update value function

The value iteration algorithm of dynamic programming is to update $V(s_t)$ toward estimated return $V(s_{t+1})$, i.e.,

$$V(s_t) \leftarrow \mathbb{E}[r_{t+1} + \gamma V(s_{t+1})]. \tag{10.14}$$

The backup diagram is depicted in Fig. 10.6.

10.4.2   Policy Iteration

Policy iteration is divided into the following two processes: policy evaluation and policy improvement, in which the policy evaluation assesses the current policy by calculating its value function, while the policy improvement process obtains its improved policy through the maximization of the value function. Therefore, the policy iteration repeats these two processes until it converges to an optimal policy. The following will detail the content of these two processes and then give the algorithm of policy iteration.

10.4.2.1   Policy Evaluation

For a given policy π, calculating its state-value function V^{π} is called policy evaluation, which is a value prediction problem based on policy (prediction problem).

Rewrite the Bellman equation for state-value function as follows:

$$V^{\pi}(s) = \sum_{s' \in S} P\left(s' \mid s, \pi(s)\right) \left(R(s, \pi(s)) + \gamma V^{\pi}(s')\right). \tag{10.15}$$

Since the dynamics of the system are fully known, that is, the MDP model is given, under the condition that the number of states $|S|$ in the state set S is unknown, Eq. 10.15 forms the simultaneous linear equations for $|S|$. In this case, linear programming can be used to solve it.

However, an iterative process is possible, and in fact, this is the case in both model-based and model-free scenarios. The Bellman equation is transformed into an update rule, using a "look-ahead" approach, i.e., extending the planning time domain by one step, updating the current value function V_k^{π} to V_{k+1}^{π}:

$$V_{k+1}^{\pi}(s) = \mathbb{E}^{\pi}\left(r_t + \gamma V_k^{\pi}(s_{t+1}) \mid S(t) = s\right)$$

$$= \sum_{s' \in S} P\left(s' \mid s, \pi(s)\right) \left(R(s, \pi(s)) + \gamma V_k^{\pi}(s')\right). \tag{10.16}$$

When k tends to infinity, the approximation sequence of V_k^{π} can show convergence. For convergence, this update rule is used for each state $s \in S$ in each iteration. It replaces the value under the old state with the value under a new state, which is weighted by the expected value of the new state, intermediate rewards and penalties, and transition probabilities. This operation is called full backup because it is based on all possible transitions from that state.

Regarding the termination problem of the policy evaluation algorithm, formally speaking, policy evaluation can only converge within the limit, but in practice, it must stop before this. A typical stopping condition for policy evaluation is to check $\Delta = \max_{s \in S} \left| V_{k+1}^{\pi}(s) - V_k^{\pi}(s) \right|$ after each iteration, if the value of Δ is small enough, the algorithm stops.

10.4.2.2   Policy Improvement

Given the value function of the policy $\pi(s)$, which is the result of policy evaluation, we can try to improve this policy. First, use the following equation to identify the value of all actions:

$$Q^{\pi}(s, a) = \mathbb{E}^{\pi}\left[r_t + \gamma V_k^{\pi}(s_{t+1}) \mid S(t) = s, A(t) = a\right]$$

$$= \sum_{s' \in S} P\left(s' \mid s, a\right) \left(R(s, a) + \gamma V^{\pi}(s')\right). \tag{10.17}$$

If for some action $a \in A$, there exists $Q^\pi(s, a)$ greater than $V^\pi(s)$, then directly choosing action a will be better. In other words, the current strategy can be improved by choosing a different and better action under a specific state. In fact, all actions under all states can be evaluated, and the optimal action under all states can be chosen. That is to say, the greedy strategy π' can be calculated by choosing the optimal action under each state, based on the current value function V^π. That is:

$$\begin{aligned}
\pi'(s) &= \arg\max_{a \in A} Q^\pi(s, a) \\
&= \arg\max_{a \in A} \mathbb{E}\left(r_t + \gamma V^\pi(s_{t+1}) \mid S(t) = s, A(t) = a\right) \\
&= \arg\max_{a \in A} \sum_{s' \in S} P(s' \mid s, a)\left(R(s, a) + \gamma V^\pi(s')\right). \quad (10.18)
\end{aligned}$$

It is called policy improvement since the process of calculating an improved policy by choosing the optimal action related to the original policy value function.

If the policy does not change through this method, it means that the policy is already optimal, and its value function satisfies the Bellman equation of the optimal value function. Similarly, these steps of a stochastic policy can also be performed by mixing the action probability into the expected operator.

10.4.2.3 Algorithm of Policy Iteration

As mentioned earlier, policy iteration consists of two processes, i.e., policy evaluation and policy improvement. The policy evaluation process calculates the value function of the current policy, the policy improvement process calculates an improved policy by maximizing the value function, and the policy iteration repeats these two processes until it converges to an optimal policy.

Main steps of the algorithm is given as follows:

(1) Initialization:
   Starting from an arbitrary initial policy π_0.
(2) Policy evaluation:
   Using Eq. 10.16 to calculate the value function of of the policy, V_k^π.
(3) Policy improvement:
   Using Eq. 10.18 to calculate π_k into π_{k+1}.
   If $\pi_{k+1}(s) = \pi_k(s)$ for all states s, then the policy tends to be stable and thus stop the algorithm, else goto 2.

The policy iteration generates a sequence of alternating policies and value functions, as shown in Fig. 10.7.

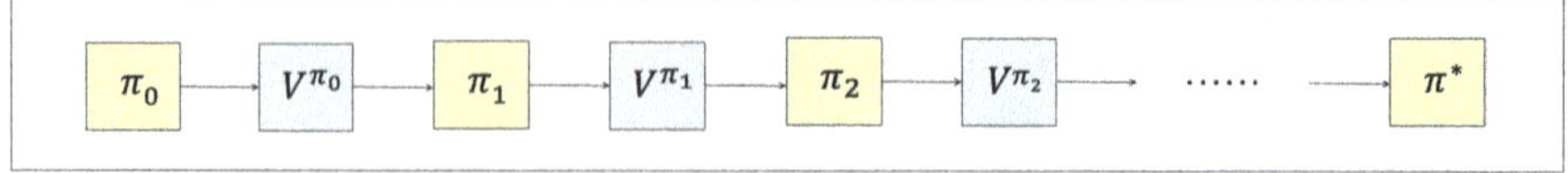

Fig. 10.7 Sequence of policies and value functions generated by policy iteration

10.4.3 Asynchronous Dynamic Programming

The dynamic programming discussed earlier has a disadvantage, i.e., it involves operations on the entire state set of the Markov decision process (MDP), that is, it requires a sweep of the state set. If the state set is very large, even a single sweep can be very time-consuming. For example, the game of backgammon has over 10^{20} states. Even if value iteration could be performed at a rate of one million states per second, it would take thousands of years to complete a single sweep.

Asynchronous dynamic programming does not systematically sweep the state set, but is an in-place algorithm for iterative dynamic programming.

The in-place algorithm refers to the conversion of input variables without the use of auxiliary storage space, but allows a small amount of additional storage space for handling auxiliary variables. During the execution of an in-place algorithm, the input variables are usually overwritten by the output variables. That is to say, the in-place algorithm only updates its input sequence within the original storage space by replacing variables, without the need for auxiliary storage space elsewhere.

Asynchronous dynamic programming backs up state-values in any order and can use any value of other states. Some state-values may be backed up multiple times before other values are backed up. However, in order to converge correctly, asynchronous algorithms must continue to back up all state-values: after calculating some points, it cannot ignore any state. Asynchronous dynamic programming has great flexibility in choosing the state to apply backup operations.

For example, an asynchronous value iteration algorithm uses the state-value iteration shown in Eq. 10.14 for backup, and only backs up the value of one state t at each step s_t. For $0 \leq \gamma < 1$, as long as all states appear in the sequence $\{s_t\}$ an infinite number of times, it can ensure the asymptotic convergence to V^*. This sequence $\{s_t\}$ can also be random. Similarly, policy evaluation and value iteration backups can be merged to produce an asynchronous truncated policy iteration. Clearly, different backup building blocks formed can be flexibly used in various sweepless dynamic programming algorithms.

Of course, avoiding sweeping does not necessarily mean that the amount of computation can be reduced. It's just that the algorithm does not need to be locked in any long-term sweeping operation during the policy evaluation process before it can improve the policy. This flexibility can be exploited by choosing the states to which backups are applied, thereby improving the execution speed of the algorithm. It is also possible to try to sort the backups, allowing value information to be transferred from one state to another effectively. Some states may not need to back up their state-values as much as other states. If they are irrelevant to the optimal behavior, you can even try to ignore backing up some states.

Asynchronous algorithms also make the mix of computation and real-time interaction easier. To solve a Markov decision process (MDP), an iterative dynamic programming algorithm can be run while the intelligent agent is actually experiencing the MDP. The experience of the intelligent agent can be used to determine the states on which the dynamic programming algorithm acts for its backups. At the same time, the latest value and policy information from the dynamic programming algorithm can guide the decisions of the intelligent agent. For example, backups can be applied to states when the intelligent agent visits. This can focus the backups of the dynamic programming algorithm on the part of the state set that is most relevant to the intelligent agent.

10.5   Monte Carlo Learning

As mentioned earlier, model-free reinforcement learning refers to the uncertainty elements in the model. That is, the intelligent agent only knows the state in the environment and the reward from the environment, but does not know the exact transition probability and reward function.

Model-free methods form a sample sequence of state, action, and reward at each moment, called experience. This sample sequence can either be the real experience of the intelligent agent interacting with the environment or the simulated experience obtained through some method.

Learning from real experiences is very important because it can make the best action decisions based on real sample sequences. Learning from simulated experiences is also useful, because it can infer what actions to take as the best based on simulated sample sequences.

Monte Carlo (MC) learning is one of the model-free methods, specifically refers to the method based on averaging sample returns, also known as unbiased estimates.

The Monte Carlo methods, which have been discussed in Chap. 3, are a set of methods of estimating the probability of a random event or the expected value of a random variable by the frequency of the event occurring through repeated random sampling and using it as an approximate numerical result.

To ensure a well-defined return, reinforcement learning using Monte Carlo learning is limited to episodic tasks. That is, its experiences can be divided into episodes, and regardless of the action chosen, all episodes will eventually terminate. Only after an episode is completed, the value assessment and policy will change. Therefore, Monte Carlo learning can continuously learn on the basis of individual episodes, rather than learning in the sense of individual steps.

An episode at time $\{0, 1, \ldots, T\}$ is a sequence consisting of states, actions, and rewards:

$$\{s_0, a_0, r_1, s_1, a_1, r_2, \ldots, s_T, a_T, r_{T+1}\}.$$

Next, we discuss how to use the Monte Carlo learning to obtain the value functions and optimize the policy.

10.5.1   Monte Carlo Prediction

The problem to be solved is how to use the Monte Carlo learning to obtain the state-value function for a given strategy.

The state-value refers to the expected return, that is, the expected cumulative rewards starting from that state. Obviously, one method to estimate the state-value from experience is to simply average the returns obtained after visiting that state. As the number of returns increases, this average will converge to the expected value. This is the basic idea of the Monte Carlo learning.

Specifically, given a set of episodes obtained by following the policy π and passing through state s, we want to estimate the value of state s according to policy π, denoted as $V^\pi(s)$. Each occurrence of state s in an episode is called a visit to s. This includes both first visits and every visit.

For the first visit, the Monte Carlo learning takes $V^\pi(s)$ as the first visit to s mean return.

The Monte Carlo learning waits until the return of the visit and then uses this return as the target of $V(s_t)$. A suitable every-visit Monte Carlo learning for non-terminating environments is as follows:

$$V(s_t) \leftarrow V(s_t) + \alpha \left(G_t - V(s_t) \right). \tag{10.19}$$

Where, G_t is the actual return at the time t, and α is the constant step-size parameter. As shown in Fig. 10.8, the return G_t needs to wait until the end of the episode to get the corresponding value, and then determine the increment of $V(s_t)$.

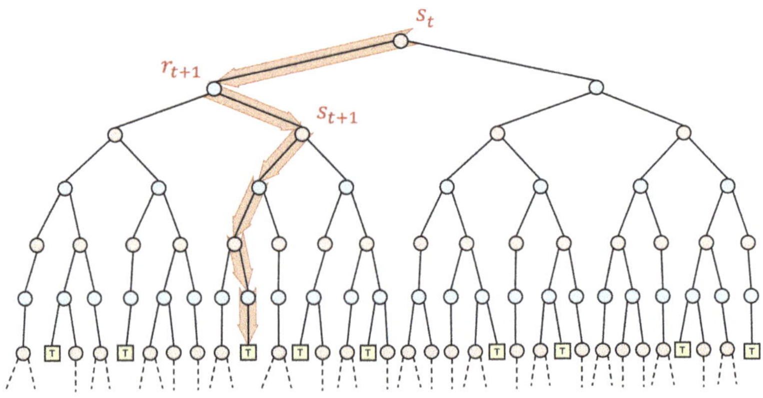

Fig. 10.8 Backup diagram for Monte Carlo methods to update value function

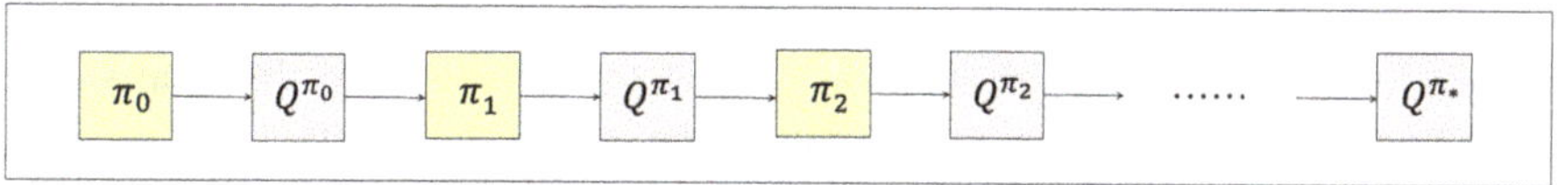

Fig. 10.9 Sequence of policies and action-value functions generated by policy iteration

Every-visit is similar to the first-visit to Monte Carlo learning but a little different: every-visit is an extension of first-visit, and $V^\pi(s)$ of every-visit is taken as the mean return after all visits.

10.5.2 Monte Carlo Control

Let's take another look at how to use the Monte Carlo learning for the optimal control of Markov decision processes, that is, how to get an approximate optimization policy. The overall idea is based on the generalized policy iteration introduced in Sect. 10.4 of this chapter.

With generalized policy iteration, an approximate policy and an approximate value function can be maintained at the same time. This value function is repeatedly modified to be closer to the current policy, and this policy is repeatedly improved relative to the current value function, as shown in Fig. 10.9. All policies and value functions are identifiable, the policy is always improved against the value function, and the value function is always driven by the policy. These two actions are somewhat contradictory because each creates a constantly moving target for the other. But because of this, both the policy and the value function approach optimality.

Consider a Monte Carlo version of policy iteration. With this method, a sequence of alternating policy evaluation and policy iteration is formed, as shown in Fig. 10.9. This sequence starts with an arbitrary policy π and ends with the optimal strategy and action-value function.

For each policy π_k, use the Monte Carlo learning to calculate the corresponding action-value function Q^{π_k}. The greedy policy corresponding to the action-value function selects an action with the maximum action-value for each $s \in S$, that is:

$$\pi(s) = \arg\max_a Q(s, a). \tag{10.20}$$

As the greedy strategy corresponding to Q^{π_k}, then by constructing each π_{k+1} to achieve policy improvement, that is:

$$Q^{\pi_k}(s, \pi_{k+1}(s)) = Q^{\pi_k}\left(s, \arg\max_a Q^{\pi_k}(s, a)\right) = \arg\max_a Q^{\pi_k}(s, a). \tag{10.21}$$

In the model-based method, i.e., dynamic programming strategy iteration, the main evaluation is of the current state's value function V, and the final strategy is obtained through the state-action-value function Q, the conversion from V to Q is direct. In the model-free methods, i.e., Monte Carlo learning and Temporal Difference learning, it cannot be directly converted, so the only choice is to directly evaluate the policy for Q.

As can be seen, whether the correct value functions can be obtained mainly depends on whether the intelligent agent's experiences are sufficient. Therefore, how to obtain sufficient experiences is the core of reinforcement learning. In the model-based method, in order to ensure the convergence of the value function, the algorithm will scan the states in the state space one by one. In model-free methods, the premise of fully evaluating the policy and value functions is that each state can be visited. To achieve this, the Monte Carlo learning often uses exploring starts.

The exploring starts referring to each episode starting from a state–action pair, and each state–action pair has a non-zero probability at startup. Then, to ensure that each state–action pair has a chance to be selected, it adopts a gentle exploration strategy, namely the ε-greedy policy:

$$\pi\,(a|s) = \begin{cases} 1 - \varepsilon + \frac{\varepsilon}{|A(s)|}, & \text{if } a = \underset{a \in A}{\arg\max}\, Q\,(s, a)\,. \\ \frac{\varepsilon}{|A(s)|}, & \text{otherwise.} \end{cases} \tag{10.22}$$

This Monte Carlo learning is an on-policy method. Using an ε-greedy strategy means that it often chooses the action with the highest estimated state-action-value and replaces random action selection with a probability of ε.

Off-policy methods are proposed to address the exploration–exploitation dilemma and are further divided into two different strategies: one that is learned and becomes the optimal strategy, and another that is more exploratory and used to generate actions; the learned strategy is called the target policy, and the strategy used to generate actions is called the behavior policy. The advantage of this separation is that the target policy can be deterministic, while the behavior policy can continue to sample all possible actions.

The main problem with Monte Carlo learning is that it must wait until the end of a sequence of episode samples before it can estimate the current value function. Temporal difference learning improves on this point.

10.6 Temporal Difference Learning

Temporal difference (TD) learning combines Monte Carlo learning with dynamic programming: it learns directly from experiences like Monte Carlo learning and updates based on learned estimates like dynamic programming.

However, temporal difference learning differs from Monte Carlo learning in that: Monte Carlo learning needs to wait for the final result of an episode before they can predict the state-value function; while temporal difference learning can predict the value functions after knowing the next result within an episode, without waiting for the final result of the episode, which is called bootstrapping.

10.6.1 Temporal Difference Prediction

Temporal difference learning, like Monte Carlo learning, also uses experiences to solve prediction problems. Given an experience based on policy π, for the non-terminal state s_t that appears in it, both methods estimate the value V of V^π.

However, as known from Eq. 10.19, the Monte Carlo learning needs to calculate the return G_t, and this return must wait for the end of the episode to get the corresponding value, before determining the increment of $V(s_t)$.

Whereas temporal difference learning only needs to wait for the next step. As shown in Fig. 10.10, at time $t + 1$, using the obtained rewards r_{t+1} and state s_{t+1} valuation $V(s_{t+1})$, we can immediately calculate the increment of $V(s_t)$. That is:

$$V(s_t) \leftarrow V(s_t) + \alpha\left(r_{t+1} + \gamma V(s_{t+1}) - V(s_t)\right). \tag{10.23}$$

Comparing Eqs. 10.19 and 10.23, the difference between them lies in their update targets. The target of the Monte Carlo learning is G_t, while the target of temporal difference learning is $r_{t+1} + \gamma V(s_{t+1})$.

The expression $r_{t+1} + \gamma V(s_{t+1}) - V(s_t)$ in Eq. 10.23 is called the *TD error*, denoted as:

$$\delta_t \doteq r_{t+1} + \gamma V(s_{t+1}) - V(s_t). \tag{10.24}$$

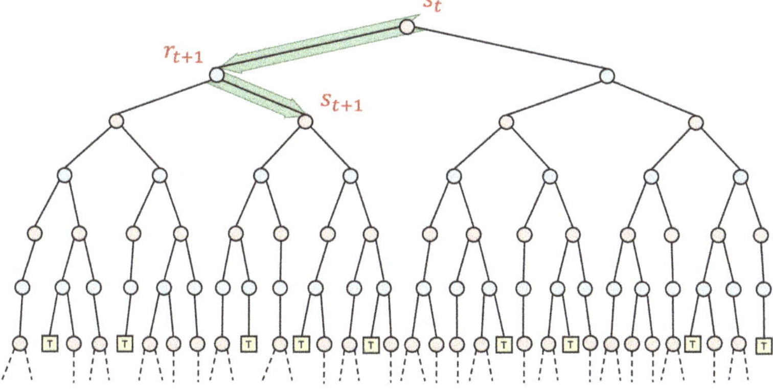

Fig. 10.10 Backup diagram of temporal-difference learning to update value function

However, each TD error δ_t is an estimation error at time t because the actual TD error depends on the next state s_{t+1} and the next reward r_{t+1}, so actually it has to wait until time $t + 1$ to know.

Equation 10.23 is known as the one-step temporal difference, i.e., $TD(0)$.

It is worth noting that one-step temporal difference $TD(0)$ is a special case of $TD(\lambda)$. More about $TD(\lambda)$ will be introduced in Sect. 10.7.

10.6.2   Temporal Difference Control

For the optimal control problem of temporal difference, similar to the Monte Carlo learning, in order to balance exploration and exploitation, there are two approaches: on-policy and off-policy.

The representative algorithms of on-policy temporal difference optimal control are SARSA, and actor-critic, while the representative algorithm of off-policy temporal difference control is Q-learning.

The following introduces SARSA and Q-learning.

10.6.2.1   SARSA Algorithm

SARSA stands for state–action–reward–state–action, named according to the algorithm process, i.e., in the current state, an action is taken, then the agent gets a reward, goes into the next state, and takes the next action.

The first step of the SARSA algorithm is to learn a state–action-value function, not the state-value function in Monte Carlo learning. Especially for on-policy methods, it is necessary to consider the current policy π, as well as all states s and actions a, to estimate $Q^{\pi}(s, a)$. For this, the algorithm converts the current state–action pair into the next state–action pair and learns the value of the state–action pair. The equation is as follows:

$$Q(s_t, a_t) \leftarrow Q(s_t, a_t) + \alpha(r_{t+1} + \gamma Q(s_{t+1}, a_{t+1}) - Q(s_t, a_t)). \qquad (10.25)$$

The aforementioned update is performed under the condition that s_{t+1} is a non-terminal state. If s_{t+1} is a terminal state, then $Q(s_{t+1}, a_{t+1})$ is defined as zero. Equation 10.25 uses each element in $(s_t, a_t, r_{t+1}, s_{t+1}, a_{t+1})$ to perform the transition from one state–action pair to the next. The full name of the SARSA is state–action–reward–state–action, which is where it comes from.

In short, the on-policy in SARSA is to generate behavior using the current estimate of the optimal policy when learning the optimal policy.

Figure 10.11a is a schematic diagram of the SARSA algorithm.

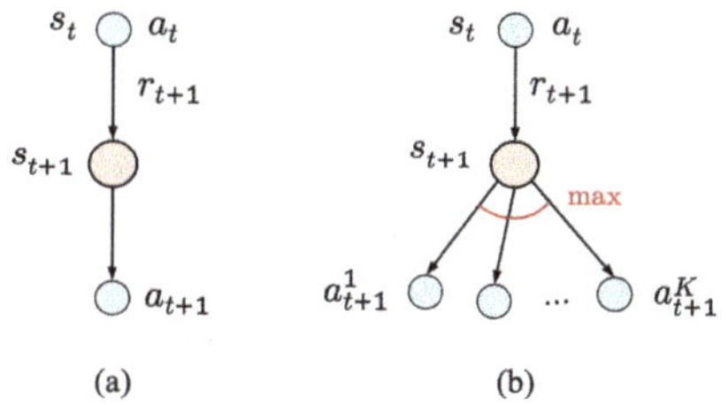

Fig. 10.11 SARSA
algorithm vs. Q-learning

(a)   (b)

10.6.2.2 *Q*-Learning Algorithm

Q-learning is an off-policy temporal difference algorithm, whose goal is to learn the
policy of what action to choose under what circumstances. The algorithm is mainly
defined by the following equation:

$$Q\left(s_t, a_t\right) \leftarrow Q\left(s_t, a_t\right) + \alpha \left(r_{t+1} + \gamma \max_{a+1} Q\left(s_{t+1}, a_{t+1}\right) - Q\left(s_t, a_t\right) \right).$$

$$(10.26)$$

Figure 10.11b is a schematic diagram of the Q-learning algorithm. The Q-
learning Eq. 10.26 updates the state–action pair, so the top node of the update,
that is, the root node of the figure, must be a small, determined action node. The
update comes from the action of this node, maximizes all possible actions in the next
state, the bottom nodes of the schematic diagram are all these action nodes, and the
figure's max represents taking the bottom node a_{t+1}^1 to a_{t+1}^K of the maximum value.

Using the Q-learning algorithm, the learned state–action-value function Q
directly approximates Q^*, this optimal state–action-value function is independent of
the followed policy. This greatly simplifies the analysis of the algorithm and makes
early convergence proofs possible. This strategy is still effective in determining
which state–action pairs to visit and update.

The Q-learning algorithm is considered an early breakthrough in reinforcement
learning.

10.7 Eligibility Trace Mechanisms

The eligibility trace is one of the basic mechanisms of reinforcement learning
(Sutton and Barto 2018).

In this section, we first learn what is the eligibility trace, then discuss the forward
view and backward view where use the eligibility trace, and have a comparative
analysis for both the forward view and backward view finally.

10.7.1   Eligibility Trace

The so-called eligibility trace is a brief record of an event, marking the parameters associated with the event and suitable for learning changes as eligible. When a temporal difference error occurs, it is only necessary to look for the cause of the error from the events with eligible marks.

The eligibility trace is a vector $z_t \in \mathbb{R}^d$, the number of components in this vector is the same as the weight vector w_t. The eligibility trace is a short-term memory, usually less than the length of an episode, while the weight vector is a long-term memory, accumulating throughout the system's lifecycle. Eligibility traces facilitate the learning process, their only effect is to influence the weight vector, which then determines the estimated value.

The events recorded in the eligibility trace refer to visiting a certain state or taking a certain action.

The eligibility trace is a basic mechanism of temporal credit assignment in reinforcement learning, which helps to bridge the gap between events and training information in reinforcement learning.

For example, the popular temporal difference algorithm $TD(\lambda)$ uses the eligibility trace mechanism, where λ $(0 \leq \lambda \leq 1)$ is the decay-rate parameter used for eligibility traces. Almost all temporal difference algorithms, including SARSA, and Q-learning, can be combined with λ to obtain a generalized method, making it more effective for reinforcement learning.

10.7.2   Forward View

The forward view refers to the expectation of all future rewards for each visited state by the reinforcement learning algorithm. Specifically, it decides how to update through all future rewards and states, like a person standing on the state flow and looking forward from the current state. After a state is updated, it will continue to move to the next state.

Further, the temporal difference algorithm $TD(\lambda)$ is used as an example to explain the forward view.

In $TD(\lambda)$, the return to be considered is called λ-return.

First, the n-step return is defined as the sum of the first n rewards plus the nth step state estimate value, each item multiplied by the discount factor. It is represented by a general form of parameterized function approximation, namely:

$$G_{t:t+n} \doteq r_{t+1} + \gamma r_{t+2} + \cdots + \gamma^{n-1} r_{t+n} + \gamma^n \hat{V}(s_{t+n}, w_{t+n-1}). \tag{10.27}$$

Where, $\hat{V}(s, w)$ represents the approximate value of state s under the given weight vector w, $0 \leq t \leq T - n$, and T is the termination time of the episode.

For $TD(\lambda)$, it is not only the n step return that needs to be considered, but also the return under all λ values, hence it is called λ-return. It is a special form of n step return, with a weight of λ^{n-1}, and by normalizing factor $1 - \lambda$ to ensure the sum of the weights is 1. Therefore, λ return is defined as:

$$G_t^{\lambda} \doteq (1 - \lambda) \sum_{n=1}^{\infty} \lambda^{n-1} G_{t:t+n}. \tag{10.28}$$

After reaching the terminal state, all subsequent returns are equal to the regular return G_t, therefore it can be separated from the summation in Eq. 10.28, so we get:

$$G_t^{\lambda} = (1 - \lambda) \sum_{n=1}^{T-t-1} \lambda^{n-1} G_{t:t+n} + \lambda^{T-t-1} G_t. \tag{10.29}$$

Since the range of λ is $0 \le \lambda \le 1$, therefore, a series of methods are generated with the change of λ value.

The eligibility traces in $TD(\lambda)$ are shown in Fig. 10.12, where:

(1) When $\lambda = 0$, the right side of Eq. 10.29 is simplified to only the first term, which is one-step temporal difference learning $TD(0)$.
(2) When $\lambda = 1$, the right side of Eq. 10.29 is simplified to only the second term, which is the Monte Carlo learning.
(3) The nth step, called n-step temporal difference (n-step TD).

As can be seen, eligibility traces serve as a bridge between temporal difference learning and Monte Carlo learning. With the addition of eligibility traces, temporal

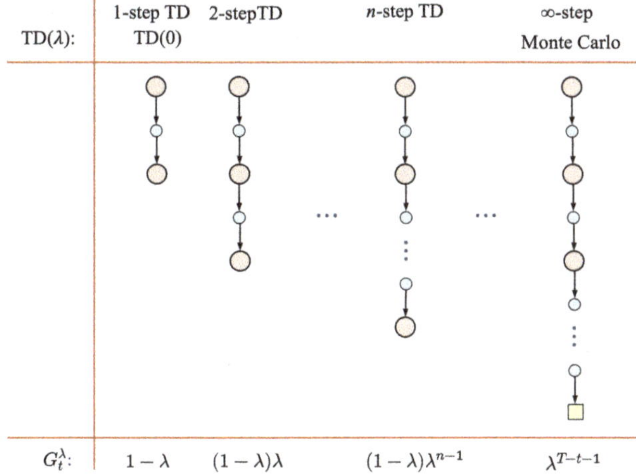

Fig. 10.12 Eligibility traces in $TD(\lambda)$

difference learning forms a lineage of reinforcement learning families: the far left is one-step temporal difference learning, the far right is the Monte Carlo learning, and the n step method in between is n-step temporal difference learning.

It can be considered that eligibility traces unify temporal difference and Monte Carlo learning, thus they are extremely valuable and enlightening.

10.7.3   Backward View

In contrast to the forward view, each update of the backward view depends on the current TD error and the eligibility traces of past events.

In $TD(\lambda)$, the eligibility trace vector z_t is initialized to 0 at the beginning of the episode, that is $z_{-1} \doteq 0$, then increasing over time, that is:

$$z_t \doteq \gamma \lambda z_{t-1} + \nabla \hat{V}(s_t, \boldsymbol{w}_t). \tag{10.30}$$

Where: $0 \le t \le T$, and γ is the discount rate, other definitions are the same as before.

Eligibility traces track the weight vector to determine which components have made positive or negative contributions to the recent state estimation, where "recent" is defined in the term $\gamma \lambda$. If a reinforcement event occurs, this tracking indicates the eligibility of each component of the weight vector for learning updates. The reinforcement event of interest is the TD error at each step. The TD error of state-value prediction is:

$$\delta_t \doteq r_{t+1} + \gamma \hat{V}(s_{t+1}, \boldsymbol{w}_t) - \hat{V}(s_t, \boldsymbol{w}_t). \tag{10.31}$$

The weight vector in $TD(\lambda)$ is updated at each step, proportional to the scalar TD error and the vector eligibility trace:

$$\boldsymbol{w}_{t+1} \doteq \boldsymbol{w}_t + \alpha \delta_t z_t. \tag{10.32}$$

10.7.4   Comparative Analysis

The comparison and analysis on forward view and the backward view of eligibility traces are depicted in Table 10.2.

Table 10.2 Forward view vs. backward view of eligibility traces

Forward view	Backward view
Cannot be directly implemented because it uses rewards and states that will occur after each step but have not actually occurred yet	Providing an incremental mechanism that can approximate the forward view and can be accurately implemented offline
Which is *acausal relations*	Which is *causal relations*
Be most useful for understanding what is computed by methods using eligibility traces	Be more appropriate for developing intuition about the algorithms themselves
Unifying TD learning and Monte Carlo learning, and has important theoretical significance	
Which is called *theoretical view*	Which is called *mechanistic view*

10.8   Deep Reinforcement Learning

Deep reinforcement learning (DRL) is the product of combining reinforcement learning with deep learning. A series of algorithms represented by deep Q-network have achieved the results comparable to or even better than the human level in video games and machine games.

This section first introduces the deep Q-network and then introduces several improved deep reinforcement learning algorithms and networks.

10.8.1   Deep Q-Network

The deep Q-network (DQN) is a deep reinforcement learning model that combines convolutional neural networks with Q-learning algorithms. It was first published by DeepMind company in 2013 (Mnih et al. 2013). The company implemented an artificial intelligence video game system on the Atari 2600 console using a deep Q-network. They conducted experiments with 49 Atari video games, of which 29 games, or 59.18%, achieved or exceeded the level of human players, achieving remarkable results (Mnih et al. 2015). This has promoted in-depth research and widespread application of deep reinforcement learning.

The convolutional neural network in DQN consists of three convolutional layers and two fully connected layers. The input is a preprocessed $84 \times 84 \times 4$ image, and the output corresponds to the effective actions of the joystick. The approximate value function of DQN is denoted as $Q(s, a; \theta_t)$, where the parameter θ_t is the weight of the Q-network in the t iteration.

Before the introduction of DQN, reinforcement learning was not stable, and when the state–action-value function approximated to nonlinear functions such as neural networks, divergence problems could even occur. However, DQN has solved these problems and made the following innovations.

Firstly, it uses experience replay for training reinforcement learning. The premise of deep network training is that the training data satisfy independent and identically distributed (i.i.d.), but the interaction data collected in reinforcement learning have a significant correlation with each other. Directly using these data for training will undoubtedly exacerbate instability. However, through experience replay, the correlation between data can be broken, and the approximation training of the state–action-value function can be stabilized. In order to implement experience replay, the intelligent agent first stores the experience at each moment t, $e_t = (s_t, a_t, r_{t+1}, s_{t+1})$ into the dataset $D = \{e_1, e_2, \ldots, e_N\}$, then uses the method of uniform random sampling to extract data from this dataset, that is $e \sim D$, and uses the extracted data to train the network.

Secondly, a target network is set up to handle the TD error in the temporal difference algorithm separately. Unlike the tabular update of the traditional Q-learning algorithm, DQN uses stochastic gradient descent to update network parameters. Its gradient update equation is:

$$\theta_{t+1} = \theta_t + \alpha \left(r_{t+1} + \gamma \max_a Q\left(s_{t+1}, a; \theta_t\right) - Q\left(s_t, a_t; \theta_t\right) \right) \nabla_{\theta_t} Q\left(s_t, a_t; \theta_t\right).$$

$$(10.33)$$

Where, $r_{t+1} + \gamma \max_a Q\left(s_{t+1}, a; \theta_t\right)$ is the TD target, and the network used to calculate the TD target is the TD network. Before the DQN algorithm, the parameters in the TD network and the network parameters corresponding to the value function to be approximated in the gradient calculation were consistent, which would cause the target function and parameters to change synchronously, making it difficult to converge; In the DQN algorithm, the two are separated: the TD network parameters are denoted as θ^-, updated once every fixed number of steps; the parameters of the overall approximated value function are denoted as θ, updated at every step. The overall gradient update equation becomes:

$$\theta_{t+1} = \theta_t + \alpha \left(r_{t+1} + \gamma \max_a Q\left(s_{t+1}, a; \theta^-\right) - Q\left(s_t, a_t; \theta_t\right) \right) \nabla_{\theta_t} Q\left(s_t, a_t; \theta_t\right).$$

$$(10.34)$$

10.8.2   *Improved Algorithms and Networks*

After DQN, the researchers at Google DeepMind continued to study deep reinforcement learning, introducing some improved algorithms and innovative network architectures.

10.8.2.1   Double Deep Q-Network

The double deep Q-network (double DQN) is mainly used to solve the overestimation of state–action-value in the Q-learning process, which leads to suboptimal solutions and stability issues in learning. The basic idea is to separate the selection and evaluation of actions and use different Q function networks. Since the original DQN has already introduced a separate Q target network, the double DQN does not need to make significant changes in architecture. The focus is on the independent training of target network parameters and online network parameters (Van Hasselt et al. 2015).

10.8.2.2   Dueling Network Architecture

The dueling network architecture is a new type of neural network architecture proposed for deep reinforcement learning, rather than using traditional architectures like CNN, LSTM. The dueling network consists of two separate evaluators, one corresponding to the state-value function, and the other is the state-dependent action advantage function. The main benefit of this decomposition is that it can generalize across action learning without any changes to the underlying reinforcement learning algorithm (Wang et al. 2015).

10.8.2.3   Deep Deterministic Policy Gradient

The deep deterministic policy gradient (DDPG) is a deep reinforcement learning algorithm based on deterministic policy gradient methods and actor-critic algorithms, enabling it to operate in continuous action spaces. DDPG combines the actor-critic algorithm with DQN. In the process of using deep networks to approximate the value function, it ensures the convergence and stability of the training process through DQN's replay buffer and target Q-network techniques, as well as batch normalization operations in deep networks (Lillicrap et al. 2016).

10.8.2.4   Asynchronous Advantage Actor-Critic

The asynchronous advantage actor-critic (A3C) is applicable not only to discrete action spaces but also to continuous action space control problems. The main feature of this algorithm is that it creates multiple intelligent agents that execute and learn in parallel and asynchronously in multiple environment instances (Mnih et al. 2016).

10.8.2.5   Unsupervised Reinforcement and Auxiliary Learning

The unsupervised reinforcement and auxiliary learning (UNREAL) is a method of reinforcement learning through unsupervised auxiliary learning tasks, which can maximize many pseudo-reward functions simultaneously (Jaderberg et al. 2016).

10.8.2.6   Neural Episodic Control

The neural episodic control (NEC) is a new type of intelligent agent for deep reinforcement learning that can significantly improve the speed of learning. Because the current speed of machine learning is much slower than humans, the results tested by DeepMind are: it takes a human 2 hours to learn an Atari game, while a machine needs 200 hours. NEC is an algorithm proposed for the speed problem, and it is said that the speed can be increased by 10 times (Pritzel et al. 2017).

10.9   Comparison of Reinforcement Learning

Some representative algorithms in reinforcement learning and their comparisons are shown in Table 10.3.

It should be emphasized that the "TD + Eligibility Trace" in Table 10.3 refers to the temporal difference (TD) algorithm after adopting the eligibility trace mechanism.

Table 10.3 Comparison of representative reinforcement learning methods and algorithms

Model	Method	Algorithm	Policy	State/Action space
Model-based	Dynamic programming	Value iteration	Optimal	Discrete/Discrete
		Policy iteration	Optimal	Discrete/Discrete
Model-free	Monte Carlo learning	Exploring starts	On-policy	Discrete/Discrete
		Importance sampling	Off-policy	Discrete/Discrete
	TD learning	SARSA	On-policy	Discrete/Discrete
		Actor-critic	On-policy	Discrete/Discrete
		Q-Learning	Off-policy	Discrete/Discrete
	TD + Eligibility traces	SARSA(λ)	On-policy	Discrete/Discrete
		Q-Learning(λ)	Off-policy	Discrete/Discrete
	Deep RL	DQN	Off-policy	Cont./Discrete
		A3C	On-policy	Cont./Cont.

Further Reading

1. Richard S. Sutton, and Andrew G. Barto. *Reinforcement Learning: An Introduction* (2nd Edition). MIT Press, 2020.
   [Notes] This book is considered a classic textbook on reinforcement learning. The first edition was published in 1998, and this is the second edition.
2. Volodymyr Mnih, Koray Kavukcuoglu, David Silver, et al. "Human-level Control through Deep Reinforcement Learning." *Nature*, 518(7540), 2015.
   [Notes] This paper introduces the deep Q-network that was proposed by some researchers from DeepMind company. The results of testing on 49 Atari video games showed that approximately 59.18% reached or exceeded the human level, hence it is called "human-level control". Google saw the significant implications of their work and acquired this small company, which was established in September 2010, for 400 million pounds in 2014.
3. David Silver, Thomas Hubert, Julian Schrittwieser, et al., "A General Reinforcement Learning Algorithm That Masters Chess, Shogi, and Go Through Self-Play." *Science*, 362(6419), 2018.
   [Notes] This paper proposes a game bot called AlphaZero, which can master three board games through self-play reinforcement learning from scratch and defeat human champions in these games.
4. Mohit Sewak. *Deep Reinforcement Learning: Frontiers of Artificial Intelligence.* Springer, 2019.
   [Notes] This book introduces reinforcement learning and deep reinforcement learning and also provides related algorithms and Python source codes.
5. Julian Schrittwieser, Ioannis Antonoglou, Thomas Hubert. "Mastering Atari, Go, Chess and Shogi by Planning with a Learned Model." *Nature*, 588(7839), 2020.
   [Notes] The paper proposes a new model-based reinforcement learning method that can control Atari video games and three board games, achieving optimal control levels for Atari and defeating human champions in Go, Chess, and Shogi.

References

Bellman, R. (1952). On the theory of dynamic programming. *Proceedings of the National Academy of Sciences of the United States of America* 38(8): 716.

Bellman, R. (1957). A Markovian decision process. *Indiana University Mathematics Journal* 6(4): 15.

Blythe, J. (1999). Decision-theoretic planning. *AI Magazine* 20.

Dasgupta, S., C. Papadimitriou, and U. Vazirani. (2006). Chapter 6 Dynamic Programming. Algorithms.

Feldman, J. A., and R. F. Sproull. (1977). Decision theory and artificial intelligence II: The hungry monkey. *Cognitive Science* 5(4): 349–371.

Howard, R. A. (1960). Dynamic programming and Markov processes. *Mathematical Gazette* 3(358): 120.

Jaderberg, M., V. Mnih, W. M. Czarnecki, T. Schaul, J. Z. Leibo, D. Silver, and K. Kavukcuoglu. (2016). Reinforcement Learning with Unsupervised Auxiliary Tasks. arXiv Preprint arXiv:1611.05397.

Lavalle, S. M. (2006). *Planning algorithms*. Cambridge: Cambridge University Press.

Lillicrap, T. P., J. J. Hunt, A. Pritzel, N. Heess, T. Erez, Y. Tassa, D. Silver, and D. Wierstra. (2016). Continuous Control with Deep Reinforcement Learning. In *International Conference on Machine Learning (ICML)*.

Mendel, J. M., and R. W. McLaren. (1970). Reinforcement learning control and pattern recognition systems. *Mathematics in Science and Engineering* 287–318.

Minsky, M. L. (1954). *Theory of neural-analog reinforcement systems and its application to the brain model problem*. Princeton: Princeton University.

Mnih, V., A. P. Badia, M. Mirza, A. Graves, T. Lillicrap, T. Harley, D. Silver, and K. Kavukcuoglu. (2016). Asynchronous Methods for Deep Reinforcement Learning. In *International Conference on Machine Learning (ICML)*.

Mnih, V., K. Kavukcuoglu, D. Silver, A. Graves, I. Antonoglou, D. Wierstra, and M. Riedmiller. (2013). Playing Atari with Deep Reinforcement Learning. arXiv preprint arXiv:1312.5602.

Mnih, V., K. Kavukcuoglu, D. Silver, A. A. Rusu, J. Veness, M. G. Bellemare, A. Graves, M. Riedmiller, A. K. Fidjeland, and G. Ostrovski. (2015). Human-level control through deep reinforcement learning. *Nature* 518(7540): 529.

Pritzel, A., B. Uria, S. Srinivasan, A. P. Badia, O. Vinyals, D. Hassabis, D. Wierstra, and C. Blundell. (2017). Neural Episodic Control.

Sutton, R. S. (1984). *Temporal credit assignment in reinforcement learning*. Amherst, MA: University of Massachusetts Amherst.

Sutton, R. S., and A. G. Barto. (2018). *Reinforcement learning: An introduction*. Cambridge: The MIT Press.

Van Hasselt, H., A. Guez, and D. Silver. (2015). Deep Reinforcement Learning with Double Q-learning. arXiv Preprint arXiv:1509.06461v3.

Waltz, M. D., and K.-s. Fu. (1965). A heuristic approach to reinforcement learning control systems. *IEEE Transactions on Automatic Control* 10(4): 390–398.

Wang, Z., T. Schaul, M. Hessel, H. van Hasselt, and M. Lanctot. (2015). Dueling Network Architectures for Deep Reinforcement Learning. arXiv Preprint arXiv:1511.06581.

Chapter 11
Other Learning Quasi-paradigm

Abstract In addition to the three learning paradigms introduced in the previous three chapters, several learning quasi-paradigms have emerged in the machine learning community. The so-called quasi-paradigm is referring each one of the quasi-paradigms having certain characteristics for a paradigm but not yet forming completely an independent paradigm. In this chapter we select some noteworthy learning quasi-paradigms to introduce, namely, ensemble learning, meta-learning, transfer learning, self-supervised learning, and n-shot learning including one-shot, few-shot, and zero-shot learning.

11.1 Ensemble Learning

The birth and development of ensemble learning have spanned over 30 years.

Ensemble learning can be traced back to the extensions of the theory of learnability, which are strong learnability and weak learnability. The technical report titled "Thoughts on Hypothesis Boosting" in 1988 provided a formal definition of strong learnability and weak learnability (Kearns 1988). The paper "The Strength of Weak Learnability" published in 1990 gave a positive answer and detailed proof to the hypothesis boosting problem and proposed a method to boost weak learning algorithms to any strong learner (Schapire 1990).

The 1990s was a prosperous period for ensemble learning, and some classic ensemble learning methods were proposed during this period. The 2000s was a period when ensemble learning was combined with neural networks, and from the 2010s onwards, ensemble deep learning became the main research trend in this field (Ganaie et al. 2022).

The following sections will introduce the combination modes, ensemble methods, averaging scheme, and voting scheme of ensemble learning, respectively.

11.1.1   Definition

Definition 11.1 (Ensemble Learning) Ensemble learning is the approach of organically combining several base learners to form a strong learner, whose performance surpasses any of the base learners before the combination.

The so-called base learners refer to the basic models used to solve machine learning tasks such as classification and regression. For each base learner, its simplicity and combinability, as well as its complementarity are usually considered, so that several base learners can complement each other.

A strong learner is relative to a weak learner. In ensemble learning, base learners are usually regarded as weak learners, because the purpose of ensemble learning is to integrate weak learners.

The ensemble learning in this section will uniformly use the following symbolic expressions:

Let X denote the input space, $\mathcal{Y}$ denote the output space,

$$ S = \left\{ (x_i, y_j) \mid x_i \in X,\, y_j \in \mathcal{Y} \right\} $$

denote the training samples, $\{h_t(x)\}_{t=1}^{T}$ denote T base learners, and $h(x)$ denote the strong learner, where $x \in X$.

11.1.2   Combination Modes

The combination modes refer to the following two modes:

- What styles of combination for base learners
- What styles of base learners to be combined

The existing modes to combine base learners include parallel and sequential combinations, while existing modes for base learners to be combined include homogeneous and heterogeneous combinations.

11.1.2.1   Parallel and Sequential

There are two styles of combination for base learners, namely, parallel combination and sequential combination.

(1) Parallel combination

It is to connect T base learners $\{h_t(x)\}_{t=1}^{T}$ parallelly, so that for a common machine learning task, it can process parallelly and independently, and then integrate the results of each base learner into a strong learner $h(x)$.

(2) Sequential combination

It is to connect T base learner $\{h_t(x)\}_{t=1}^{T}$ sequentially, so that for a common machine learning task, it learns in stages and iteratively, finally resulting in a strong learner $h(x)$.

11.1.2.2 Homogeneous and Heterogeneous

There are also two styles of base learners to be combined, namely, homogeneous combination and heterogeneous combination.

(1) Homogeneous combination

It refers to a strong learner $h(x)$ that is composed of base learners of the same type and used for the same learning task.

For example, T base learners $\{h_t(x)\}_{t=1}^{T}$ all use same logistic regression algorithm for binary classification.

(2) Heterogeneous combination

It refers to the strong learner $h(x)$ that can be composed of different types of base learners, but used for the same learning task.

For example, T base learners $\{h_t(x)\}_{t=1}^{T}$ respectively use logistic regression, naive Bayes, decision tree, and support vector machine for binary classification.

11.1.3 Ensemble Methods

There are some methods with different combination styles for ensemble learning. Mainly divided into three types: boosting, bagging, and stacking.

11.1.3.1 Boosting

Boosting is not the name of a specific ensemble learning algorithm, but a collective name for a class of ensemble algorithms, so it is considered a meta-algorithm of ensemble learning.

Boosting uses the sequential combination of ensemble learning, aiming to reduce bias and variance through the iteration of base learners. For the bias-variance problem, please see Chap. 8 of this book.

The earliest boosting method was proposed in 1990, which did not consider adaptability (Schapire 1990). AdaBoost, short for adaptive boosting, was presented in 1995, supporting different types of base learners to combine their outputs into a boosted strong learner (Freund and Schapire 1995). AnyBoost was presented in 1999, which is a general algorithm framework, showing that boosting performs gradient descent in a function space using a convex cost function (Mason et al. 1999); thus, many boosting algorithms can be fitted into the AnyBoost framework.

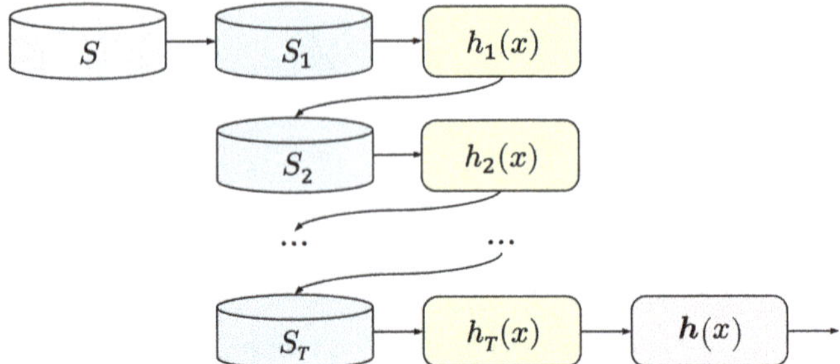

Fig. 11.1 Schematic diagram of the boosting

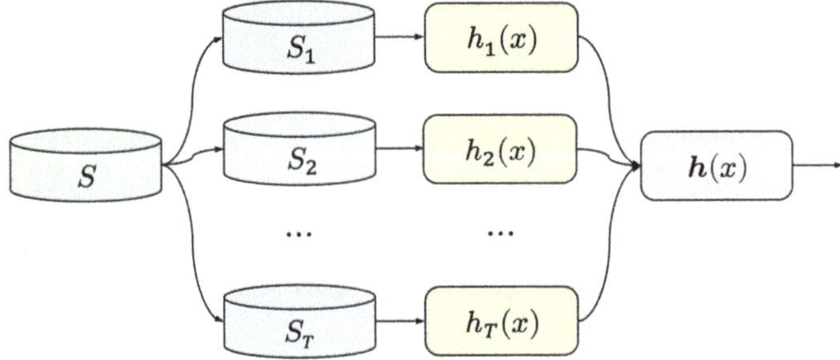

Fig. 11.2 Schematic diagram of the bagging

The schematic diagram of boosting is shown in Fig. 11.1. First, it assigns the same weight to each sample in the original samples S and generates boosted samples S_1 to be used to train the base learner $h_1(x)$. Then, the samples mis-predicted by $h_1(x)$ are weighted and used to generate boosted samples S_2, and it is used to train the base learner $h_2(x)$. Repeat the above process, and finally, the prediction results of each base learner are combined by weighted voting to obtain the strong learner $h(x)$.

11.1.3.2   Bagging

The "bagging" is an abbreviation of "bootstrap aggregating."

Bagging is not a specific algorithm, but a general term for a class of ensemble algorithms, so it is also considered a meta-algorithm of ensemble learning.

Bagging was first proposed by Leo Breiman in 1996, using the parallel combination of ensemble learning, aiming to improve the stability and accuracy of the learner, mainly to reduce variance and avoid overfitting (Breiman 1996).

The schematic diagram of bagging is shown in Fig. 11.2. This method first randomly selects samples from the original samples S using sampling with replacement and generates T bootstrap samples $\{S_t\}_{t=1}^{T}$. And second, each bootstrap sample S_t

is used to train each base learner $h_t(x)$. Then, by aggregating the prediction results of each base learner, the required strong learner $h(x)$ is obtained.

Although each sample in the bootstrap samples S_t is randomly selected from the original samples S, there are two characteristics: first, the number of samples in S_t is the same as that in S, i.e., $|S_t| = |S|$; second, a sample in S may be repeatedly selected or not selected.

Random forest is an ensemble learning algorithm for classification and regression that combines the bagging strategy with random feature selection of decision trees (Breiman 2001).

11.1.3.3 Stacking

Stacking, also known as stacked generalization, was proposed by David H. Wolpert in 1992 to reduce generalization error (Wolpert 1992).

Stacking allows several heterogeneous base learners to be combined into a strong learner.

The schematic diagram of the stacking method is depicted in Fig. 11.3. First, this method uses the original samples as the *level*-0 samples S_0, and for the samples S_0 to use the *level*-0 base learner $\{h_{0,t}(x)\}_{t=1}^{T}$ for learning. Then, the output of the *level*-0 base learner forms the *level*-1 samples S_1, which are then learned by the *level*-1 meta-learner $h_1(x)$. After that, the result of the *level*-1 meta-learner forms a strong learner $h(x)$.

11.1.4 Averaging Scheme

In ensemble learning, the strong learner $h(x)$ is an ensemble of the learning results from T base learners $\{h_t(x)\}_{t=1}^{T}$. For the regression task based on ensemble learning, there are two type of averaging scheme, namely, simple averaging and weighted averaging.

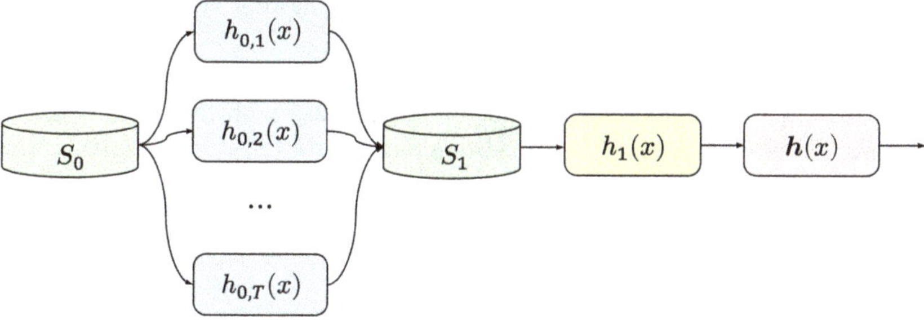

Fig. 11.3 Schematic diagram of the stacking

11.1.4.1   Simple Averaging

The simple averaging scheme calculates the sum of T base learners $\{h_t\,(x)\}_{t=1}^{T}$ and then divides by the number of base learners T to get a strong learner $h\,(x)$. Its expression is as follows:

$$h\,(x) = \frac{1}{T}\sum_{t=1}^{T} h_t\,(x).\tag{11.1}$$

11.1.4.2   Weighted Averaging

In the weighted averaging scheme, it calculates the sum of T *weighted* base learners $\{w_t h_t(x)\}_{t=1}^{T}$ to derive a strong learner $h\,(x)$. The calculation equation is as follows:

$$h\,(x) = \sum_{t=1}^{T} w_t h_t\,(x),\tag{11.2}$$

where $w_t \geq 0$, and $\sum_{t=1}^{T} w_t = 1$.

11.1.5   Voting Scheme

For the classification task based on ensemble learning, there are two type of voting scheme, namely, hard voting and soft voting.

11.1.5.1   Hard Voting

Given the training samples $S = \{(x_i, c_j)\}$ for classification, let $\{c_j\}_{j=1}^{k}$ denote the k labeled classes. Therefore, for the data point x_i, the base classifier $h_t\,(x_i)$ may have k classification results.

Hard voting also has two subtypes, namely, majority voting and weighted majority voting.

In the majority voting scheme, each base classifier $h_t\,(x_i)$ predicts if the data point x_i belongs to target class c_j and the target class with the most votes wins. Therefore, the prediction result of the strong classifier $h\,(x)$ is:

$$h\,(x_i) = c_j = \arg\max_{j} \sum_{t=1}^{T} \mathbb{I}\left(h_t\,(x_i) = c_j\right),\tag{11.3}$$

where $\mathbb{I}\,(\omega)$ is an indicator function that returns 1 if ω is true; otherwise 0.

For example: there are three base classifiers $\{h_t(x)\}_{t=1}^3$, performing binary classification $\{c_1 = 0, c_2 = 1\}$. Therefore, we have $h_t(x) = c_j \in \{0, 1\}$. Let the classification results of the three base classifiers be $h_1(x) = 0$, $h_2(x) = 0$, and $h_3(x) = 1$, then they are calculated based on Eq. 11.3, which can be represented in the following:

	$h_1(x)$	$h_2(x)$	$h_3(x)$	$\arg\max_j \sum_{t=1}^3 \mathbb{I}(h_t(x) = c_j)$	$h(x)$
$c_1 = 0$	0	0	1	poll = 2 for $j = 1$	c_1
$c_2 = 1$	0	0	1	poll = 1 for $j = 2$	–

Since the number of votes for $j = 1$ is 2, which is greater than the number of votes for $j = 2$, the result of the strong classifier $h(x)$ is $c_1 = 0$.

In statistical terms, the target class with the most votes in majority voting is equivalent to the mode of the target class predicted by each base classifier.

Let $\{w_t\}_{t=1}^T$ denote the weights of T base learners, then the prediction result of the strong classifier $h(x_i)$ based on the weighted majority voting scheme is:

$$h(x_i) = c_j = \arg\max_j \sum_{t=1}^T w_t \mathbb{I}(h_t(x_i) = c_j), \tag{11.4}$$

where $w_t \geq 0$, $\sum_{t=1}^T w_t = 1$, and $\mathbb{I}(\omega)$ is an indicator function.

11.1.5.2 Soft Voting

In soft voting, there are also two subtypes, namely, probabilistic voting and weighted probabilistic voting.

In the probabilistic voting scheme, given a data point x_i, calculate the probability value for each base classifier $h_t(x_i)$ belonging to a target class c_j, i.e., $P(h_t(x_i) = c_j | x_i)$, the target class with the highest sum of probabilities wins. Therefore, we have:

$$h(x_i) = c_j = \arg\max_j \sum_{t=1}^T P(h_t(x_i) = c_j | x_i). \tag{11.5}$$

Similarly, the result of weighted probabilistic voting is:

$$h(x_i) = c_j = \arg\max_j \sum_{t=1}^T w_t P(h_t(x_i) = c_j | x_i), \tag{11.6}$$

where $w_t \geq 0$, and $\sum_{t=1}^T w_t = 1$.

11.2　Meta-Learning

The prefix "meta-" in meta-learning originates from Greek, meaning "above" or "beyond." In epistemology, it is defined in an abstract recursive way, that is, "*X* about *X*." For example, the meaning of meta-cognition is "cognition about cognition."

Meta-learning is a branch of meta-cognition, known as "learning about one's own learning and learning processes."[1]

It is well known that humans can learn and need to continue learning, not only learning new concepts and skills but also learning their inductive bias, that is, learning how to obtain a hypothesis and how to make a generalization.

The term meta-learning first appeared in cognitive psychology in 1979. Years later, meta-learning was introduced into the field of machine learning and gradually became a noteworthy research direction. Since 2017, the *Workshop on Meta-Learning* has been held annually as one of the workshops of *Conference on Neural Information Processing Systems* (NeurIPS).

Meta-learning is also considered "learning to learn." During the 2021 *International Conference on Learning Representations* (ICLR), a special *Workshop on Learning to Learn* was held.

11.2.1　Definition

Different researchers have given different definitions of meta-learning.

In 2002, two researchers from the IBM Thomas J. Watson Research Center published an article titled "A Perspective View and Survey of Meta-Learning" in the journal *Artificial Intelligence Review* (Vilalta and Drissi 2002). They proposed that the goal of meta-learning is to build self-adaptive learners, that is, to build algorithms that can dynamically improve their preferences through the accumulation of meta-knowledge.

In 2008, Pavel Brazdil et al. from the University of Porto in Portugal published a book titled *Metalearning: Applications to Data Mining*. They believe that meta-learning is a principled method of using meta-knowledge and adjusting machine learning and data mining processes to obtain effective models and solutions (Brazdil et al. 2008).

In 2015, a review article titled "Metalearning: A Survey of Trends and Technologies" was published in the journal *Artificial Intelligence Review* (Lemke et al. 2015). It states: (i) A meta-learning system must include a learning subsystem, which adapts with experience. (ii) Experience is gained by exploiting extracted meta-knowledge.

---

[1] https://en.wikipedia.org/wiki/Meta-learning

In 2018, Joaquin Vanschoren from the Eindhoven University of Technology in the Netherlands gave a definition as follows: meta-learning, or learning to learn, is systematically observing how different machine learning methods perform in a wide range of learning tasks, and learning from this experience or meta-data, in order to learn new tasks more quickly (Vanschoren 2018).

Combining the above standpoints, here we give the following definition.

Definition 11.2 (Meta-Learning) Meta-learning, also known as "learning to learn" or "learning how to learn," is a quasi-paradigm of machine learning, which is dedicated to acquiring meta-knowledge, using it to train the meta-model, and then applying the meta-model to solve new problems and tasks in machine learning.

Based on the recursive definition of meta-learning, it can be thought of as "learning about learning"; therefore, it is logical to call meta-learning "learning to learn" or "learning how to learn."

After the term "learning to learn" appeared, several terms derived from the phrase "learning to do" have also emerged, such as: learning to think, learning to reason, learning to rank, and learning to index.

11.2.2 Related Discourses

Due to the broad definition of meta-learning, its connotation and extension are constantly being explored and supplemented. Here are two related discourses from international academic workshops.

11.2.2.1 Workshop on Meta-Learning

The statement about meta-learning on the homepage of the 2022 *Workshop on Meta-Learning*[2] is as follows:

> Recent years have seen rapid progress in meta-learning methods, which transfer knowledge across tasks and domains to efficiently learn new tasks, optimize the learning process itself, and even generate new learning methods from scratch. Meta-learning can be seen as the logical conclusion of the arc that machine learning has undergone in the last decade, from learning classifiers, to learning representations, and finally to learning algorithms that themselves acquire representations, classifiers, and policies for acting in environments. In practice, meta-learning has been shown to yield new state-of-the-art automated machine learning methods, novel deep learning architectures, and substantially improved one-shot learning systems. Moreover, improving one's own learning capabilities through experience can also be viewed as a hallmark of intelligent beings, and neuroscience shows a strong connection between human and reward learning and the growing subfield of meta-reinforcement learning.

---

[2] https://meta-learn.github.io/2022/

It is worth noting that in the above description the phrase "transfer knowledge" is used.

The homepage of the above Workshop also proposes several fundamental questions aimed to address, which are as below:

- What are the meta-learning processes in nature (e.g., in humans), and how can we take inspiration from them?
- What is the relationship between meta-learning, continual learning, and transfer learning?
- What interactions exist between meta-learning and large pre-trained/foundation models?
- What principles can we learn from meta-learning to help us design the next generation of learning systems?
- What kind of theoretical principles can we develop for meta-learning?
- How can we exploit our domain knowledge to effectively guide the meta-learning process and make it more efficient?
- How can we design better benchmarks for different meta-learning scenarios?

11.2.2.2   Workshop on Learning to Learn

The 2021 *Workshop on Learning to Learn*[3] started like this on its homepage:

> Recent years have seen a lot of interest in the use and development of learning-to-learn algorithms. Research on learning-to-learn, or meta-learning, algorithms is often motivated by the hope to learn representations that can be easily transferred to the learning of new skills, and lead to faster learning. Yet, current meta-learned representations often struggle to generalize to novel task settings. In this workshop, we'd like to discuss how humans meta-learn, and what we can and should expect from learning to learn in the field of machine learning.

It is worth emphasizing that this workshop also includes a discussion on "how humans meta-learn," which can be considered as an attempt to sort out meta-learning from the origin and apply it to learning to learn.

11.2.3   Related Terminologies

Several machine learning terms prefixed with "meta-" have been mentioned earlier, namely: meta-data, meta-knowledge, meta-model, meta-reinforcement learning, etc.

---

[3] https://sites.google.com/view/learning-2-learn/

11.2.3.1 Meta-Data

The so-called meta-data, in cognitive psychology, refers to data about data, but not the content of text, images, etc. Meta-data is also known as meta-information, which is data that provides information about other data.

In the field of machine learning, meta-data is the data for machine learning, such as datasets, data configurations, annotated samples, hyperparameters, and performance metrics.

Meta-data is also an important part of meta-knowledge.

11.2.3.2 Meta-Knowledge

Meta-knowledge, in cognitive psychology, refers to knowledge about types, domains, functions, or knowledge premises. It is knowledge independent of system issues and domain issues.

In the field of machine learning, meta-knowledge refers to knowledge about machine learning model knowledge, including datasets, meta-data, learning algorithms, hardware configurations, training techniques, evaluation methods, and ablation experiments.

11.2.3.3 Meta-Model

A meta-model is a model that can be used to build other machine learning models, including the model's documentation, source code, datasets, and evaluation metrics.

In a sense, a meta-model is similar to the baseline of machine learning and is a reference model or foundational model for further development of machine learning.

11.3 Transfer Learning

Transfer learning, originating from the theory of transfer of learning in psychology, is also one of the research contents in cognitive science. Learning transfer occurs when people apply the information, strategies, and skills they have learned to new scenes or environments.

In machine learning, the theory of transfer learning was initially used for neural network training. In 1976, a paper written in the Croatian language proposed to train perceptron using transfer learning; the first author was Stevo Bozinovski. He reviewed this work in a paper published in English in 2020 (Bozinovski 2020).

In 1992, Lorien Y. Pratt proposed an algorithm based on discriminability-based transfer, transferring the weights trained by one neural network to another neural network (Pratt 1992).

In 1995, the *Workshop on Learning to Learn: Knowledge Consolidation and Transfer in Inductive Systems*,[4] as one of the post-conference workshops held at NIPS-95, discussed how to transfer knowledge from other source systems to the current inductive system.

In 2005, the Information Processing Technology Office (IPTO) of the Defense Advanced Research Projects Agency (DARPA) issued the "Transfer Learning Program" in its Broad Agency Announcement (BAA) 05-29, also known as "DARPA BAA 05-29," making transfer learning one of the key capabilities of cognitive systems (Gasser et al. 2005).

In 2013, a method called propagated semantic transfer was proposed for image classification (Rohrbach et al. 2013).

Since then, transfer learning has increasingly attracted attention in the field of machine learning, and many methods of transfer learning as well as several review papers have been published (Pan and Yang 2010; Weiss et al. 2016; Zhuang et al. 2021).

In the following, we will first define transfer learning and then introduce the working principle, types, and applications of transfer learning.

11.3.1   Definition

Generally speaking, transfer learning is to apply the knowledge or skills learned in solving a certain problem to another different but related task. In machine learning, transfer learning occurs between the source model and the target model.

Definition 11.3 (Transfer Learning) Transfer learning in machine learning is a quasi-paradigm that transfers the knowledge or functionality of a source model of machine learning to a different but related target model.

In the classic paradigm of supervised learning, there are usually some basic conditions: that is, the training data and test data must be in the same feature space, with the same distribution, and with the same data type, and a machine learning model can only solve a specific task. In addition, supervised learning requires a large number of manually labeled training samples for extensive training. However, these basic conditions are often difficult to meet the actual needs, and their limitations have become an urgent problem to be solved. Therefore, transfer learning came into being.

For ease of understanding, a conceptual comparison of the differences between classic supervised learning and transfer learning is depicted in Fig. 11.4, where $\mathcal{X}_x$

---

[4] https://plato.acadiau.ca/courses/comp/dsilver/NIPS95_LTL/transfer.workshop.1995.html

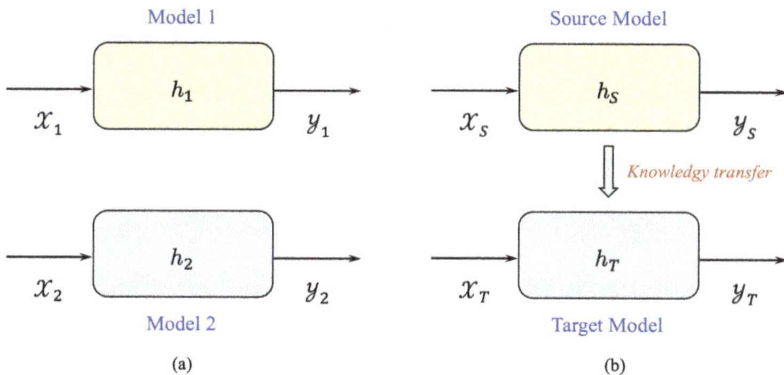

Fig. 11.4 Comparison between traditional supervised learning and transfer learning

denotes the input space, $\mathcal{Y}_x$ denotes the output space, h_x denotes the hypothesis. We can see:

- In supervised learning, as shown in Fig. 11.4a, the hypotheses h_1 and h_2 in "Model 1" and "Model 2" are trained separately and employ their respective tasks independently, without any relation each other. And $\mathcal{X}_1$ with $\mathcal{X}_2$, and $\mathcal{Y}_1$ with $\mathcal{Y}_2$ are also without any relation too.
- In transfer learning shown in Fig. 11.4b, the knowledge of h_S in "Source model" can be transferred to h_T in "Target model", while $\mathcal{X}_S$ with $\mathcal{X}_T$, and $\mathcal{Y}_S$ with $\mathcal{Y}_T$ are different but related, thereby reducing the sample annotation cost and shortening the training time for the target model.

11.3.2 Working Principle

Here we discuss the working principle of transfer learning in a formalized way. Without loss of generality, the dimensions of the input and output spaces and parameterization issues are omitted.

The source model (SM) in transfer learning can be represented as a 5-tuple:

$$SM = \langle \mathcal{X}_S, \mathcal{Y}_S, S_S, H_S, \mathcal{L}_S \rangle,$$

where $\mathcal{X}_S$ and $\mathcal{Y}_S$ are the input and output spaces of the source model, respectively, S_S is a set of training samples of the source model, $S_S = \{(x_i, y_i) \mid i = 1, \ldots, n_S\}$, H_S is the hypothesis set of the source model, and $\mathcal{L}_S$ is the loss function of the source model.

The target model (TM) in transfer learning is also represented as a 5-tuple:

$$TM = \langle \mathcal{X}_T, \mathcal{Y}_T, S_T, H_T, \mathcal{L}_T \rangle,$$

where $S_T = \{(\boldsymbol{x}_i, y_i) \mid i = 1, \ldots, n_T\}$ and other elements in this 5-tuple are the same as that in the source model. The number of training samples between the target model and source model is satisfied $n_T \ll n_S$.

Through the above formalization of the source model and the target model, it can be known that the following four situations may occur:

$\mathcal{X}_S \neq \mathcal{X}_T$ that is, the input space of the source model is different from that of the target model.

$\mathcal{Y}_S \neq \mathcal{Y}_T$ that is, the output space of the source model is different from that of the target model.

$S_S \neq S_T$ that is, the training samples of the source model are different from the target model.

$H_S \neq H_T$ that is, the hypothesis set of the source model is different from that of the target model.

The definition of transfer learning is to transfer the knowledge of the source model to a different but related target model, which is also the "premise" of transfer learning. The so-called different but related means—the input and output spaces of the source model and the target mode—can be different, but they should be domain-related. This can be formally expressed as: $\mathcal{X}_S \neq \mathcal{X}_T$ but $\mathcal{X}_S \sim \mathcal{X}_T$, $\mathcal{Y}_S \neq \mathcal{Y}_T$ but $\mathcal{Y}_S \sim \mathcal{Y}_T$.

For example, in computer vision, an object recognition model, after being extensively trained to recognize cats, can recognize tigers when given a detailed description of tigers. Similarly, the knowledge of a source model in natural language processing can only be transferred to a different but domain-related target model.

Being domain-related has a specific meaning, that is, within the same domain and related. Otherwise, even if the domains are the same but unrelated, the transfer cannot be performed.

For example: In the field of computer vision, there are various subtasks, including image classification and 3D reconstruction. Since the two are unrelated, it is impossible to transfer the source model of image classification to the target model of 3D reconstruction. For natural language processing, it is also impossible to transfer the source model of machine translation to the target model of text generation.

$\mathcal{X}_S$ and $\mathcal{Y}_S$ are the input and output spaces of the source model, respectively. S_S is a set of training samples of the source model, $S_S = \{(\boldsymbol{x}_i, y_i) \mid i = 1, \ldots, n_S\}$. $\boldsymbol{x}_i \in \mathcal{X}_S$ is the observed data collected from the input space based on the probability distribution $P(\boldsymbol{x})$, and $y_i \in \mathcal{Y}_S$ is the artificial annotation based on the conditional probability $P(y|\boldsymbol{x})$. H_S is the hypothesis set of the source model, and h_S is one of the hypotheses in it, that is, $h_S \in H_S$. Adopting S_S is trained on it, and the number of its training samples n_S is often very large. $\mathcal{L}_S$ is the loss function of the source model. The goal is to minimize the loss $\mathcal{L}_S(h_S(\boldsymbol{x}), y)$ through extensive training. That is the best hypothesis h_S obtained through training.

The most common approach in transfer learning is to transfer the hypothesis $h_S \in H_S$ of the source model to the hypothesis $h_T \in H_T$ of the target model; this can greatly shorten the training time of h_T. That is to say, the hypothesis h_T of the target model is created based on the pre-trained hypothesis h_S of the source model and is trained slightly to reuse sample set S_T. Therefore, only a small number n_T of training samples are needed in the target model.

11.3.3   Categorization

In the field of machine learning, transfer learning can be divided into different categories, and different researchers have proposed different dividing methods from different viewpoints; there are mainly two categories as follows.

The first is considered to be the label-setting-based categorization (Zhuang et al. 2021), namely: inductive transfer learning, transductive transfer learning, and unsupervised transfer learning (Pan and Yang 2010).

The second is considered to be the space-setting-based categorization (Zhuang et al. 2021), namely: homogeneous transfer learning and heterogeneous transfer learning. The former is for the case where the input spaces of the source model and the target model are homogeneous, that is, $\mathcal{X}_S = \mathcal{X}_T$. The latter is for the case where the input spaces of the source model and the target model are heterogeneous, that is, $\mathcal{X}_S \neq \mathcal{X}_T$ (Weiss et al. 2016). Among them, homogeneous transfer learning is further divided into instance-based, feature-based (asymmetric, symmetric), parameter-based, and relation-based transfer learning; and heterogeneous transfer learning is mainly feature-based transfer learning (symmetric, asymmetric).

In summary, the categories of transfer learning can be represented in Fig. 11.5.

The application fields of transfer learning are quite extensive, such as the discovery of cancer subtypes, the utilization rate of rooms in buildings, text classification, and spam email filtering.

A learning technique related to transfer learning is called multitask learning (Caruana 1997), which attempts to learn multiple tasks simultaneously, even if they are different. A typical method of multitask learning is to reveal the latent features that can benefit each task.

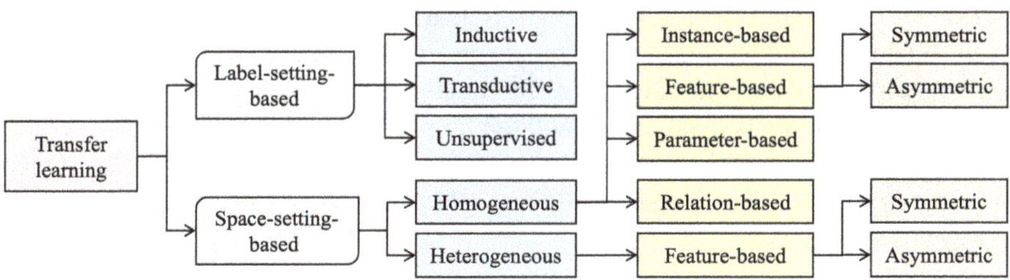

Fig. 11.5 Categories of transfer learning

11.4   Self-Supervised Learning

The supervised learning paradigm represented by classification and regression has achieved great success. For example, in ILSVRC (ImageNet large scale visual recognition challenge) 2015, the deep residual network (ResNet) achieved an outstanding result with an error rate of 3.57%, which is lower than the human error rate of 5.1%.

However, supervised learning often requires thousands or tens of thousands of training samples, and these training samples are often manually annotated. Some fields, such as medical images, require domain experts to annotate, so the annotation cost is laborious and expensive, becoming the bottleneck of supervised learning. Especially some data in the real world are almost impossible to annotate, that is, they cannot be labeled.

As we all know, humans mainly acquire general and common sense knowledge through self-study. Therefore, self-supervised learning came into being.

11.4.1   Definition

Self-supervised learning is an approach to machine learning that has emerged in recent years that does not require manually annotated training data, becoming a new type of machine learning.

Definition 11.4 (Self-Supervised Learning) Self-supervised learning (SSL) is a quasi-paradigm of machine learning that automatically extracts *pseudo-labels* from raw data and then uses them for supervised learning in the next stage.

From the above definition, it is known that self-supervised learning is usually divided into two stages, namely, the pretext stage of extracting pseudo-labels and the downstream stage of supervised learning with these pseudo-labels (Jing and Tian 2021).

The "pseudo-labels" in self-supervised learning are different from the "true labels" in supervised learning. The former are automatically extracted in the pretext stage of self-supervised learning and different labels from supervised learning, while the latter are real labels manually annotated by professionals.

The "pseudo-labels" in self-supervised learning can be seen as the "representation" of original data and therefore can also be referred to as self-supervised representation learning.

11.4.2 Related Discourses

On May 6, 2020, during the *International Conference on Learning Representation* (ICLR) 2020, there is one of *ACM TechNews* published on *Communications of the ACM*:

> Turing Award recipients Yann LeCun and Yoshua Bengio say that self-supervised learning could lead to the creation of artificial intelligence (AI) programs that are more humanlike in their reasoning.[5]

On March 4, 2021, Yann LeCun, chief scientist of Facebook AI, and researcher Ishan Misra published a blog post titled "Self-Supervised Learning: The Dark Matter of Intelligence" on the *Meta AI Blog* (LeCun and Misra 2021), in which they stated the following arguments:

> A working hypothesis is that generalized knowledge about the world, or common sense, forms the bulk of biological intelligence in both humans and animals. This common sense ability is taken for granted in humans and animals, but has remained an open challenge in AI research since its inception. In a way, common sense is the dark matter of artificial intelligence.

The so-called dark matter is a type of invisible matter in the universe that has been indirectly detected, accounting for about 85% of the total matter in the universe, and is an important part of the universe. LeCun and Misra compare self-supervised learning to the dark matter of intelligence, which shows their understanding of the importance of self-supervised learning.

LeCun and Misra also wrote on the blog:

> Common sense helps people learn new skills without requiring massive amounts of teaching for every single task. For example, if we show just a few drawings of cows to small children, they'll eventually be able to recognize any cow they see....... The short answer is that humans rely on their previously acquired background knowledge of how the world works.

Their viewpoint is:

> We believe that self-supervised learning (SSL) is one of the most promising ways to build such background knowledge and approximate a form of common sense in AI systems.

11.4.3 Typical Methods

The main characteristic of self-supervised learning is the pretext stage of automatically extracting pseudo-labels, which often combines with other machine learning methods to complete the subtask in the pretext stage. Its major methods include generative, contrastive, adversarial, and also general framework.

---

[5] https://cacm.acm.org/news/244720-yann-lecun-yoshua-bengio-self-supervised-learning-is-key-to-human-level-intelligence/fulltext

11.4.3.1   Generative

Generative self-supervised learning is the generative model-based self-supervised learning. The generative models adopted include autoregressive models, flow-based models, and autoencoders (Jing and Tian 2021; Liu et al. 2021).

11.4.3.2   Contrastive

Contrastive self-supervised learning is the contrastive learning model-based self-supervised learning. It uses positive instances and negative instances to train the model, and its loss function is to minimize the distance between positive instances while maximizing the distance between negative instances. The contrastive models adopted include context-instance contrast and instance-instance contrast (Liu et al. 2021).

11.4.3.3   Adversarial

Adversarial self-supervised learning is adversarial model-based self-supervised learning. The approaches adopted include adversarial self-supervised contrastive learning (Kim et al. 2020), adversarial self-supervised learning for semi-supervised learning (Si et al. 2020), and self-supervised learning based on adversarial enhanced neural networks (Zeng et al. 2021).

11.4.3.4   General Framework

The general framework refers to a unified self-supervised learning framework that can support multiple modalities. In February 2022, Mate AI proposed a general framework for self-supervised learning—Data2vec—which can be used for computer vision, speech, and natural language processing (Baevski et al. 2022). In December 2022, Meta AI published an upgraded version of this framework—Data2vec 2.0—and the related paper was also published at the *International Conference on Machine Learning* (ICML) in 2023 (Baevski et al. 2023).

11.5 n-Shot Learning

The classic classification task in computer vision belongs to supervised learning, which often requires tens of thousands of manually labeled samples. However, the ability to quickly learn object classification from a very small number of samples has been confirmed in humans, who can also use existing knowledge to distinguish some new object categories without any training samples.

It is estimated that a child around the age of 6 has basically learned the tens of thousands of object categories that exist in the world. This is not only due to the recognition ability of the child's brain at this age, but also because the child possesses the remarkable ability to rapidly learn new visual concepts by observing only one or a few visual instances (Li et al. 2002; Lake et al. 2011).

11.5.1 Meaning of the Term

The *n*-shot learning is an umbrella term that covers *few*-shot learning, *one*-shot learning, and even *zero*-shot learning.

Since *n*-shot learning does not require a lot of manual annotations, it has expanded from computer vision to natural language processing and other fields.

By the way, there are similar terms in the field of machine learning, namely, "few sample learning," "one sample learning," and "zero sample learning."

11.5.2 Related Subtasks

The field of machine learning initially introduced *n*-shot learning as an object classification task in computer vision (Romaniuk 1995; Li et al. 2003, 2006).

Below, we introduce the three subtasks in the order in which they appeared, i.e., one-shot learning, few-shot learning, and zero-shot learning.

11.5.2.1 One-Shot Learning

One-shot learning (OSL) can be traced back to the paper titled "A Bayesian Approach to Unsupervised One-Shot Learning of Object Categories" by Fei-Fei Li et al. at the *IEEE International Conference on Computer Vision* (ICCV) in 2003 (Li et al. 2003). Subsequently, these researchers published a paper titled "One-Shot Learning of Object Categories" in the *IEEE Transactions on Pattern Analysis and Machine Intelligence* (TPAMI) in 2006 (Li et al. 2006).

In the above papers, one-shot learning refers to having only one or a very few of training samples. The number of training samples used by their experiments is 1 to 5.

The method of one-shot learning proposed by the above papers was based on the variational Bayesian framework, in which object categories were characterized by a corresponding number of probability models, prior category knowledge was represented as a probability density function, and posterior category probability was obtained by updating prior knowledge with one or several observed data. Therefore, this was a research approach based on probability models.

With the rise of deep learning, some one-shot learning methods based on deep neural networks have emerged.

Due to the effectiveness of one-shot learning, in addition to being used for object classification, it is also used for other tasks in computer vision, such as pedestrian re-identification, and semantic segmentation.

One-shot learning has also been extended to other research fields, such as natural language processing (Xia et al. 2020) and speech processing (Lake et al. 2014).

11.5.2.2   Few-Shot Learning

Following one-shot learning, the few-shot learning (FSL) also emerged in computer vision community.

Some researchers regard one-shot learning as a special case of few-shot learning. From the logical relationship between the two, their view is valid, although they appeared in a different chronological order.

Fei-Fei Li et al. published a paper titled "Learning Generative Visual Models from Few Training Examples" in 2004 (Li et al. 2004), which did not mention the term few-shot learning, but it involved computer vision problems with a few number of samples.

Some researchers proposed low-shot learning, attempting to unify few-shot learning and one-shot learning (Xia et al. 2020).

11.5.2.3   Zero-Shot Learning

Zero-shot learning (ZSL) means being able to solve a task without any training samples. The premise is to use many training samples from similar tasks for training, forming prior knowledge. Then, with some additional information, it can solve new tasks.

In 2008, the team of Joshua Bengio published a paper titled "Zero-data Learning of New Tasks" at the annual *AAAI Conference on Artificial Intelligence* (Larochelle et al. 2008). This paper can be considered as the precursor of zero-shot learning, as the term used in the paper is "zero-data learning."

The first academic paper titled with zero-shot was published in 2009, titled "Zero-Shot Learning with Semantic Output Codes," presented at the 2009 *Conference on Neural Information Processing Systems* (NIPS) (Palatucci et al. 2009). The paper proposed a zero-shot learning algorithm called the semantic output code classifier, which can use a semantic knowledge base to encode the common features of new categories and the training set, thereby predicting new categories that are ignored in the training set.

11.5.3 Working Principle

Let $\mathcal{X}$ denote the input space, $\mathcal{Y}$ denote the output space, and H denote the hypothesis set. Without loss of generality, assume that the input data $x \in \mathcal{X}$ is sampled through the probability distribution $P(x)$, and the output data $y \in \mathcal{Y}$ is labeled through conditional probability $P(y|x)$.

n-shot can be regarded as a set of labeled samples with m training data, derived based on the joint probability distribution $P(x, y) = P(y|x) P(x)$, and represented as:

$$S_m = \{(x_i, y_i) \mid i = 1, \ldots, m\}.$$

Therefore, n-shot learning (NSL) can be represented as the following 2-tuple:

$$\text{NSL}_n = \langle S_m, H \rangle.$$

Our aim is to find a hypothesis $h \in H$, so that we can use $h(x) = y$ for n-shot learning.

Therefore, the relationship of n-shot learning (NSL) with zero-shot learning (ZSL), one-shot learning (OSL), and few-shot learning (FSL) is as follows:

- If n is "zero" and $m = 0$, it is zero-shot learning $\text{NSL}_{\text{zero}} = \text{ZSL}$.
- If n is "one" and $m = 1$, it is one-shot learning $\text{NSL}_{\text{one}} = \text{OSL}$.
- If n is "few" and m is very small, it is few-shot learning $\text{NSL}_{\text{few}} = \text{FSL}$.

11.5.4 Case Study of One-Shot Learning

Here we select a paper published in the journal *Science* in December 2015, titled "Human-Level Concept Learning Through Probabilistic Program Induction" (Lake et al. 2015). At a time when deep learning is in full swing, this paper presents us with a different viewpoint.

The three authors of the paper are Brenden Lake from the Data Science Center at New York University, Ruslan Salakhutdinov from the Department of Computer Science and Statistics at the University of Toronto, and Joshua Tenenbaum from the Department of Brain and Cognitive Sciences at MIT. Among them, Ruslan Salakhutdinov had co-authored a paper titled "Reducing the Dimensionality of Data with Neural Networks" with Professor Hinton in the journal *Science* in 2006.

This paper addresses the learning problem of one-shot classification. By modeling the learning ability of human visual concepts through a Bayesian model, a framework called Bayesian program learning is designed, which achieves human-level performance in one-shot classification problems. In addition, the method proposed in this paper can also recognize new data, create new categories, etc., that is, the model can automatically induce and abstract high-level information in the training data. The paper validates the model's creative generalization abilities through several visual Turing tests, showing that the model's behavior is close to human behavior in many cases.

Supervised learning, including supervised deep learning, often requires a large amount of training data, yet humans can learn rich concepts from limited data. The Bayesian program learning framework proposed in this paper can learn many visual concepts from a single example and generalize them in a way that is almost indistinguishable from humans. In this framework, concepts are represented as simple probabilistic programs, i.e., probabilistic generative models of structured processes in abstract description languages. The framework combines the following three important ideas: compositionality, causality, and learning to learn. These ideas are very important in cognitive science and machine learning.

The Bayesian program learning learns concepts through simple random programs, each new type can also be represented as a generative model, and this low-level generative model produces new examples or tokens of the concept. The steps are as below:

(a) Start with primitive tokens.
(b) Combine them to sub-parts.
(c) Synthesize parts.
(d) Generate object templates with relations.
(e) Combine and generate new token exemplars.
(f) Finally, render them as raw token data.

To compare Bayesian program learning with several deep learning models and humans, this paper proposes the following two tasks: the first is one-shot classification, i.e., given a character image, human testers and each machine learning model are required to select the same type of image from 20 images; the second is to generate new samples, including standard, dynamic, and new concept samples. The effectiveness of the proposed method was verified through these two tasks.

This paper also proposes several visual Turing tests to detect the creative generalization abilities of the model. The results show that in many cases, the model is almost indistinguishable from human behavior.

Further Reading

1. Zhi-Hua Zhou. *Ensemble Methods: Foundations and Algorithms*. CRC Press, 2012.
   [Notes] This book introduces the core knowledge of ensemble learning such as boosting and bagging and also introduces clustering ensembles and some advanced topics.
2. Timothy Hospedales, Antreas Antoniou, Paul Micaelli, and Amos Storkey. "Meta-Learning in Neural Networks: A Survey." *IEEE Transactions on Pattern Analysis and Machine Intelligence* (TPAMI), 44(9), 2022.

[Notes] This is a review paper on meta-learning. It was submitted to arXiv in 2020 and then officially published in TPAMI in 2022. The paper comprehensively discusses the taxonomy, methodologies, applications, challenges, and open questions of meta-learning in neural networks.
3. Fuzhen Zhuang, Zhiyuan Qi, Keyu Duan, Dongbo Xi, Yongchun Zhu, Hengshu Zhu, Hui Xiong, and Qing He. "A Comprehensive Survey on Transfer Learning." *Proceedings of the IEEE*, 109(1) 2021.
[Notes] This is a review paper on transfer learning. It systematically discusses the mechanisms, strategies, methods, and applications of transfer learning.
4. Xiao Liu, Fanjin Zhang, Zhenyu Hou, Li Mian, Zhaoyu Wang, Jing Zhang, and Jie Tang. "Self-Supervised Learning: Generative or Contrastive." *IEEE Transactions on Knowledge and Data Engineering*, 35(1), 2023.
[Notes] This is a review paper on self-supervised learning. It comprehensively summarizes the methods, categories, open problems, and future directions for self-supervised learning.
5. Maxime Oquab, Timothée Darcet, Théo Moutakanni, Huy Vo, Marc Szafraniec, Vasil Khalidov, Pierre Fernandez, et al. "DINOv2: Learning Robust Visual Features Without Supervision." *arXiv preprint* arXiv:2304.07193, 2023.
[Notes] This is a paper submitted by researchers at Meta AI, introducing their latest developed computer vision model based on self-supervised learning, and they have opened their source code.
6. Yisheng Song, Ting Wang, Puyu Cai, Subrota K. Mondal, and Jyoti Prakash Sahoo. "A Comprehensive Survey of Few-Shot Learning: Evolution, Applications, Challenges, and Opportunities." *ACM Computing Surveys*, 2023.
[Notes] This is a comprehensive review paper on few-shot learning. It was published by the authors after a broad survey of over 200 papers on few-shot learning over the past 3 years.
7. Brenden Lake, Ruslan Salakhutdinov, Joshua Tenenbaum. "Human-Level Concept Learning Through Probabilistic Program Induction." *Science*, 350(6266), 2015.
[Notes] This paper delves into one-shot learning and organically combines the concepts of compositionality, causality, and learning to learn. The AI system they developed can quickly learn to write unfamiliar characters.

References

Baevski, A., W.-N. Hsu, Q. Xu, A. Babu, J. Gu, and M. Auli. (2022). Data2vec: A General Framework for Self-supervised Learning in Speech, Vision and Language. arXiv preprint arXiv:2202.03555.

Baevski, A., A. Babu, W.-N. Hsu, and M. Auli. (2023). Efficient self-supervised learning with contextualized target representations for vision, speech and language. In *International Conference on Machine Learning*, PMLR.

Bozinovski, S. (2020). Reminder of the first paper on transfer learning in neural networks,1976. *Informatica* 44(3): 291–302.

Brazdil, P., C. G. Carrier, C. Soares, and R. Vilalta. (2008). *Metalearning: Applications to data mining*. Berlin: Springer Science & Business Media.

Breiman, L. (1996). Bagging predictors. *Machine Learning* 24: 123–140.

Breiman, L. (2001). Random forests. *Machine Learning* 45(1): 5–32.

Caruana, R. (1997). Multitask learning. *Machine Learning* 28(1): 41–75.

Freund, Y., and R. E. Schapire. (1995). A decision-theoretic generalization of on-line learning and an application to boosting. In *Computational Learning Theory: Second European Conference, EuroCOLT '95.*

Ganaie, M. A., M. Hu, A. Malik, M. Tanveer, and P. Suganthan. (2022). Ensemble deep learning: A review. *Engineering Applications of Artificial Intelligence* 115: 105151.

Gasser, L., K. Lakkaraju, S. Ray, and S. Swarup. (2005). DARPA BAA 05-29: Transfer Learning Issues and Potential Contributions. University of Illinois at Urbana-Champaign.

Jing, L., and Y. Tian. (2021). Self-supervised visual feature learning with deep neural networks: A survey. *IEEE Transactions on Pattern Analysis and Machine Intelligence* 43(11): 4037–4058.

Kearns, M. (1988). Thoughts on Hypothesis Boosting. Technical report, Machine Learning Class Project.

Kim, M., J. Tack, and S. J. Hwang. (2020). Adversarial self-supervised contrastive learning. In *Conference on Neural Information Processing Systems (NIPS)* 2983–2994.

Lake, B., R. Salakhutdinov, J. Gross, and J. Tenenbaum. (2011). One shot learning of simple visual concepts. In *Proceedings of the Annual Meeting of the Cognitive Science Society.*

Lake, B., C.-y. Lee, J. Glass, and J. Tenenbaum. (2014). One-shot learning of generative speech concepts. In *Proceedings of the Annual Meeting of the Cognitive Science Society.*

Lake, B. M., R. Salakhutdinov, and J. B. Tenenbaum. (2015). Human-level concept learning through probabilistic program induction. *Science* 350(6266): 1332.

Larochelle, H., D. Erhan, and Y. Bengio. (2008). *Zero-data learning of new tasks*. Washington: AAAI.

LeCun, Y., and I. Misra. (2021). Self-supervised learning: The dark matter of intelligence. *Meta AI* 23: 2.1.

Lemke, C., M. Budka, and B. Gabrys. (2015). Metalearning: A survey of trends and technologies. *Artificial Intelligence Review* 44(1): 117–130.

Li, F.-F., R. VanRullen, C. Koch, and P. Perona. (2002). Rapid natural scene categorization in the near absence of attention. *Proceedings of the National Academy of Sciences* 99(14): 9596–9601.

Li, F.-F., R. Fergus, and P. Perona. (2003). A bayesian approach to unsupervised one-shot learning of object categories. In *IEEE International Conference on Computer Vision (ICCV)* Piscataway: IEEE.

Li, F.-F., R. Fergus, and P. Perona. (2004). Learning generative visual models from few training examples: An incremental Bayesian approach tested on 101 object categories. In *2004 Conference on Computer Vision and Pattern Recognition Workshop*. Piscataway: IEEE.

Li, F.-F., R. Fergus, and P. Perona. (2006). One-shot learning of object categories. *IEEE Transactions on Pattern Analysis and Machine Intelligence (TPAMI)* 28(4): 594–611.

Liu, X., F. Zhang, Z. Hou, L. Mian, Z. Wang, J. Zhang, and J. Tang. (2021). Self-supervised learning: Generative or contrastive. *IEEE Transactions on Knowledge and Data Engineering* 35(01): 857–876.

Mason, L., J. Baxter, P. Bartlett, and M. Frean. (1999). Boosting algorithms as gradient descent. In *Conference on Neural Information Processing Systems (NIPS)*.

Palatucci, M., D. Pomerleau, G. E. Hinton, and T. M. Mitchell. (2009). Zero-shot learning with semantic output codes. In *Conference on Neural Information Processing Systems (NIPS)*.

Pan, S. J., and Q. Yang. (2010). A survey on transfer learning. *IEEE Transactions on Knowledge & Data Engineering* 22(10): 1345–1359.

Pratt, L. Y. (1992). Discriminability-based transfer between neural networks. In *Conference on Neural Information Processing Systems (NIPS)* 5.

Rohrbach, M., S. Ebert, and B. Schiele. (2013). *Transfer learning in a transductive setting*. In *Conference on Neural Information Processing Systems (NIPS)*.

Romaniuk, S. G. (1995). *Application of learning to learn to real-world pattern recognition.* Artificial Neural Nets and Genetic Algorithms. Berlin: Springer.

Schapire, R. E. (1990). The strength of weak learnability. In *Symposium on Foundations of Computer Science.*

Si, C., X. Nie, W. Wang, L. Wang, T. Tan and J. Feng. (2020). Adversarial self-supervised learning for semi-supervised 3d action recognition. In *European Conference on Computer Vision (ECCV).*

Vanschoren, J. (2018). Meta-learning: A survey. arXiv preprint arXiv:1810.03548.

Vilalta, R., and Drissi, Y. (2002). A perspective view and survey of meta-learning. *Artificial Intelligence Review* 18: 77–95. https://doi.org/10.1023/A:1019956318069.

Weiss, K., T. M. Khoshgoftaar, and D. Wang. (2016). A survey of transfer learning. *Journal of Big Data* 3(9): 1–40.

Wolpert, D. H. (1992). Stacked generalization. *Neural Networks* 5(2): 241–259.

Xia, C., C. Zhang, J. Zhang, T. Liang, H. Peng, and S. Y. Philip. (2020). Low-shot learning in natural language processing. In *2020 IEEE Second International Conference on Cognitive Machine Intelligence (CogMI).* Piscataway: IEEE.

Zeng, Z., K. He, Y. Yan, H. Xu, and W. Xu. (2021). Adversarial self-supervised learning for out-of-domain detection. In *Proceedings of the 2021 Conference of the North American Chapter of the Association for Computational Linguistics: Human Language Technologies.*

Zhuang, F., Z. Qi, K. Duan, D. Xi, Y. Zhu, H. Zhu, H. Xiong, and Q. He. (2021). A comprehensive survey on transfer learning. *Proceedings of the IEEE* 109(1): 43–76.

Part IV
Tasks

Here the tasks refer to the learning tasks that are accomplished by machine learning. A task has the relation with a common problem to be solved by machine learning, in which the common problem is abstracted from some domain-specific problems and makes it to be a domain-independent problem. Such a common problem is viewed as a task that needs to be fulfilled by machine learning and can solved by the same machine learning algorithms with the same functions.

Machine learning has been playing a significant role in various domain problems, in which each domain has also some tasks. In computer vision, there are tasks such as image segmentation, object detection tracking, visual understanding, and generation. In natural language processing, there are tasks such as text summarization, machine translation, text understanding, and generation. In addition, in computer speech, data mining, robotics, computer games, autonomous driving, and so on, they all have their own tasks. However, those tasks are all domain-specific tasks that are different from the domain-independent tasks we discuss in this part.

Therefore, it is necessary to discern the above domain-specific and domain-independent tasks, find out their common characteristics, and then sort out their respective solutions.

The tasks in machine learning include what to do for the input data, what form the output data appears in, and how to decide the next action based on current state and feedback from the environment.

The representative tasks of machine learning include classification, regression, ranking, clustering, dimensionality reduction, association, and decision-making.

This part will mainly introduce classification, regression, clustering, and dimensionality reduction. While the association has been discussed in Chap. 6, decision-making and sequential decision-making have been introduced in Chaps. 7 and 10, respectively.

Chapter 12
Classification Task

Abstract Classification is the most common task in supervised learning paradigm. In this chapter, we first present the problems and definition, and the working principle using formal and illustrated descriptions. We second discuss the basic types including one-class, two-class, multi-class, multi-label, and imbalanced classifications and related elements such as linear and nonlinear, hard and soft, as well as lazy and eager classifications. Next, we focus on several typical classification algorithms including logistic regression, naive Bayes, AdaBoost, support vector machine, and artificial neural network. We then introduce the loss functions covering 0–1 loss, logistic loss, exponential loss, hinge loss, and cross-entropy loss. After that, we discuss some evaluation metrics such as accuracy, precision and recall, sensitivity and specificity, F1 score, precision-recall curve (PR curve), receiver operating characteristic curve (ROC curve), and Matthew's correlation coefficient (MCC). Finally, we introduce the application fields of classification briefly.

12.1 Problem and Definition

In this section, we first review the classification problem and then give a definition of the classification.

12.1.1 Classification Problems

Classification is the process of categorizing something. Many problems involve classification, for example: classifying emails, judging whether they belong to "spam" or "non-spam" emails, classifying handwritten Arabic numerals, and judging which one of "0" to "9" they are. In addition, handwritten character recognition, face recognition, fingerprint recognition, speaker recognition, and image classification all need to be processed by classification.

W. Wang, *Principles of Machine Learning*,
https://doi.org/10.1007/978-981-97-5333-8_12

12.1.2 Definition

Definition 12.1 (Classification) Classification in machine learning is the task that trains a classifier with the labeled samples whose categories are known, and then uses it to identify which categories the other data belongs to.

From the above definition, we can see the three elements of classification:

(1) 12:Identifying data into categories.
(2) 12:The categories are known.
(3) 12:Training with samples.

In short, classification is identifying data into known categories by training.

12.2   Working Principle

For understanding of the working principle of classification, here we first give a formal description and then explain it through an illustration.

12.2.1   Formal Description

Let $\mathbb{R}^m (m \geq 1)$ denote a set of m-dimensional real-valued vectors, $X \subseteq \mathbb{R}^m$ be the input space, and $C = \{c_j \mid j = 1, \ldots, k\}$ be the output space which consists of k known categories.

Through an unknown probability distribution $P(x)$ we obtain n independent and identically distributed (i.i.d.) data:

$$D = \{x_i \mid x_i \in X \text{ and } i = 1, \ldots, n\}. \tag{12.1}$$

Also, let the following target function of classification

$$f : X \rightarrow C \tag{12.2}$$

be a conditional probability distribution $P(c|x)$, where $x \in X$, and $c \in C$.

Based on the joint probability distribution $P(x, c) = P(c|x) P(x)$, each data $x_i (i = 1, \ldots, n)$ is labeled with its target category $c_j (j = 1, \ldots, k)$ by $P(c|x)$, resulting in a set of training samples S with n elements, where each element is represented as a data pair $(x_i, c_j) \in X \times C$, i.e.,

$$S = \{(x_i, c_j) \mid i = 1, \ldots, n \text{ and } j = 1, \ldots, k\}. \tag{12.3}$$

Let H be a set of hypothesis functions for classification mapping from $\mathcal{X}$ to C, i.e.,

$$H : \mathcal{X} \to C . \tag{12.4}$$

The goal is to obtain $h \in H$, one of the hypothesis functions through the training on S, and $h(\boldsymbol{x}) = \hat{c}$ has the smallest expected error with the target function $f(\boldsymbol{x}) = c$:

$$\underset{h \in H}{\arg\min}\, R(h) = \underset{h \in H}{\arg\min}\, \mathbb{E}\left[\mathcal{L}(h(\boldsymbol{x}), f(\boldsymbol{x}))\right] = \underset{h \in H}{\arg\min}\, \mathbb{E}\left[\mathcal{L}(\hat{c}, c)\right], \tag{12.5}$$

where $\hat{c}$ is referred to as the predicted category, c is the target category, and $\mathcal{L}(\cdot, \cdot)$ represents the loss function of the classification.

After obtaining the hypothesis function h, it is tested with actual unknown data for classification. Given n' test data $\boldsymbol{x}_i \in \mathcal{X}$ with unknown categories, that is:

$$T = \left\{ \boldsymbol{x}_i \mid i = 1, \ldots, n' \right\} \subseteq \mathcal{X}.$$

Use the trained hypothesis function $h(\boldsymbol{x})$ to classify the n' test data, that is: $h(\boldsymbol{x}_i) = \hat{c}_j \in C$; the result is a set of data pairs, that is:

$$T_{\text{Output}} = \left\{ (\boldsymbol{x}_i, \hat{c}_j) \mid i = 1, \ldots, n' \text{ and } j = 1, \ldots, k \right\}.$$

12.2.2 Illustrated Description

Here, the working principle of classification is further explained with two figures, respectively elaborating the training process and the testing process of the classification model.

The training process of the classification algorithm is depicted in Fig. 12.1. Here, the training data is labeled data, and its category is known. For each input data $\boldsymbol{x}_i \in \mathcal{X}$, the target category is obtained through labeling $c_j \in C$. The purpose is to create a training dataset S, which is used for training the set of hypothesis functions H, and then we obtain a hypothesis $h(\boldsymbol{x})$ that is closest to the target output c, i.e., a classifier, which satisfies the smallest expected error between $h(\boldsymbol{x}) = \hat{c}$ and labeled category c.

The testing process of the classification model is depicted in Fig. 12.2. After the classifier hypothesis $h(\boldsymbol{x})$ is trained, it can be used for actual classification. For some input data whose categories are unknown (also called unseen data), use the trained classifier hypothesis $h(\boldsymbol{x}) = \hat{c}$ and get the predicted classification result $\hat{c}$.

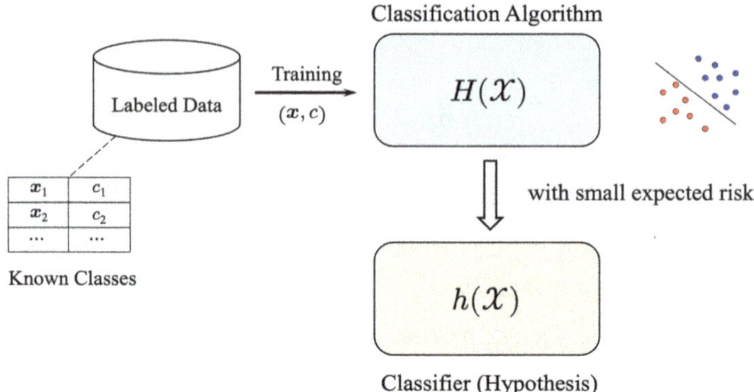

Fig. 12.1 The training process of the classification model

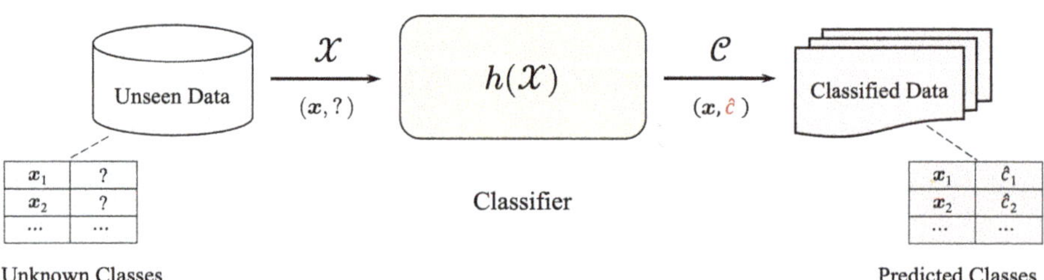

Fig. 12.2 The testing process of the classification model

12.2.3 Classification Functions

There are mainly two types of classification functions: one is probabilistic classification, and the other is statistical classification.

12.2.3.1 Probabilistic Classification

Probabilistic classification is based on the probability framework of machine learning, and its classification result is the probability of which category the unknown data belongs to.

This type of classifier can classify independently, and it can also combine several weak probabilistic classifiers to form a strong classifier.

As introduced before, classification is to obtain a hypothesis $h(x)$ that has the smallest expected error with the target category c. In probabilistic classification, this hypothesis is a probability distribution.

Assume there exist k known categories, and $k \geq 2$. The classification function in probabilistic classification is a conditional probability distribution $P(c|x)$. It means to measure the probability that c belongs to the jth category, given input data $x \in X$, where $j \in \{1, \ldots, k\}$. Its formal representation is as follows:

$$h_j(x) = \arg\max_j P\left(c = c_j \mid x\right). \tag{12.6}$$

That is, its conditional probability $P\left(c = c_j \mid x\right)$ is obtained, then the best decision method to get the final classification result $h_j(x)$ is used.

In other words, probabilistic classification first measures the probability that the input data belongs to one of the k different categories, and then obtains the category with the highest probability as the best category.

The characteristic of the probabilistic classification function is that it can output confidence values associated with its classification, usually referred to as confidence-weighted classification; when the confidence value of a category is below a certain threshold, it can be abandoned; in addition, it can be more effectively used in large-scale machine learning tasks.

In machine learning, the algorithms, such as logistic regression, naive Bayes, and the softmax function used for classification in the last layer of the multilayer perceptron, all use probabilistic classification functions. Details about the above algorithms will be discussed in Sect. 12.5 of this chapter.

12.2.3.2 Statistical Classification

Statistical classification is based on the statistical framework of machine learning, and the result of its classification is what category the unknown data belongs to.

Like probabilistic classification, statistical classification is also trained through labeled samples to obtain a hypothesis $h(x)$ with the smallest expected error for the target category c; but unlike probabilistic classification, the hypothesis of statistical classification is a linear or nonlinear function.

Assume there exist k known categories, and $k \geq 2$. Let $x \in X$ be an input data, w_j be the weight parameters of the jth category, and $j \in \{1, \ldots, k\}$. Without loss of generality, the classification function of statistical classification is expressed as follows:

$$h_j(x) = \arg\max_j f\left(c_j = w_j^{\mathsf{T}} \cdot x + b\right). \tag{12.7}$$

In machine learning, those algorithms, such as linear discriminant analysis, k-nearest neighbor, and support vector machine (SVM), all use statistical methods of classification functions, where the SVM algorithm will be discussed in detail in Sect. 12.5 of this chapter.

12.3   Basic Types

There are several basic types in the classification task, namely: one-class classification, two-class classification, multi-class classification, multi-label classification, and imbalanced classification.

We discuss these basic types as follows.

12.3.1   One-Class Classification

The one-class classification (OCC), also known as unary classification or class modeling, is a special type of classification that trains on a set of samples containing only a specific category, enabling to identify of whether unknown data belongs to that specific category.

Let $h : X \rightarrow C$ be the function of one-class classification, then the output space of this classification function is $C = \{c_j \mid j = 1\}$.

One-class classifiers are mainly used to identify the data points on the outside of a specific category in the input data, as shown in Fig. 12.3. Its corresponding subtasks include outlier detection, anomaly detection, and novelty detection.

Classic methods of one-class classification include:

- Density methods, such as the Gaussian model and Parzen density estimation;
- Boundary methods, such as k-center algorithm

The reference (Perera et al. 2021) provides a systematic review of one-class classification and also introduces the method using deep learning for one-class classification.

12.3.2   Two-Class Classification

Two-class classification (TCC), also known as binary classification, is the type of dividing unknown input data into only two categories.

Let $h : X \rightarrow C$ be the function of two-class classification, then its output space is $C = \{c_j \mid j = 1,\ 2\}$.

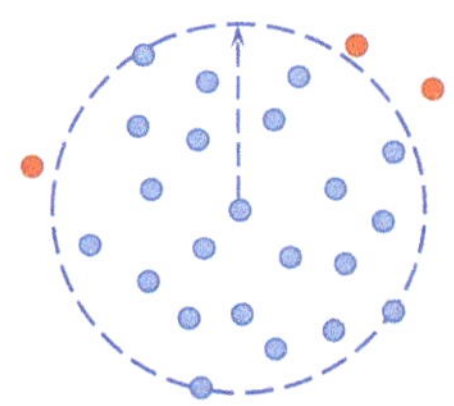

Fig. 12.3 Schematic diagram of one-class classification

Two-class classification is used to predict whether input data has a certain specific property and qualitative characteristics. Its application scenarios include email detection (spam or non-spam), disease diagnosis (whether the disease is present), and credit fraud detection (fraud or legal).

Commonly used algorithms of two-class classification include logistic regression, k-nearest neighbors, decision trees, support vector machine (SVM), and naive Bayes.

12.3.3 Multi-class Classification

Multi-class classification (MCC) is the method of dividing unknown input data into multiple categories through training.

Let $h : X \to C$ be the function of multi-class classification, then its output space is $C = \{c_j \mid j = 1, \ldots, k\}$, where $k > 2$.

There are several ways to solve multi-class classification problems, such as: 1) transforming multi-class classification into two-class classification, 2) extending existing algorithms of two-class classification to multi-class classifiers, and 3) hierarchical classification.

The softmax function is an effective method for solving multi-class classification. It normalizes a vector with k real values into a probability distribution composed of k probability values.

In addition, there are some methods that combine deep neural networks with multi-class classifiers. For example, a paper titled "Deep Neural Decision Forests" (Kontschieder et al. 2015) presented at the 2015 *IEEE International Conference on Computer Vision* (ICCV) combines a deep neural network with decision forests for image classification. This paper won the Marr Prize for the best paper of the year.

The upper part of this model is a deep convolutional neural network with a variable number of layers, with parameters Θ; the middle part is a fully connected layer, represented as an inner product function $f_n(\cdot; \Theta)$; each output unit of the fully connected layer is connected to a node in a tree of the decision forest, $d_n(x) = \sigma(f_n(x))$, making routing decisions; the order of allocation from the output units of the fully connected layer to the decision tree nodes can be arbitrary, and the order marked in the figure is just for visualization; the bottom is the leaf node, as a result of solving the convex optimization problem, maintaining the probability distribution π_l.

The model of deep neural decision forest has surpassed existing methods in experiments on the MNIST handwritten digit dataset and the ImageNet image classification dataset, verifying the effectiveness of the method of deep neural decision forest.

12.3.4 *Multi-label Classification*

Multi-label classification (MLC) is a variant of classification problems, characterized by the ability to assign multiple labels to a single data point. The labels in multi-label classification are non-exclusive, and there is no limit to how many can be assigned to a single class.

Let $h : \mathcal{X} \to \{0, 1\}^C$ be the function of multi-label classification, where $C = \{c_j \mid j = 1, \ldots, k\}$ and $k > 1$.

Therefore, multi-label classification is to construct a model that maps the input $x \in \mathcal{X}$ to a binary vector $y \in \{0, 1\}^C$. That is, it assigns 0 or 1 to each label in y.

Multi-label classification is different from multi-class classification, because the latter is a single-label problem, which is to divide the data into one of several categories.

The schematic diagram of multi-label classification and multi-class classification is as depicted in Fig. 12.4, where:

- Figure 12.4a is a training sample, there are three objects in the picture, and the result after labeling is strawberry, banana, and grapes.
- Figure 12.4b belongs to multi-label classification, there are three data in the picture, each data has two or three objects, and the results of multi-label classification are

$$\{1, 1, 0\}, \{0, 1, 1\}, \{1, 1, 1\}$$

respectively.

- Figure 12.4c is multi-class classification, there are three data in the picture, each data has only one object, and the results of multi-class classification are

$$\{1, 0, 0\}, \{0, 0, 1\}, \{0, 1, 0\}$$

respectively.

Common algorithms of multi-label classification include multi-label k-nearest neighbor, multi-label decision tree, and artificial neural networks (ANNs). Ensemble learning methods can also be used to combine several multi-class classifiers into a multi-label classifier.

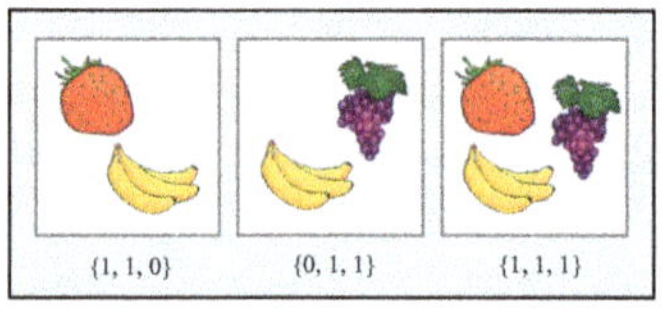

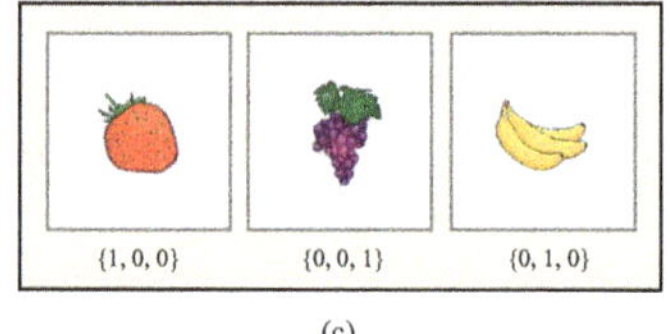

Fig. 12.4 Schematic diagram of multi-label classification and multi-class classification

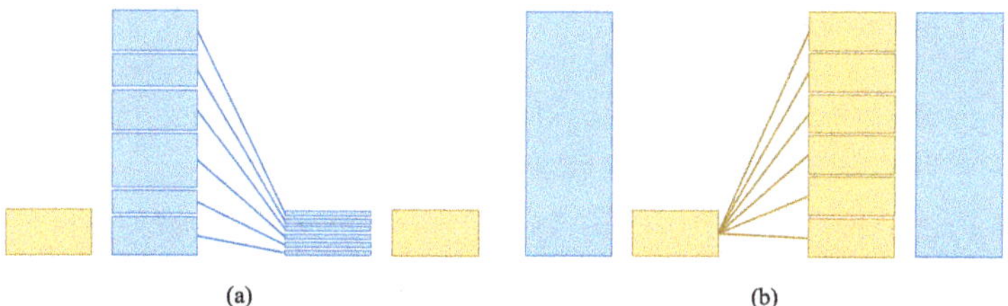

Fig. 12.5 Schematic diagram of imbalanced classification

12.3.5 Imbalanced Classification

Imbalanced classification is the classification of imbalanced data.

The imbalanced data refers to a dataset where the distribution of categories is extremely imbalanced, which usually belongs to two-class classification problems. For example, the ratio of data between two categories is 1:10, 1:100, or even 1:1000. This makes it difficult for some machine learning models to learn from such imbalanced data and achieve good classification performance.

An effective way to solve the problem of imbalanced classification is to use resampling techniques to transform imbalanced data into balanced data, and then use two-class classification algorithms for classification.

There are mainly three resampling techniques for imbalanced data:

(1) Down-sampling for data with a high ratio, as shown in Fig. 12.5a
(2) Up-sampling for data with a low ratio, as shown in Fig. 12.5b
(3) A combination of down-sampling and up-sampling

There are many application scenarios for imbalanced classification, including defect detection, medical diagnosis, fraud detection, spam email filtering, text categorization, and land use classification.

12.4 Related Elements

12.4.1 Linear and Nonlinear

Linear and nonlinear classifications are closely related to the data to be classified. In this subsection, we will first give the definition on linearly separability of data and then introduce linear and nonlinear classifications.

12.4.1.1 Linear Separable

Linear separability refers to the data points in binary classification problems which can be separated using linear decision boundary.

Definition 12.2 (Linear Separable) Let $\mathbb{R}^m$ be a set of m-dimensional real-valued vectors, $\mathcal{X} \subseteq \mathbb{R}^m$ be an input space, $D \subset \mathcal{X}$ and $D' \subset \mathcal{X}$ be two sets of data points, $\boldsymbol{w} \in \mathbb{R}^m$ be weight vector, and $b \in \mathbb{R}$ be bias. Then D and D' are called linear separable, if there exists a linear function

$$u(\boldsymbol{x}) = \boldsymbol{w}^\mathsf{T}\boldsymbol{x} - b$$

such that every data point $\boldsymbol{x} \in D$ satisfies $u(\boldsymbol{x}) = \boldsymbol{w}^\mathsf{T}\boldsymbol{x} - b > 0$ and every data point $\boldsymbol{x}' \in D'$ satisfies $u(\boldsymbol{x}') = \boldsymbol{w}^\mathsf{T}\boldsymbol{x}' - b < 0$.

12.4.1.2 Linear Classification

If the data is linearly separable, it can be classified using linear classification, the function of which is called linear classification function.

Without loss of generality, a linear classification for two-class classification can be expressed as:

$$f(\boldsymbol{x}) = \mathrm{sign}\,(u(\boldsymbol{x})) = \mathrm{sign}(\boldsymbol{w}^\mathsf{T}\boldsymbol{x} - b), \tag{12.8}$$

where $\mathrm{sign}(\cdot)$ is a sign function. And the linear classification function is defined as follows:

$$f(\boldsymbol{x}) = \begin{cases} +1 & \boldsymbol{w}^\mathsf{T}\boldsymbol{x} - b > 0 \\ 0 & \boldsymbol{w}^\mathsf{T}\boldsymbol{x} - b = 0 \\ -1 & \boldsymbol{w}^\mathsf{T}\boldsymbol{x} - b < 0. \end{cases}$$

A schematic diagram of two-class linear classification in two-dimensional input data space is depicted in Fig. 12.6. As can be seen from the figure, $\boldsymbol{w} \cdot \boldsymbol{x} - b = 0$ is equivalent to a linear decision boundary, classifying the data on both sides of this line: the one side is the data $\boldsymbol{w} \cdot \boldsymbol{x} - b > 0$, and the other side is the data $\boldsymbol{w} \cdot \boldsymbol{x} - b < 0$.

Fig. 12.6 Schematic diagram of two-class linear classification in two-dimensional input data space

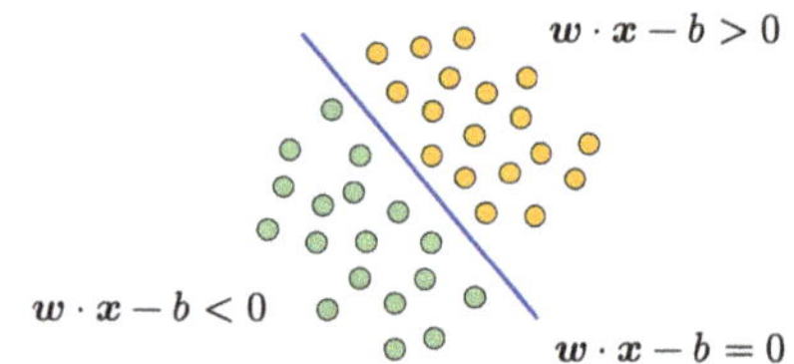

If the linear decision margins with no data can be found on either side of a linear classification boundary and maximizes it, then the classifier has the best classification effect. The classifier of support vector machine (SVM) has the largest explicit linear decision margin ((Boser et al. 1992), (Cortes and Vapnik 1995)).

It is worth pointing out that the linear classification is closely related to the dimension of the input data space and its hyperplane.

Let $\mathbb{R}^m$ be a set of m-dimensional vectors, $\mathcal{V} \subseteq \mathbb{R}^m$ denote input data space, and $\boldsymbol{w}, \boldsymbol{x} \in \mathcal{V}$ and $\boldsymbol{w} \neq 0$, $b \in \mathbb{R}$ be the bias, then the hyperplane $\mathcal{H}$ of linear classification is defined as follows:

$$\mathcal{H} = \left\{ \boldsymbol{x} \mid \boldsymbol{w}^\mathsf{T}\boldsymbol{x} - b = 0 \right\}. \tag{12.9}$$

From the above definition, it can be seen that the hyperplane satisfied $\boldsymbol{w}^\mathsf{T}\boldsymbol{x} - b = 0$ is a set of points.

In geometry, the input data space $\mathcal{V}$ with the hyperplane $\mathcal{H}$ has the following properties:

(1) In the m-dimensional input data space $\mathcal{V}$, the dimension of its hyperplane $\mathcal{H}$ is $m - 1$. Therefore:

  - When the input data space $\mathcal{V}$ is three-dimensional, its hyperplane $\mathcal{H}$ appears as a two-dimensional plane.
  - When the input data space $\mathcal{V}$ is two-dimensional, its hyperplane $\mathcal{H}$ appears as a one-dimensional line.
  - When the input data space $\mathcal{V}$ is one-dimensional, its hyperplane $\mathcal{H}$ is a point with 0 dimension (zero dimension).

(2) Any $(m - 1)$-dimensional hyperplane $\mathcal{H}$ will divide the m-dimensional input data space $\mathcal{V}$ into two half-spaces. They are $\boldsymbol{w}^\mathsf{T}\boldsymbol{x} - b > 0$ and $\boldsymbol{w}^\mathsf{T}\boldsymbol{x} - b < 0$.

Linear classifiers in machine learning, especially support vector machine (SVM), utilize the above properties: according to the dimension of the input data space, find a hyperplane of one dimension lower to classify the data.

The schematic diagram, using a two-dimensional plane for linear classification when dealing with three-dimensional data, is depicted in Fig. 12.7.

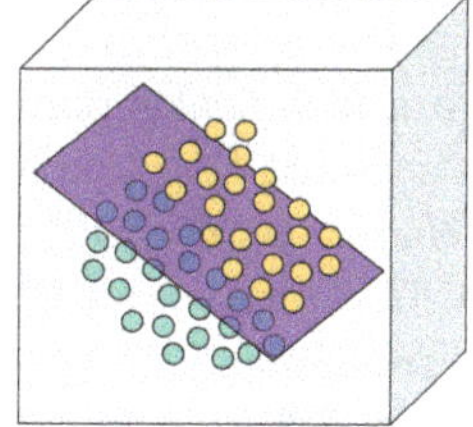

Fig. 12.7 The two-dimensional plane in three-dimensional input data space for linear classification

12.4.1.3 Nonlinear Classification

If the data is nonlinearly separable, it needs to be classified by the methods of nonlinear classification, and their functions are called nonlinear classification functions.

Unlike linear classification, a nonlinear classification function is a nonlinear combination of a model's parameters and depends on one or more independent variables.

For instance, a nonlinear classification function is expressed as follows:

$$f(\boldsymbol{x}) = w_0 + w_1 x + w_2 x^2 + \cdots + w_m x^m = w_0 + \sum_{i=1}^{m} w_i x^i. \tag{12.10}$$

After normalizing the above equation, we get the following nonlinear classification expression:

$$f(\boldsymbol{x}) = w_0 + \sum_{i=1}^{m} w_i \phi_i(\boldsymbol{x}) = \boldsymbol{w}^\mathsf{T} \Phi(\boldsymbol{x}) + b, \tag{12.11}$$

where $\Phi(\boldsymbol{x}) = \left(x, x^2, \ldots, x^m\right)^\mathsf{T}$ is called the basis function of the classification, $w_0 = b$, and $\sum_{i=1}^{m} w_i \phi_i(\boldsymbol{x}) = \boldsymbol{w}^\mathsf{T} \Phi(\boldsymbol{x})$.

Some nonlinear classification functions may have several nonlinear decision boundaries, and these decision boundaries may be discontinuous.

Common nonlinear classification algorithms include naive Bayes classifiers, k-nearest neighbors, kernel support vector machines, decision trees, random forests, and artificial neural networks, in which naive Bayes, support vector machines, and artificial neural networks are discussed in more detail in Sect. 12.5 of this chapter.

12.4.2 Hard and Soft

Classification methods in machine learning can also be divided into hard classification and soft classification.

The concept of hard classification and soft classification can be traced back to a technical report in 1993 titled "Soft Classification, aka Risk Estimation, via Penalized Log Likelihood and Smoothing Spline Analysis of Variance" by Grace Wahba et al. from the University of Wisconsin-Madison (Wahba et al. 1993), which interpreted the concepts of hard classification and soft classification from a probabilistic viewpoint.

In 1996, Yoram Baram from the Israel Institute of Technology proposed the concept of firm classification and soft classification in a technical report titled "Consistent Classification, Firm and Soft." In 1997, he published a paper with the same title at the *Conference on Neural Information Processing Systems* (NIPS)

(Baram 1997), studying these two types of classification problems using geometric methods.

In another technical report in 1998 titled "Support Vector Machines, Reproducing Kernel Hilbert Spaces and The Randomized GACV," Grace Wahba further described hard classification and soft classification. This technical report later was published as a chapter in the book *Advances in Kernel Methods: Support Vector Learning* (Wahba 1998).

In 2002, Grace Wahba published a paper titled "Soft and Hard Classification by Reproducing Kernel Hilbert Space Methods" (Wahba 2002), proposing a method that can handle soft classification and hard classification in a unified manner.

The concept of hard classification and soft classification gradually gained recognition from some researchers. In 2011, several researchers from the University of North Carolina at Chapel Hill published a paper titled "Hard or Soft Classification? Large-Margin Unified Machines" (Liu et al. 2011).

Some researchers have applied the concepts of hard classification and soft classification to image classification (Choodarathnakara et al. 2012; Pan et al. 2012; Sharma et al. 2016). Some other researchers have also explored the combination model of hard classification and soft classification (Hu et al., 2014; Cao et al. 2018).

Let's consider why do we need to distinguish classification methods with "hard" and "soft":

Consider a problem in medical data classification. Given n samples:

$$S = \left\{ (x_i, \ c_j) \mid i = 1, \ldots, n \text{ and } c_j = 1, \ldots, k \right\},$$

where x_i is the symptom of some disease, and c_j is the category of the disease.

There are two possible methods of classification:

(1) The first classification method: Design a classification algorithm $f_c(x)$; after training with aforementioned samples, the new symptom x_i' is clearly divided into some disease class c_j.
(2) The second classification method: Design a classification algorithm $p(c|x)$; after training with aforementioned samples, the new symptom x_i' is predicted as the likelihood

Clearly, among the two classification methods mentioned above, the first one is hard classification, while the second one is soft classification.

In summary, hard classification explicitly divides the input data into a known category, while soft classification gives the probability value of the input data being a known category.

For example, the support vector machine classifier belongs to hard classification, while logistic regression belongs to soft classification.

12.4.3   Lazy and Eager

The classification methods of machine learning can be divided into lazy classification and eager classification. These two classification properties are closely related to the two stages, the training stage and testing stage, of classification.

12.4.3.1   Lazy Classification

The work style of lazy classification is as follows. During the training phase, it only stores the training data. When entering the testing phase, the classifier checks whether the input is the most relevant object in the training data, and then classifies it.

The characteristic of the lazy classifier is less training time, but more time in classification.

A typical lazy classification algorithm is the k-nearest neighbor.

12.4.3.2   Eager Classification

The work style of eager classification is as follows. During the training phase, it undergoes rigorous debugging and training. When entering the testing phase, the trained classifier directly classifies the input data.

The eager classifier is the opposite of the lazy classifier, characterized by a long time for training, but less time for classification.

Representative eager classification algorithms include decision trees, naive Bayes, and artificial neural networks (ANNs).

12.5   Typical Algorithms

There are many classification algorithms.

In *Wikipedia*, there is a webpage, namely, "Category: Classification algorithms[1]" When I wrote this book, the last update of this page was on December 6, 2016, and the listed classification algorithms had already reached 83.

Representative algorithms include AdaBoost, artificial neural network classifier, decision trees, k-nearest neighbors, logistic regression, linear discriminant analysis, naive Bayes classifier, random forests, and support vector machine classifier.

---

[1] https://en.wikipedia.org/wiki/Category:Classification_algorithms

Here we will introduce some of those representative classification algorithms that are logistic regression, naive Bayes, adaptive boosting, support vector machine (classifier), and artificial neural networks.

12.5.1 Logistic Regression

Logistic regression, also known as logistic regression analysis, is a statistical analysis method that predicts the corresponding binary results based on the relationship between the independent variable and the dependent variable and therefore can be used for two-class classification.

Classification is essentially a regression problem with discrete output; the algorithm of logistic regression reflects this nature.

Let X denote the input space of classification and $C = \{0, 1\}$ denote the output space of two-class classification. For input data $x \in X$, the logistic regression expression $P(c|x)$ for predicting its category as $c \in C$ is as follows:

$$P(c|x) = \sigma(z) = \frac{1}{1 + e^{-z}} = \frac{1}{1 + \exp(-z)} \tag{12.12}$$

where $\sigma(\cdot)$ is referred to as the logistic function, also known as the Sigmoid function, which is also used as one of the activation functions in artificial neural networks.

Logistic regression has the following property, i.e., $P(c = 0|x) = 1 - P(c = 1|x)$. This is because the probability distribution of a two-class classification should satisfy $P(c = 0|x) + P(c = 1|x) = 1$. Also, let $z = w^{\mathsf{T}}x - b$; therefore, we have:

$$P(c = 1|x) = \sigma\left(w^{\mathsf{T}}x - b\right) = \frac{1}{1 + \exp\left(-\left(w^{\mathsf{T}}x - b\right)\right)},$$

$$P(c = 0|x) = 1 - \sigma\left(w^{\mathsf{T}}x - b\right) = 1 - \frac{1}{1 + \exp\left(-\left(w^{\mathsf{T}}x - b\right)\right)}$$

$$= \frac{\exp\left(-\left(w^{\mathsf{T}}x - b\right)\right)}{1 + \exp\left(-\left(w^{\mathsf{T}}x - b\right)\right)}. \tag{12.13}$$

A schematic diagram of the logistic regression classification algorithm is depicted in Fig. 12.8.

The above logistic regression method is only applicable to two-class classification.

Later, some extended versions of logistic regression appeared for multi-class classification problems, including multinormal logistic regression, ordered logistic regression, and conditional logistic regression.

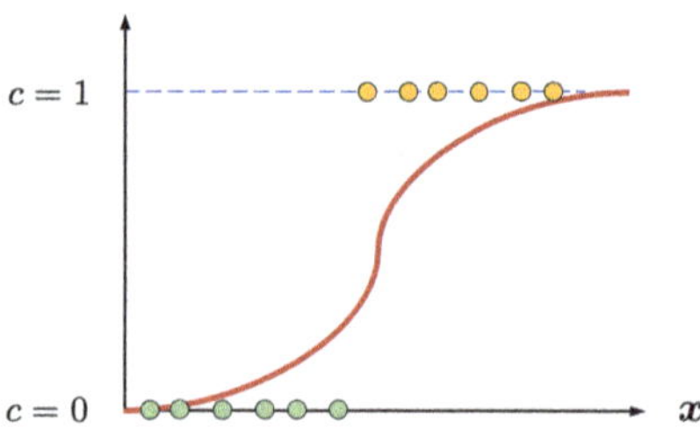

Fig. 12.8 Schematic diagram of the logistic regression classification algorithm

12.5.2   Naive Bayes

Naive Bayes classifier belongs to probability classification, which is based on the Bayes theorem, but the premise is that the features have strong independence.

The Bayes theorem classifier based on strong independence between features is a simplification of the theorem conditions; this is why it is called naive Bayes.

Given the data vector $x = (x_1, \ldots, x_n) \in X$ to be classified in the input space, where the output space is a set of categories $C = \{c_j \mid j = 1, \ldots, k\}$, naive Bayes classification is the conditional probability of the category c_j given the data vector x, i.e., $P(c_j|x)$.

According to Bayes theorem, the following equation exists:

$$P\left(c_j|x\right) = \frac{P\left(x|c_j\right)P\left(c_j\right)}{P(x)} = \frac{P\left(x|c_j\right)P\left(c_j\right)}{\sum_k P\left(x|c_j\right)P\left(c_j\right)}. \tag{12.14}$$

And, according to the chain rule of joint probability distribution (see Sect. 3.2.4 of this book), we have:

$$P\left(c_j, x\right) = P\left(c_j, x_1, \ldots, x_n\right) = P\left(c_j\right)\prod_{i=1}^{n} P\left(x_i|x_{i-1}, \ldots, , c_j\right). \tag{12.15}$$

Based on the strong independence between features in naive Bayes, it should satisfy $P\left(x_i|x_{i-1}, \ldots, x_1, c_j\right) = P\left(x_i|c_j\right)$; hence:

$$P(c_j, x_1, \ldots, x_n) = P\left(c_j\right)\prod_{i=1}^{n} P\left(x_i|c_j\right). \tag{12.16}$$

Therefore, the following proportional relationship exists, expressed as:

$$P\left(c_j|x_1, \ldots, x_n\right) \propto P\left(c_j\right)\prod_{i=1}^{n} P\left(x_i|c_j\right). \tag{12.17}$$

This implies that the conditional probability is based on the strong independence between features $P\left(c_j|x\right)$, and the following equation holds:

$$P\left(c_j|x_1,\ldots,x_n\right) = \frac{1}{Z}\,P\left(c_j\right)\prod_{i=1}^{n}P\left(x_i|c_j\right), \tag{12.18}$$

where $Z = P(x) = \sum_j P\left(x|c_j\right)P\left(c_j\right)$ is a scaling factor that only depends on $x = (x_1,\ldots,x_n)$. If its feature variables are known, then Z appears as a constant.

Naive Bayes is one of the simplest Bayesian networks, but when combined with kernel density estimation, it can achieve high accuracy. It also has high scalability, with the number of parameters required being linearly related to the number of variables. Maximum likelihood training can be completed by evaluating a closed-form expression, requiring only linear time, unlike some other types of classifiers that require expensive iterative approximation to complete.

Naive Bayes can be used for binary and multi-class classification.

12.5.3 Adaptive Boosting

The development history of the boosting method was introduced in Sect. 1.3 of this book. The basic idea of this method is to combine several weak learners into a strong learner.

The adaptive boosting (AdaBoost) algorithm was proposed in 1995 (Freund and Schapire 1995) and is a representative work of boosting methods. AdaBoost can be combined with many other types of classifiers, that is, the outputs of other weak classifiers are combined into a weighted sum and boosted to the final output of the strong classifier. The adaptability of AdaBoost is reflected in the fact that the weak classifiers used will be appropriately adjusted to improve the performance of the classification.

Let X denote the input space of the classification data and C denote the output space of the classification. Given a set of labeled training samples $S = \{(x_i,\ c_i)\mid i = 1,\ldots,n\}$, where $x_i \in X$, and only considering the case of two-class classification, that is $c_i \in C = \{-1,+1\}$.

Given T weak classifiers, $h_t : X \to C$, where $t = 1,\ldots,T$.

The AdaBoost algorithm iteratively calls the given weak classifier; its main idea is to maintain a set of weighted distributions of training samples during the T iteration processes, i.e., $D = \{d_t\left(i\right)\mid t = 1,\ldots,T \text{ and } i = 1,\ldots,n\}$. $\forall i$, its initial value, is $d_1\left(i\right) = \frac{1}{n}$.

For $t = 1,\ldots,T$, the iterative process of the AdaBoost algorithm is as follows.

First, in order to find the weak classifier h_t suitable for the weighted distribution d_t, calculate the weighted distribution error ϵ_t of the training process, that is:

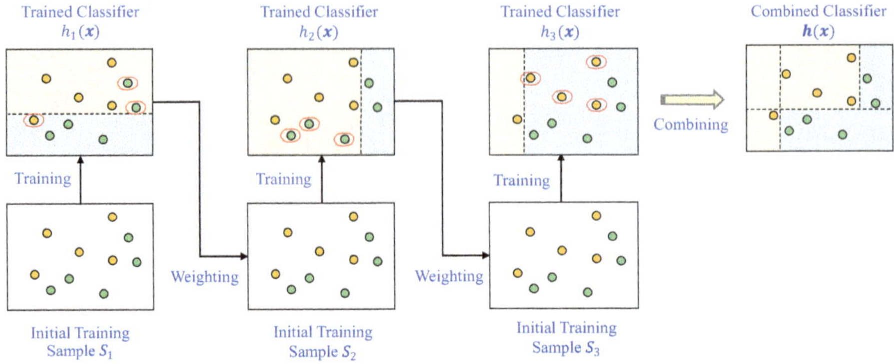

Fig. 12.9 Schematic diagram of the AdaBoost algorithm

$$\epsilon_t = \sum_{i=1}^{n} d_t(i)\mathbb{I}(h_t(\boldsymbol{x}_i) \neq c_i), \tag{12.19}$$

where $\mathbb{I}(\omega)$ is an indicator function that equals 1 if ω is true; otherwise, it is 0.

Next, calculate its adaptive parameter:

$$\alpha_t = \frac{1}{2}\ln\left(\frac{1-\epsilon_t}{\epsilon_t}\right). \tag{12.20}$$

This α_t is used to measure the importance of h_t.

Then, $\forall i$, the weighted distribution, is updated:

$$d_{t+1}(i) = \frac{1}{Z_t}d_t(i)\exp\left(-\alpha_t c_i h_t(\boldsymbol{x}_i)\right), \tag{12.21}$$

where Z_t is the normalization factor.

Thus, after T iterations, we obtain the following combined strong classifier:

$$h_T(\boldsymbol{x}) = \text{sgn}\left(\sum_{t=1}^{T} \alpha_t h_t(\boldsymbol{x})\right). \tag{12.22}$$

A schematic diagram of the AdaBoost algorithm is depicted in Fig. 12.9, where the data points marked with circles are the places where classification errors occurred during the iteration training process.

12.5.4 Support Vector Machine

Support vector machine (SVM) was proposed in the 1990s, and its development history is introduced in Sect. 1.3 of this book. The following discusses its processing methods for linear separable and nonlinear separable data.

12.5.4.1 Linear Separable Data

The support vector machine is originally designed for linear separable data, known as linear support vector machine (linear SVM).

Given is a set of labeled training samples:

$$S = \{(x_i, \ c_i) \mid i = 1, \ldots, n \text{ and } c_i \in \{-1, +1\}\}.$$

which can be linearly separated. The hyperplane is an important factor for the support vector machine to perform linear classification, as described in 12.4.1; a hyperplane $\mathcal{H}$ is a set of points that satisfy the following expression:

$$\mathcal{H} = \{x \mid w^\mathsf{T}x - b = 0\},$$

where w is the normal vector of the hyperplane. The parameter $\frac{b}{\|w\|}$ is the offset of this hyperplane from the origin along the normal vector w.

Since the set of labeled training sample S is linear separable, we can choose two hyperplanes parallel to the hyperplane $\mathcal{H}$ to distinguish these two types of data, making the distance between them as large as possible, and $\mathcal{H}$ is the hyperplane located in the middle of them.

These two hyperplanes are denoted as $\mathcal{H}_+$ and $\mathcal{H}_-$; their expressions are respectively as follows:

$$\mathcal{H}_+ = \{x \mid w^\mathsf{T}x - b = +1\},$$
$$\mathcal{H}_- = \{x \mid w^\mathsf{T}x - b = -1\}. \tag{12.23}$$

Therefore, for each data point located above the hyperplane $\mathcal{H}_+$, its category label is $c = +1$; for each data point located below the hyperplane $\mathcal{H}_-$, its category label is $c = -1$, and each data point on hyperplanes $\mathcal{H}_+$ and $\mathcal{H}_-$ is called the support vector.

From a geometric viewpoint, the distance between the hyperplanes $\mathcal{H}_+$ and $\mathcal{H}_-$ is $\frac{2}{\|w\|}$, which is called the margin. Therefore, maximizing the margin is equivalent to minimizing $\|w\|$.

Use the set of samples S to train the support vector machine, so that it satisfies the following constraints:

$$w^\mathsf{T}x_i - b \geq +1, \quad \text{if } c_i = +1,$$

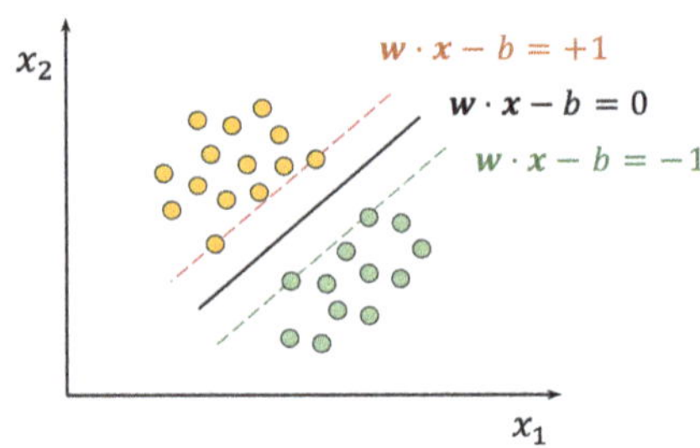

Fig. 12.10 Schematic diagram of support vector machine classification

$$w^{\mathsf{T}} x_i - b \leq -1, \quad \text{if } c_i = -1. \tag{12.24}$$

That is to say, the above two constraints ensure that each data point is on the correct side of the margin. It can also be written in the following form:

$$c_i \left(w^{\mathsf{T}} x_i - b \right) \geq +1, \quad \forall i \in 1, \ldots, n. \tag{12.25}$$

A schematic diagram of a support vector machine used for two-class linear classification is depicted in Fig. 12.10.

12.5.4.2   Nonlinear Separable Data

When dealing with nonlinear separable data, support vector machines do not directly classify, but linearize the data. That is, a kernel function is introduced to map the nonlinear separable data in the input space to a linear separable high-dimensional space, and then find the hyperplane with the maximum margin in this high-dimensional space for linear classification.

Kernel functions have already been introduced in detail in Sect. 4.7 of this book. It is known from the expression of the kernel function

$$\kappa \left(x_i, x_j \right) = \left\langle \varphi \left(x_i \right), \varphi \left(x_j \right) \right\rangle = \varphi \left(x_i \right) \cdot \varphi \left(x_j \right)$$

that the kernel method performs $\varphi \left(x_i \right)$ transformation. The value of w is also in the transformed space:

$$w = \sum_i a_i c_i \varphi \left(x_i \right). \tag{12.26}$$

The inner product of w used for classification can be calculated again using the kernel trick, that is:

$$w \cdot \varphi \left(x_i \right) = \sum_{i=1}^{n} a_i c_i \kappa \left(x_i, x \right). \tag{12.27}$$

There are several types of kernel functions, and their selection depends on the representation of nonlinear separable data. Five commonly used kernel functions have been introduced in Table 4.6 of Chap. 4.

12.5.5 Artificial Neural Networks

Chapter 5 of this book has already introduced artificial neural networks. For multi-class classification neural networks, after the fully connected layer outputs the results of deep representation learning, some multi-class classification method is used for classification.

The softmax function is an effective method for handling multi-class classification in artificial neural networks. It normalizes a vector with k real numbers into a probability distribution composed of k probability values. That is to say, each real number in this vector is mapped to a value in the range of $\{0, 1\}$, and the sum of k values equals 1.

Let $c = (c_1, \ldots, c_k) \in \mathbb{R}^k$, and $j = 1, \ldots, k$, then calculate the jth expression of the softmax function is as follows:

$$f_{c_j \in c}\left(c_j\right) = \frac{e^{c_j}}{\sum_{i=1}^{k} e^{c_i}}. \tag{12.28}$$

Subsequently, let $c_j = \boldsymbol{x}^\mathsf{T} \boldsymbol{w}_j$, and $\boldsymbol{x} = (x_1, \ldots, x_m) \in \mathbb{R}^m$, and if the softmax function is expressed in the form of conditional probability, it is a probability distribution covering k different outcomes, that is:

$$P_{c_j \in C}\left(c_j \middle| \boldsymbol{x}\right) = \frac{e^{\boldsymbol{x}^\mathsf{T} \boldsymbol{w}_j}}{\sum_{i=1}^{k} e^{\boldsymbol{x}^\mathsf{T} \boldsymbol{w}_i}}. \tag{12.29}$$

A schematic diagram of the softmax function performing multi-class classification in artificial neural networks is shown in Fig. 12.11.

The softmax classification function also plays an important role in other multi-class classification algorithms, such as multinormal logistic regression, multi-class linear discriminant analysis (multi-class LDA), and multi-class naive Bayes.

12.6 Loss Functions

A loss function for classification, also known as the error function, is used to measure the error between the predicted class output by the classifier and the target class labeled in the training samples.

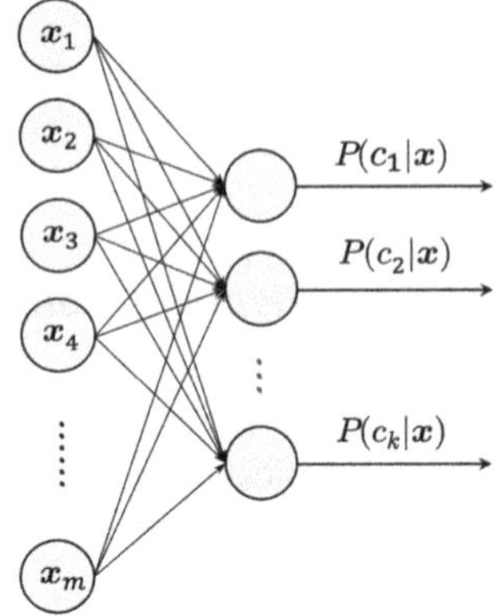

Fig. 12.11 The softmax classifier for multi-class classification

The loss function for classification is used to train and optimize the classifier and is also the cost paid for measuring classification errors, so it is also called the cost function.

This section will introduce several commonly used loss functions for classification, namely: 0–1 loss, logistic loss, exponential loss, hinge loss, and cross-entropy loss, where we use the following same notations:

$x \in X$ denotes the input data, $c \in C$ denotes the labeled target class, $h(x)$ denotes the classification hypothesis, $\hat{c} = h(x)$ is the predicted class, and the loss function is denoted as $\mathcal{L}(h(x), c) = \mathcal{L}(\hat{c}, c)$.

12.6.1   0–1 Loss

In two-class classification, the most commonly used loss function is the 0–1 loss function. It is defined as follows:

$$\mathcal{L}_{0-1}(\hat{c}, c) = \sum_{i=1}^{n} \mathbb{I}(h(x_i) \neq c_i) \tag{12.30}$$

where $\mathbb{I}(\omega)$ is an indicator function that equals 1 if ω is true, otherwise 0. The meaning of the above expression is: if the predicted class is different from the target class, the value is 1; otherwise, it is 0.

Generally, according to the nature of two-class classification, and assuming that the cost of false positives and false negatives is the same, the 0–1 loss is naturally chosen as its loss function. However, the problem with the 0–1 loss is that it is non-differentiable, so it cannot be used for methods such as gradient descent.

12.6.2 Logistic Loss

The logistic loss function is mainly used for the training of logistic regression classifiers.

The hypothesis of logistic regression is represented as $\hat{c} = P(h(x)|x)$, the labeled target output is $c \in C = \{0, 1\}$, then the expression of its logical loss function is as follows:

$$\mathcal{L}_{\mathrm{Log}}(\hat{c}, c) = -\left(c \log(h(x)) + (1 - c) \log(1 - h(x))\right). \tag{12.31}$$

That is:

$$\mathcal{L}_{\mathrm{Log}}(\hat{c}, c) = \begin{cases} -\log(h(x)) & \text{if } c = 1 \\ -\log(1 - h(x)) & \text{if } c = 0. \end{cases} \tag{12.32}$$

12.6.3 Exponential Loss

The exponential loss function is mainly used for the training of AdaBoost. Let the AdaBoost classifier be $\hat{c} = h_T(x)$, the labeled target output is $c \in C$, then its loss function is as follows:

$$\mathcal{L}_{\mathrm{Exp}}(\hat{c}, c) = \exp\left(-c \cdot \hat{c}\right) = \exp\left(-c \cdot h_T(x)\right). \tag{12.33}$$

12.6.4 Hinge Loss

Hinge loss is a specific type of loss function, mainly used for training two-class classification of support vector machines.

The hinge loss takes into account the distance of data points from the classification decision boundary as a cost, meaning that even if the input data is correctly classified, if the distance from the decision boundary is not large enough, it will also be considered as a loss.

For a linear support vector machine classifier, its classification function is a hyperplane, $\hat{c} = h(x) = w \cdot x + b$, where w is the weight vector, x is the input variable, and b is the bias. Let the target output of the classification $c \in \{-1, +1\}$, then its hinge loss is defined as follows:

$$\mathcal{L}_{\mathrm{Hinge}}(\hat{c}, c) = \max(0, \ 1 - c \cdot h(x)). \tag{12.34}$$

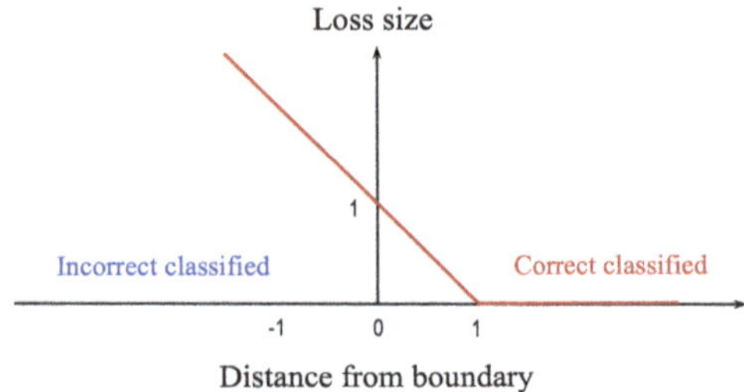

Fig. 12.12 Schematic diagram of hinge loss

According to the above equation, we know: if $c \cdot h(x) \geq 1$ then the hinge loss $\mathcal{L}_{\text{Hinge}}(\hat{c}, c) = 0$; else if $c \cdot h(x) < 1$ then the value of $\mathcal{L}_{\text{Hinge}}(\hat{c}, c)$ increases linearly with the change of $c \cdot h(x)$.

A schematic diagram of hinge loss is depicted in Fig. 12.12, where the red line is the result of $\mathcal{L}_{\text{Hinge}}(\hat{c}, c) = \max(0, \ 1 - c \cdot h(x))$. The horizontal axis is equivalent to the distance between the data points and the boundary, and the vertical axis represents the size of the loss value. The right side of the horizontal axis is the correctly classified part: when $c \cdot h(x) \geq 1$, it indicates that $h(x)$ and c have the same positive and negative signs and $h(x)$ predicts the correct category, and its hinge loss is 0, while when $0 < c \cdot h(x) < 1$, it will also be a certain loss even if the prediction of $h(x)$ is correct but the distance from the boundary is not large enough. The left side of the horizontal axis is the part of misclassification, that is, $c \cdot h(x) \leq 0$.

12.6.5 Cross-Entropy Loss

The cross-entropy loss function, also known as the logarithmic loss function, is mainly used for training the softmax classifier.

In information theory, the entropy of a set of random variables is defined as the average level of inherent information or uncertainty in the possible outcomes of these variables. Suppose the discrete information space is $\mathcal{X}$, there exists $x \in \mathcal{X}$, $P(x)$ is its probability distribution, then the calculation equation for entropy is as follows:

$$H(P) = \mathbb{E}_{x \sim P}[-\log P(x)] = -\sum_{x \in \mathcal{X}} P(x) \log P(x), \tag{12.35}$$

where $\mathbb{E}[\cdot]$ is the expectation.

Cross-entropy is built on the basis of entropy and used to calculate the difference between the two probability distributions of a set of random variables. These two probability distributions are denoted as $P(x)$ and $Q(x)$, and the expression of their cross-entropy is as follows:

$$H(P, Q) = \mathbb{E}_{x \sim Q}[-\log P(x)] = -\sum_{x \in \mathcal{X}} Q(x) \log P(x), \qquad (12.36)$$

where $Q(x)$ is referred to as the empirical probability distribution and $P(x)$ for the predicted probability distribution.

When there are only two random variables x_1 and x_2:

$$H(P, Q) = -\sum_{i=1}^{2} Q(x_i) \log P(x_i) = -(Q(x_1) \log P(x_1) + Q(x_2) \log P(x_2)).$$

And its probability distribution should satisfy $Q(x_1) + Q(x_2) = 1$, and $P(x_1) + P(x_2) = 1$. Therefore, we have:

$$H(P, Q) = -(Q(x_1) \log P(x_1) + (1 - Q(x_1)) \log(1 - P(x_1)))$$
$$= -((1 - Q(x_2)) \log(1 - P(x_2)) + Q(x_2) \log P(x_2)).$$

In machine learning, cross-entropy is often used as the loss function for classification. Its expression is as follows:

$$\mathcal{L}_{\text{Cross}}(\hat{c}, c) = -\sum_{i=1}^{n} c_i \log h(x_i), \qquad (12.37)$$

where c_i denote the target class of x_i, $h(x_i) = \hat{c}$ denotes the predicted class of x_i, and $n \geq 2$. In Eq. 12.37:

- If $n = 2$, it can be used to train a two-class classifier; else
- If $n \geq 3$, it can be used to train a multi-class classifier.

It is worth noting that the cross-entropy loss of two-class classification is actually equivalent to the logistic loss function.

12.7 Evaluation Metrics

Without loss of generality, here we will introduce several evaluation metrics for the two-class classification.

The terms involved in two-class classifier evaluation are shown in Fig. 12.13 that is known as a confusion matrix, also called a contingency table. The meanings of these terms are as follows:

(1) There are following four basic categories:

- Actual:
  The labeled actual samples

Actual		
	Positive	Negative
Predicted — Positive	True Positive (TP)	False Positive (FP)	Predicted Positive (TP+FP)
Predicted — Negative	False Negative (FN)	True Negative (TN)	Predicted Negative (FN+TN)
	Actual Positive (TP+FN)	Actual Negative (FP+TN)

Fig. 12.13 Basic terms for classifier evaluation

- Predicted:
  The predicted results of the classification model
- Positive:
  The positive cases in the actual samples or predicted results
- Negative:
  The negative cases in the actual samples or predicted results

(2) Therefore, there are following four basic terms based on the above categories:

- True positive (TP):
  The actual sample is positive, and the predicted result is also positive.
- False positive (FP):
  The actual sample is negative, but the predicted result is positive.
- False negative (FN):
  The actual sample is positive, but the predicted result is negative.
- True negative (TN):
  The actual sample is negative, and the predicted result is also negative.

(3) Consequently, there are following four synthesized terms:

- Actual positive (AP):
  The actual positive samples, composed of true positive (TP) and false negative (FN)
- Actual negative (AN):
  The actual negative samples, composed of false positive (FP) and true negative (TN)
- Predicted positive (PP):
  The predicted positive results, composed of true positive (TP) and false positive (FP)
- Predicted negative (PN):
  The predicted negative results, composed of false negative (FN) and true negative (TN)

To further understand these terms, the physical view of the terms for classifier evaluation is depicted in Fig. 12.14, where:

- "Actual" refers to the samples inside the black square.
- "Predicted" refers to the samples inside the red circle.

Fig. 12.14 Physical view of
the terms for classifier
evaluation

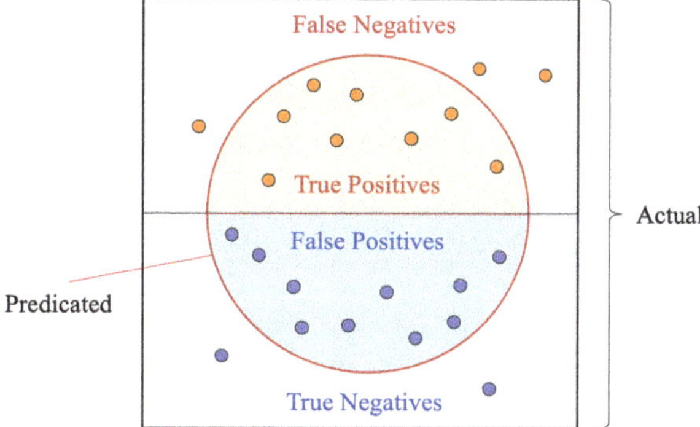

- "True positives" refers to the samples inside the red up half circle.
- "False positives" refers to the samples inside the red down half circle.
- "False negatives" refers to the samples inside the black up half square.
- "True negatives" refers to the samples inside the black down half square.

The following will introduce several evaluation metrics for two-class classifiers, namely, accuracy, precision and recall, sensitivity and specificity, F1 score, P-R curve, ROC curve, and Matthews correlation coefficient.

12.7.1   Accuracy

Accuracy is an important metric reflecting the classification results, the proportion of correct predictions among the total number of samples. It is defined as follows:

$$\text{Accuracy} = \frac{\text{True Positive (TP)} + \text{TrueNegative (TN)}}{\text{Actural Positive (AP)} + \text{Actural Negative (AN)}}. \qquad (12.38)$$

In the above formula, (TP+TN) are correctly predicted positive and negative cases, and (AP+AN) are actual positive and negative cases in the samples. Therefore, accuracy is the proportion of correctly predicted cases in the actual positive and negative cases.

Accuracy is the most intuitive performance evaluation metric, useful when all classes are of equal importance. However, for imbalanced datasets, the ratio of true positive to true negative is unbalanced, and accuracy can be misleading.

For example, for a dataset containing 95 negative cases and 5 positive cases, if all values are classified as negative cases, an accuracy of 95% can be achieved, but none of the 5 positive cases are detected. Conversely, if it is a dataset containing 95 positive cases and 5 negative cases, then none of the 5 negative cases are detected. If it is fraud detection, a fraud rate of 5% is a very serious problem!

Therefore, in practical applications, we often consider the following indicators to be introduced in the next two sections, namely: precision and recall, and sensitivity and specificity.

12.7.2 Precision and Recall

Precision and recall are important concepts for evaluating classifiers and important metrics reflecting classification predicted results.

The precision is defined as follows:

$$\text{Precision} = \frac{\text{True Positive (TP)}}{\text{True Positive (TP)} + \text{False Positive (FP)}}. \tag{12.39}$$

In other words, the precision is how many of the total number of predicted positive cases are actual positive. Therefore, it is also called the positive predictive value (PPV).

And, recall is defined as follows:

$$\text{Recall} = \frac{\text{True Positive (TP)}}{\text{True Positive (TP)} + \text{False Negative (FN)}}. \tag{12.40}$$

In other words, the recall is how many of the true positives are predicted, so it is also known as the true positive rate (TPR).

12.7.3 Sensitivity and Specificity

Sensitivity and specificity are used more in the medical field.

The sensitivity, also known as the true positive rate (TPR), is equivalent to the recall introduced in the previous subsection. For comparison, the definition of sensitivity is given here again, that is, the ratio of true positives to actual positives:

$$\begin{aligned}
\text{Sensitivity} &= \text{True Positive Rate (TPR)} \\
&= \frac{\text{True Positive (TP)}}{\text{True Positive (TP)} + \text{False Negative (FN)}}.
\end{aligned} \tag{12.41}$$

Sensitivity assesses the proportion of true positives that are correctly identified, for example, the percentage of people correctly identified as having a certain disease.

Corresponding to sensitivity is specificity, which is defined as the ratio of true negatives to actual negatives:

$$\text{Specificity} = \text{True Negative Rate (TNR)}$$

$$= \frac{\text{True Negatives (TN)}}{\text{True Negatives (TN)} + \text{False Positives (FP)}}. \tag{12.42}$$

Specificity, also known as the true negative rate (TNR), assesses the proportion of true negatives that are correctly identified, for example, the percentage of people correctly identified as not having a certain disease.

12.7.4 F1 Score

F1 score is the harmonic mean of recall and precision. That is:

$$\text{F1 Score} = \frac{2}{\text{Precision}^{-1} + \text{Recall}^{-1}}$$

$$= 2 \times \frac{(\text{Precision} \times \text{Recall})}{(\text{Precision} + \text{Recall})}.$$

$$= \frac{\text{True Positive (TP)}}{\text{True Positive (TP)} + \frac{1}{2}(\text{False Positive (FP)} + \text{False Negative (FN)})}. \tag{12.43}$$

Therefore, the F1 score takes into account both false positives and false negatives.

Intuitively, the F1 score is not as easy to understand as precision, but it is usually more useful, especially when the class distribution is not uniform. If the number of false positives and false negatives is similar, precision will perform best. If the number of false positives and false negatives is very different, it is best to consider both recall and precision.

12.7.5 P-R Curve

Based on the results of precision and recall, the corresponding precision-recall curve (P-R curve) can be drawn. Its vertical axis is precision, the horizontal axis is recall, and the curve is used to judge the quality of classification results.

A schematic diagram of a P-R curve is depicted in Fig. 12.15, from which the following conclusions can be drawn.

- The line near the precision of 0.5 in the figure is equivalent to the baseline. If the P-R curve is above the baseline, its classifier is superior; otherwise, it is inferior.
- For a classifier, its recall and precision cannot be optimized at the same time. Because the higher the recall, the lower the precision, and vice versa. Although

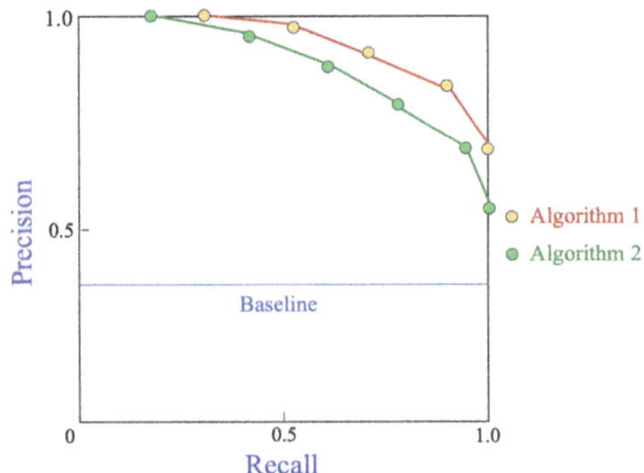

Fig. 12.15 Schematic diagram of a P-R curve

we hope that the classifier has a high recall and precision, we can only make a trade-off between the two.

- But for two classifiers, their merits can be compared through the P-R curve. There are two curves in the figure: the larger the area under the curve (AUC), the better the classification performance. Therefore, the classifier corresponding to the blue line in the figure has better performance than the classifier corresponding to the green line.

12.7.6 ROC Curve

The ROC curve stands for the receiver operating characteristic curve. When the ROC curve is used for classifier evaluation, the receiver refers to the classifier.

The ROC curve is not an evaluation curve unique to machine learning, as can be seen from its name. It was invented by electronic and radar engineers during World War II, originally used for radar signal analysis and later used in signal detection theory. In the 1950s, psychophysics introduced the ROC, used for the detection of weak signals in humans or animals. Since then, the ROC has been used in medicine, radio, biology, and criminal psychology. It has only been introduced into the field of machine learning in recent years.

For the performance evaluation of classifiers, the ROC curve is used to depict its predictive ability when the recognition threshold changes. Its vertical axis is the true positive rate (TPR), and the horizontal axis is the false positive rate (FPR).

The true positive rate (TPR), also known as sensitivity or recall, has been defined in the above subsections. The false positive rate (FPR) is defined as the ratio of false positives to true negatives, that is:

$$\text{False Positive Rate (FPR)} = \frac{\text{False Positive (FP)}}{\text{False Positive (FP)} + \text{True Negative (TN)}}.$$

$$(12.44)$$

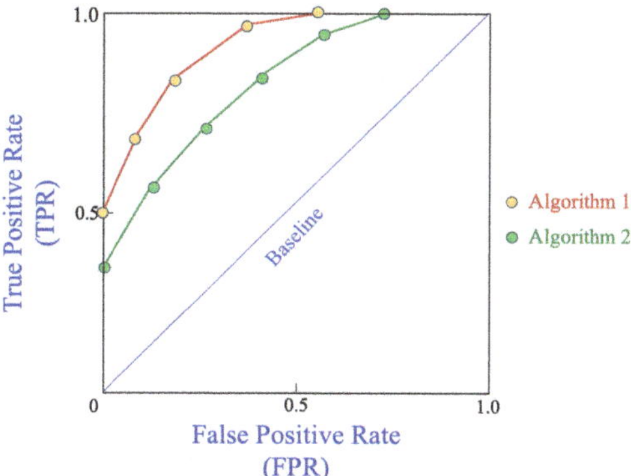

Fig. 12.16 Schematic diagram of ROC curve

In other words, the false positive rate is how many of the actual negatives are incorrectly predicted as positives. Moreover, the false positive rate also satisfies the following relationship:

False Positive Rate (FPR) $= 1 -$ True Negative Rate (TNR) $= 1 -$ Specificity.

A schematic diagram of an ROC curve is depicted in Fig. 12.16, from which the following conclusions can be drawn:

- The diagonal line in the figure, that is, connecting $(0, 0)$ to $(1, 1)$ line, is called the baseline. If the ROC curve is above the baseline, its classifier is superior; otherwise, it is inferior.
- For a classifier, its true positive rate (sensitivity) and false positive rate $(1 -$ specificity) cannot be optimized at the same time. Because the higher the true positive rate (sensitivity), the higher the false positive rate (i.e., the lower the specificity). Although we hope that the classifier has a higher true positive rate and a lower false positive rate, a balance must be struck between the two. According to the ROC curve of the classifier, the best threshold is selected through trade-offs.
- But for two classifiers, their merits can be compared through the ROC curve. There are two curves in the figure: the larger the area under the curve (AUC), the better its classification performance. Therefore, the classifier corresponding to the blue line in the figure has better performance than the green line.

12.7.7 Matthew's Correlation Coefficient

The Matthew's correlation coefficient (MCC) is a more reliable method for evaluating the performance of a classifier, and its expression is as follows:

$$MCC = \frac{TP \times TN - FP \times FN}{\sqrt{(TP + FP)\,(TP + FN)\,(TN + FP)\,(TN + FN)}}. \tag{12.45}$$

As can be seen from the above expression, the Matthew's correlation coefficient will only score high when all four metrics in the contingency table (confusion matrix), i.e., true positives (TP), false negatives (FN), false positives (FP), and true negatives (TN), achieve good results, which is proportional to the number of positive elements and negative elements in the data sample.

12.8 Application Fields

Classification is the most common task in machine learning, and its application fields are very wide.

For example, in the field of computer vision (CV) research, it is necessary to perform handwritten digit and text recognition, optical character recognition (OCR), action recognition and content understanding in images and videos, medical imaging analysis, as well as face recognition, fingerprint recognition, and other biometric recognition.

In recent years, intelligent surveillance systems have developed rapidly, and high attention has been paid to the classification recognition of moving objects in surveillance videos, including the detection of vehicle types, sizes, speeds, in public transportation, re-identification of pedestrians in surveillance videos, and detection of abnormal scenes.

Classification algorithms are also often applied to natural language processing (NLP). Among them, text classification refers to the process of automatically dividing the text in a document collection into one or several categories according to predefined topic categories and certain rules. It can help people organize and manage a large amount of text information well and achieve personalized information acquisition. Common applications of text classification include email classification, web page classification, text indexing, automatic summarization, information retrieval, and information push.

Voice and audio also need to be classified, which include speech recognition and audio scene classification. Speech recognition is a subfield of computational linguistics, which converts speech waves into text. For a given target speech data, a statistical model of speech recognition is obtained through training. Audio scene classification is a specific task in the field of computer auditory scene analysis. It recognizes the specific scene semantic labels corresponding to the acoustic content of the audio stream, thereby achieving the purpose of perceiving and understanding the surrounding environment.

The development of classification in the field of artificial intelligence in medicine is also rapid. For example, in 2016, the *Journal of the American Medical Association* published an article on diagnosing diabetic retinopathy using deep learning (Gulshan et al. 2016). In 2017, *Nature* published a paper on classifying skin cancer using deep neural networks (Esteva et al. 2017). On February 2, 2018, *Cell* magazine published a cover article proposing a method for screening blinding retinal diseases through optical coherence tomography (OCT) images of the retina (Kermany et al. 2018). On March 14, 2018, *Nature* published an AI system that can accurately distinguish nearly 100 different central nervous system tumors based on methylation data of tumor tissue DNA and can also find some disease classifications not in clinical guidelines (Wong and Yip 2018).

Further Reading

1. Corinna Cortes and Vladimir Vapnik. "Support-Vector Networks." *Machine learning*, 20, 1995.

   [Notes] The paper's title "Support-Vector Networks" was later also known as support vector machines. Its solid theoretical foundation and excellent experimental results have had attracted widespread attention in the machine learning community. Google Scholar shows that this paper has been cited nearly 60,000 times.
2. Pramuditha Perera, Poojan Oza, and Vishal M. Patel. "One-class Classification: A Survey." *arXiv preprint* arXiv:2101.03064, 2021.

   [Notes] This is a review paper that describes one-class classification methods based on statistics and neural networks, discusses the advantages and disadvantages of existing methods, and proposes research directions in this field. It also provides commonly used datasets and evaluation metrics.
3. Weiwei Liu, Haobo Wang, Xiaobo Shen, and Ivor W. Tsang. "The Emerging Trends of Multi-label Learning." *IEEE Transactions on Pattern Analysis and Machine Intelligence* (TPAMI), 44(11), 2021.

   [Notes] This review paper argues that with the large, complex, and growing multimodal data, traditional multi-label classification can no longer meet practical needs; thus, some emerging research trends and methods have appeared, and new multi-label learning paradigms are needed. The paper also conducts a comprehensive survey of these emerging trends and methods and discusses potential valuable research directions.

References

Baram, Y. (1997). Consistent classification, firm and soft. In *Conference on Neural Information Processing Systems (NIPS)*.

Boser, B. E., I. M. Guyon, and V. N. Vapnik. (1992). A training algorithm for optimal margin classifiers. In *Fifth Annual Workshop on Computational Learning Theory*.

Cao, S., D. Hu, Z. Hu, W. Zhao, Y. Mo, and K. Qiao. (2018). An integrated soft and hard classification approach for evaluating urban expansion from multisource remote sensing data: A case study of the Beijing–Tianjin–Tangshan Metropolitan Region, China. *International Journal of Remote Sensing* 39(11): 3556–3579.

Choodarathnakara, A. L., T. A. Kumar, S. Koliwad, and C. G. Patil. (2012). Soft classification techniques for RS data. *International Journal of Computer Science Engineering & Technology* 2(11): 1468–1471.

Cortes, C., and V. Vapnik. (1995). Support vector networks. *Machine Learning* 20(3): 273–297.

Esteva, A., B. Kuprel, R. A. Novoa, J. Ko, S. M. Swetter, H. M. Blau, and S. Thrun. (2017). Dermatologist-level classification of skin cancer with deep neural networks. *Nature* 542(7639): 115–118.

Freund, Y., and R. E. Schapire. (1995). A decision-theoretic generalization of on-line learning and an application to boosting. In *Computational Learning Theory: Second European Conference, EuroCOLT '95.*

Gulshan, V., L. Peng, M. Coram, M. C. Stumpe, D. Wu, A. Narayanaswamy, S. Venugopalan, K. Widner, T. Madams, and J. Cuadros. (2016). Development and validation of a deep learning algorithm for detection of diabetic retinopathy in retinal fundus photographs. *The Journal of the American Medical Association (JAMA)* 316(22): 2402–2410.

Kermany, D. S., M. Goldbaum, W. Cai, C. C. S. Valentim, and H. Liang. (2018). Identifying medical diagnoses and treatable diseases by image-based deep learning. *Cell* 172(5): 1122–1131.

Kontschieder, P., M. Fiterau, A. Criminisi, and S. R. Bulo. (2015). Deep neural decision forests. In *IEEE International Conference on Computer Vision (ICCV).*

Liu, Y., H. H. Zhang, and Y. Wu. (2011). Hard or soft classification? Large-margin unified machines. *Journal of the American Statistical Association* 106(493): 166–177.

Pan, Y., T. Hu, X. Zhu, J. Zhang, and X. Wang. (2012). Mapping cropland distributions using a hard and soft classification model. *IEEE Transactions on Geoscience and Remote Sensing* 50(11): 4301–4312.

Perera, P., P. Oza, and V. M. Patel. (2021). One-class classification: A survey. arXiv preprint arXiv:2101.03064.

Sharma, R., A. K. Goyal, and R. K. Dwivedi. (2016). *A review of soft classification approaches on satellite image and accuracy assessment.* Berlin: Springer.

Wahba, G. (1998). Support vector machines, Reproducing Kernel Hilbert spaces and the randomized GACV. In *Advances in Kernel Methods: Support Vector Learning* 69–87.

Wahba, G. (2002). Soft and hard classification by reproducing Kernel Hilbert space methods. *Proceedings of the National Academy of Sciences* 99(26): 16524–16530.

Wahba, G., C. Gu, and Y. Wang. (1993). *Soft classification, Aka risk estimation, via penalized log likelihood and smoothing spline analysis of variance.* University of Wisconsin.

Wong, D., and S. Yip. (2018). Machine learning classifies cancer. *Nature* 555(7697): 446–448.

Chapter 13
Regression Task

Abstract Regression, also known as regression analysis, is also a task that belongs to supervised learning paradigm. We first in this chapter present the regression problem and definition and explain the working principle using formal and illustrated descriptions. Next, we discuss the relevant elements of regression, including the number and relationship of independent variables and dependent variables, linear or nonlinear regressions, parametric or nonparametric regressions, and interpolation or extrapolation of regression fitting lines. We then focus on several commonly used regression algorithms, such as multiple linear regression, polynomial regression, ridge regression, Lasso regression, and Bayesian regression. After that, we explain the loss functions, namely, mean squared error (MSE), mean absolute error (MAE), mean bias error (MBE), relative absolute error (RAE), relative squared error (RSE), Huber loss, log-cash loss, and coefficient of determination. Finally, the application fields of regression are introduced briefly.

13.1 Problem and Definition

In this section, we first discuss the regression problem, then give a definition of the regression, and next propose a formal definition of regression model, and after that present the underlying assumptions for regression task.

13.1.1 Regression Problems

Regression is to statistically analyze the relationship between sample data and use it to predict unknown input data. Many fields such as economics, finance, marketing, healthcare, environment, social sciences, and engineering involve regression problems, for example, predicting future stock trends based on past stock trading data and predicting house prices based on data such as the size, structure, location, and construction year of the house.

However, the literal meaning of "regression" is somewhat puzzling.

W. Wang, *Principles of Machine Learning*,
https://doi.org/10.1007/978-981-97-5333-8_13

It is said that the term "regression" was coined by Sir Francis Galton in the 1880s. Galton was a British polymath who made important contributions in many fields, including statistics, psychology, and biology. He first used the term "regression" in a paper published in 1886, titled "Regression Towards Mediocrity in Hereditary Stature," because he found such a biological phenomenon that the heights of descendants of tall ancestors tend to regress down towards a normal average.

Since then, Galton's work has been extended by some researchers to more general statistical scenarios, making the term regression gradually widely used in statistics to describe the relationship between variables, especially to describe the trend of a variable's value associated with another variable's value.

From another viewpoint, the essence of regression analysis is to find a curve that fits the relationship between multiple sample data points, represented as a regression function, manifested as a regression model. In this way, another coordinate value corresponding to one coordinate value can be calculated. However, in the process of fitting multiple data points, a certain compromise needs to be made to find an appropriate curve, which is a product of "regression" towards the average value of multiple data points.

13.1.2   Definition

Definition 13.1 (Regression)   Regression in machine learning is the process of training a regression algorithm based on samples of corresponding values between labeled input and output data, obtaining an optimal regression hypothesis, and then using it to predict unknown input data. The output is a continuous corresponding value.

From the view of machine learning, regression belongs to the supervised learning paradigm, so there are training samples. The values of the input data in the above definition are equivalent to the features of the observed data in the training samples, and the values of the output data are equivalent to the labels of the observed data.

Both the regression and the classification belong to the supervised learning paradigm, and both need to be trained based on labeled samples. But the difference lies in the output that is shown in Table 13.1.

Table 13.1 Differences between regression and classification

	Regression	Classification
Difference	The output is a continuous value	The output is a discrete category
Example	Sales forecasting, risk analysis	{sunny, cloudy, rainy}, $\{0, 1, \ldots, 9\}$

13.1.3 Regression Model

Definition 13.2 (Regression Model) A regression model can be formalized as the following equation:

$$y = f(x, \theta) + \varepsilon, \tag{13.1}$$

where y is the *dependent variable*, x is the *independent variable*, θ is the parameter, ε is the error term, and $f(x, \theta)$ is referred to as the regression function.

In the field of regression analysis, there are some synonyms as below:

- Independent variable:
  Also called predictor variable, or explanatory variable
- Dependent variable:
  Also called outcome variable, or response variable
- Parameter
  Also known as the coefficient, or weight

Furthermore, the error term in the regression model refers to the noise from random sampling or the influence of variables not included in the model.

The goal of the regression model is to find the best-fitting regression function based on labeled sample data.

13.1.4 Underlying Assumptions

As for the regression task itself, it is based on training samples to obtain the best regression function. In order for regression to be an effective method of estimating the relationship between independent and dependent variables, it often relies on some underlying assumptions, namely:

- The sample is representative of the population at large.
- The independent variables are measured without error.
- The error ε of the model has an expected value of zero, given the independent variables, i.e., $\mathbb{E}[\varepsilon|x] = 0$.
- The variance of error ε is constant across all input data, i.e., homoscedasticity.
- The errors ε are uncorrelated. Mathematically, this satisfies the diagonal property of the variance-covariance matrix of the errors.

13.2   Working Principle

For further understanding of the regression task, here we first give a formal description and then explain it through an illustration. We will see that the formal descriptions between regression and classification are similar, but the main difference lies in the output space.

13.2.1   Formal Description

Without loss of generality, let $\mathbb{R}^m$ denote a set of m-dimensional real vectors, $m \geq 1$, $X \subseteq \mathbb{R}^m$ denote input space, and $\mathcal{Y} \subseteq \mathbb{R}$ denote output space with real numbers.

Through a probability $P(x)$, we obtain n independent and identically distributed (i.i.d.) observed data:

$$D = \{x_i \mid x_i \in X \text{ and } i = 1, \ldots, n\}. \tag{13.2}$$

Let the target function of regression be:

$$f : X \to \mathcal{Y}. \tag{13.3}$$

It can be represented as a conditional probability $P(y|x)$, where $x \in X$, $y \in \mathcal{Y}$.

Based on the joint probability distribution $P(x, y) = P(y|x) P(x)$ to label the actual output value y_i for each input data x_i, we obtain a set of training sample S with n elements, where each element is represented as a data pair $(x_i, y_i) \in X \times \mathcal{Y}$, that is:

$$\arg\min_{h \in H} R(h) = \arg\min_{h \in H} \mathbb{E}\left[\mathcal{L}(h(x), f(x))\right] = \arg\min_{h \in H} \mathbb{E}\left[\mathcal{L}(\hat{y}, y)\right], \tag{13.4}$$

where $\hat{y}$ is referred to as the predicted value, y is the target value, and $\mathcal{L}(\cdot, \cdot)$ represents the loss function of regression.

After obtaining the hypothesis function h, use the actual data to test the regression of h. Given the test data with n' unknown outputs $x_i \in X$, that is:

$$T = \left\{x_i \mid i = 1, \ldots, n'\right\} \subseteq X, \tag{13.5}$$

use the well-trained regression function $h(x)$ to predict; the result is a set of data pairs, namely:

$$T_{\text{Output}} = \left\{(x_i, \hat{y}_i) \mid \hat{y}_i = h(x_i) \text{ and } i = 1, \ldots, n'\right\} \subseteq \mathcal{Y}. \tag{13.6}$$

13.2.2   Illustrated Description

Here, the working principle of regression is further explained in conjunction with two figures, which respectively elaborate on the training process and testing process of the regression model.

The training process of the regression model is depicted in Fig. 13.1. We will notice that its training process is similar to classification. But the difference is that the hypothesis function for regression is called the regression function, which can be regarded as the fitting function of the training data. This is the key to understanding the regression method.

The testing process of the regression model is depicted in Fig. 13.2. It is also similar to classification; the difference is that for a given input x_i, through the regression function $h(x_i)$, what we obtained is an output value $\hat{y}_i$, but not a category.

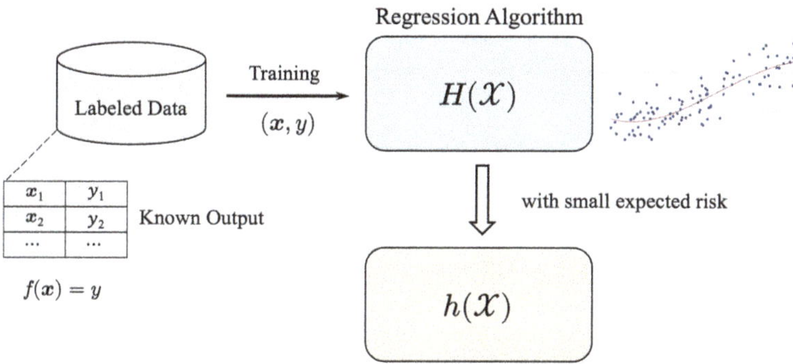

Fig. 13.1 Training process of the regression model

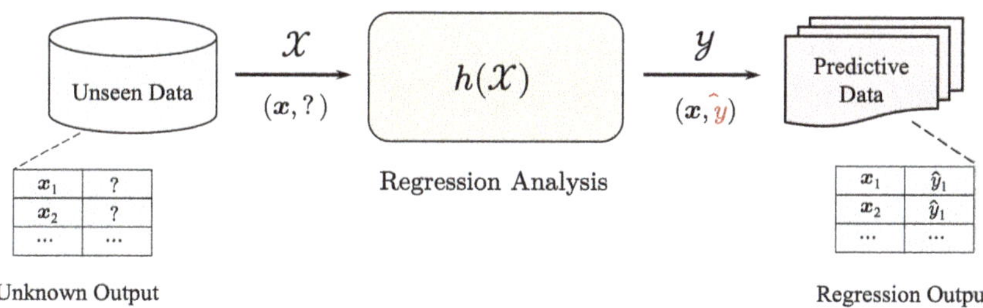

Fig. 13.2 Testing process of the regression model

13.3   Related Elements

This section discusses several elements in regression analysis, namely, unique and multiple, linear and nonlinear, parametric and nonparametric, as well as interpolation and extrapolation of fitting lines.

13.3.1   Unique and Multiple

Regression is a statistical process for estimating the relationship between independent and dependent variables, but different regression problems have different numbers of independent and dependent variables.

13.3.1.1   Uni- and Multi-independent Variable

In the regression model $y = f(x, \theta) + \varepsilon$, the x is the independent variable, and $x = (x_1, x_2, \ldots, x_m) \in X \subseteq \mathbb{R}^m$, where m denotes the number of the independent variables and $m \geq 1$. After substituting the independent variable back into the regression model, it becomes:

$$y = f(x_1, x_2, \ldots, x_m, \theta) + \varepsilon. \tag{13.7}$$

Therefore, there are two situations regarding the number of variables in the regression model.

(1) Uni-independent variable
    There is only one independent variable in the regression model, that is, $m = 1$.
(2) Multi-independent variable
    The number of independent variables is two or more, i.e., $m \geq 2$.

13.3.1.2   Uni- and Multi-dependent Variable

In the regression model $y = f(x, \theta) + \varepsilon$, the y is the dependent variable and $y = (y_1, \ldots, y_n) \in Y \subseteq \mathbb{R}$, where n is the number of the dependent variables and $n \geq 1$. After substituting the dependent variables into the regression model, it becomes:

$$y = (y_1, \ldots, y_n) = f(x, \theta) + \varepsilon. \tag{13.8}$$

Table 13.2 The four types of regression submodels

	Uni-independ var	Multi-independ var
Uni-depend var	Uni-independ, uni-depend var	Multi-independ, uni-depend var
Multi-depend var	Uni-independ, multi-depend var	Multi-independ, multi-depend var

Therefore, there are two situations for the number of dependent variables in the regression model.

(1) Uni-dependent variable
    There is only one dependent variable in the regression model, that is, $n = 1$.
(2) Multi-dependent variable
    The number of dependent variables is two or more, i.e., $n \geq 2$.

13.3.1.3   Four Submodels

Since each of independent and dependent variables has two situations, the following four types of regression submodels can be derived, as shown in Table 13.2.

The regression problems with multi-independent and uni-dependent variables are common and easy to understand.

But the regression problems with multi-independent and multi-dependent variables are also not uncommon. For example, a doctor collected some data on the cholesterol, blood lipids, blood pressure, and weight of some subjects, and some data on the subjects' dietary habits per week, such as intake of red meat, fish, eggs, dairy products, and chocolate, so that those data are used to analyze the relationship between four health indicators and dietary habits.

13.3.2   *Linear and Nonlinear*

Linearity and nonlinearity of the regression model are two of the important concepts in regression analysis. First, we will introduce the definition of linear combination and then discuss linear and nonlinear regression models.

13.3.2.1   Linear Combination

Linear combination is an important concept in mathematics including linear algebra.

Definition 13.3 (Linear Combination of Parameters)   A linear combination of parameters is an expression composed of a set of terms, each of which is multiplied by parameter and independent variable and then added together. The value of the first independent variable is usually 1.

13.3.2.2   Linear Regression

Definition 13.4 (Linear Regression Model)   A regression model $y = f(x, \boldsymbol{\theta}) + \varepsilon$ is referred to as a linear regression model, if the expression $f(x, \boldsymbol{\theta})$ is a linear combination of parameters.

It is worth noting that linear regression relies on its expression as a linear combination of parameters, but does not depend on the fitted line presented by the expression being a straight line, as the fitted line of linear regression may also be a curve.

For example, the following equation is a linear regression model, which appears as a straight line:

$$y = \theta_0 + \theta_1 x + \varepsilon.$$

The following equation is also a linear regression model, although it appears as a parabola:

$$y = \theta_0 + \theta_1 x + \theta_2 x^2 + \varepsilon.$$

13.3.2.3   Nonlinear Regression

Definition 13.5 (Nonlinear Regression Model)   A regression model $y = f(x, \boldsymbol{\theta}) + \varepsilon$ is referred to as a nonlinear regression, if the expression $f(x, \boldsymbol{\theta})$ is not a linear combination of parameters.

In other words, all regression models that do not meet the definition of linear regression model are nonlinear regression models.

For example, in the following function,

$$y = \frac{\theta_1 x}{\theta_0 + x} + \varepsilon,$$

although there are only two parameters and one independent variable, because the two parameters θ_0 and θ_1 are not linearly combined, so the function is still nonlinear.

13.3.3   *Parametric and Nonparametric*

Depending on the relationship between the independent and dependent variables in the regression model, it can be divided into parametric regression models, nonparametric regression models, and semiparametric regression models.

13.3.3.1 Parametric Regression

Definition 13.6 (Parametric Regression Model) A regression model $y = f(x, \theta) + \varepsilon$ is referred to as a parametric regression model, if the θ is a finite and fixed number of parameters.

In other words, if the relationship between the independent and dependent variables in the regression model is known and can be defined using a finite number of parameters, it is a parametric regression model.

Both linear regression models and nonlinear regression models are parametric regression models.

For parametric regression models, the task is to estimate the values of the parameters.

13.3.3.2 Nonparametric Regression

Definition 13.7 (Nonparametric Regression Model) A regression model $y = f(x, \theta) + \varepsilon$ is referred to as a nonparametric regression model, if the number of parameters of θ is not predetermined but adjusted according to the size of the dataset.

Nonparametric regression models are not parameter-free, but the number of parameters is not fixed.

13.3.3.3 Semiparametric Regression

Definition 13.8 (Semiparametric Regression Model) A regression model $y = f(x, \theta) + \varepsilon$ is referred to as a semiparametric regression model, if it is a mixture of parametric and nonparametric regression.

13.3.4 Interpolation and Extrapolation

Regression models are statistical processes that estimate the relationship between independent and dependent variables based on training samples and obtain their best-fit line. Therefore, this process can be divided into two types: interpolation and extrapolation.

13.3.4.1 Interpolation

Interpolation is based on the best-fit line to estimate the value of the dependent variable for a given value of the independent variable, provided it is within the data

Fig. 13.3 Interpolation and extrapolation in regression

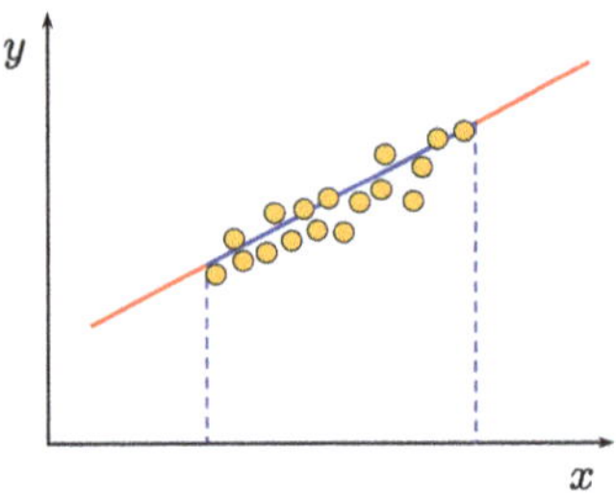

range. The blue line within the range of the two dashed lines in Fig. 13.3 belongs to the part of interpolation.

13.3.4.2   Extrapolation

Unlike interpolation, extrapolation is based on the projection, expansion, or extension of the best-fit line to make predictions, beyond the range used for the training samples. That is, extrapolation is the estimation of the value of the dependent variable for a given value of the independent variable outside the data range.

The red line outside the range of the two dashed lines in Fig. 13.3 belongs to the part of extrapolation.

It should be noted that extrapolation carries certain risks. Because the premise of extrapolation is to assume that the relationship between the dependent variable and the independent variable remains unchanged outside the data range, it may lead to inaccurate predictions. Therefore, be cautious when extrapolating, and only extrapolate under certain conditions.

13.4   Typical Algorithms

Typical regression algorithms fall into the following two types:

One type is based on classification algorithms, which are modified to be used for regression tasks, including AdaBoost, naive Bayes, decision trees, random forests, k-nearest neighbors, support vector machines (SVM), and artificial neural networks. This is because classification and regression both belong to the supervised learning paradigm.

The other type is dedicated regression algorithms, such as multiple linear regression, polynomial regression, ridge regression, lasso regression, multilinear interpolation, quantile regression, and Bayesian regression.

Here we introduce a few typical dedicated regression algorithms, namely, multiple linear regression, polynomial regression, ridge regression, lasso regression, and Bayesian regression.

13.4.1   Multiple Linear Regression

Multiple linear regression, also known as multiple regression, is the earliest proposed and widely used linear regression analysis algorithm. This is because linear models are easier to fit, and the statistical characteristics of their linear models are easier to determine.

For the multiple linear regression algorithm, the premise is that there is multi-independent and uni-dependent variables, and its expression $f(x, \theta)$ is manifested as a linear combination of parameters.

Given the labeled n training samples $S = \{(x_i, y_i) \mid i = 1, \ldots, n\}$, where $x_i \in \mathbb{R}^m$, $y_i \in \mathbb{R}$, for the regression model $y_i = f(x_i, \theta) + \varepsilon_i$, let the multi-independent variable of the multiple linear regression be $x_i = (x_{i1}, \ldots, x_{im})$ and parameter $\theta = (\theta_0, \theta_1, \ldots, \theta_m)$; therefore, the objective function of multiple linear regression is expressed as:

$$y_i = f(x_i, \theta) + \varepsilon_i = \theta_0 + \theta_1 x_{i1} + \cdots + \theta_m x_{im} + \varepsilon_i = \sum_{j=0}^{m} \theta_j x_{ij} + \varepsilon_i, \qquad (13.9)$$

where $m \geq 1$, and $x_{i0} = 1$.

It is called simple linear regression or simple regression for the special case of multiple linear regression, i.e., $m = 1$.

Since the parameters in the multiple linear regression function are unknown, it is necessary to obtain the following prediction model of multiple linear regression based on the labeled n training samples S:

$$\hat{y}_i = \hat{\theta}_0 + \hat{\theta}_1 x_{i1} + \cdots + \hat{\theta}_m x_{im} = \sum_{j=0}^{m} \hat{\theta}_j x_{ij}. \qquad (13.10)$$

Let $\varepsilon_i = y_i - \hat{y}_i$ be the difference between actual value y_i of dependent variable in objective function and the predicted value $\hat{y}_i$ of dependent variable in prediction model, and ε_i is called the residual, depicted in Fig. 13.4.

Fig. 13.4 Residual ε_i between the actual value y_i and the predicted value $\hat{y}_i$

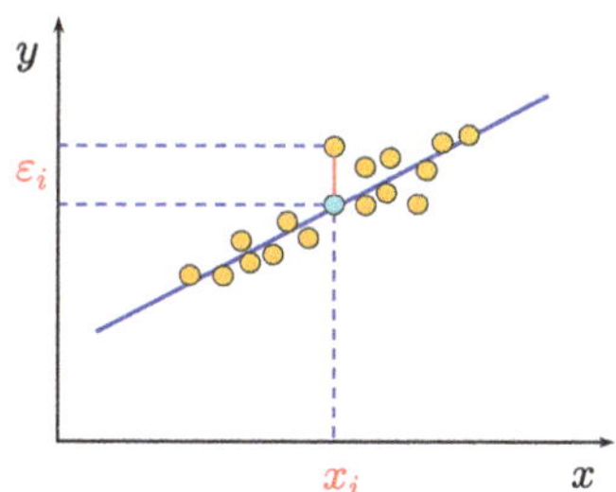

The commonly used parameter estimation is the ordinary method of least squares, which calculates its parameters $\hat{\boldsymbol{\theta}}$ by minimizing the residual sum of squares (RSS). The equation of residual sum of squares for multiple linear regression $\text{RSS}_{\text{multi}}\,(\cdot)$ is as follows:

$$\text{RSS}_{\text{multi}}\left(\hat{\boldsymbol{\theta}}\right) = \sum_{i=1}^{n}\varepsilon_i^2 = \sum_{i=1}^{n}\left(y_i - \hat{y}_i\right)^2 = \sum_{i=1}^{n}\left(y_i - \sum_{j=0}^{m}\hat{\theta}_j x_{ij}\right)^2. \tag{13.11}$$

To simplify the derivation process of $\text{RSS}\,(\cdot)$, for n training samples, it can be denoted as the following vectors and matrices:

$$\boldsymbol{y} = \begin{bmatrix} y_1 \\ y_2 \\ \vdots \\ y_n \end{bmatrix},\; \boldsymbol{x} = \begin{bmatrix} 1 & x_{11} & \cdots & x_{1m} \\ 1 & x_{21} & \cdots & x_{2m} \\ \vdots & \vdots & \ddots & \vdots \\ 1 & x_{n1} & \cdots & x_{nm} \end{bmatrix},\; \boldsymbol{\theta} = \begin{bmatrix} \theta_0 \\ \theta_1 \\ \vdots \\ \theta_m \end{bmatrix},\; \boldsymbol{\varepsilon} = \begin{bmatrix} \varepsilon_1 \\ \varepsilon_2 \\ \vdots \\ \varepsilon_n \end{bmatrix}, \tag{13.12}$$

where $\boldsymbol{x}$ is referred to as the design matrix of multiple linear regression, with dimensions $n \times (m + 1)$.

Therefore, the objective function and prediction model of multiple linear regression can be expressed as:

$$\boldsymbol{y} = \boldsymbol{x}\boldsymbol{\theta} + \boldsymbol{\varepsilon},$$
$$\hat{\boldsymbol{y}} = \boldsymbol{x}\hat{\boldsymbol{\theta}}. \tag{13.13}$$

Thus, the residual sum of squares for multiple linear regression can be represented as follows:

$$\begin{aligned}
\text{RSS}_{\text{multi}}\left(\hat{\boldsymbol{\theta}}\right) &= \|\boldsymbol{y} - \hat{\boldsymbol{y}}\|^2 \\
&= \|\boldsymbol{y} - \boldsymbol{x}\hat{\boldsymbol{\theta}}\|^2 \\
&= (\boldsymbol{y} - \boldsymbol{x}\hat{\boldsymbol{\theta}})^{\mathsf{T}}(\boldsymbol{y} - \boldsymbol{x}\hat{\boldsymbol{\theta}}) \\
&= \boldsymbol{y}^{\mathsf{T}}\boldsymbol{y} - \boldsymbol{y}^{\mathsf{T}}(\boldsymbol{x}\hat{\boldsymbol{\theta}}) - (\boldsymbol{x}\hat{\boldsymbol{\theta}})^{\mathsf{T}} + (\boldsymbol{x}\hat{\boldsymbol{\theta}})^{\mathsf{T}}(\boldsymbol{x}\hat{\boldsymbol{\theta}}) \\
&= \boldsymbol{y}^{\mathsf{T}}\boldsymbol{y} - 2\boldsymbol{y}^{\mathsf{T}}\boldsymbol{x}\hat{\boldsymbol{\theta}} + \hat{\boldsymbol{\theta}}^{\mathsf{T}}\boldsymbol{x}^{\mathsf{T}}\boldsymbol{x}\hat{\boldsymbol{\theta}}.
\end{aligned} \tag{13.14}$$

Because $\text{RSS}_{\text{multi}}\left(\hat{\boldsymbol{\theta}}\right)$ is a convex function, its optimal solution is at the point where the gradient is zero. By calculating its derivative and setting it equal to zero, we get:

$$\frac{\partial \text{RSS}_{\text{multi}}\left(\hat{\boldsymbol{\theta}}\right)}{\partial \hat{\boldsymbol{\theta}}} = \frac{\partial(\boldsymbol{y}^{\mathsf{T}}\boldsymbol{y} - 2\boldsymbol{y}^{\mathsf{T}}\boldsymbol{x}\hat{\boldsymbol{\theta}} + \hat{\boldsymbol{\theta}}^{\mathsf{T}}\boldsymbol{x}^{\mathsf{T}}\boldsymbol{x}\hat{\boldsymbol{\theta}})}{\partial \hat{\boldsymbol{\theta}}}$$

Fig. 13.5 Schematic diagram
of multiple linear regression

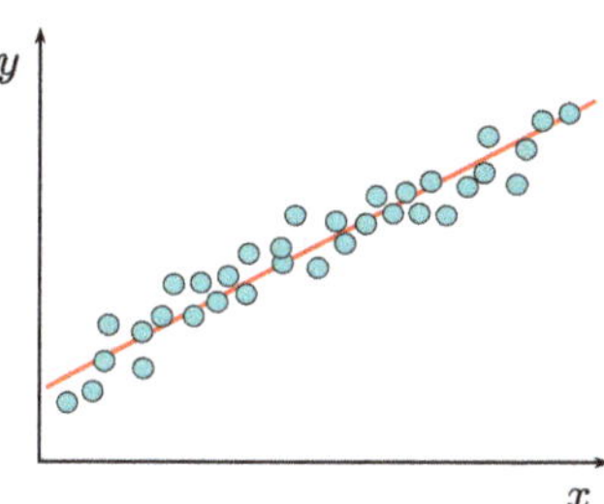

$$= -2x^\mathsf{T} y + 2x^\mathsf{T} x\hat{\theta}$$

$$= 0. \tag{13.15}$$

Thus far, the equation for solving the parameters of multiple linear regression using the method of least squares (i.e., minimizing the sum of residual squares) is as follows:

$$\hat{\theta}_{\text{multi}} = (x^\mathsf{T} x)^{-1} x^\mathsf{T} y. \tag{13.16}$$

A schematic diagram of multiple linear regression is depicted in Fig. 13.5. The red line in the middle of the data points is called the regression line, which is the best-fit line for these data points obtained through training samples.

13.4.2 Polynomial Regression

A regression model $y = f(x, \theta) + \varepsilon$ is referred to as a polynomial regression model, if the relationship between the independent variable x and the dependent variable y is modeled as a pth degree polynomial in x.

Given the labeled n training samples $S = \{(x_i, y_i) \mid i = 1, \ldots, n\}$, where $x_i \in \mathbb{R}^m$, $y_i \in \mathbb{R}$, for the regression model $y_i = f(x_i, \theta) + \varepsilon_i$, let $x_i = (x_i, x_i^2, \ldots, x_i^m)$ be the independent variable of the polynomial regression and $\theta = (\theta_0, \theta_1, \ldots, \theta_m)$ be the parameters, then the objective function of polynomial regression is expressed as:

$$
\begin{aligned}
y_i &= f(x_i, \theta) + \varepsilon_i \\
&= \theta_0 + \theta_1 x_i + \theta_2 x_i^2 + \cdots + \theta_m x_i^m + \varepsilon_i \\
&= \sum_{j=0}^{m} \theta_j x_i^j + \varepsilon_i,
\end{aligned}
\tag{13.17}
$$

where $m \geq 2$, and $x_i^0 = 1$.

It should be pointed out that, although the relationship between the independent and dependent variables in the above equation is nonlinear, polynomial regression is a linear regression model as defined by Definition 13.4.

The polynomial regression function can be represented in the following matrix form:

$$
\begin{bmatrix} y_1 \\ y_2 \\ y_3 \\ \vdots \\ y_n \end{bmatrix} = \begin{bmatrix} 1 & x_1 & x_1^2 & \cdots & x_1^m \\ 1 & x_2 & x_2^2 & \cdots & x_2^m \\ 1 & x_3 & x_3^2 & \cdots & x_3^m \\ \vdots & \vdots & \vdots & \ddots & \vdots \\ 1 & x_n & x_n^2 & \cdots & x_n^m \end{bmatrix} \begin{bmatrix} \theta_0 \\ \theta_1 \\ \theta_2 \\ \vdots \\ \theta_m \end{bmatrix} + \begin{bmatrix} \varepsilon_1 \\ \varepsilon_2 \\ \varepsilon_3 \\ \vdots \\ \varepsilon_n \end{bmatrix}.
\tag{13.18}
$$

Therefore, the objective function and prediction model of polynomial regression can be represented as:

$$
y = \mathbf{X}_{\text{poly}}\theta + \varepsilon
$$

$$
\hat{y} = \mathbf{X}_{\text{poly}}\hat{\theta}.
\tag{13.19}
$$

The ith row of the design matrix in the above equation, $\mathbf{X}_{\text{poly}}[i, :]$ is an mth degree polynomial.

Similar to the method of least squares introduced in the previous section, calculate its residual sum of squares and calculate its gradient to obtain the parameter estimation of polynomial regression. The equation is as follows:

$$
\hat{\theta} = (\mathbf{X}_{\text{poly}}{}^{\mathsf{T}}\mathbf{X}_{\text{poly}})^{-1}\mathbf{X}_{\text{poly}}{}^{\mathsf{T}}y.
\tag{13.20}
$$

A schematic diagram of polynomial regression is depicted in Fig. 13.6.

13.4.3   Ridge Regression

Ridge regression is a method for estimating the parameters of a linear regression model when the independent variables are highly correlated.

Fig. 13.6 Schematic diagram of polynomial regression

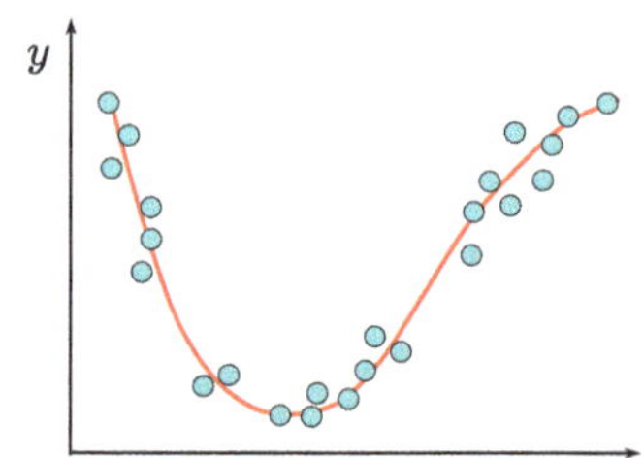

As known from Sect. 13.4.1, the parameter estimation commonly used in multiple linear regression is the method of least squares, which estimates its parameters by minimizing the residual sum of squares:

$$\text{RSS}_{\text{multi}}\left(\hat{\boldsymbol{\theta}}\right) = \sum_{i=1}^{n}\left(y_i - \sum_{j=0}^{m}\hat{\theta}_j x_{ij}\right)^2.$$

However, when *multicollinearity* occurs in multiple linear regression, it can lead to unreliable parameter estimates and high variance.

Multicollinearity often occurs when two or more independent variables in multiple linear regression are highly correlated. In other words, multicollinearity is when one independent variable can be linearly predicted from the others.

If there is an exact multicollinearity relationship between two independent variables, it is called *perfect collinearity*, for example, given n training samples $S = \{(\boldsymbol{x}_i, y_i) \mid i = 1, \ldots, n\}$, where $\boldsymbol{x}_i = (x_{i1}, x_{i2}) \in \mathbb{R}^m$ and $y_i \in \mathbb{R}$. If for all training samples, there exists parameters $\hat{\theta}_0$ and $\hat{\theta}_1$, such that the equation $x_{i2} = \hat{\theta}_0 + \hat{\theta}_1 x_{i1}$ holds, then it is perfect collinearity.

For multicollinearity, ridge regression adds a regularization term, also known as a shrinkage penalty term, to the residual sum of squares RSS $(\cdot)$.

Here, the residual sum of squares with the regularization term is called the generalized residual sum of squares, denoted as $\widehat{\text{RSS}}_{\text{ridge}}(\cdot)$, that is:

$$\widehat{\text{RSS}}_{\text{ridge}}\left(\hat{\boldsymbol{\theta}}\right) = \sum_{i=1}^{n}\left(y_i - \sum_{j=0}^{m}\hat{\theta}_j x_{ij}\right)^2 + \lambda \sum_{j=1}^{m}\left(\hat{\theta}_j\right)^2$$

$$= \text{RSS}_{\text{multi}}(\hat{\boldsymbol{\theta}}) + \lambda \sum_{j=1}^{m}\left(\hat{\theta}_j\right)^2, \tag{13.21}$$

where λ is the tuning parameter to be determined.

Minimizing the above $\widehat{\text{RSS}}_{\text{ridge}}\left(\hat{\boldsymbol{\theta}}\right)$ requires a trade-off between minimizing the two terms on the right side of Eq. 13.21,

Here, the first term $\text{RSS}(\hat{\boldsymbol{\theta}})$ is the residual sum of squares of linear regression, which is minimized using the method of least squares to estimate the parameters. And the second term is the regularization term, which is minimized by making $\lambda \sum_{j=1}^{m}(\hat{\theta}_j)^2 \leq c$ for some constant $c > 0$. As $\hat{\theta}_1, \hat{\theta}_2, \ldots, \hat{\theta}_m$ approaching zero, this regularization term is minimized, thus having the effect of reducing the estimates of $\hat{\theta}_j$ ($j = 1, 2, \ldots, p$) to approach zero. The adjustment parameter λ is used to control the relative impact of these two terms on the regression parameter estimation: when $\lambda = 0$, the regularization term equals zero, and ridge regression is equivalent to linear regression; however, as $\lambda \to \infty$, the impact of the shrinkage regularization term increases, and the parameter estimates of ridge regression will approach zero.

Therefore, the most important thing is to choose a suitable λ value, that is, under the premise of fully balancing the λ value, provide accurate ridge parameter estimates.

The matrix representation of ridge regression parameter estimation is given below. Similar to multiple linear regression, the residual sum of squares of ridge regression is represented as follows:

$$\widehat{\mathrm{RSS}}_{\mathrm{ridge}}\left(\hat{\boldsymbol{\theta}}\right) = (\boldsymbol{y} - \mathbf{X}\hat{\boldsymbol{\theta}})^{\mathsf{T}}(\boldsymbol{y} - \mathbf{X}\hat{\boldsymbol{\theta}}) + \lambda\hat{\boldsymbol{\theta}}^{\mathsf{T}}\hat{\boldsymbol{\theta}}$$

$$= \boldsymbol{y}^{\mathsf{T}}\boldsymbol{y} - 2\boldsymbol{y}^{\mathsf{T}}\mathbf{X}\hat{\boldsymbol{\theta}} + \hat{\boldsymbol{\theta}}^{\mathsf{T}}\mathbf{X}^{\mathsf{T}}\mathbf{X}\hat{\boldsymbol{\theta}} + \lambda\hat{\boldsymbol{\theta}}^{\mathsf{T}}\hat{\boldsymbol{\theta}}. \qquad (13.22)$$

Calculate the gradient of $\widehat{\mathrm{RSS}}_{\mathrm{ridge}}(\cdot)$ and set it to zero, that is:

$$\frac{\partial\widehat{\mathrm{RSS}}_{\mathrm{ridge}}(\hat{\boldsymbol{\theta}})}{\partial\hat{\boldsymbol{\theta}}} = \frac{\partial(\boldsymbol{y}^{\mathsf{T}}\boldsymbol{y} - 2\boldsymbol{y}^{\mathsf{T}}\mathbf{X}\hat{\boldsymbol{\theta}} + \hat{\boldsymbol{\theta}}^{\mathsf{T}}\mathbf{X}^{\mathsf{T}}\mathbf{X}\hat{\boldsymbol{\theta}} + \lambda\hat{\boldsymbol{\theta}}^{\mathsf{T}}\hat{\boldsymbol{\theta}})}{\partial\hat{\boldsymbol{\theta}}}$$

$$= -2\mathbf{X}^{\mathsf{T}}\boldsymbol{y} + 2\mathbf{X}^{\mathsf{T}}\mathbf{X}\hat{\boldsymbol{\theta}} + 2\lambda\hat{\boldsymbol{\theta}}$$

$$= 0. \qquad (13.23)$$

Therefore, the parameter estimation of ridge regression is:

$$\hat{\boldsymbol{\theta}}_{\mathrm{ridge}} = (\mathbf{X}^{\mathsf{T}}\mathbf{X} - \lambda\mathbf{I})^{-1}\mathbf{X}^{\mathsf{T}}\boldsymbol{y}, \qquad (13.24)$$

where $\mathbf{I}$ is the identity matrix.

13.4.4   Lasso Regression

Lasso regression is another method to solve the problem of multicollinearity, wherein the "lasso" is an abbreviation for "least absolute shrinkage and selection operator."

For multicollinearity, lasso regression also adds a regularization term to the residual sum of squares $\mathrm{RSS}(\cdot)$, but the regularization term is somewhat different from the one in ridge regression. The generalized sum of squares is denoted as $\widehat{\mathrm{RSS}}_{\mathrm{lasso}}(\cdot)$, that is:

$$\widehat{\mathrm{RSS}}_{\mathrm{lasso}}(\hat{\boldsymbol{\theta}}) = \sum_{i=1}^{n}\left(y_i - \sum_{j=0}^{m}\hat{\theta}_j x_{ij}\right)^2 + \lambda\sum_{j=1}^{m}|\hat{\theta}_j|$$

$$= \mathrm{RSS}_{\mathrm{multi}}(\hat{\boldsymbol{\theta}}) + \lambda\sum_{j=1}^{m}|\hat{\theta}_j|. \qquad (13.25)$$

Similarly, lasso regression is also for a certain constant $c > 0$, making its regularization term $\sum_{j=1}^{m} \left| \hat{\theta}_j \right| \leq c$. However, when the parameter λ is sufficiently large, the L_1 regularization of lasso regression has the effect of forcing some parameter estimates to be exactly zero. Therefore, lasso regression is usually easier for variable selection and exclusion of irrelevant variables than ridge regression.

Comparing Eq. 13.21 in ridge regression and Eq. 13.25 in lasso regression, we may find both of the generalized residual sum of squares are same; the only difference lies in the regularization term. Lasso regression replaces $\left(\hat{\theta}_j \right)^2$ with $\left| \hat{\theta}_j \right|$.

The regularization term of ridge regression $\sum_{j=1}^{m} \left(\hat{\theta}_j \right)^2$ is L_2 regularization, and the regularization term of lasso regression $\sum_{j=1}^{p} \left| \hat{\theta}_j \right|$ is L_1 regularization, because the L_1 norm of parameter vector $\hat{\boldsymbol{\theta}}$ is $\|\hat{\boldsymbol{\theta}}\|_1 = \sum |\hat{\theta}_j|$. This is a subtle but important change, as some parameters of the lasso regression can be precisely reduced to zero.

Below, given a dataset with only two variables, that is, $m = 2$, let's compare the differences between lasso regression and ridge regression through two diagrams.

The two diagrams in Fig. 13.7 are schematic diagrams of lasso regression and ridge regression with two variables, i.e., in two dimensions. Their residual sum of squares $\sum_{i=1}^{n} \left(y_i - \sum_{j=0}^{2} \hat{\theta}_j x_{ij} \right)^2$ is the same, represented as three-layer red elliptical curves. In two-dimensional space, the regularization term of lasso regression $\sum_{j=1}^{2} |\theta_j| = |\theta_1| + |\theta_2|$ becomes a diamond, while the regularization term of ridge regression $\sum_{j=1}^{2} \left(\theta_j \right)^2 = (\theta_1)^2 + (\theta_2)^2$ is a circle. Comparing Fig. 13.7a and b, it can be seen that the L_1 regularization of lasso regression has the effect of forcing a parameter estimate to be zero, while the L_2 regularization of ridge regression can only make it approach zero, but cannot be equal to zero.

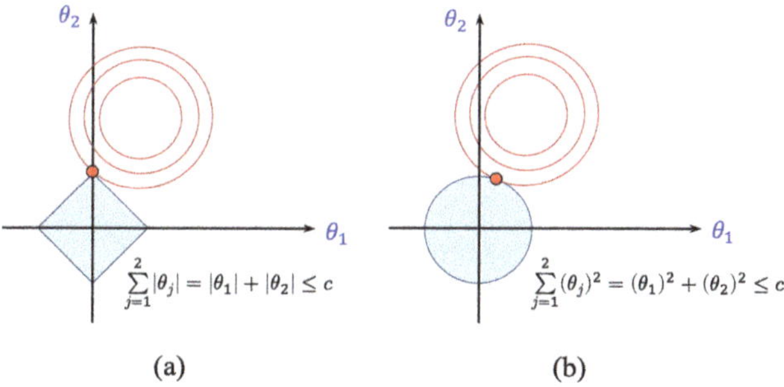

Fig. 13.7 Comparison of lasso regression and ridge regression

The matrix representation of the lasso regression parameter estimate is given below. Similar to linear regression, the residual sum of squares of lasso regression is represented as follows:

$$\widehat{\text{RSS}}_{\text{lasso}}(\hat{\boldsymbol{\theta}}) = (\boldsymbol{y} - \mathbf{X}\hat{\boldsymbol{\theta}})^{\mathsf{T}}(\boldsymbol{y} - \mathbf{X}\hat{\boldsymbol{\theta}}) + \lambda\hat{\boldsymbol{\theta}}$$
$$= \boldsymbol{y}^{\mathsf{T}}\boldsymbol{y} - 2\boldsymbol{y}^{\mathsf{T}}\mathbf{X}\hat{\boldsymbol{\theta}} + \hat{\boldsymbol{\theta}}^{\mathsf{T}}\mathbf{X}^{\mathsf{T}}\mathbf{X}\hat{\boldsymbol{\theta}} + \lambda\hat{\boldsymbol{\theta}}. \tag{13.26}$$

Solve its gradient and set it to zero:

$$\frac{\partial \widehat{\text{RSS}}_{\text{lasso}}(\hat{\boldsymbol{\theta}})}{\partial \hat{\boldsymbol{\theta}}} = \frac{\partial(\boldsymbol{y}^{\mathsf{T}}\boldsymbol{y} - 2\boldsymbol{y}^{\mathsf{T}}\mathbf{X}\hat{\boldsymbol{\theta}} + \hat{\boldsymbol{\theta}}^{\mathsf{T}}\mathbf{X}^{\mathsf{T}}\mathbf{X}\hat{\boldsymbol{\theta}} + \lambda\hat{\boldsymbol{\theta}})}{\partial \hat{\boldsymbol{\theta}}}$$
$$= -2\mathbf{X}^{\mathsf{T}}\boldsymbol{y} + 2\mathbf{X}^{\mathsf{T}}\mathbf{X}\hat{\boldsymbol{\theta}} + \lambda$$
$$= 0. \tag{13.27}$$

Therefore, the parameter estimation for lasso regression is:

$$\hat{\boldsymbol{\theta}}_{\text{lasso}} = (2\mathbf{X}^{\mathsf{T}}\mathbf{X})^{-1}(2\mathbf{X}^{\mathsf{T}}\boldsymbol{y} - \lambda). \tag{13.28}$$

13.4.5   *Bayesian Regression*

The regression algorithms introduced earlier (i.e., multiple linear regression, polynomial regression, ridge regression, and lasso regression) are all based on the method of least squares (i.e., minimizing the residual sum of squares) to solve the parameters of linear regression. Ridge regression and lasso regression just add a regularization term. These methods are based on the law of large numbers, requiring a large number of training samples, equivalent to the frequency method of linear regression, so they are called frequentist linear regressions.

Bayesian regression, also known as Bayesian linear regression, is a linear regression analysis model based on Bayesian inference. It infers regression parameters based on prior and likelihood information, thus providing an effective way for regression problems with insufficient training samples.

The most commonly used in Bayesian regression is the normal linear model, that is, given the independent variables, the dependent variables follow a Gaussian distribution.

Given a limited number of n training samples $S = \{(\boldsymbol{x}_i, y_i) \mid i = 1, \ldots, n\}$, using the expression of Eq. 13.13, its objective function is expressed as $\boldsymbol{y} = \mathbf{X}\boldsymbol{\theta} + \boldsymbol{\varepsilon}$. Based on the Gaussian prior distribution, the parameters $\boldsymbol{\theta}$ and error term under the condition of zero mean are $\boldsymbol{\varepsilon}$ can be represented as:

$$\boldsymbol{\theta} \sim \mathcal{N}(0, \Sigma)$$

$$\boldsymbol{\varepsilon} \sim \mathcal{N}(0, \sigma^2 \mathbf{I}),$$

where the covariance Σ and variance σ^2 are known and $\mathbf{I}$ represents n order identity matrix.

Bayesian regression calculates the conditional probability of the parameter $\boldsymbol{\theta}$ under the condition of the training sample S, that is, $p(\boldsymbol{\theta}|S) = p(\boldsymbol{\theta}|\mathbf{X}, \boldsymbol{y})$. Based on Bayes' theorem, we have:

$$p(\boldsymbol{\theta}|\mathbf{X}, \boldsymbol{y}) = \frac{p(\boldsymbol{y}|\mathbf{X}, \boldsymbol{\theta})p(\boldsymbol{\theta})}{p(\boldsymbol{y}|\mathbf{X})}, \tag{13.29}$$

where the likelihood function $p(\boldsymbol{y}|\mathbf{X}, \boldsymbol{\theta})$ and the prior $p(\boldsymbol{\theta})$ are as follows:

$$p(\boldsymbol{y}|\mathbf{X}, \boldsymbol{\theta}) = \mathcal{N}\left(\boldsymbol{y}\middle|\mathbf{X}\boldsymbol{\theta}, \sigma^2 \mathbf{I}\right),$$

$$p(\boldsymbol{\theta}) = \mathcal{N}(\boldsymbol{\theta}|0, \Sigma).$$

Below, based on Eq. 13.29, the conditional probability of its regression $p(\boldsymbol{\theta}|\mathbf{X}, \boldsymbol{y})$ is derived as follows:

$$p(\boldsymbol{\theta}|\mathbf{X}, \boldsymbol{y}) \propto p(\boldsymbol{y}|\mathbf{X}, \boldsymbol{\theta})p(\boldsymbol{\theta})$$

$$= \mathcal{N}\left(\boldsymbol{y}\middle|\mathbf{X}\boldsymbol{\theta}, \sigma^2 \mathbf{I}\right)\mathcal{N}(\boldsymbol{\theta}|0, \Sigma)$$

$$\propto \exp\left(-\frac{1}{2\sigma^2}(\boldsymbol{y} - \mathbf{X}\boldsymbol{\theta})^{\mathsf{T}}(\boldsymbol{y} - \mathbf{X}\boldsymbol{\theta})\right)\exp\left(-\frac{1}{2}\boldsymbol{\theta}^{\mathsf{T}}\Sigma^{-1}\boldsymbol{\theta}\right)$$

$$= \exp\left(-\frac{1}{2}\left(\frac{1}{\sigma^2}(\boldsymbol{y} - \mathbf{X}\boldsymbol{\theta})^{\mathsf{T}}(\boldsymbol{y} - \mathbf{X}\boldsymbol{\theta}) + \boldsymbol{\theta}^{\mathsf{T}}\Sigma^{-1}\boldsymbol{\theta}\right)\right).$$

Analyzing the exponential part of the above equation separately:

$$\frac{1}{\sigma^2}(\boldsymbol{y} - \mathbf{X}\boldsymbol{\theta})^{\mathsf{T}}(\boldsymbol{y} - \mathbf{X}\boldsymbol{\theta}) + \boldsymbol{\theta}^{\mathsf{T}}\Sigma^{-1}\boldsymbol{\theta} = \frac{1}{\sigma^2}\left(\boldsymbol{\theta}^{\mathsf{T}}\mathbf{X}^{\mathsf{T}}\mathbf{X}\boldsymbol{\theta} - 2\boldsymbol{\theta}^{\mathsf{T}}\mathbf{X}^{\mathsf{T}}\boldsymbol{y} + \boldsymbol{y}^{\mathsf{T}}\boldsymbol{y}\right)$$

$$+ \boldsymbol{\theta}^{\mathsf{T}}\Sigma^{-1}\boldsymbol{\theta}$$

$$= \boldsymbol{\theta}^{\mathsf{T}}\left(\frac{1}{\sigma^2}\mathbf{X}^{\mathsf{T}}\mathbf{X} + \Sigma^{-1}\right)\boldsymbol{\theta} - 2\frac{1}{\sigma^2}\boldsymbol{y}^{\mathsf{T}}\mathbf{X}\boldsymbol{\theta}.$$

To simplify the above expression, let $M = \frac{1}{\sigma^2}\mathbf{X}^\mathsf{T}\mathbf{X} + \Sigma^{-1}$, $b = \frac{1}{\sigma^2}\mathbf{X}^\mathsf{T}\mathbf{y}$, then the expression becomes:

$$\frac{1}{\sigma^2}(\mathbf{y} - \mathbf{X}\boldsymbol{\theta})^\mathsf{T}(\mathbf{y} - \mathbf{X}\boldsymbol{\theta}) + \boldsymbol{\theta}^\mathsf{T}\Sigma^{-1}\boldsymbol{\theta} = \boldsymbol{\theta}^\mathsf{T} M\boldsymbol{\theta} - 2b^\mathsf{T}\boldsymbol{\theta}.$$

By applying the following identity to complete the *quadratic form*:[1]

$$\boldsymbol{\theta}^\mathsf{T} M\boldsymbol{\theta} - 2b^\mathsf{T}\boldsymbol{\theta} = \left(\boldsymbol{\theta} - M^{-1}b\right)^\mathsf{T} M(\boldsymbol{\theta} - M^{-1}b) - b^\mathsf{T} M^{-1}b.$$

Substitute the above exponential part, so we have:

$$p(\boldsymbol{\theta}|\mathbf{X}, \mathbf{y}) \propto \exp\left(-\frac{1}{2}\left(\left(\boldsymbol{\theta} - M^{-1}b\right)^\mathsf{T} M\left(\boldsymbol{\theta} - M^{-1}b\right) - b^\mathsf{T} M^{-1}b\right)\right)$$

$$= \exp\left(-\frac{1}{2}\left(\left(\boldsymbol{\theta} - M^{-1}b\right)^\mathsf{T} M\left(\boldsymbol{\theta} - M^{-1}b\right)\right)\right) \exp\left(-\frac{1}{2}(-b^\mathsf{T} M^{-1}b)\right)$$

$$\propto \exp\left(-\frac{1}{2}\left(\left(\boldsymbol{\theta} - M^{-1}b\right)^\mathsf{T} M\left(\boldsymbol{\theta} - M^{-1}b\right)\right)\right).$$

The Gaussian density of the known parameter $\boldsymbol{\theta}$ satisfies the following expression:

$$\mathcal{N}(\boldsymbol{\theta}|\mu, \Sigma) = |2\pi\Sigma|^{-\frac{1}{2}} \exp\left(-\frac{1}{2}(\boldsymbol{\theta} - \mu)^\mathsf{T}\Sigma^{-1}(\boldsymbol{\theta} - \mu)\right).$$

Therefore, we have:

$$p(\boldsymbol{\theta}|\mathbf{X}, \mathbf{y}) \propto \mathcal{N}\left(\boldsymbol{\theta}|M^{-1}b, M^{-1}\right). \tag{13.30}$$

Substituting $M = \frac{1}{\sigma^2}\mathbf{X}^\mathsf{T}\mathbf{X} + \Sigma^{-1}$ and $b = \frac{1}{\sigma^2}\mathbf{X}^\mathsf{T}\mathbf{y}$, the posterior mean $M^{-1}b$ is:

$$M^{-1}b = \left(\frac{1}{\sigma^2}\mathbf{X}^\mathsf{T}\mathbf{X} + \Sigma^{-1}\right)^{-1}\frac{1}{\sigma^2}\mathbf{X}^\mathsf{T}\mathbf{y}$$

$$= \left(\mathbf{X}^\mathsf{T}\mathbf{X} + \sigma^2\Sigma^{-1}\right)^{-1}\mathbf{X}^\mathsf{T}\mathbf{y}$$

$$= \left(\mathbf{X}^\mathsf{T}\mathbf{X} + \sigma^2\Lambda\right)^{-1}\mathbf{X}^\mathsf{T}\mathbf{y},$$

---

[1] https://en.wikipedia.org/wiki/Quadratic_form

where $\Lambda = \Sigma^{-1}$. Therefore, M^{-1} is:

$$M^{-1} = \left(\frac{1}{\sigma^2}\mathbf{X}^{\mathsf{T}}\mathbf{X} + \Sigma^{-1}\right)^{-1}$$

$$= \sigma^2 \left(\mathbf{X}^{\mathsf{T}}\mathbf{X} + \sigma^2\Lambda\right)^{-1}.$$

13.5 Loss Functions

For a regression model, since the actual values of the labels are real numbers, it is unrealistic to expect the model to accurately predict the actual values. Instead, we hope that its predicted values are as close as possible to the actual values. This is the essential difference between regression and classification: the loss function of a regression model measures the error between the predicted value and the actual value, instead of judging whether these two values are same just like a classification model.

The loss function of regression is used to evaluate the performance of the regression model, that is, by calculating the error between the predicted value $\hat{y}_i$ and the actual value y_i, to measure the accuracy of the regression model. The loss function is also known as evaluation metric, used to optimize the regression model.

Let the labeled training sample set be $S = \{(x_i, y_i) \mid i = 1, \ldots, n\}$, where: x_i is the ith sample and $x_i \in \mathbb{R}^m$, y_i is the actual value corresponding to this sample, and $y_i \in \mathbb{R}$, n represents the number of training sample data pairs. Also, let the regression model's prediction for the ith sample x_i be $\hat{y}_i = h(x_i)$, where $h(x_i)$ is referred to as the hypothesis function of the regression model.

The following introduces several commonly used loss functions, i.e., mean square error, root mean square error, mean absolute error, mean absolute percentage error, mean bias error, relative absolute error, relative squared error, Huber loss, log-cosh loss, and coefficient of determination.

13.5.1 Mean Square Error

Mean square error (MSE), also known as quadratic loss, L_2 loss, is one of the most common regression loss functions. Its expression is as follows:

$$\text{MSE} = \frac{1}{n}\sum_{i=1}^{n}(y_i - \hat{y}_i)^2. \tag{13.31}$$

The optimization of regression models benefits from this regularization of large errors, as it helps to find the best values of the parameters. Since the square of the

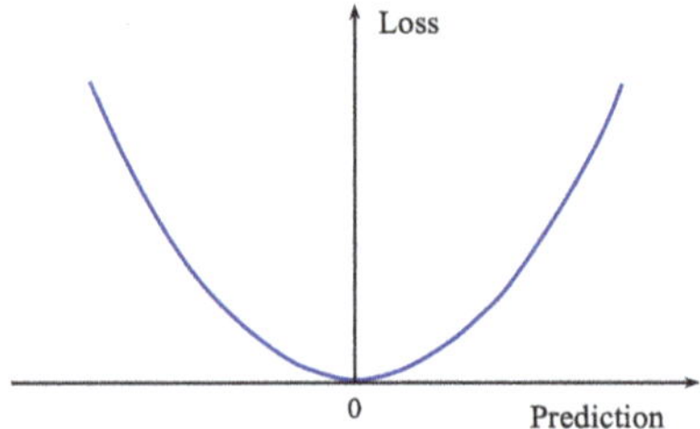

Fig. 13.8 Schematic diagram of mean square error

error is taken, the MSE loss will never be a negative value. The range of error is from zero to infinity. The MSE increases exponentially with the increase in error. A good regression model will have an MSE value closer to zero.

The mean square error is the most commonly used loss function in regression because it is easy to differentiate, making model optimization easy, and it has stable properties.

The schematic diagram of the mean square error is shown in Fig. 13.8.

The horizontal axis in the figure represents the predicted value, and the vertical axis represents the mean square error (MSE) loss.

The loss measured by the mean square error is proportional to the square of the error, and the loss of outliers will become very large, that is, it gives higher weight to outliers. Therefore, when there are fewer outliers in the data, the mean square error is preferred. Using normalized data can effectively use MSE loss to better optimize the regression model.

The root mean square error (RMSE) is also referred to as the root mean square deviation (RMSD).

$$\text{RMSE} = \sqrt{\frac{1}{n} \sum_{i=1}^{n} (y_i - \hat{y}_i)^2}. \tag{13.32}$$

This metric calculates the root mean square error between the actual values and the predicted values. Because it takes the square root of the mean square error, it restores the measurement unit to its original scale.

RMSE is the most commonly used metric to measure the error of a regression model. It measures the average size of the error, which is related to the deviation from the actual value. An RMSE value of 0 indicates a perfect fit of the model. The lower the RMSE, the better the prediction of the model. A higher RMSE indicates a larger deviation between the residuals and the actual values. RMSE can be used for different features because it helps to determine whether the feature helps to improve the predictive performance of the model.

Compared with the mean square error (MSE), the root mean square error (RMSE) is often used to evaluate how well the model fits the dataset because the unit of RMSE is the same as the dependent variable. In contrast, MSE is measured in the square of the unit of the dependent variable.

Fig. 13.9 Schematic diagram
of mean absolute error

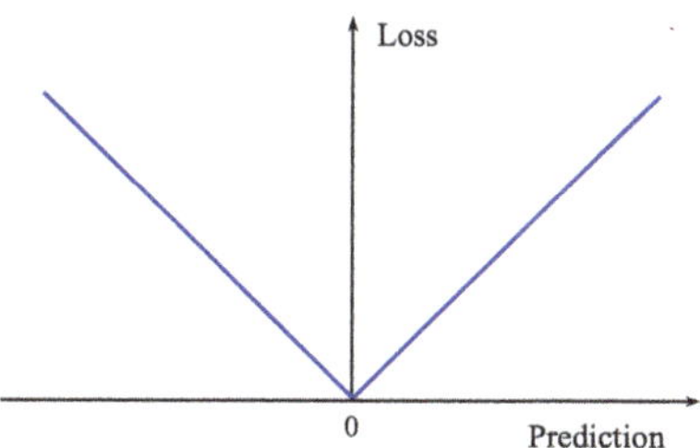

13.5.2   Mean Absolute Error

The mean absolute error (MAE), also known as L_1 loss, is one of the simplest
loss functions used to evaluate regression models and is also an easy-to-understand
evaluation metric.

Mathematically speaking, it is the average of the absolute errors, used to evaluate
the effectiveness of a regression model. MAE only measures the size of the error,
not the direction of the error. The lower the MAE, the higher the accuracy of the
model. The definition of mean absolute error is as follows:

$$\text{MAE} = \frac{1}{n} \sum_{i=1}^{n} |y_i - \hat{y}_i|. \tag{13.33}$$

A schematic diagram of the mean absolute error (MAE) is depicted in Fig. 13.9.

The mean absolute percentage error (MAPE), also known as the mean absolute
percentage deviation (MAPD), measures the error in percentage.

Its calculation method is the absolute error divided by the average of the actual
values, multiplied by 100, to get a percentage error. That is:

$$\text{MAPE} = \frac{1}{n} \sum_{i=1}^{n} \frac{|y_i - \hat{y}_i|}{y_i} \times 100. \tag{13.34}$$

The condition for MAPE is that the actual value $y_i \neq 0$. Furthermore, when y_i is
close to 0, it will produce a very large value.

13.5.3   Mean Bias Error

The bias in the mean bias error (MBE) is a tendency to overestimate or underesti-
mate the parameter value in the measurement process. Bias has only one direction:
it can be positive or negative. Positive bias means that the error of the data is
overestimated, and negative bias means that the error is underestimated. The mean
bias error is the average of the difference between the actual value and the predicted

Fig. 13.10 Schematic
diagram of mean bias error

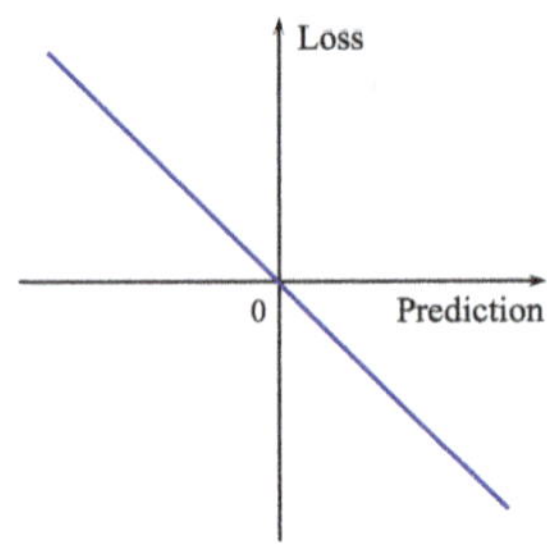

value. This evaluation metric quantifies the overall bias and captures the average
bias in the prediction.

The definition of the mean bias error is as follows:

$$\text{MBE} = \frac{1}{n} \sum_{i=1}^{n} (y_i - \hat{y}_i). \tag{13.35}$$

And the schematic diagram is depicted in Fig. 13.10.

The mean bias error (MBE) is similar to the mean absolute error (MAE), the
only difference being that MBE does not take the absolute value. This evaluation
metric should be handled with caution, as positive and negative errors may cancel
each other out.

13.5.4   Relative Absolute Error

The relative absolute error (RAE) is calculated by dividing the total absolute error
by the sum of the absolute differences between the actual values and the average
value. Its expression is as follows:

$$\text{RAE} = \frac{\sum_{i=1}^{n} |y_i - \hat{y}_i|}{\sum_{i=1}^{n} |y_i - \bar{y}|}, \tag{13.36}$$

where $\bar{y}$ is the average of n actual values y_i and its calculation equation is as
follows:

$$\bar{y} = \frac{1}{n} \sum_{i=1}^{n} y_i. \tag{13.37}$$

RAE measures the performance of the regression model and is expressed as a
ratio. The range of RAE is $0 \sim 1$. A better regression model is close to 0, and the
best regression model is equal to 0.

13.5.5 Relative Squared Error

The relative squared error (RSE) is calculated by dividing the mean squared error (MSE) by the square of the difference between the actual value and the average value of the data. Its expression is as follows:

$$\text{RSE} = \frac{\sum_{i=1}^{n} \left(y_i - \hat{y}_i\right)^2}{\sum_{i=1}^{n} \left(y_i - \bar{y}\right)^2},$$

(13.38)

where $\bar{y}$ is the average of n actual values y_i, the calculation of which is shown in Eq. 13.37.

13.5.6 Huber Loss

Huber loss is also known as smooth mean absolute error (SMAE).

Huber loss is a combination of linear and quadratic scoring methods. It balances and combines the characteristics of mean absolute error (MAE) and mean square error (MSE). It has a hyperparameter δ, which can be adjusted according to the data. For values greater than δ, the loss will be linear (L_1 loss), and for values less than δ, the loss will be quadratic (L_2 loss). In other words, for loss values less than δ, MSE will be used, and for loss values greater than δ, MAE will be used.

Its expression is as follows:

$$\mathcal{L}_\delta(y, \hat{y}) = \begin{cases} \frac{1}{2}\left(y_i - \hat{y}_i\right)^2, & \text{for } |y_i - \hat{y}_i| \leq \delta \\ \delta\left(|y_i - \hat{y}_i| - \frac{1}{2}\delta\right) & \text{otherwise.} \end{cases}$$

The choice of δ is crucial as it defines the tolerance for outliers. The Huber loss reduces the sensitivity to outliers for larger loss values by using MAE, while for smaller loss values, it maintains a quadratic function using MSE.

A schematic diagram of the Huber loss is depicted in Fig. 13.11.

Fig. 13.11 Schematic diagram of Huber loss

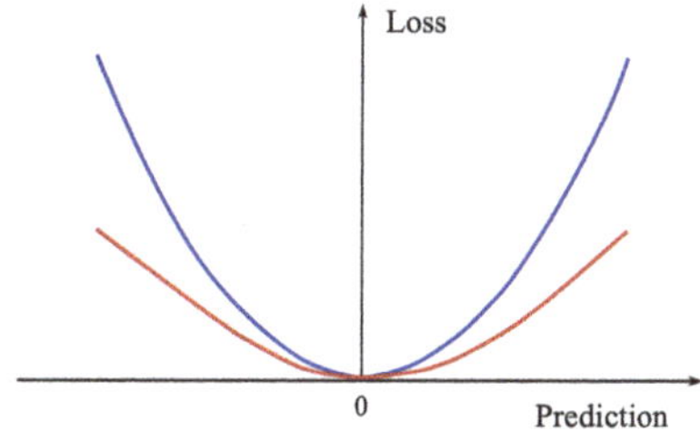

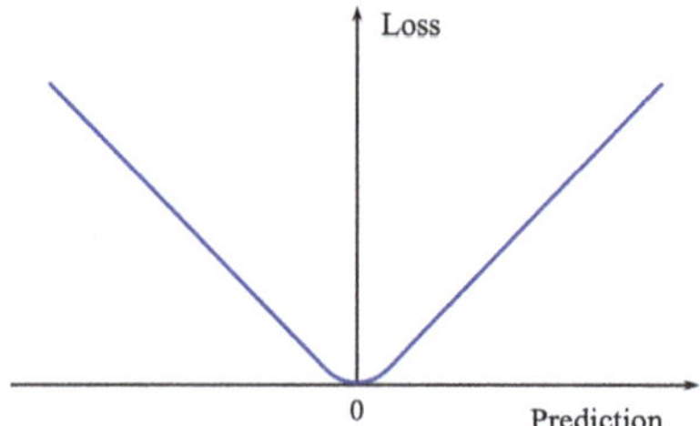

Fig. 13.12 Schematic diagram of log-cosh loss

13.5.7 Log-Cosh Loss

The log-cosh loss calculates the logarithm of the hyperbolic cosine of the error. This function is smoother than the quadratic loss function. It works similarly to the mean squared error (MSE), but is not affected by larger prediction errors.

Its formula is as follows:

$$\mathcal{L}(y, \hat{y}) = \sum_{i=1}^{n} \log\left(\cosh(y_i - \hat{y}_i)\right). \tag{13.39}$$

The log-cosh loss is very similar to the Huber loss, as it is a combination of linear and quadratic scoring methods. Fig. 13.12 is a schematic diagram of it.

13.5.8 Coefficient of Determination

The coefficient of determination, also known as R-squared (R^2), represents the proportion of the variance in the dependent variable that can be predicted from the independent variable(s), and is therefore also referred to as explained variance.

In other words, R-squared represents the degree of fit of the regression model to the training samples. Its formula is as follows:

$$R^2 = 1 - \text{RSE} = 1 - \frac{\sum_{i=1}^{n}(y_i - \hat{y}_i)^2}{\sum_{i=1}^{n}(y_i - \bar{y})^2}. \tag{13.40}$$

From the above equation, it can be seen that R-squared can take any value in the range of 0 to 1, but the meaning varies with different values:

- If the value is 0, it means that the dependent variable cannot be explained by the independent variable.
- If the value is 1, it means that the independent variable can perfectly explain the dependent variable, with no error.

13.6 Application Fields

Regression is used to estimate the relationship between independent and dependent variables, so its application fields are very wide.

In the economic field, regression can be used to predict GDP, inflation rate, unemployment rate, consumer spending, fixed asset investment spending, demand for liquid assets, etc.

In the financial field, regression is used to predict stock prices, analyze market trends, evaluate the performance of investment portfolios, predict quantitative investment risks, etc.

In the field of marketing, regression is used to analyze the relationship between marketing variables such as advertising expenditure, sales volume, and customer satisfaction.

In the field of healthcare, regression is used to predict epidemics, analyze disease causality, analyze the effectiveness of treatments, and identify risk factors for diseases.

In the field of environmental science, regression can be used for environmental prediction, environmental monitoring, and environmental evaluation.

In the field of social sciences, regression is used to study the relationship between social variables such as education, income, and health.

Further Reading

1. Ludwig Fahrmeir, Thomas Kneib, Stefan Lang, and Brian Marx. *Regression: Models, Methods and Applications*, Springer, 2013.
   [Notes] This book discusses several important models and methods in regression and also introduces some applications of regression through actual data and cases.
2. Samprit Chatterjee, and Ali S. Hadi. *Regression Analysis by Example*. John Wiley & Sons, 2013.
   [Notes] This book introduces several important methods in regression analysis, provides several typical examples from the real world, and discusses them in detail.

Chapter 14
Clustering Task

Abstract Clustering, or cluster analysis, is a task that belongs to unsupervised learning paradigm. This chapter starts from the clustering problem and definition and studies the working principle employing formal and illustrated descriptions. Second, we discuss the relevant elements such as hard and soft, as well as linear and nonlinear clustering. We next divide clustering algorithms into classical and neoclassical methods and discuss them respectively. Then, we introduce several typical clustering algorithms, including k-means, Gaussian mixture clustering, DBSCAN, and the density peak clustering. After that, we explain evaluation metrics including adjusted Rand index (ARI), adjusted mutual information (AMI), Fowlkes-Mallows index (FMI), V-measure, Calinski-Harabasz index (CHI), Davies-Bouldin index (DBI), and silhouette coefficient (SC). Finally, the application fields of clustering are briefly introduced.

14.1 Problem and Definition

In this section, we first discuss the clustering problem and then give a definition of the clustering.

14.1.1 Clustering Problems

Clustering, also known as cluster analysis, is to divide data into several clusters.

The clustering problems exist in many fields. For example: in the medical field, clustering is used for medical image analysis, medical image segmentation, and three-dimensional reconstruction; in the business and marketing field, customers or shopping goods are grouped through cluster analysis; in the World Wide Web, clustering methods are used for social network analysis and search result grouping; in the field of information filtering and retrieval, clustering can be used for scene segmentation and building recommendation systems.

© The Author(s), under exclusive license to Springer Nature Singapore Pte Ltd. 2025
W. Wang, *Principles of Machine Learning*,
https://doi.org/10.1007/978-981-97-5333-8_14

Table 14.1 Similarities and differences between clustering and classification

Clustering	Classification
Dividing input data into clusters according to certain criteria	Dividing input data into predefined classes
The clusters and their numbers are unknown beforehand	The categories and their numbers are known in advance
Without labeled training samples	With labeled training samples

14.1.2   Definition

Here we first give the definition of clustering.

Definition 14.1 (Clustering)   Clustering in machine learning is a task that is based on certain criteria to analyze input data and divide it into several groups called clusters. These clusters are unknown in advance, but there is some correlation between the data within the same cluster.

From the above definition, it can be seen that clustering has two factors: one is to divide the data into clusters, and the other is the correlation within the same cluster.

Clustering is essentially different from the classification introduced in Chap. 12, although both tasks divide data into corresponding categories or several clusters. This is because classification has labeled training samples, and the categories and their numbers are determined before classification. Clustering does not have labeled training samples, and the clusters and their numbers are obtained by analyzing some correlation of input data according to certain criteria. This is why clustering is called cluster analysis.

The similarities and differences between clustering and classification are shown in Table 14.1.

14.2   Working Principle

For further understanding of the clustering task, here we first give a formal description and then explain it through an illustration.

14.2.1   Formal Description

Let $\mathbb{R}^m$ ($m \geq 1$) denote a set of m-dimensional real vectors, $\mathcal{X} \subseteq \mathbb{R}^m$ be the input space, and $\mathcal{G} \subseteq \mathbb{R}^m$ be the output space. Assume there is a dataset with n independent and identically distributed (i.i.d.) data, and unknown their clusters and number, represented as:

$$D = \{\boldsymbol{x}_i \mid i = 1, \ldots, n\} \subseteq \mathcal{X}. \tag{14.1}$$

Let H denote a set of hypotheses for the clustering; the goal is to find an optimal clustering hypothesis $h \in H$, which divides the input data into k clusters based on certain criteria, that is:

$$h : \mathcal{X} \to \mathcal{G} . \tag{14.2}$$

Using the above clustering hypothesis $h\,(\mathcal{X})$ to cluster the dataset D, we get a set of k clusters G, that is:

$$G = \left\{G_j \mid j = 1, \ldots, k\right\} \subseteq \mathcal{G} \tag{14.3}$$

where $G_j \subseteq D$, and $k \leq n$.

14.2.2 Illustrated Description

The clustering process includes two important steps: (i) cluster analysis and (ii) result verification.

The dividing process of cluster analysis is depicted in Fig. 14.1, which divides the input data into several clusters based on certain criteria. As can be seen, clustering does not have a training phase, meaning there are no labeled samples as training data. It uses the clustering hypothesis to directly analyze the original dataset, and its output is the divided clusters.

The verifying process of the clustering model is also shown in Fig. 14.1. Its purpose is to evaluate the quality and performance of the clustering, optimize it if necessary, and make the clustering results more reasonable using various evaluation metrics.

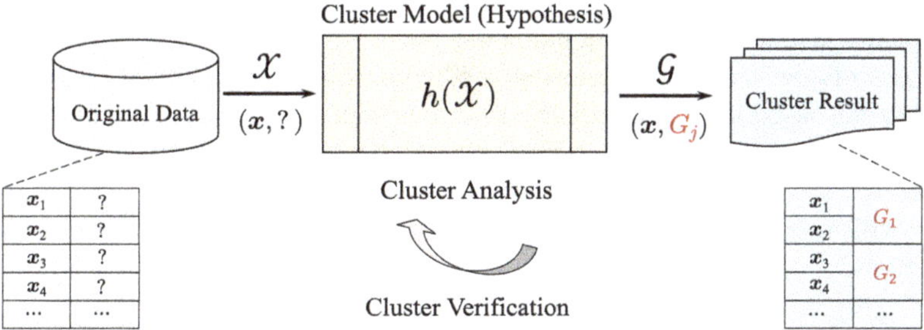

Fig. 14.1 Dividing and verifying processes of the clustering model

14.3   Related Elements

14.3.1   Hard and Soft

Just as classification methods are divided into hard classification and soft classification, clustering methods can also be divided into hard clustering and soft clustering.

- Hard clustering:
  It determines whether an input data point belongs to a certain cluster and the result is a definite value.
- Soft clustering:
  It determines the degree to which an input data point belongs to a certain cluster, and the result is expressed as a likelihood or fuzzy value.

14.3.2   Linear and Nonlinear

Based on the data distribution of clustering, clustering problems can be divided into linearly separable clustering and nonlinearly separable clustering.

- Linearly separable clustering:
  Clustering performed on a linearly separable dataset.
- Nonlinearly separable clustering:
  There are no restrictions similar to linearly separable clustering.

The definitions of linear and nonlinear separability have been given in Sect. 12.4.1 of Chap. 12.

14.4   Classical Methods

Classical clustering methods, also known as traditional clustering methods, refer to some earlier proposed clustering methods.

Representative classical clustering methods include hierarchical-based clustering, centroid-based clustering, distribution-based clustering, density-based clustering, and grid-based clustering. These five classical clustering methods and their representative algorithms are shown in Fig. 14.2.

We will introduce these five classical clustering methods in the following subsections.

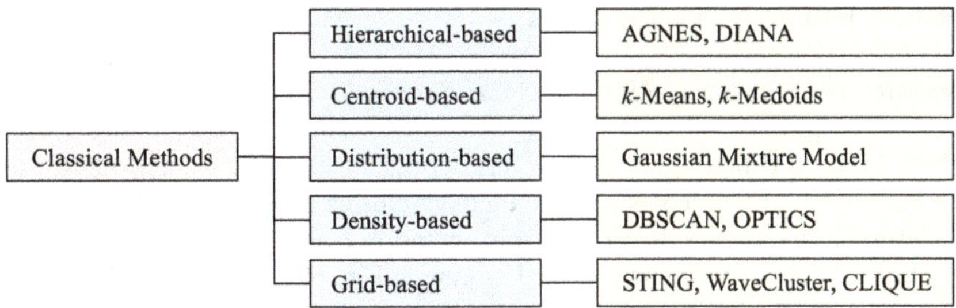

Fig. 14.2 Classical clustering methods and their representative algorithms

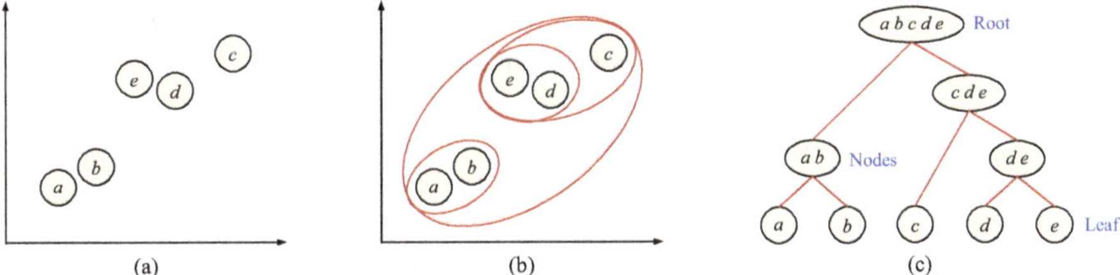

Fig. 14.3 Schematic diagram of hierarchical-based clustering

14.4.1 Hierarchical-Based Clustering

14.4.1.1 Core Idea

The core idea of hierarchical-based clustering is that there is a closer relationship between data points that are closer to each other.

This type of algorithm divides the input data points into several clusters based on the distance between them. It creates a hierarchical decomposition of a set of data points according to some criteria. A cluster is mainly described by connecting the maximum distance between the data points in the cluster. Different distances form different clusters, which can be represented by a dendrogram, this is why it is called "hierarchical-based clustering."

As shown in Fig. 14.3a are several original data points, and (b) is a schematic diagram divided into several clusters according to their distance, which can also be organized into a dendrogram as shown in (c) where the vertical represents the merged clusters and their distances, and the input data points are placed horizontally, so that a clear dendritic hierarchical structure is formed intra-cluster and inter-cluster.

14.4.1.2   Bottom-Up vs. Top-Down

There are mainly two ways of hierarchical-based clustering: one is bottom-up, and the other is top-down.

Bottom-up is called agglomerative clustering; its algorithm is as follows: first, each data point is regarded as a single-element cluster, then the two closest clusters are merged into a larger cluster, and this operation is repeated until all data points become members of a largest cluster. Looking at Fig. 14.3c, bottom-up can be seen as the process of clustering from leaf nodes to root node, and its representative algorithm is AGNES (AGglomerative NESting).

Top-down, on the other hand, is the opposite of bottom-up; it is the clustering process from root node to leaf nodes; therefore, it is called divisive clustering, and its representative algorithm is DIANA (DIvisive ANAlysis).

14.4.1.3   Linkage Criteria

There are several linkage criteria for hierarchical-based clustering, which can be distinguished according to different ways of calculating distance. In addition to the chosen distance function, it is also necessary to determine the partitioning criteria used. This is because a cluster consists of multiple data points, and its distance is calculated from multiple candidate data points.

Let G and G' denote two different clusters; the clustering function $\mathrm{dist}\left(x, x'\right)$ represents the distance between the elements $x \in G$ and $x' \in G'$ in the two clusters. Then, hierarchical clustering has the following three representative linkage criteria:

(1) Single-linkage clustering:
    Calculates the minimum distance between each clustering element:

$$\min \left\{ \mathrm{dist}\left(x, x'\right) \mid x \in G, x' \in G' \right\}. \tag{14.4}$$

(2) Complete-linkage clustering:
    Calculates the maximum distance between the elements of each cluster:

$$\max \left\{ \mathrm{dist}\left(x, x'\right) \mid x \in G, x' \in G' \right\}. \tag{14.5}$$

(3) Average-linkage clustering:
    Calculates the average distance between the elements of each cluster:

$$\frac{1}{|G| \cdot |G'|} \sum_{x \in G} \sum_{x' \in G'} \mathrm{dist}\left(x, x'\right). \tag{14.6}$$

14.4.2 *Centroid-Based Clustering*

Centroid-based clustering is also known as partition-based clustering.

This type of clustering method usually follows these steps: it randomly generates the k initial number of clusters based on the input data and involves finding the centroid of each cluster, then assigning the input data points to the cluster whose centroid is closest, minimizing the distance between the data in the cluster and the centroid. The position of the centroid is then adjusted, and the process is iterated until it converges to the optimal clustering result.

The so-called centroid refers to the average value of the data points in the cluster. Representative centroid-based clustering algorithms are as below:

- k-means:
  Its centroid is the *center* of the cluster, i.e., the geometric center of the cluster, but not necessarily a data point in the cluster.
- k-medoids:
  Its centroid is the *medoid* of the cluster, and is a data point in the cluster, but not necessarily the geometric center of the cluster.

The schematic diagram of a cluster in k-means clustering is depicted in Fig. 14.4a, where the red cross symbol is the center of the cluster. And the schematic diagram of a cluster in k-medoids clustering is depicted in Fig. 14.4b, where the data point behind red cross symbol is the medoid of the cluster.

We will in detail introduce k-means clustering algorithm in Sect. 14.6.1 of this chapter.

14.4.3 *Distribution-Based Clustering*

Distribution-based clustering is the clustering based on the probability distribution of the data. This method is based on the assumption that the data satisfies a certain

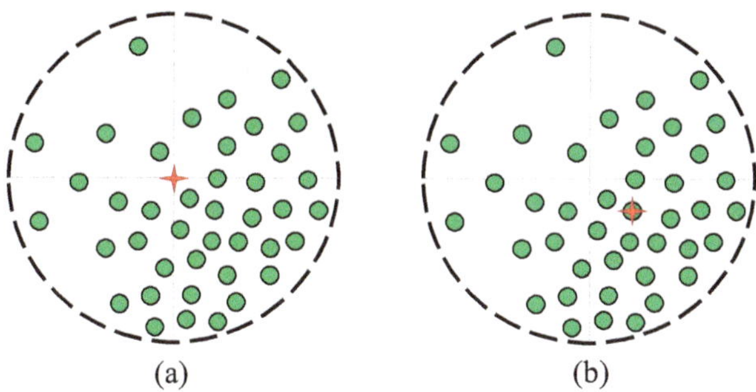

(a)   (b)

Fig. 14.4 Schematic diagrams of k-means clustering vs. k-medoids clustering

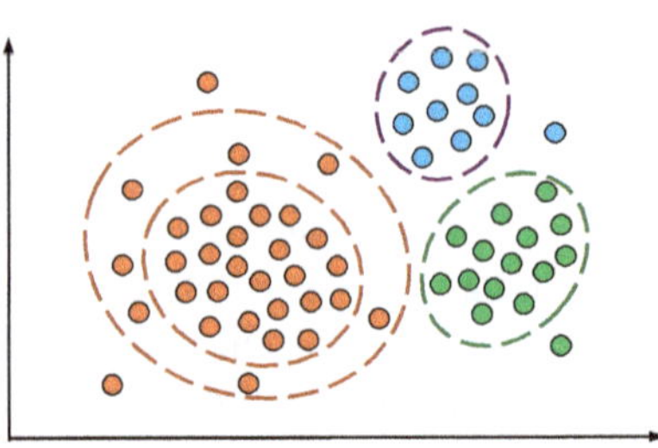

Fig. 14.5 Schematic diagram of Gaussian mixture model clustering

probability distribution during the sampling process, such as normal distribution or binomial distribution.

The Gaussian mixture model (GMM) using the expectation-maximization (EM) algorithm is commonly used in distribution-based clustering.

In the Gaussian mixture model, a fixed number of Gaussian distributions are usually used to model the dataset to avoid overfitting. These Gaussian distributions are randomly initialized, and their parameters are optimized iteratively to better partition the dataset. Multiple iterations may produce different results and will eventually converge to a local optimum. As a result, the data in the same cluster all belong to the same distribution.

A schematic diagram of distribution-based clustering using the Gaussian mixture model is depicted in Fig. 14.5. And we will introduce Gaussian mixture clustering algorithm in Sect. 14.6.2 of this chapter in detail.

If we compare the Gaussian mixture model with the k-means algorithm, both allocate input data to a cluster according to a certain criterion, the difference is that k-means minimizes the square distance of data points in each cluster, while the Gaussian mixture model gives the probability of these data points being assigned to a cluster.

14.4.4   Density-Based Clustering

Density-based clustering accords to the density of the areas where the data is located, and divides each connected high-density data area into corresponding clusters. The scattered low-density data areas distinguish these clusters, and the data points located in these areas are considered noise or outliers.

Density-based clustering can be seen as a nonparametric method because it does not make any assumptions about the number or distribution of clusters. Moreover, due to the local nature of high-density data areas, each high-density data area can have any shape.

Density-based clustering algorithms need to solve two key problems:

(1) How to define the density of data points
(2) How to determine the connectivity of data points

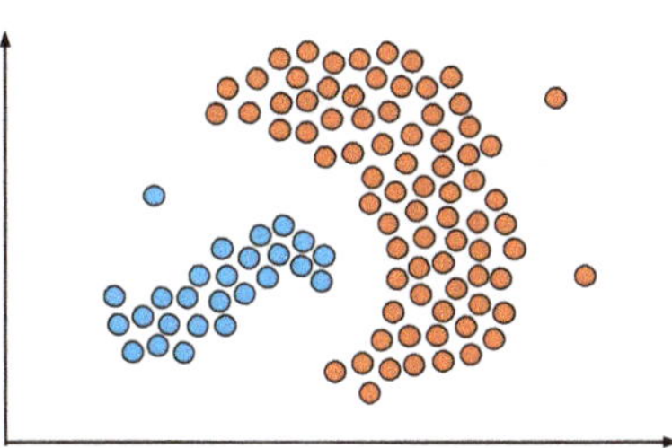

Fig. 14.6 Schematic diagram of density-based clustering

For example, in the DBSCAN (density-based spatial clustering of applications with noise) method based on density clustering, the minimum number of data points in high-density areas is defined by the density threshold, and the maximum radius of high-density areas is determined by the distance threshold, which is called the density-reachability clustering model. Another algorithm called OPTICS (ordering points to identify the clustering structure) is a variant of DBSCAN, which improves the handling of different density clusters (Fig. 14.6).

We will introduce DBSCAN clustering algorithm in Sect. 14.6.3 of this chapter in detail.

14.4.5 Grid-Based Clustering

Grid-based clustering methods are used for multidimensional datasets. This method divides the multidimensional data space into a finite number of cells to form a grid structure, and then forms clusters based on the cells in the grid structure. Each cluster corresponds to a different area, and the data points in it are denser than their surrounding environment.

The biggest advantage of grid-based clustering is that it significantly reduces time complexity, especially for very large datasets. This method does not cluster data points directly but clusters the data point neighborhoods represented by cells. Since the number of cells is significantly smaller than the number of data points, the performance of grid-based clustering is significantly higher than that of methods that directly cluster data points.

Grid-based clustering algorithms usually involve the following five steps: (i) create a grid structure, that is, partition the data space into a finite number of cells; (ii) calculate the density for each cell; (iii) sort the cells by cell density; (iv) identify the center of each cluster; and (v) traverse adjacent cells.

Representative grid-based clustering algorithms include:

- STING:
  Stands for "statistical information grid," a clustering algorithm based on statistical information grid

- WaveCluster:
  Uses wavelet transform for multi-resolution clustering
- CLIQUE:
  Stands for "clustering in ques," for high-dimensional data space clustering based on grid and density

14.5   Neo-classical Methods

Different from classical clustering methods, some researchers use quantum theory, spectral graph theory, metaheuristics, and deep learning to explore new clustering methods. These methods are depicted in Fig. 14.7, where we call them neo-classical clustering methods, and some researchers call them modern clustering methods.

Below we introduce some representative neo-classical clustering methods.

14.5.1   Quantum Clustering

Quantum clustering (QC), also known as quantum theory-based clustering, is a clustering algorithm that uses quantum mechanics concepts and mathematical tools. Its basic idea is to study the distribution of data in scale space based on the distribution of particles in the energy field.

The quantum clustering method was first proposed in 2001. Later, an extended version of the quantum clustering algorithm appeared in 2009, that is, dynamic quantum clustering.

14.5.1.1   Original Quantum Clustering

Given a set of data points in n-dimensional data space, quantum clustering uses a multidimensional Gaussian distribution to represent each data point, with its width (standard deviation) σ, centered at each data point location in the space. Then these

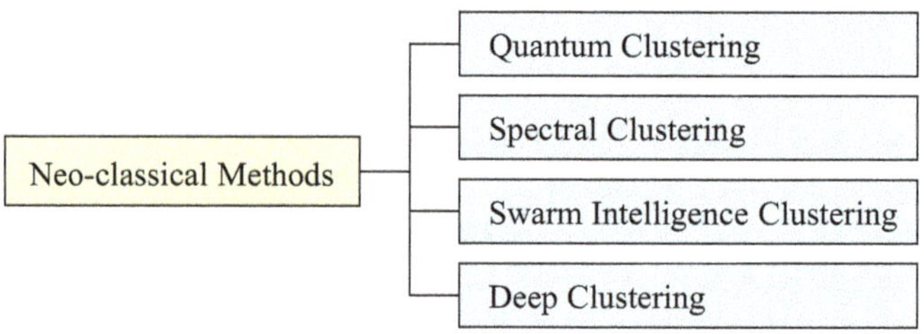

Fig. 14.7 Neo-classical clustering methods

Gaussian coefficients are added together to create a single distribution for the entire dataset. This step is a special case of kernel density estimation, often referred to as the Parzen-Rosenblatt window estimator. This distribution is considered to be the quantum mechanical wave function of the dataset. Simply put, the wave function is a generalized description of the possible locations of data points in space.

Quantum clustering introduces the concept of quantum potential: using the time-independent Schrödinger equation to construct a potential surface and using the wave function of the dataset as a stable solution. Compared to the wave function, the potential surface is more robust to changes in σ (Gaussian width). This feature was a motivation for developing quantum clustering.

The low points in the potential surface correspond to areas of high data density, and then the gradient descent method is used to move each data point in the potential surface downwards, causing the data points to cluster in the nearby minimum area, thus revealing the clusters in the dataset.

14.5.1.2   Dynamic Quantum Clustering

Dynamic quantum clustering (DQC) uses the same potential surface as the original quantum clustering, but replaces the original gradient descent with a quantum evolution method. The time-dependent Schrödinger equation is used to calculate the evolution of each wave function in a given quantum potential over time, establishing a trajectory for each data point in the data space. Repeat the above evolution process until all data points stop moving.

Dynamic quantum evolution is equivalent to a data point moving downwards in the potential surface, acting as a form of non-local gradient descent within the potential surface, thereby avoiding the problem of numerous useless local minima brought about by gradient descent. This method introduces two new hyperparameters: the time step and mass of each data point. However, compared to the original quantum clustering, dynamic quantum evolution increases time complexity.

14.5.2   Spectral Clustering

Spectral clustering, also known as spectral graph theory-based clustering, represents the dataset as a spectral graph, where nodes represent data points and edges represent the similarity between data points. Then, the similarity matrix is extracted from the spectral graph, and the eigenvectors in the matrix are mapped to low-dimensional space for clustering.

By the definition of clustering, the clustering problem is to divide data into several groups called clusters according to certain criteria. Classic clustering methods, such as k-means, are often aimed at convex, compact datasets. However, when the dataset presents a clear non-convex shape, such as a winding spiral, it is

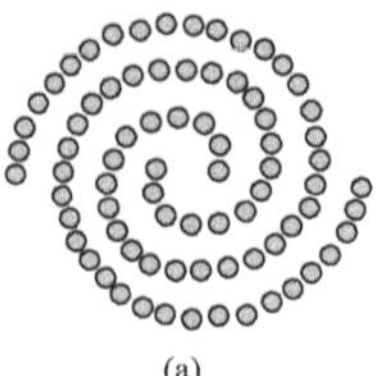 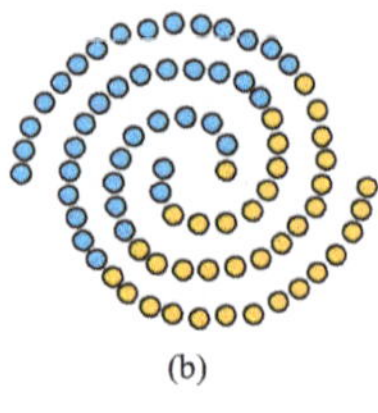 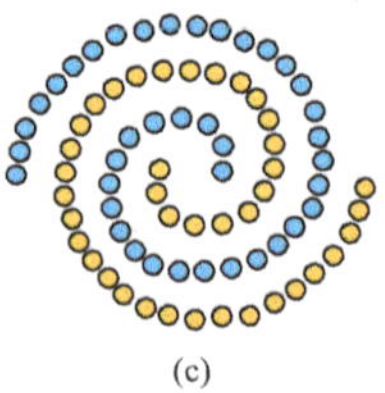

(a)   (b)   (c)

Fig. 14.8 Schematic diagrams of non-convex dataset, k-means clustering, and spectral clustering

necessary to cluster the data based on the connectivity of the data points rather than their compactness. Therefore, clustering methods based on spectral graph theory were born.

There are three schematic diagrams in Fig. 14.8, which are: (a) the raw non-convex dataset, (b) the result of k-means clustering, and (c) the result of spectral clustering.

Since spectral clustering is based on the connectivity of data points in the spectral graph, it is not only suitable for non-convex datasets but also for convex datasets.

14.5.3   Swarm Intelligence Clustering

Swarm intelligence clustering is the clustering analysis using swarm intelligence.

Swarm intelligence[1] refers to a series of artificial intelligence algorithms inspired by nature, especially the biological collective behavior, which can provide sufficiently good solutions for combinatorial optimization problems, especially in the cases that information is incomplete and imperfect or computational resources are limited.

Some natural or biological phenomena are often random, local, decentralized, and self-organizing, but follow simple rules, through the interaction between individuals or the influence of the environment, leading to group optimization.

Swarm intelligence is used for clustering because clustering is regarded as a kind of combinatorial optimization problem, and swarm intelligence is effective in solving complicated combinatorial optimization problems.

The following will introduce some typical swarm intelligence clustering methods.

(1) Clustering based on ant colony optimization (ACO)
     Data are regarded as *ants*:

- The data is distributed randomly on the grid of two dimensions.
- Be selected or not for further operation based on the decision of an ant.
- This process is iterated, until a satisfactory clustering result is got.

---

[1] https://en.wikipedia.org/wiki/Swarm_intelligence

(2) Clustering based on particle swarm optimization (PSO)
    Data are regarded as *particles*:

   - Initial clusters of particles are got by the other clustering algorithm.
   - The clusters are updated continuously based on the centers of clusters.
   - This process is iterated, until a satisfactory clustering result is got.

(3) Clustering based on shuffled frog-leaping algorithm (SFLA)
    To simulate the information interaction of *frogs*:

   - Taking advantage of the local search and the global information interaction.

(4) Clustering based on artificial bee colony (ABC)
    To simulate the foraging behavior of three types of *bee*: queen, drone, and
    worker:

   - The duty is to determine the food source, in a bee population and making use
     of the exchange of local information and global information for clustering.

It is worth noting that metaheuristic clustering algorithms are very suitable for
solving non-convex clustering problems, because they can find global solutions,
while classical clustering algorithms (such as k-means) can only guarantee convergence to local solutions.

14.5.4   Deep Clustering

Deep clustering, also known as deep learning-based clustering, is the use of deep
neural networks for clustering.

Successful methods of deep clustering all follow the same principle, that is,
using deep neural networks for representation learning and then using these
representations as inputs for specific clustering methods (Ren et al. 2022).

Deep clustering has been proposed since the mid-2010s, along with the rapid
development of deep learning. Classical clustering methods often struggle when
dealing with large, high-dimensional, implicit patterns, and complex datasets. Deep
clustering effectively solves the clustering problems of these types of datasets.

Representative deep clustering methods can be divided into the following
categories (Zhou et al. 2022).

(1) Multi-stage deep clustering
    The multi-stage deep clustering is a method with two modules, sequentially
    connected and processed in stages. The two modules respectively complete
    the tasks of representation learning and clustering analysis, that is, using
    deep neural networks for data representation learning and then inputting this
    representation into the classical clustering module to obtain the final clustering
    analysis results.

(2) Iterative deep clustering

In iterative deep clustering, for the two steps of representation learning and clustering analysis, an iterative processing method is used, that is, calculating the clustering results given the representation and updating the representation given the clustering results. Through multiple iterations, good representations are beneficial for clustering, and the clustering results in turn provide supervised samples for representation learning.

(3) Generative deep clustering

The generative deep clustering uses generative models, among which variational autoencoders (VAE) and generative adversarial networks (GAN) are commonly used generative models in generative deep clustering that can capture, represent, and reconstruct data points, propose hypotheses about the potential cluster structure, and then infer the division of clusters by estimating data density.

(4) Simultaneous deep clustering

The simultaneous deep clustering is another method in deep clustering, where the representation learning module and the clustering module are processed simultaneously in an end-to-end manner.

Although most iterative deep clustering methods also optimize the two modules under a single objective, these two modules are optimized in an explicit iterative manner, so they are different from simultaneous deep clustering and cannot be updated simultaneously.

14.6   Typical Algorithms

There are many clustering algorithms; below we introduce some representative clustering algorithms that are k-means, Gaussian mixture clustering, DBSCAN, and density peak-based clustering.

14.6.1   k-Means

The k-means is the most commonly used clustering algorithm, aiming to divide n input data into k clusters, where each data point belongs to the cluster with the nearest mean.

The k-means method originated in the mid-1950s, used for pulse code modulation in signal processing. The term "k-means" began in 1967, as a method of cluster analysis for multivariate observational data.

Below is the explanation to the working principle of k-means clustering algorithm.

Given n input data $D = \{x_i \mid i = 1, \ldots, n\}$, where each data $x_i \in \mathbb{R}^m$ is a m-dimensional real vector, the purpose of the k-means algorithm is to divide the dataset D into k clusters $G = \{G_j \mid j = 1, \ldots, k\}$. Let μ_j be the mean of all data in the jth cluster, also known as the jth cluster center, then its objective function f is:

$$f = \arg\min_{G} \sum_{j=1}^{k} \sum_{x_i \in G_j} \|x_i - \mu_j\|^2. \tag{14.7}$$

Let t denote the number of iterations, and $\mu_1^{(1)}, \mu_2^{(1)}, \ldots, \mu_k^{(1)}$ be the initial (i.e., $t = 1$) means of the k clusters, then the k-means algorithm performs the following operations:

(1) Assignment: Each input data is assigned to a cluster such that the squared Euclidean distance of that cluster is minimized. That is: $\forall j'$, $1 \leq j' \leq k$, then

$$G_j^{(t)} = \left\{ x_i \ \middle| \ \|x_i - \mu_j^{(t)}\|^2 \leq \|x_i - \mu_{j'}^{(t)}\|^2 \right\}. \tag{14.8}$$

(2) Update: Recalculate the mean of the input data assigned to each cluster to obtain the new cluster center. That is:

$$\mu_j^{(t+1)} = \frac{1}{\left|G_j^{(t)}\right|} \sum_{x_i \in G_j^{(t)}} x_i. \tag{14.9}$$

(3) Repeat the above two steps, until the results of assignment and update no longer change, and convergence is reached, and the k-means algorithm ends.

An example of clustering process of k-means algorithm is depicted in Fig. 14.9; the algorithm is as below:

(a) Initial means, in this case $k = 3$, are randomly generated within the data domain.

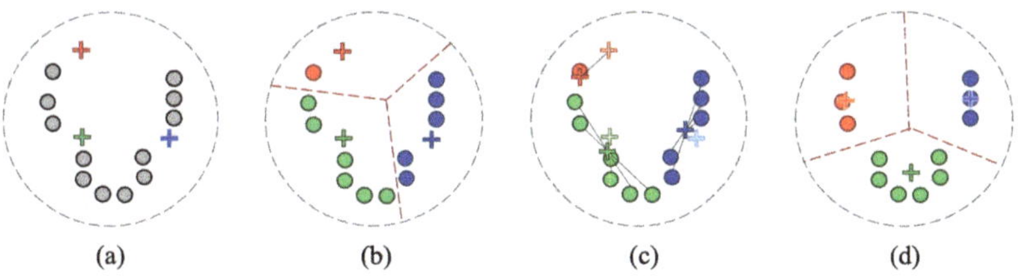

Fig. 14.9 Example of k-means clustering process diagram

(b) The $k = 3$ clusters are created by associating every observation with the nearest mean. The partitions here represent the *Voronoi diagram*[2] generated by the means.

(c) The center of each of the $k = 3$ clusters becomes the new mean.

(d) Repeat steps (b) and (c), until convergence has been reached.

It should be pointed out that the number of clusters k in the k-means algorithm can be preset, or different k values can be tried to get the best clustering results.

The characteristic of the k-means algorithm is its relative effectiveness, with a complexity of $O\,(tkn)$, where n is the number of input data, k is the number of clusters, and t is the number of iterations.

k-means algorithm iterates until convergence, so it is likely to end at a local optimum, which means it cannot guarantee a global optimum. If techniques such as annealing algorithms or genetic algorithms in artificial intelligence are used, a global optimum solution can be obtained.

k-means algorithm belongs to *hard clustering*.

14.6.2　Gaussian Mixture Clustering

Mixture of Gaussians (MoG) is a complex probabilistic model, often referred to as Gaussian mixture model (GMM), which assumes that all data points are generated by a mixture of finite Gaussians with unknown parameters and can classify data into different clusters based on probability distribution.

Classical clustering algorithms, such as k-means, have certain limitations in identifying clusters of different shapes and sizes, and this is where the Gaussian mixture model comes in handy.

The parameters of Gaussian mixture clustering are determined by maximum likelihood, usually using the expectation-maximization (EM) algorithm.

The Gaussian mixture model superimposed by k Gaussian distributions is defined as follows:

$$p(x|\theta) = \sum_{j=1}^{k} \pi_j \mathcal{N}\left(x|\mu_j, \sigma_j\right), \tag{14.10}$$

where each Gaussian distribution $\mathcal{N}\left(x|\mu_j, \sigma_j\right)$ is called a component of the mixture model, and it has its own mean μ_j and variance σ_j, while π_j is referred to as the mixture coefficients, satisfying $0 \le \pi_j \le 1$ and $\sum_{j=1}^{k} \pi_j = 1$.

---

[2] https://en.wikipedia.org/wiki/Voronoi_diagram

Let there be n input data $D = \{x_i \mid i = 1, \ldots, n\}$, using the Gaussian mixture model (GMM) to divide the input dataset D into k clusters $G = \{G_j \mid j = 1, \ldots, k\}$.

Here we introduce a discrete latent variable $z_i \in \{1, \ldots, k\}$, used to specify x_i which cluster G_j it belongs to; the prior probability of this latent variable is $p(z_i = j|\boldsymbol{\theta}) = \pi_j$.

During clustering, first calculate the posterior probability of the input data x_i belonging to the jth cluster G_j, denoted as $p(z_i = j|x_i, \boldsymbol{\theta})$. The following Bayes' theorem can be used for calculation:

$$p(z_i = j|x_i, \boldsymbol{\theta}) = \frac{p(z_i = j|\boldsymbol{\theta})\, p(x_i|z_i = j, \boldsymbol{\theta})}{\sum_{j'=1}^{k} p(z_i = j'|\boldsymbol{\theta}) p(x_i|z_i = j', \boldsymbol{\theta})}. \tag{14.11}$$

The above method is called *soft clustering*.

Let $1 - \max_j p(z_i = j|x_i, \boldsymbol{\theta})$ be the measure of uncertainty. To make it very small, we can use the mean average precision (mAP) to calculate the clustering results, that is:

$$z_i^* = \arg\max_j p(z_i = j|x_i, \boldsymbol{\theta})$$

$$= \arg\max_j (\log p(x_i|z_i = j, \boldsymbol{\theta}) + \log p(z_i = j|\boldsymbol{\theta})). \tag{14.12}$$

The above method is called *hard clustering*.

14.6.3 DBSCAN

DBSCAN (density-based spatial clustering of applications with noise) is the most representative density-based clustering algorithm.

This algorithm was published in 1996 under the title "A Density-Based Algorithm for Discovering Clusters in Large Spatial Databases with Noise" at the annual *ACM SIGKDD International Conference on Knowledge Discovery and Data Mining (KDD)* (Ester et al. 1996).

The DBSCAN algorithm provides the following basic definitions, where D denotes the noise space dataset:

- Two parameters:

  - Density threshold minPts:
    Denotes the minimum number of data points in a high-density area
  - Distance threshold ε:
    Denotes the radius of the high-density area

- Three types of data points:

  - Core points:
    If there is a data point $q \in D$, and $|N_\varepsilon (q)| \geq$ minPts, then q is called a core point.
  - Neighbors: If there is a data point $p \in D$, and $N_\varepsilon (p) : \{p \mid \text{dist}(p, q) \leq \varepsilon\}$, then p is called a neighbor.
  - Outliers:
    The data points are called outliers if they are not core points or neighbors.

- Density-reachable:

  If there is a chain of data points $p_1, \ldots, p_n$, such that $p_1 \rightarrow q$, $p_n \rightarrow p$, and p_{i+1} can be directly density-reachable from p_i, then the data point p is density-reachable from the data point q.

- Density-connected:

  If there exists a data point o, such that data points p and q can be density-reachable from o, then the data point p is density-connected to the data point q.

The above basic definitions are shown in Fig. 14.10, where the density threshold is minPts $= 5$ and distance threshold $\varepsilon = 1$ cm.

The main steps of the DBSCAN algorithm are as follows:

- Retrieve all the neighbors of each point and identify the core points among them.
- Find the data points that are density-connected with the core points on the neighborhood graph, ignoring all non-core points.
- Form the corresponding clusters with all the neighbors of each core point, and treat the outliers as noise points.
- If there are no more density-reachable points for this core point, retrieve the next data point in the dataset.

In summary, DBSCAN can divide the data in high-density areas greater than the density threshold into a cluster and mark the scattered data points in non-high-density areas as noise.

It is worth noting that the DBSCAN algorithm won the *Test of Time Award* at the 2014 *ACM SIGKDD International Conference on Knowledge Discovery and Data Mining*. This award is given by ACM SIGKDD to algorithms in the field of data mining that have received great attention in theory and practice.

Fig. 14.10 Schematic diagram of the basic definition of DBSCAN

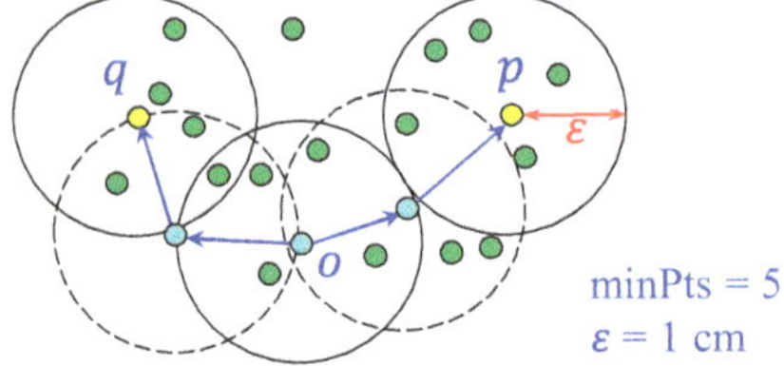

More than 20 years after the publication of the DBSCAN algorithm, in 2017, a paper titled "DBSCAN Revisited, Revisited: Why and How You Should (Still) Use DBSCAN" was published in the *ACM Transactions on Database Systems*, which included all the authors who proposed the DBSCAN algorithm (Schubert et al. 2017).

14.6.4 Density Peak Clustering

One of the main problems of clustering is how to find the centroid of the cluster, that is, the center or medoid.

In the classic clustering algorithms, k-means is to find the center of the cluster, that is, the geometric center of the cluster, but it may not be a data point in the cluster, while the k-medoids is to find the medoid of the cluster, that is, a data point in the cluster, but it may not be the geometric center of the cluster. The commonality of these two algorithms is that they need to find the locally optimal cluster centroid through multiple iterations, but they cannot guarantee global optimality.

Density peak clustering is a new method for quickly finding cluster centers.

In June 2014, a paper titled "Clustering by Fast Search and Find of Density Peaks" (Rodriguez and Laio 2014) was published in the journal *Science*. This paper proposed the following two basic ideas for the selection of cluster centers:

- In terms of density: a cluster center is characterized by a higher density than their neighbors.
- In terms of distance: a cluster center is characterized by a larger distance from points with higher densities.

Let's understand how this method calculates the cluster centers.

For an input dataset $D = \{x_i \mid i = 1, \ldots, n\}$, it defines two variables ρ_i and δ_i, which correspond to the above two basic ideas.

The ρ_i denotes the local density of ith data point, and its calculation equation is:

$$\rho_i = \sum_j \mathbb{I}\left(d_{ij} - d_c\right), \tag{14.13}$$

where $d_{ij} = \text{dist}\left(x_i, x_j\right)$ is the Euclidean distance between data points x_i and x_j, and d_c is the cut-off distance; $\mathbb{I}(\omega)$ is an indicator function, which equals 1 if $\omega < 0$ and 0 otherwise.

Therefore, as defined by Eq. 14.13, ρ_i is equal to the number of data points in the dataset D that are closer to the data point x_i than d_c.

The δ_i is the minimum distance between ith data point and any other data points with higher density, calculated by the following equation:

$$\delta_i = \min_{j:\rho_j > \rho_i}\left(d_{ij}\right). \tag{14.14}$$

For the point with the highest density, it conventionally takes:

$$\delta_i = \min_j \left(d_{ij} \right).$$

Note that δ_i is much larger than the typical nearest neighbor distance only for points that are local or global maxima in the density. Thus, cluster center is recognized as the point for which the value of δ_i is anomalously large.

This paper presents two figures to illustrate their idea that cluster centers are characterized by a higher density than their neighbors and by a relatively large distance from points with higher densities.

This method has the following characteristics:

- It intuitively obtains the number of clusters and their center points.
- It automatically discovers and excludes outliers.
- It does not need to consider the shape and dimensions of the data when identifying clusters.

The above method was validated by cluster analysis on the Olivetti face dataset (Rodriguez and Laio 2014).

14.7   Evaluation Metrics

In supervised learning, labels are known, so the predictions of the supervised learning model can be compared with the labels to evaluate their correctness. However, in unsupervised learning, labels are unknown, which makes it difficult to evaluate the correctness of the unsupervised learning model.

Clustering is a task of unsupervised learning. Some researchers have proposed several methods for evaluating clustering models, mainly divided into the following two types:

(1) Extrinsic evaluation measures:
    That is, given manually labeled clustering benchmarks (ground-truth), the grouping results of the clustering algorithm are compared with them. For situations where clustering benchmarks cannot be given, this evaluation method is not applicable in practice.
(2) Intrinsic evaluation measures:
    That is, without any manually labeled clustering benchmarks, directly evaluate the clustering algorithm and its clustering results. For this, it is necessary to evaluate based on the characteristics of the clustering algorithm and the characteristics of the data.

These two types each contain several clustering evaluation methods, as shown in Table 14.2.

Table 14.2 Measures of extrinsic evaluation vs. intrinsic evaluation for clustering

	Extrinsic evaluation measures	Intrinsic evaluation measures
Evaluation metrics	• Adjusted Rand index	• Calinski-Harabasz index
	• Adjusted mutual information	• Davies-Bouldin index
	• Fowlkes-Mallows index	• Silhouette coefficient
	• V-Measure	

The following introduces these clustering evaluation metrics. Extrinsic evaluation measures come first, followed by intrinsic evaluation measures.

14.7.1 Adjusted Rand Index

The adjusted Rand index (ARI) is a function based on the Rand index, which can be used to measure the similarity between clustering algorithms and clustering benchmarks.

14.7.1.1 Rand Index

The Rand index (RI) originated from a paper published in 1971 titled "Objective Criteria for the Evaluation of Clustering Methods" (Rand 1971).

Given the dataset $D = \{x_i \mid i = 1, \ldots, n\}$, the labeled cluster set of clustering benchmark $U = \{U_k \mid k = 1, \ldots, r\}$, and the predicted cluster set of clustering algorithm $V = \{V_l \mid l = 1, \ldots, s\}$, let

- $a = \left| \hat{D} \right|$, and $\hat{D} = \{\langle x_i, x_{i'} \rangle \mid x_i, x_{i'} \in U_k, x_i, x_{i'} \in V_l\}$.
- $b = \left| \hat{D} \right|$, and $\hat{D} = \{\langle x_i, x_{i'} \rangle \mid x_i \in U_k, x_{i'} \in U_{k'}, x_i \in V_l, x_{i'} \in V_{l'}\}$,

where $\langle x_i, x_{i'} \rangle$ is a pair of data and $i \neq i'$.

The Rand index (RI) is defined as:

$$\mathrm{RI} = \frac{a + b}{\binom{n}{2}}. \tag{14.15}$$

In Eq. 14.15,

$$\binom{n}{2} = C_2^n = \frac{n(n-1)}{2}$$

is the number of data pairs that can be enumerated in n data.

The range of RI is $[0, 1]$, where 0 indicates that the predicted cluster set V is completely different from the labeled cluster set U, and 1 indicates that V is completely identical to U.

Exercise 14.1 (Example of Rand Index Calculation) Given the dataset $D = \{x_1, x_2, x_3, x_4, x_5, x_6\}$, the labeled cluster set U and the predicted cluster set V are as follows:

$$U = \{U_1, U_2, U_3\} = \{(x_1, x_2), (x_3, x_4), (x_5, x_6)\},$$

$$V = \{V_1, V_2\} = \{(x_1, x_2, x_3), (x_4, x_5, x_6)\}.$$

By comparing U and V, we know that:

- For the data pair $\langle x_1, x_2 \rangle$, there exist $x_1, x_2 \in U_1$ and $x_1, x_2 \in V_1$; and for $\langle x_5, x_6 \rangle$, there exist $x_5, x_6 \in U_3$ and $x_5, x_6 \in V_2$, so that $a = 2$.
- For the data pair $\langle x_1, x_4 \rangle$, $\langle x_1, x_5 \rangle$, $\langle x_1, x_6 \rangle$, $\langle x_2, x_4 \rangle$, $\langle x_2, x_5 \rangle$, $\langle x_2, x_6 \rangle$, $\langle x_3, x_5 \rangle$, $\langle x_3, x_6 \rangle$ are not assigned to any clusters in U and V, so that $b = 8$.

Therefore, the Rand index of this example is as follows:

$$\mathrm{RI} = \frac{2 + 8}{\binom{6}{2}} = \frac{10}{15} = 0.67.$$

14.7.1.2   Adjusted Rand Index

Based on the Rand index, a paper titled "Comparing Partitions" (Hubert and Arabic 1985) introduced the adjusted Rand index (ARI) in 1985, which is defined as follows:

$$\mathrm{ARI} = \frac{\mathrm{RI} - \mathbb{E}[\mathrm{RI}]}{\max(\mathrm{RI}) - \mathbb{E}[\mathrm{RI}]}, \tag{14.16}$$

where $\mathbb{E}[\mathrm{RI}]$ is the expected value of RI and $\max(\mathrm{RI})$ is the maximum value of RI.

The range of the adjusted Rand index is $[-1, 1]$; the larger the value, the better the similarity between the clustering algorithm and the clustering benchmark.

14.7.2   Adjusted Mutual Information

Adjusted mutual information (AMI) is also a function that can be used to measure the consistency between the clustering algorithm and the clustering benchmark. The adjusted mutual information originates from the concept of mutual information.

14.7.2.1 Mutual Information

Mutual information (MI) refers to the measure of the interdependence between two random variables. The concept of mutual information is closely related to the entropy of random variables, which is a basic concept in information theory.

Information theory began with Claude Shannon's landmark paper "A Mathematical Theory of Communication" (Shannon 1948), while "mutual information" was named by Robert Fano.

Given a dataset with n data, and the labeled cluster set of clustering benchmarks $U = \{U_k \mid k = 1, \ldots, r\}$, and the predicted cluster set of the clustering algorithm $V = \{V_l \mid l = 1, \ldots, s\}$, then the entropy of U and V is defined as follows:

$$\mathrm{H}(U) = -\sum_{k=1}^{r} P(k) \log(P(k)),$$

$$\mathrm{H}(V) = -\sum_{l=1}^{s} P'(l) \log\big(P'(l)\big), \tag{14.17}$$

where $P(k) = |U_k|/n$ is the probability that randomly selected data belongs to cluster U_k. Similarly, $P(l) = |V_l|/n$ is the probability that randomly selected data belongs to cluster V_l.

The mutual information between U and V can be calculated by the following formula:

$$\mathrm{MI}(U, V) = \sum_{k=1}^{r}\sum_{l=1}^{s} P(k, l) \log\left(\frac{P(k, l)}{P(k)P'(l)}\right), \tag{14.18}$$

where $P(k, l) = |U_k \cap V_l|/n$ is the probability that randomly selected data belongs to cluster U_k and cluster V_l.

$\mathrm{MI}(\cdot)$ is a non-negative value, with entropy $\mathrm{H}(U)$ and $\mathrm{H}(V)$ as the upper bound. It quantifies the information shared by two clusters, so it can be used to measure the similarity of clusters.

14.7.2.2 Normalized Mutual Information

In 2002, in a paper titled "Cluster Ensembles: A Knowledge Reuse Framework for Combining Multiple Partitions" (Strehl and Ghosh 2002), the concept of normalized mutual information (NMI) was proposed. Its definition is as follows:

$$\mathrm{NMI} = \frac{\mathrm{MI}(U, V)}{\mathrm{mean}(\mathrm{H}(U), \mathrm{H}(V))}. \tag{14.19}$$

Regardless of the actual number of "mutual information" between the benchmark clusters, the values of mutual information and normalized information will not be adjusted due to randomness, but will tend to increase with the increase in the number of different benchmark clusters.

The range of normalized mutual information is $[0, 1]$.

14.7.2.3   Adjusted Mutual Information

In 2009, in a paper titled "Information Theoretic Measures for Clusterings Comparison: Is a Correction for Chance Necessary?" (Vinh et al. 2009), a form similar to the adjusted Rand index was adopted, and the adjusted mutual information (AMI) was proposed. Its definition is:

$$\text{AMI} = \frac{\text{MI} - \mathbb{E}[\text{MI}]}{\text{mean}\left(\text{H}\left(U\right), \text{H}\left(V\right)\right) - \mathbb{E}[\text{MI}]}. \tag{14.20}$$

The AMI is equal to 1 if the two clustering algorithms are the same, and AMI is 0 if the MI of the two clustering algorithms is equal to the expected value.

Upon reading this, one might ask a question: the expressions of adjusted Rand index and adjusted mutual information are similar, and both can be used to measure the similarity between the clustering algorithm and the clustering benchmark, but what is the difference?

We can compare the differences from the following three aspects:

- From the temporal view:
  The adjusted Rand index was proposed in 1985, while the adjusted mutual information was published in 2009.
- From the methodological view:
  The adjusted Rand index uses pair-counting method, while adjusted mutual information is based on information theory.
- From the usage scenario:
  The adjusted Rand index will be used when the clustering benchmarks have large clusters of almost same size, and the adjusted mutual information will be used when the clustering benchmarks are unbalanced and there are small clusters (Romano et al. 2016).

14.7.3   Fowlkes-Mallows Index

The Fowlkes-Mallows index (FMI), also known as the Fowlkes-Mallows scores, is used to measure the similarity of two clustering algorithms.

This index can be used to measure the similarity of two hierarchical clustering algorithms, as well as the similarity between two non-hierarchical clustering algorithms and the clustering benchmark. This method was proposed in a paper titled "A Method for Comparing Two Hierarchical Clusterings" published in 1983 (Fowlkes and Mallows 1983).

Given n data and two known as hierarchical clustering trees C and C', the trees C and C' can be cut to produce $k = 2, \ldots, n - 1$ clusters by selecting clusters at a specific height of the tree or setting different levels of clustering strength. Therefore, for each specific value of k, the following confusion matrix can be generated: $M = (m_{i,j})$, where: $i = 1, \ldots, k$ and $j = 1, \ldots, k$.

Therefore, the Fowlkes-Mallows index for a specific value k is defined as follows:

$$\text{FMI}_k = \frac{T_k}{\sqrt{P_k \cdot Q_k}}, \tag{14.21}$$

where

$$T_k = \sum_{i=1}^{k} \sum_{j=1}^{k} m_{i,j}^2 - n,$$

$$P_k = \sum_{i=1}^{k} \left(\sum_{j=1}^{k} m_{i,j} \right)^2 - n,$$

$$Q_k = \sum_{j=1}^{k} \left(\sum_{i=1}^{k} m_{i,j} \right)^2 - n.$$

In summary, by calculating the FMI_k for different values of k, we can judge the similarity of two hierarchical clustering algorithms. The FMI_k can also be generalized to measure the similarity of two clustering algorithms with different numbers of clusters or non-hierarchical clustering algorithms.

There is another method to calculate the Fowlkes-Mallows index, which is to count the number of data point pairs that appear in the same cluster in both clusterings. Let U be the set of labeled clusters in the clustering benchmark, and V be the set of predicted clusters in the clustering algorithm, then the Fowlkes-Mallows index is defined as follows:

$$\text{FMI} = \frac{\text{TP}}{\sqrt{(\text{TP} + \text{FP})(\text{TP} + \text{FN})}}, \tag{14.22}$$

where:

TP: true positive, that is, the number of data point pairs that appear in the same cluster of U and V

FP: false positive, that is, the number of data point pairs that appear in the same cluster of U but not V

FN: false negative, that is, the number of data point pairs that appear in the same
      cluster of V but not U

The range of the Fowlkes-Mallows index is $[0, 1]$; the higher the score, the better
the similarity between the two clusters.

14.7.4   V-Measure

V-Measure was proposed in 2007 by a paper titled "V-Measure: A Conditional
Entropy-Based External Cluster Evaluation Measure" (Rosenberga and Hirschberg
2007), which can be used to measure the similarity between clustering algorithms
and clustering benchmarks.

V-Measure is based on two metrics of cluster assignments, namely:

- Homogeneity:
  Each cluster contains the members of one category only.
- Completeness:
  All members of the same category are assigned to the same cluster.

Given a dataset with n data, the labeled cluster set of the clustering benchmark
$U = \{U_k \mid k = 1, \ldots, r\}$, and the predicted cluster set of the clustering algorithm
$V = \{V_l \mid l = 1, \ldots, s\}$, also let $M = (m_{k,l})$ be a confusion matrix, also known as a
contingency table, generated by U and V.

Therefore, the formal definitions of homogeneity and completeness are as
follows:

$$h = 1 - \frac{H(U|V)}{H(U)},$$

$$c = 1 - \frac{H(V|U)}{H(V)}, \tag{14.23}$$

where $H(U|V)$ and $H(V|U)$ are conditional entropies, defined as follows:

$$H(U|V) = -\sum_{k=1}^{r}\sum_{l=1}^{s} \frac{m_{k,l}}{n} \log\left(\frac{m_{k,l}}{m_{\cdot,l}}\right),$$

$$H(V|U) = -\sum_{l=1}^{s}\sum_{k=1}^{r} \frac{m_{k,l}}{n} \log\left(\frac{m_{k,l}}{m_{k,\cdot}}\right).$$

And the entropies of $H(U)$ and $H(V)$ can be calculated by the following equation:

$$H(U) = -\sum_{k=1}^{r} \frac{m_{k,\cdot}}{n} \log\left(\frac{m_{k,\cdot}}{n}\right),$$

$$H(V) = -\sum_{l=1}^{s} \frac{m_{\cdot,l}}{n} \log\left(\frac{m_{\cdot,l}}{n}\right).$$

In the above equations, $m_{\cdot,l} = \sum_{k=1}^{r} m_{k,l}$, and $m_{k,\cdot} = \sum_{l=1}^{s} m_{k,l}$.

Therefore, the V-measure is defined as the harmonic mean of homogeneity and completeness, that is:

$$V = 2 \cdot \frac{h \cdot c}{h + c}. \tag{14.24}$$

14.7.5 Calinski-Harabasz Index

The Calinski-Harabasz index (CHI), also known as variance ratio criterion (VRC), can be used to evaluate clustering models. This method was proposed in a paper titled "A Dendrite Method for Cluster Analysis" (Caliński and Harabasz 1974) in 1974.

CHI is a measure of cohesion and separation of data points, where cohesion refers to the distance from the data points to the centroid of the cluster, while separation is the distance from the cluster centroid to the global centroid.

Given a dataset $D = \{x_i \mid i = 1, \ldots, n\}$ and the assignment of the n data points to k clusters, the definition of the Calinski-Harabasz index is as follows:

$$\text{CHI} = \left[\frac{B_k}{(k-1)}\right] / \left[\frac{W_k}{(n-k)}\right], \tag{14.25}$$

where B_k is the between-cluster dispersion matrix and W_k is the within-cluster dispersion matrix. Their equations are as follows:

$$B_k = \sum_{j=1}^{k} n_k \|c_k - c\|^2$$

$$W_k = \sum_{j=1}^{k} \sum_{i=1}^{n_k} \|x_i - c_k\|^2,$$

where n_k and c_k are the number of data points and the centroid in the kth cluster, respectively, and c is the centroid of all data points in the dataset. The higher the Calinski-Harabasz index, the better the clustering model.

14.7.6 Davies-Bouldin Index

The Davies-Bouldin index (DBI) was proposed in 1979 by a paper titled "A Cluster Separation Measure" (Davies and Bouldin 1979). This index evaluates the clustering model by calculating the average similarity between clusters.

Given N clusters, let G_j denote the jth cluster, x_i be a data point in G_j, and r_j be the centroid of G_j, then the average distance S_j between each point in G_j and its centroid is:

$$S_j = \left(\frac{1}{|G_j|} \sum_{i=1}^{|G_j|} \|x_i - r_j\|_p^q \right)^{1/q},$$

where if $q = 1$, then S_j is the average Euclidean distance, while the value of p is usually set to 2, making it a Euclidean distance function.

The equation for the deviation rate between clusters G_j and $G_{j'}$ is as follows:

$$M_{j,j'} = \|G_j - G_{j'}\|_p,$$

where p is usually set to 2.

Also, let $R_{j,j'}$ be a measure of the quality of the clustering scheme. By definition, this measure must consider:

- The divergence between the clusters j and j', which ideally should be as large as possible
- The cohesion of the cluster j, which should be as low as possible

The $R_{j,j'}$ that satisfies the above conditions is defined as follows:

$$R_{j,j'} = \frac{S_j + S_{j'}}{M_{j,}}.$$

Therefore, the Davies-Bouldin index (DBI) is:

$$\text{DBI} = \frac{1}{N} \sum_{j=1}^{N} \max_{j \neq j'} R_{j,j'}. \tag{14.26}$$

The lower the Davies-Bouldin index, the better the clustering model.

14.7.7   Silhouette Coefficient

The silhouette refers to a method of interpreting and validating the consistency of data within clusters and provides a concise graphical representation of the clustering situation for each data point. This method was published in 1987 (Rousseeuw 1987).

The silhouette coefficient (SC), also called silhouette score, is used for each data point to measure the cohesion with its cluster, and the separation rate from neighboring cluster. The range of the silhouette coefficient is $[-1, +1]$, a high value indicates that the data point has a high degree of cohesion with its own cluster, and its deviation from the neighboring cluster is not bad. If most data points have high silhouette coefficients, it indicates that the cluster configuration is good; if many data points have low or negative silhouette coefficients, the clustering result is not good.

Let x_i and $x_{i'}$ denote data points, and let G_{own} be the owner cluster and G_{next} be the neighbor cluster.

For the data point $x_i \in G_{\text{own}}$, its average distance to all other data points $x_{i'} \in G_{\text{own}}$ is as follows:

$$a(i) = \frac{1}{|G_{\text{own}}| - 1} \sum_{x_{i'} \in G_{\text{own}}} d(x_i, x_{i'}).$$

The $a(i)$ indicates the cohesion of x_i assigned to G_{own}; the smaller the value, the higher the cohesion.

In addition, for each data point $x_i \in G_{\text{own}}$, its minimum average distance to all data points of any neighbor cluster $x_{i'} \in G_{\text{next}}$ is:

$$b(i) = \min_{G_{\text{own}} \neq G_{\text{next}}} \frac{1}{|G_{\text{next}}|} \sum_{x_{i'} \in G_{\text{next}}} d(x_i, x_{i'}).$$

Therefore, the silhouette coefficient (SC) of the data point x_i is defined as follows:

$$\text{SC}(i) = \frac{b(i) - a(i)}{\max\{a(i), b(i)\}}. \tag{14.27}$$

The range of $\text{SC}(i)$ is $[0, 1]$.

14.8   Application Fields

Clustering has a wide range of applications.

In the field of data mining, clustering is used to analyze data and discover potential patterns.

In the medical field, clustering is used for medical image analysis, medical image segmentation, and three-dimensional reconstruction.

In the business and marketing field, it is used to group customers or shopping goods.

On the World Wide Web, it can perform social network analysis and search result grouping.

In the field of computer science, it can be used for scene segmentation and recommendation system formulation.

Further Reading

1. Olfa Nasraoui, and Chiheb-Eddine Ben N'Cir, *Clustering Methods for Big Data Analytics: Techniques, Toolboxes and Applications.* Springer, 2019.
   [Notes] This book consists of seven chapters that were edited and co-authored by Nasraoui and N'Cir. The chapters in the book include balanced coverage of big data clustering theory, methods, tools, frameworks, applications, representation, visualization, and clustering validation.
2. Satyasai J. Nanda, and Ganapati Panda. "A Survey on Nature Inspired Metaheuristic Algorithms for Partitional Clustering." *Swarm and Evolutionary Computation*, 16, 2014.
   [Notes] This survey paper systematically reviews the main metaheuristic clustering algorithms that have emerged since the 1990s, and also discusses the key issues of various metaheuristic clustering problems and major application areas.
3. Sheng Zhou, et al. "A Comprehensive Survey on Deep Clustering: Taxonomy, Challenges, and Future Directions." *arXiv preprint* arXiv:2206.07579, 2022.
   [Notes] This survey paper, based on a comprehensive survey of various deep clustering methods, proposes a taxonomy of deep clustering and also lists the challenges faced by deep clustering and future research directions.
4. Alex Rodriguez and Alessandro Laio. "Clustering by Fast Search and Find of Density Peaks" . *Science*, 344(6191), 2014.
   [Notes] The paper proposes a new method for determining the center of clusters. Sect. 14.6.4 of this book provides an introduction to it, and interested readers are recommended to read this paper in depth.

References

Caliński, T., and J. Harabasz, (1974). A dendrite method for cluster analysis. *Communications in Statistics-Theory and Methods* 3(1): 1–27.

Davies, D. L., and D. W. Bouldin. (1979). A cluster separation measure. *IEEE Transactions on Pattern Analysis and Machine Intelligence* 1(2): 224–227.

Ester, M., H.-P. Kriegel, J. Sander, and X. Xu. (1996). A density-based algorithm for discovering clusters in large spatial databases with noise. In *International Conference on Knowledge Discovery and Data Mining (KDD)*.

Fowlkes, E., and C. Mallows. (1983). A method for comparing two hierarchical clusterings. *Journal of the American Statistical Association* 78(383): 553–569.

Hubert, L., and P. Arabic, (1985). Comparing partitions. *Journal of Classification* 2: 193–218.

Rand, W. M. (1971). Objective criteria for the evaluation of clustering methods. *Journal of the American Statistical Association* 66(336): 846–850.

Ren, Y., J. Pu, Z. Yang, J. Xu, G. Li, X. Pu, P. S. Yu, and L. He. (2022). Deep clustering: A comprehensive survey. arXiv preprint: arXiv: 2210.04142.

Rodriguez, A., and A. Laio. (2014). Clustering by fast search and find of density peaks. *Science* 344(6191): 1492–1496.

Romano, S., N. X. Vinh, J. Bailey, and K. Verspoor. (2016). Adjusting for chance clustering comparison measures. *Journal of Machine Learning Research* 17: 1–32.

Rosenberg, A., and J. Hirschberg. (2007). V-measure: A conditional entropy-based external cluster evaluation measure. In *Joint Conference on Empirical Methods in Natural Language Processing and Computational Natural Language Learning (EMNLP-CoNLL)*.

Rousseeuw, P. J. (1987). Silhouettes: A graphical aid to the interpretation and validation of cluster analysis. *Journal of Computational and Applied Mathematics* 20: 53–65.

Schubert, E., J. Sander, M. Ester, H.-P. Kriegel, and X. Xu (2017). DBSCAN revisited, revisited: Why and how you should (still) use DBSCAN. *ACM Transactions on Database Systems* 42(3): 1–21.

Shannon, C. E. (1948). A mathematical theory of communication. *The Bell System Technical Journal* 27(3): 379–423.

Strehl, A., and J. Ghosh. (2002).Cluster ensembles: A knowledge reuse framework for combining multiple partitions. *Journal of Machine Learning Research* 3: 583–617.

Vinh, N. X., J. Epps, and J. Bailey. (2009). Information theoretic measures for clusterings comparison: Is a correction for chance necessary. In *International Conference on Machine Learning*.

Zhou, S., H. Xu, Z. Zheng, J. Chen, J. Bu, J. Wu, X. Wang, W. Zhu and M. Ester. (2022). A comprehensive survey on deep clustering: Taxonomy, challenges, and future directions. arXiv preprint arXiv:2206.07579.

Chapter 15
Dimensionality Reduction Task

Abstract Dimensionality reduction is to map high-dimensional data into low-dimensional representation while retaining the intrinsic properties in data. This chapter first elaborates on the problem and definition, the working principle employing formal and illustrated descriptions, and the related elements of dimensionality reduction. Next, we discuss linear dimensionality reduction, nonlinear dimensionality reduction, and deep dimensionality reduction separately, in which the linear dimensionality reduction can be divided to linear subspace learning and multilinear subspace learning. Nonlinear dimensionality reduction includes distance matrix method, kernel method, manifold learning, and neighborhood embedding. Moreover, deep dimensionality reduction is mainly based on deep autoencoders and variational autoencoders. We also expound several typical algorithms for dimensionality reduction, such as principal component analysis (PCA), multidimensional scaling (MDS), isometric feature mapping (Isomap), local linear embeddings (LLE), stochastic neighbor embedding (SNE), and t-distributed stochastic neighbor embedding (t-SNE). Finally, the application fields of dimensionality reduction are briefly introduced.

15.1 Problem and Definition

The dimensionality reduction is also called dimension reduction by some researchers. In this section, we first discuss the dimensionality reduction problem and then give a definition of the dimensionality reduction.

15.1.1 Dimensionality Reduction Problems

15.1.1.1 Dimension

For the term dimension, it is defined differently in different fields. The dimensions involved in dimensionality reduction are mainly the dimensions of mathematical space and data space.

W. Wang, *Principles of Machine Learning*,
https://doi.org/10.1007/978-981-97-5333-8_15

- Dimension of mathematical space:
  It refers to the minimum number of coordinates required for the representation of any point. For example, for Euclidean space in mathematics, the point in one-dimensional space is represented by one coordinate, in two-dimensional space by two coordinates, and in three-dimensional space by three coordinates.
- Dimension of data space:
  It refers to the number of features, also called attributes, in the data, which can be represented by variables or vectors. The data is called high-dimensional data if it is in high-dimensional data space.

15.1.1.2   High-Dimensional Data

Dimensionality reduction is for high-dimensional data.

Definition 15.1 (High-Dimensional Data)
Given a dataset, let n denote the number of its data points and p denote the number of its features (attributes); if n is very large and $p \gg 3$, then the dataset is referred to as high-dimensional data, where 3 is the dimension of Euclidean space.

We are faced with a large amount of high-dimensional data, such as human gene distribution, global climate patterns, stellar spectra, voice signals, digital images and videos, and natural language text, all containing a large amount of high-dimensional data.

Processing high-dimensional data is very difficult; it needs to be denoised through dimensionality reduction, especially for visualization. Dimensionality reduction can also serve as a preprocessing step for other machine learning tasks, used to reduce the number of features in the input data, improve the accuracy of feature learning, and reduce model training time.

By the way, high-dimensional data is different from big data: High-dimensional data refers to datasets with very high dimensions of data features, while big data refers to datasets with very large sizes and hard to process by traditional methods.

15.1.2   Definition

Definition 15.2 (Dimensionality Reduction)
Dimensionality reduction in machine learning is a task that maps the data in high-dimensional space to a low-dimensional space while retaining the basic features of the original high-dimensional data.

Why do we need dimensionality reduction? The main reasons are as follows.

(1) Curse of dimensionality problem:
   As the dimensionality of the data increases, the workload required for data processing or analysis grows exponentially, forming a curse of dimensionality.

(2) Data sparsity problem:
   The representation of high-dimensional data leads to the inflation of useless information, making meaningful features in the data-scarce, with numerous meaningless features.
(3) Intrinsic dimension problem:
   The intrinsic dimension refers to the minimum number of feature vectors required to represent high-dimensional data. This is because there are numerous redundant features and noise in the data.
(4) Data visualization problem:
   Mapping high-dimensional data to a low-dimensional space can represent the complex relationships and patterns in the data in an easy-to-understand visual way.

15.2 Working Principle

For further understanding the task of dimensionality reduction, here we first give a formal description and then explain it through an illustration.

15.2.1 Formal Description

Let $\mathbb{R}^h$ and $\mathbb{R}^l$ denote the sets of high-dimensional and low-dimensional real vectors, respectively, $h \gg l$ and $1 \le l \le 3$. Also, let $\mathcal{X}$ and $\mathcal{Y}$ denote the high-dimensional input space and low-dimensional output space, $\mathcal{X} \subseteq \mathbb{R}^h$ and $\mathcal{Y} \subseteq \mathbb{R}^l$, respectively.

Given m high-dimensional dataset:

$$D_{\text{high}} = \{x_i \mid i = 1, 2, \ldots, m\}. \tag{15.1}$$

Where $x_i \in \mathcal{X}$.

Design a hypothesis set H that maps the data in the high-dimensional input space $\mathcal{X}$ to the low-dimensional output space $\mathcal{Y}$:

$$H : \mathcal{X} \to \mathcal{Y}. \tag{15.2}$$

The goal is to find the best hypothesis $h \in H$ of dimensionality reduction, mapping the high-dimensional dataset $D_{\text{high}} \subseteq X$ to the low-dimensional dataset $D_{\text{low}} \subseteq Y$, i.e.,

$$h : D_{\text{high}} \to D_{\text{low}},$$

while preserving the basic properties of the high-dimensional data.

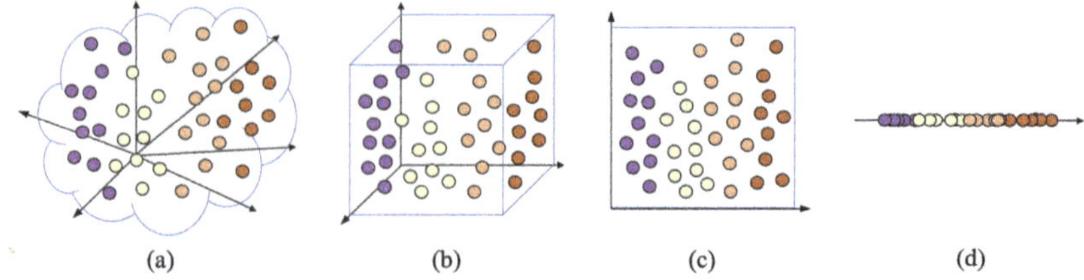

Fig. 15.1   Schematic diagrams of high-dimensional space and low-dimensional space

15.2.2   Illustrated Description

Four schematic diagrams are given in Fig. 15.1

Among them, Fig. 15.1a is a schematic diagram of the high-dimensional space $\mathcal{X} \subseteq \mathbb{R}^h$ before dimensionality reduction, while Fig. 15.1b, c, and d shows schematic diagrams of the three-dimensional space $l = 3$, two-dimensional space $l = 2$, and one-dimensional space $l = 1$ after dimensionality reduction, respectively.

The difference before and after dimensionality reduction is that the dimension of the data space has changed, but the basic properties of high-dimensional data still remain unchanged in the low-dimensional space.

15.3   Related Elements

Dimensionality reduction is the mapping of high-dimensional data to a low-dimensional space. The elements involved include:

- Methods of feature processing:
  That is, feature selection and feature extraction
- Types of dimensionality reduction:
  That is, linear dimensionality reduction, nonlinear dimensionality reduction, and deep dimensionality reduction

15.3.1   Feature Selection and Extraction

Feature selection is to select a subset from the original feature pool. Specifically, it is based on the correlation of the original features of the input data to filter out irrelevant or redundant features, while retaining the feature subset with the highest correlation, but the features themselves do not change. The main techniques of feature selection include filter methods, wrapper methods, and embedded methods.

Feature extraction, also known as feature projection, is to extract useful features from existing data. Specifically, it is to transform the features of the input data and then map them to a new low-dimensional feature space. Some common techniques used for feature extraction are principal component analysis (PCA) and linear discriminant analysis (LDA).

15.3.2   Linear and Nonlinear

Linear dimensionality reduction uses linear mapping to map the data in high-dimensional space input data to a low-dimensional space. Linear dimensionality reduction can be further divided into two subtypes, namely: linear subspace learning and multilinear subspace learning.

Nonlinear dimensionality reduction uses nonlinear mapping to map the data in high-dimensional space to a low-dimensional space. It includes the following subtypes, namely: distance matrix, kernel methods, manifold learning, and neighborhood embedding methods.

15.3.3   Shallow and Deep

Shallow dimensionality reduction refers to traditional methods for dimensionality reduction.

Deep dimensionality reduction uses deep artificial neural networks to map the data from high-dimensional space to a low-dimensional space. The algorithms for deep dimensionality reduction include deep autoencoders and variational autoencoders.

15.3.4   Hierarchical Relationship

Due to the nature of the dimensionality reduction task of mapping high-dimensional data to a low-dimensional space, various algorithms are often divided according to their dimensionality reduction processing methods. Therefore, the subtypes and their hierarchical relationship between linear dimensionality reduction, nonlinear dimensionality reduction, and deep dimensionality reduction are explained here in Fig. 15.2.

The following will provide a detailed introduction to the three types and their subtypes in dimensionality reduction separately.

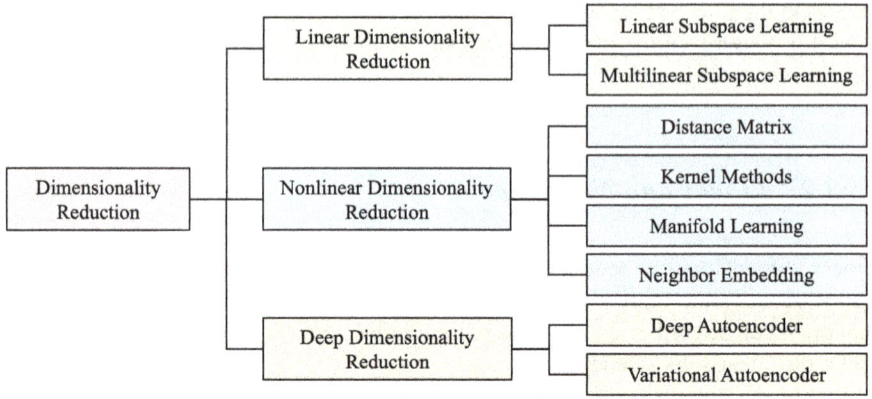

Fig. 15.2 Hierarchical relationship of linear, nonlinear, and deep dimensionality reduction

15.4   Linear Dimensionality Reduction

Linear dimensionality reduction has two subtypes, referred to as linear subspace learning and multilinear subspace learning.

The two subtypes of linear dimensionality reduction and their representative algorithms are depicted in Fig. 15.3.

15.4.1   Linear Subspace Learning

Linear subspace learning is a traditional dimensionality reduction technique that represents the original high-dimensional data as vectors and maps them to a lower-dimensional space using an optimal linear mapping method.

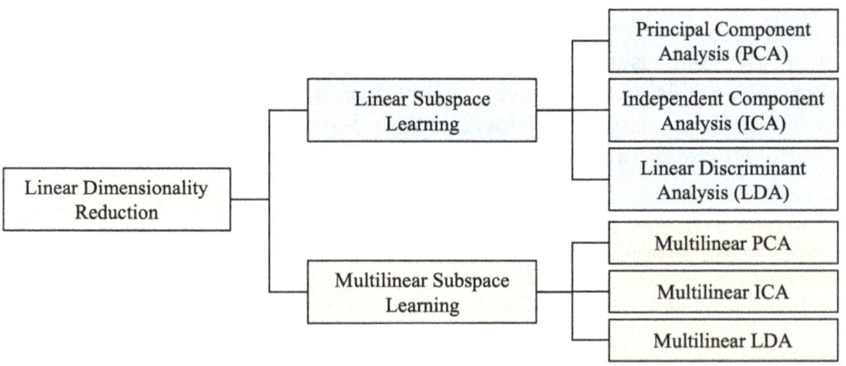

Fig. 15.3 Subtypes of linear dimensionality reduction and their representative algorithms

Typical linear subspace learning algorithms include principal component analysis, independent component analysis, and linear discriminant analysis.

- Principal component analysis (PCA)

  It is a commonly used dimensionality reduction method. It uses an orthogonal transformation to linearly transform high-dimensional data to a new coordinate system, describing it with fewer dimensions. This retains the maximum amount of information while improving the interpretability of the data and achieving visualization of multidimensional data.
- Independent component analysis (ICA)

  It assumes that each high-dimensional data is a mixture of independent components, decomposes it into independent non-Gaussian distributions, and identifies the components with the least correlation with other components. Independent component analysis is a blind source separation method, a classic example of which is the "cocktail party problem," i.e., focusing on listening to a person's speech in a noisy room at a cocktail party.
- Linear discriminant analysis (LDA)

  It is a generalization of Fisher discriminant analysis. It reduces dimensions by finding the projection matrix in the dataset, ensuring that the difference between the same class data is minimized and the difference between different class data is maximized in the low-dimensional space.

15.4.2 *Multilinear Subspace Learning*

When dealing with a large amount of high-dimensional data, linear subspace learning can produce high-dimensional vectors, leading to the need to estimate numerous parameters, thus becoming difficult to apply.

Multilinear subspace learning maps high-dimensional vector spaces to a set of low-dimensional vector spaces.

Multilinear subspace learning uses different types of data tensor analysis tools to reduce dimensions, so it can be used for such high-dimensional data, for example, its metrics are vectorized and organized into tensors or viewed as matrices and connected into tensors.

Multilinear subspace learning is a higher-order generalization of linear subspace learning, and linear subspace learning algorithms can be generalized to multilinear space learning algorithms. Therefore, following algorithms have been proposed.

- Multilinear principal component analysis (MPCA)
- Multilinear independent component analysis (MICA)
- Multilinear linear discriminant analysis (MLDA)

15.5   Nonlinear Dimensionality Reduction

Nonlinear dimensionality reduction uses nonlinear methods to map high-dimensional data to low-dimensional space.

Nonlinear dimensionality reduction mainly has four subtypes, namely: distance matrix, kernel methods, manifold learning, and neighborhood embedding. Each subtype contains several dimensionality reduction algorithms, as shown in Fig. 15.4.

15.5.1   Distance Matrix Method

The distance matrix, also known as the dissimilarity matrix, is a square matrix that contains the pairwise distances between data points in the dataset.

Assume there are n data points $\{x_1, x_2, \ldots, x_n\}$, the distance matrix is $D = [d_{ij}]$, where $d_{ij} = d(x_i, x_j)$, $i, j = 1, 2, \ldots, n$, and if $i = j$ then $d_{ij} = 0$, that is, the elements on the diagonal of the matrix are zero.

A conceptual diagram of a distance matrix is depicted in Fig. 15.5.

Depending on the application involved, it can be divided into metric distance matrices, non-metric distance matrices, and multidimensional scaling.

The characteristic of a metric distance matrix is that each of the non-diagonal elements, $d_{ij} = d(x_i, x_j)$ and $i \neq j$, represents the distance between x_i and x_j, such as the Euclidean distance.

The non-diagonal elements of the non-metric distance matrix, $d_{ij} = d(x_i, x_j)$ and $i \neq j$, are the distance-like information between x_i and x_j, such as the rank order between x_i and x_j.

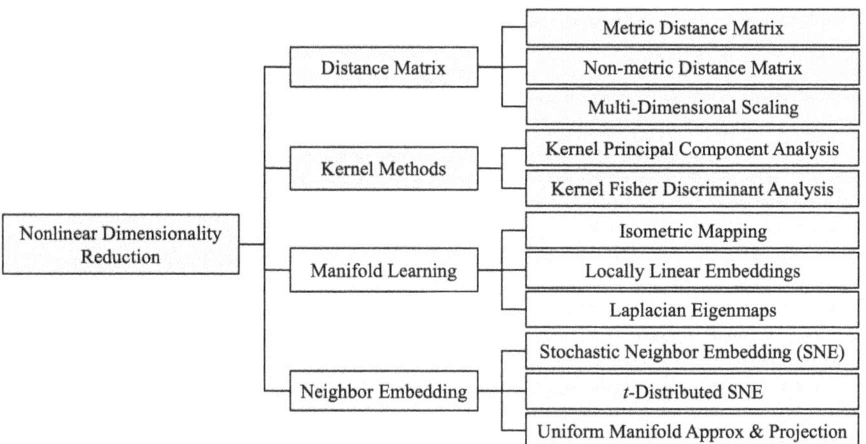

Fig. 15.4 Subtypes of nonlinear dimensionality reduction and their representative algorithms

Fig. 15.5 Conceptual
diagram of a distance matrix

0	x_1	x_2	x_3	$\cdots$	x_n
x_1	0	d_{12}	d_{13}	$\cdots$	d_{1n}
x_2	d_{21}	0	d_{23}	$\cdots$	d_{2n}
x_3	d_{31}	d_{32}	0	$\cdots$	d_{3n}
$\vdots$	$\vdots$	$\vdots$	$\vdots$	$\ddots$	$\vdots$
x_n	d_{n1}	d_{n2}	d_{n3}	$\cdots$	0

Multidimensional scaling (MDS) is a nonlinear dimensionality reduction method based on distance matrix; this will be discussed in detail in Sect. 15.7.2 of this chapter.

15.5.2   Kernel Methods

Kernel methods are a class of algorithms for pattern analysis in machine learning, which have been introduced in Sect. 4.7 of Chap. 4. Kernel methods are also used in dimensionality reduction algorithms.

Using kernel methods, that is, using kernel functions to replace linear functions in linear algorithms, can convert any linear model into a nonlinear model. Using kernel methods for dimensionality reduction is to use kernel function to convert linear dimensionality reduction into nonlinear dimensionality reduction.

The main nonlinear dimensionality reduction algorithms using kernel methods include kernel principal component analysis and kernel Fisher discriminant analysis.

- Kernel principal component analysis (kernel PCA)
  It is an extension of principal component analysis (PCA) based on kernel methods. After using kernel methods, the original linear operations of principal component analysis (PCA) run in the reproducing kernel Hilbert space.
- Kernel Fisher discriminant analysis (kernel FDA)
  It is also known as generalized discriminant analysis and kernel discriminant analysis is an extension of Fisher discriminant analysis (FDA) based on kernel methods.

15.5.3 *Manifold Learning*

Manifold learning is a type of machine learning method for nonlinear dimensionality reduction based on manifold.

Manifold, this term originates from the German term "mannigfaltigkeit" named by German mathematician Bernhard Riemann.

Definition 15.3 (Manifold)
An n-dimensional manifold is a topological space with the property that each point has a neighborhood that is homeomorphic to an open subset of n-dimensional Euclidean space.

Manifolds in mathematics are used to describe geometric bodies and can be equipped with additional structures, resulting in different types of manifolds, such as topological manifolds, differentiable manifolds, Riemannian manifolds, and Finsler manifolds.

Manifold learning is based on the manifold hypothesis. The manifold hypothesis assumes that many high-dimensional datasets in the real world are actually located on low-dimensional latent manifolds in high-dimensional space. Based on this hypothesis, a large amount of data that requires many variables to describe can be described with relatively fewer variables, similar to the local coordinate system of low-dimensional manifolds. This hypothesis becomes the theoretical basis for manifolds in nonlinear dimension reduction.

Therefore, manifold learning is a technique for finding low-dimensional non-linear manifolds in high-dimensional space, and the basic features of this low-dimensional data are consistent with the original features of high-dimensional data.

Representative manifold learning algorithms include isometric mapping, locally linear embedding, Laplacian eigenmaps, local tangent space alignment, inductive manifold learning, and symmetric positive definite manifold. The following is a brief introduction to these algorithms.

- Isometric mapping (Isomap)
  It was proposed in a paper titled "A Global Geometric Framework for Nonlinear Dimension Reduction" published in the journal *Science* in 2000 (Tenenbaum et al. 2000). This algorithm calculates the distance between data points on a high-dimensional manifold, not using the traditional Euclidean distance, but the geodesic distance in differential geometry.
- Locally linear embedding (LLE)
  It was also proposed in another paper published in the journal *Science* in 2000, titled "Nonlinear Dimensionality Reduction by Locally Linear Embeddings" (Roweis and Saul 2000). The feature of this algorithm is to find the best linear reconstruction in a small neighborhood and then embed the data points in a low-dimensional space.
- Laplacian eigenmaps (LE)
  It is a geometrically driven algorithm for high-dimensional data representation based on the correspondence between the graph Laplacian operator, the Laplace-

Beltrami operator on the manifold, and the heat equation. The algorithm was published in 2003, with the paper titled "Laplacian Eigenmaps for Dimensionality Reduction and Data Representation" (Belkin and Niyogi 2003).

- Local tangent space alignment (LTSA)

  It was proposed in a paper titled "Principal Manifolds and Nonlinear Dimensionality Reduction via Local Tangent Space Alignment" in 2005 (Zhang and Zha 2004). This algorithm learns the local geometry of the manifold based on a set of data points sampled from the parameterized manifold, by constructing an approximation of the tangent space at each data point and then aligning these tangent spaces to give the global coordinates of the data points that are relative to the underlying manifold.

- Inductive manifold learning

  It was published in 2014, with the paper titled "Inductive Manifold Learning Using Structured Support Vector Machine" (Kim and Lee 2014). As the title suggests, an inductive manifold learning algorithm is proposed, using structured support vector machines to learn the mapping from high-dimensional data representation to low-dimensional coordinates.

- Symmetric positive definite manifold (SPD manifold)

  It was published in *IEEE Transactions on Pattern Analysis and Machine Intelligence* in 2017, with the title "Dimensionality Reduction on SPD Manifolds: The Emergence of Geometry-Aware Methods" Harandi et al. (2017). This paper introduces an algorithm for dealing with high-dimensional SPD matrices by constructing low-dimensional SPD manifolds for dimensionality reduction.

In the above algorithms, the isometric mapping (Isomap) and local linear embedding (LLE) will be further detailed in Sect. 15.7 of this chapter.

15.5.4 Neighborhood Embedding

Neighborhood embedding is a class of nonlinear dimensionality reduction methods evolved from neighborhood graphs, including stochastic neighbor embedding, t-distributed stochastic neighbor embedding, and uniform manifold approximation and projection. Brief introductions to these three algorithms are as follows.

Stochastic neighbor embedding (SNE) was proposed by Geoffrey Hinton's team in a paper presented at *Conference on Neural Information Processing Systems* (NIPS) in 2002 (Hinton and Roweis 2002). Stochastic neighbor embedding is a probabilistic method that converts the Euclidean distance between data points into the similarity of Gaussian neighborhoods, embedding high-dimensional data into low-dimensional space in a way that preserves their neighborhood identity. Its loss function is the sum of KL divergences, minimized by gradient descent.

The t-distributed stochastic neighbor embedding (t-SNE) was published in 2008 and is a variant of stochastic neighbor embedding (Van der Maaten and Hinton 2008). In the stochastic neighbor embedding (SNE) algorithm, both the high-dimensional input space and the low-dimensional space use Gaussian distributions. This is not a problem for the high-dimensional input space, but when embedding high-dimensional data into low-dimensional space, it is difficult to fit the information of all points in the same neighborhood, resulting in the so-called crowding problem. In response to this, t-SNE uses a t-distribution for the low-dimensional embedding space. The bell curve of this distribution has much larger tails than the Gaussian distribution, making it easier to fit high-dimensional data information in low-dimensional space.

Uniform manifold approximation and projection (UMAP) was proposed in 2018, with the paper titled "UMAP: Uniform Manifold Approximation and Projection for Dimension Reduction" (McInnes et al. 2018). It can also be considered as an algorithm in the category of neighborhood embedding, as it has many similarities with t-SNE in terms of data visualization. But the biggest difference of UMAP lies in three assumptions about the data:

- The data is uniformly distributed on the Riemannian manifold.
- The Riemannian metric is locally constant or approximately locally constant.
- The manifold is locally connected.

15.6   Deep Dimensionality Reduction

Deep dimensionality reduction is a class of dimensionality reduction methods based on deep artificial neural networks.

Representative algorithms for deep dimensionality reduction include deep autoencoders and variational autoencoders, as shown in Fig. 15.6.

15.6.1   Deep Autoencoder

The deep autoencoder was proposed in the paper "Reducing the Dimensionality of Data with Neural Networks" published in the journal *Science* by Geoffrey Hinton et al. in 2006 (Hinton and Salakhutdinov 2006). This was the first method for dimensionality reduction based on deep autoencoders.

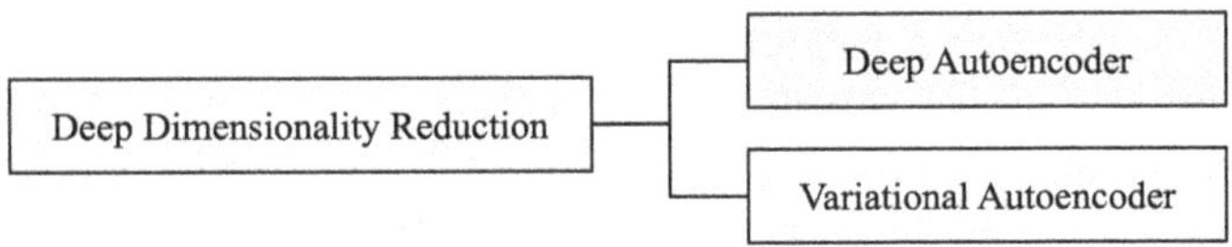

Fig. 15.6 Representative algorithms for deep dimensionality reduction

Section 9.6 of this book provides a detailed introduction to several representative autoencoders. The above deep autoencoder for dimensionality reduction is an undercomplete autoencoder (UAE). In the network architecture of this undercomplete autoencoder, the dimension of the latent space layer is less than the dimensions of the input and output spaces, so that is divided logically into an encoder and a decoder. Therefore, the encoder can convert high-dimensional input data into low-dimensional encoding in the latent space layer, and then the decoder can recover the low-dimensional encoding, thereby achieving the purpose of data dimensionality reduction.

The deep autoencoder can be seen as a nonlinear generalization of principal component analysis (PCA), but it is far superior to classical dimensionality reduction methods such as PCA and can effectively extract low-dimensional features of data.

15.6.2   Variational Autoencoder

The variational autoencoder (VAE) is a generative model based on probabilistic graphical models (Kingma and Welling 2013), because its basic architecture is an encoder-decoder, which can also be used for dimensionality reduction.

The variational autoencoder introduces variational Bayesian methods, and its latent space is composed of probability distributions rather than fixed vectors, so it can be used for both generation and dimensionality reduction, and its dimensionality reduction function is superior to traditional autoencoders.

15.7   Typical Algorithms

There are many dimensionality reduction algorithms, and here we introduce a few representative ones: principal component analysis in linear dimensionality reduction, multidimensional scaling based on distance matrices in nonlinear dimensionality reduction, isometric feature mapping and local linear embedding in manifold learning, and stochastic neighborhood embedding and t-distributed stochastic neighborhood embedding in neighbor embedding. The deep autoencoder and variational autoencoder in deep dimensionality reduction have been introduced in Sect. 9.6 of this book and will not be repeated here.

15.7.1   Principal Component Analysis

Principal component analysis (PCA) is a commonly used linear dimensionality reduction method. This method was invented by British mathematician Karl Pearson in 1901 as a simulation of the principal axis theorem in mechanics. It was independently developed and named by American mathematical statistician Harold Hotelling in the 1930s, to solve the problem of multiple indicators of educational ability.

Ian T. Jolliffe, a professor of statistics at the University of Edinburgh, published *Principal Component Analysis* in 1986, which is the first monograph on principal component analysis. The second edition, which was released in 2002, has been updated and significantly expanded to 488 pages, almost double the size of the first edition, including core content, research progress, and its applications (Jolliffe 2002).

The formal description of principal component analysis is as follows.

Let $\mathbb{R}^h$ and $\mathbb{R}^l (h \gg l)$ denote the sets of high-dimensional and low-dimensional real vectors, respectively, and given a dataset with n data points $D_{\text{high}} = \{x_i \mid i = 1, \ldots, n\} \subseteq \mathbb{R}^h$. The purpose of the principal component analysis is to find a low-dimensional linear subspace $D_{\text{low}} \subseteq \mathbb{R}^l$ and map the high-dimensional dataset D_{high} to the low-dimensional linear subspace D_{low}, so as to form a new coordinate system, where each coordinate axis represents a principal component (PC). These principal components are orthogonal linear transformations of the original high-dimensional data, where the first principal component is the orthogonal axis with the largest variance, and so on.

The kth row $(k = 1, \ldots, n)$ of data points and the linear combination of coefficients are represented as:

$$z_k = \sum_{j=1}^{n} w_{kj} x_j = \mathbf{X} w_k, \qquad (15.3)$$

where $\mathbf{X} = [x_1 x_2 \cdots x_n]$ is a $h \times n$ data matrix and $w_k = [w_{k1} w_{k2} \cdots w_{kn}]^\mathsf{T}$ is a weight vector of length n.

Let U_k be the kth principal component; its maximum variance is:

$$U_k = \max\left(\text{var}\left(z_k\right)\right) = \max\left(\text{var}\left(\mathbf{X} w_k\right)\right) = \max\left(w_k{}^\mathsf{T} \mathbf{S} w_k\right), \qquad (15.4)$$

where $\mathbf{S}$ represents the $n \times n$ covariance matrix.

Obviously, in order to maximize the variance $\text{var}\left(w_k{}^\mathsf{T} \mathbf{S} w_k\right)$, it can be achieved by increasing the value of the vector w_k, while restricting w_k to be a unit-norm vector, that is, $w_k{}^\mathsf{T} w_k = 1$. Therefore, the maximization of U_k can be rewritten as:

$$U_k = \max\left(w_k{}^\mathsf{T} \mathbf{S} w_k\right) = \underset{w_k{}^\mathsf{T} w_k = 1}{\arg\max}\left(w_k{}^\mathsf{T} \mathbf{S} w_k - \lambda_k \left(w_k{}^\mathsf{T} w_k - 1\right)\right) \qquad (15.5)$$

where λ is the Lagrange multiplier. For the vector w_k in the brackets of $\arg\max(\cdot)$, differentiate it and set it to zero, that is:

$$\frac{d}{d w_k}(w_k{}^\mathsf{T} \mathbf{S} w_k - \lambda_k(w_k{}^\mathsf{T} w_k - 1)) = 0.$$

Hence, we have:

$$\mathbf{S}\mathbf{w}_k - \lambda_k \mathbf{w}_k = 0,$$

$$\mathbf{S}\mathbf{w}_k = \lambda_k \mathbf{w}_k, \tag{15.6}$$

where $\mathbf{w}_k$ is the unit norm eigenvector of the covariance matrix $\mathbf{S}$ and λ_k is the corresponding eigenvalue.

Therefore, the kth principal component is:

$$U_k = \max\left(\mathbf{w}_k{}^\mathsf{T}\mathbf{S}\mathbf{w}_k\right) = \max\left(\mathbf{w}_k{}^\mathsf{T}\lambda_k\mathbf{w}_k\right) = \max\left(\lambda_k\mathbf{w}_k{}^\mathsf{T}\mathbf{w}_k\right) = \max\left(\lambda_k\right). \tag{15.7}$$

That is the kth principal component U_k equals the maximum value of the eigenvalue λ_k.

It should be emphasized that: the first principal component is $U_1 = \max\left(\mathrm{var}\left(X\mathbf{w}_1\right)\right)$, while the second principal component is $U_2 = \max\left(\mathrm{var}\left(X\mathbf{w}_2\right)\right)$, but the U_2 is not related to U_1 and so on.

The first two principal components are usually used to draw a two-dimensional data graph for visualizing the clusters of data points after dimension reduction. There are two schematic diagrams in Fig. 15.7 which respectively represent before and after dimension reduction using principal component analysis.

15.7.2 Multidimensional Scaling

Multidimensional scaling (MDS) is a nonlinear dimension reduction method that uses a set of related ranking techniques for information visualization; specifically, it represents high-dimensional data information based on a distance matrix.

The term "scaling" in multidimensional scaling comes from psychometrics, which assigns numbers to some abstract concepts (objects) according to certain rules. A representative example is quantifying customer satisfaction with numbers, where "5" means "very satisfied" and "1" means "very dissatisfied."

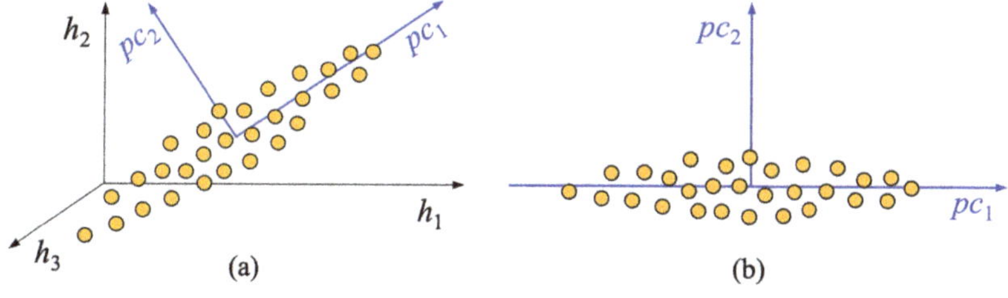

Fig. 15.7 Schematic diagram of principal component analysis

Given a distance matrix, which contains the distances between each pair of data in the dataset, multidimensional scaling selects the dimension l of the low-dimensional space and maps each data to this l-dimensional space while trying to maintain the distances between the data. When dimension l equals 1, 2, and 3, the data points obtained can be visualized on a scatter plot.

This subsection first introduces the types of multidimensional scaling methods and then introduces a commonly used multidimensional scaling method.

15.7.2.1   Types of Multidimensional Scaling

There are mainly four types of multidimensional scaling, depending on the differences in the input matrix.

(1) Classic multidimensional scaling (classic MDS):
     Also known as principal coordinates analysis (PCoA). This method is based on the metric distance matrix, uses the original distances between data points, and tries to preserve these distances in the dimensions after dimension reduction. This method is often used when there is an accurate dissimilarity matrix (such as geographical distance).
(2) Metric multidimensional scaling (metric MDS):
     Similar to classic MDS, it is based on the metric distance matrix and tries to preserve the metric distance as much as possible. The difference is that it uses an iterative algorithm, dealing with ratio data and interval data at the same time.
(3) Non-metric multidimensional scaling (non-metric MDS):
     It focuses on the rank order between data points, but not the actual distance between data points. This method is suitable for non-metric dissimilarities and unknown or nonlinear dissimilarity transformations. It models similar or dissimilar data as distances in geometric space.
(4) Generalized multidimensional scaling (generalized MDS):
     It is an extension of metric multidimensional scaling, and its target space is any smooth non-Euclidean space. When dissimilarity is a distance on a surface and the target space is another surface, generalized MDS can find a minimum distortion embedding from one surface to another. Generalized multidimensional scaling is designed to handle different types of data, such as matrices with missing values.

15.7.2.2   Classic Multidimensional Scaling

Classic multidimensional scaling takes an input matrix, in which each item is the distance between two data points, and outputs a coordinate matrix, whose configuration minimizes the loss function.

For example, given the Euclidean distances between several cities, represented by a matrix $D = [d_{ij}]$, where:

$$d_{ij} = \sqrt{(x_i - x_j)^2 + (y_i - y_j)^2} \tag{15.8}$$

represents the distance between the ith and jth cities, the goal is to find the coordinates between cities.

The loss function in classical multidimensional scaling is called strain, and its expression is as follows:

$$\text{Strain}_D (x_1, x_2, \ldots, x_n) = \left(\frac{\sum_{i,j} (b_{ij} - x_i^\mathsf{T} x_j)^2}{\sum_{i,j} b_{ij}^2} \right)^{1/2}, \tag{15.9}$$

where x_i represents a vector in the one-dimensional space, $x_i^\mathsf{T} x_j$ represents the inner product between x_i and x_j, and b_{ij} is an element in the matrix $\mathbf{B}$.

Multidimensional scaling is based on the fact that the distance matrix can be derived from the eigenvalue decomposition of $\mathbf{B} = \mathbf{X}\mathbf{X}'$, and the matrix $\mathbf{B}$ can be calculated from the distance matrix $\mathbf{D}$ through double centering (Wickelmaier 2003).

Assume there are four cities City_1, City_2, City_3 and City_4, and the distances between these four cities have been measured; the example matrix $\mathbf{D}$ is shown in Fig. 15.8a.

The square matrix of the distance matrix $\mathbf{D}^{(2)}$ is:

$$\mathbf{D}^{(2)} = \begin{bmatrix} 0 & 8649 & 6724 & 17689 \\ 8649 & 0 & 2704 & 3600 \\ 6724 & 2704 & 0 & 12321 \\ 17689 & 3600 & 12321 & 0 \end{bmatrix}.$$

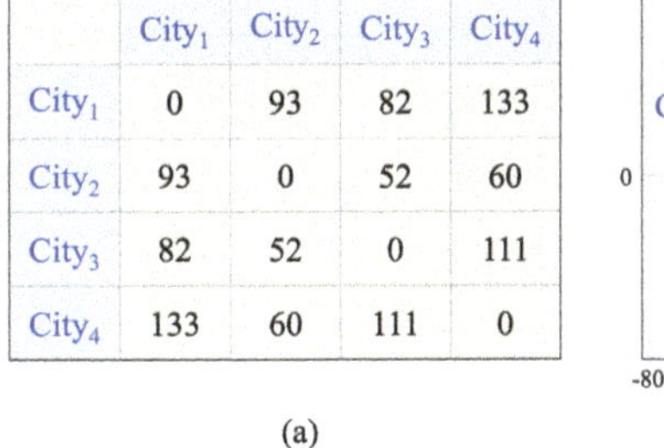

	City_1	City_2	City_3	City_4
City_1	0	93	82	133
City_2	93	0	52	60
City_3	82	52	0	111
City_4	133	60	111	0

(a)

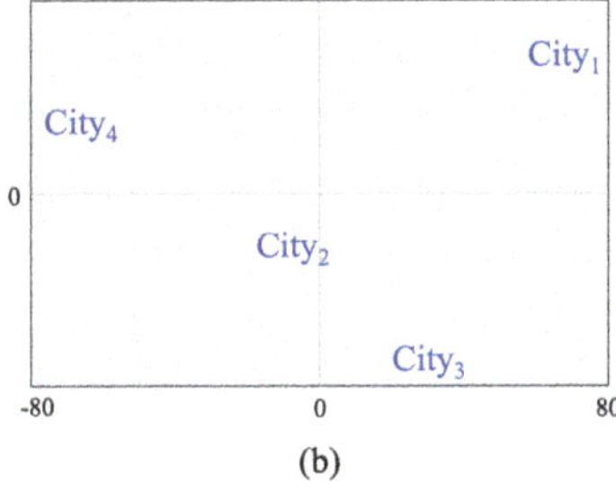

(b)

Fig. 15.8 Example of classical multidimensional scaling

Since there are $n = 4$ cities, the matrix calculation is as follows:

$$\mathbf{J} = \begin{bmatrix} 1 & 0 & 0 & 0 \\ 0 & 1 & 0 & 0 \\ 0 & 0 & 1 & 0 \\ 0 & 0 & 0 & 1 \end{bmatrix} - \frac{1}{4} \times \begin{bmatrix} 1 & 1 & 1 & 1 \\ 1 & 1 & 1 & 1 \\ 1 & 1 & 1 & 1 \\ 1 & 1 & 1 & 1 \end{bmatrix} = \begin{bmatrix} 0.75 & -0.25 & -0.25 & -0.25 \\ -0.25 & 0.75 & -0.25 & -0.25 \\ -0.25 & -0.25 & 0.75 & -0.25 \\ -0.25 & -0.25 & -0.25 & 0.75 \end{bmatrix}.$$

Applying $\mathbf{J}$ to $\mathbf{D}^{(2)}$ results in the double-centered matrix B:

$$\mathbf{B} = -\frac{1}{2}\mathbf{J}\mathbf{D}^{(2)}\mathbf{J} = \begin{bmatrix} 5035.0625 & -1553.0625 & 258.9375 & -3740.938 \\ -1553.0625 & 507.8125 & 5.3125 & 1039.938 \\ 258.9375 & 5.3125 & 2206.8125 & -2471.062 \\ -3740.9375 & 1039.9375 & -2471.0625 & 5172.062 \end{bmatrix}.$$

For the two-dimensional representation of the four cities, it is necessary to extract the first two largest eigenvalues and corresponding eigenvectors from matrix $\mathbf{B}$:

$$\lambda_1 = 9724.168, \quad \lambda_2 = 3160.986,$$

$$e_1 = \begin{pmatrix} -0.637 \\ 0.187 \\ -0.253 \\ 0.704 \end{pmatrix}, \quad e_2 = \begin{pmatrix} -0.586 \\ 0.214 \\ 0.706 \\ -0.334 \end{pmatrix}.$$

Finally, the coordinates of these cities are obtained by multiplying the eigenvectors with the eigenvalues:

$$\mathbf{X} = \begin{bmatrix} -0.637 & -0.586 \\ 0.187 & 0.214 \\ -0.253 & 0.706 \\ 0.704 & -0.334 \end{bmatrix} \begin{bmatrix} \sqrt{9724.168} & 0 \\ 0 & \sqrt{3160.986} \end{bmatrix} = \begin{bmatrix} -62.831 & -32.97448 \\ 18.403 & 12.02697 \\ -24.960 & 39.71091 \\ 69.388 & -18.76340 \end{bmatrix}.$$

Figure 15.8b is a graphical representation of the four cities obtained through the above multidimensional scaling method. This figure is calculated from the distances between the city points and therefore cannot represent information about the cardinal directions.

15.7.3   Isometric Feature Mapping

Isometric feature mapping (Isomap) is one of the representative algorithms of manifold learning.

This algorithm was published in a paper titled "A Global Geometric Framework for Nonlinear Dimensionality Reduction" in *Science* magazine in December 2000 (Tenenbaum et al. 2000). Among the three authors of this paper, Joshua Tenenbaum and Vin Silva respectively come from the department of psychology and the department of mathematics at Stanford University, while John Langford comes from the department of mathematics at Carnegie Mellon University.

The paper uses the method of manifold learning to learn and obtain a global geometric framework from local metric information to solve the problem of nonlinear dimensionality reduction.

The innovation of this paper is that it does not use the traditional Euclidean distance when calculating the distance between data points on the high-dimensional manifold, but uses the geodesic distances in differential geometry.

Isometric mapping, as the name suggests, is to map the geodesic distances between the original data that appear as nonlinear manifolds isometrically to the distances on the two-dimensional linear manifold.

The Isomap algorithm is explained using a *Swiss roll* dataset to illustrate how it uses geodesic distances for nonlinear dimensionality reduction.

15.7.4   Local Linear Embedding

Local linear embedding (LLE) is another one of the representative algorithms of manifold learning.

This algorithm was also published in another paper in *Science* magazine in December 2000, titled "Nonlinear Dimensionality Reduction by Locally Linear Embedding" (Roweis and Saul 2000). The two authors are Sam Roweis from University College London and Lawrence Saul from AT&T Labs.

The local linear embedding proposed in this paper is an unsupervised learning algorithm that calculates the low-dimensional, neighborhood-preserving embedding contained in high-dimensional input data.

The algorithm finds the best linear reconstruction in small neighborhoods and then embeds the data points in a low-dimensional space. It calculates the reconstruction weights of each point and then minimizes the embedding cost by calculating eigenvalues.

The LLE algorithm can discover the global internal coordinates of the manifold without explicitly representing how the data should be embedded in two-dimensional graphics.

15.7.5   Stochastic Neighbor Embedding

Stochastic neighbor embedding (SNE) was proposed by Geoffrey Hinton and Sam Roweis of the Department of Computer Science at the University of Toronto, in a paper published at the *Conference on Neural Information Processing Systems* (NIPS) in 2002 (Hinton and Roweis 2002).

Let $D_{\text{high}} = \{x_i \mid i = 1, 2, \ldots, m\} \subseteq \mathbb{R}^h$ and $D_{\text{low}} = \{y_j \mid j = 1, 2, \ldots, m\} \subseteq \mathbb{R}^l$ denote the high-dimensional input data and low-dimensional output data, respectively.

In the high-dimensional input space, stochastic neighbor embedding takes a probabilistic method based on Gaussian distribution. For the data point x and its mean $\mu = 0$, the probability density function of its Gaussian distribution is:

$$p(x) = \frac{1}{\sqrt{2\pi\sigma^2}} \exp\left(-\frac{(x-\mu)^2}{2\sigma^2}\right) = \frac{1}{\sqrt{2\pi\sigma^2}} \exp\left(-\frac{x^2}{2\sigma^2}\right).$$

The first step of SNE is to convert the Euclidean distance between high-dimensional data points into a conditional probability representing their similarity in the high-dimensional space. For each Gaussian distribution centered on $x_i \in D_{\text{high}}$, select its neighborhood data point $x_j \in D_{\text{high}}$ according to the proportion of probability density, then the asymmetric conditional probability $p_{j|i}$ is used as the similarity between x_i and x_j. The value of $p_{j|i}$ is relatively high for the data points near x_i; otherwise the value of $p_{j|i}$ will be very small. The calculation equation for the similar conditional probability $p_{j|i}$ in high-dimensional space is as follows:

$$p_{j|i} = \frac{\exp\left(-\frac{\|x_i - x_j\|^2}{2\sigma_i^2}\right)}{\sum_{k \neq i} \exp\left(-\frac{\|x_i - x_k\|^2}{2\sigma_i^2}\right)}, \tag{15.10}$$

where $p_{i|i} = 0$ and σ_i^2 is the variance of the data point x_i which can be manually set or found by a binary search for the value that makes the entropy of the distribution over neighbors equal to $\log k$. Here, the k is the effective number of local neighborhood data points which is manually selected.

In the low-dimensional space, the Gaussian neighborhood method is also used to calculate the similar conditional probability $q_{j|i}$ of the low-dimensional data points $y_i \in D_{\text{low}}$ and $y_j \in D_{\text{low}}$ corresponding to the high-dimensional data points x_i and x_j; the difference is that a fixed variance value is used. Without loss of generality, let $\sigma_i^2 = \frac{1}{2}$, then the calculation equation of the similar conditional probability in the low-dimensional space is as follows:

$$q_{j|i} = \frac{\exp\left(-\|y_i - y_j\|^2\right)}{\sum_{k \neq i} \exp\left(-\|y_i - y_k\|^2\right)}. \tag{15.11}$$

Similarly, $q_{i|i} = 0$.

A schematic diagram of stochastic neighbor embedding is depicted in Fig. 15.9, in which Fig. 15.9a represents the similarity calculation between the data points

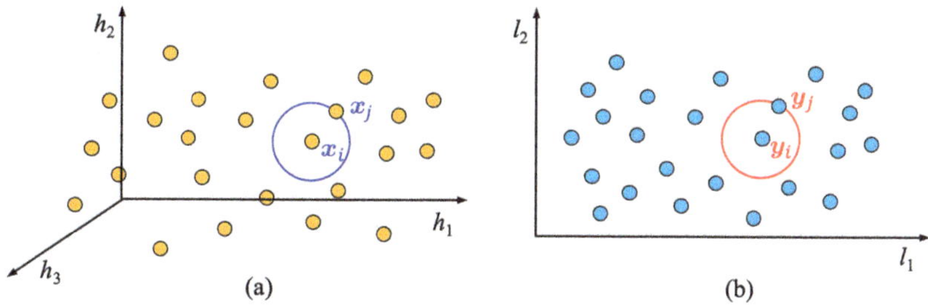

Fig. 15.9 Schematic diagram of stochastic neighbor embedding (SNE)

x_i and x_j in the high-dimensional space and Fig. 15.9b represents the similarity calculation between the corresponding data points y_i and y_j in the low-dimensional space.

In order to maintain the same similarity between the data pairs corresponding to the high-dimensional and low-dimensional spaces, that is, to make the data pairs corresponding to the two spaces maintain the same conditional probability as much as possible, let its loss function C equal to the sum of KL divergence between the conditional probability distribution of each data point neighborhood in the high-dimensional and low-dimensional spaces $p_{j|i}$ and $q_{j|i}$:

$$C = \sum_i D_{KL}\left(P_i \| Q_i\right) = \sum_i \sum_j p_{j|i} \log \frac{p_{j|i}}{q_{j|i}}. \tag{15.12}$$

The loss function C is minimized by using gradient descent, and then we have:

$$\frac{\partial C}{\partial y_i} = 2 \sum_j (p_{j|i} - q_{j|i} + p_{i|j} - q_{i|j})(y_i - y_j). \tag{15.13}$$

15.7.6 *t-Distributed Stochastic Neighbor Embedding*

The t-distributed stochastic neighbor embedding (t-SNE) is an improved version of stochastic neighbor embedding (SNE). This method was published in the *Journal of Machine Learning Research* in 2008, with the title "Visualizing Data using t-SNE." The first author of the paper, Laurens van der Maaten, at the time was a Ph.D. student from Tilburg University in the Netherlands, and the second author was Geoffrey Hinton from the University of Toronto in Canada (Van der Maaten and Hinton 2008).

Although stochastic neighbor embedding (SNE) has reasonably good visualizations, it uses Gaussian distribution for both the input high-dimensional space and the embedded low-dimensional space. This is not a problem for high-dimensional space, but when embedding high-dimensional data into low-dimensional space, it is difficult to fit the information of all points in the same neighborhood area—that is the so-called crowding problem. In addition, the loss function of SNE is also difficult to optimize.

To address the above issues, t-SNE has made the following important improvements:

(1) In the high-dimensional space, it uses symmetric stochastic neighbor embedding based on Gaussian distribution, making its loss function have a simpler gradient.
(2) In the low-dimensional space, it uses the t-distribution to calculate the similarity between two data points.

t-SNE uses a symmetric conditional probability based on Gaussian distribution in high-dimensional space, that is, the joint probability distribution p_{ij}:

$$p_{ij} = \frac{p_{j|i} + p_{i|j}}{2m}. \tag{15.14}$$

This ensures that for all data points x_i, $\sum_j p_{ij} > \frac{1}{2m}$. As a result, even if x_i is an outlier, it will still make a significant contribution to the loss function.

To address the crowding problem in low-dimensional space, the similarity between pairs of data points is calculated based on the t-distribution, rather than the Gaussian distribution used by SNE.

The full name of the t-distribution is Student's t-distribution. This distribution belongs to the heavy-tailed distribution, that is, the distribution at the tail ends of its bell-shaped curve is much larger than the Gaussian distribution, so it is convenient to fit high-dimensional data in low-dimensional space and can effectively solve the crowding problem.

By the way, the Student's t-distribution was proposed by British statistician William S. Gosset in a paper in 1908, he used "Student" as a pseudonym for this paper, and then the t-distribution was named under his pseudonym.[1]

In low-dimensional space, t-SNE employs a Student's t-distribution with one degree of freedom, and its joint probability distribution q_{ij} is defined as follows:

$$q_{ij} = \frac{\left(1 + \|y_i - y_j\|^2\right)^{-1}}{\sum_{k \neq l} \left(1 + \|y_i - y_j\|^2\right)^{-1}}. \tag{15.15}$$

A Student's t-distribution with one degree of freedom is equivalent to a standard Cauchy distribution.[2] This distribution has very superior characteristics, that is: for the distance between data pairs in low-dimensional mapping $\|y_i - y_j\|$, the expression $\left(1 + \|y_i - y_j\|^2\right)^{-1}$ is inversely proportional to its square. This makes

---

[1] https://en.wikipedia.org/wiki/Student's_t-distribution

[2] https://en.wikipedia.org/wiki/Cauchy_distribution

the map's representation of the joint probability (almost) unaffected by the ratio changes of data pairs that are far apart. It also means that the processing of distant data point clusters is the same as that of individual data points, so optimization can be carried out in the same way except for the smallest scale. The theoretical basis for choosing Student's t-distribution is that it is closely related to the Gaussian distribution, because Student's t-distribution is an infinite mixture of Gaussian distributions. A convenient feature for calculation is that the density of points under the Student's t-distribution is much faster than under the Gaussian distribution because it does not involve exponential operations.

The loss function of t-distribution stochastic neighbor embedding C equals the sum of the KL divergence between the joint probability distribution in high-dimensional space P and the joint probability distribution in low-dimensional space Q:

$$C = D_{KL}(P||Q) = \sum_i \sum_j p_{ij} \log \frac{p_{ij}}{q_{ij}}. \tag{15.16}$$

Minimize the above loss function C using the gradient descent method:

$$\frac{\partial C}{\partial y_i} = 4 \sum_j (p_{ij} - q_{ij})(y_i - y_j)(1 + \|y_i - y_j\|^2)^{-1}. \tag{15.17}$$

To facilitate further understanding of the evolution from stochastic neighbor embedding (SNE) to t-distributed stochastic neighbor embedding (t-SNE), we list several key equations in SNE and t-SNE in Table 15.1.

Three researchers from Google Brain and Google Cloud published an article titled "How to Use t-SNE Effectively" on *Distill.pub*, which vividly introduces the use of t-distributed stochastic neighbor embedding and provides some examples (Wattenberg et al. 2016).

Table 15.1 Comparison between stochastic neighbor embedding (SNE) and t-distributed stochastic neighbor embedding (t-SNE)

SNE	t-SNE
$p_{j\|i} = \dfrac{\exp\left(-\dfrac{\|x_i - x_j\|^2}{2\sigma_i^2}\right)}{\sum_{k \neq i} \exp\left(-\dfrac{\|x_i - x_k\|^2}{2\sigma_i^2}\right)}$	$p_{ij} = \dfrac{p_{j\|i} + p_{i\|j}}{2m}$
$q_{j\|i} = \dfrac{\exp(-\|y_i - y_j\|^2)}{\sum_{k \neq i} \exp(-\|y_i - y_k\|^2)}$	$q_{ij} = \dfrac{(1 + \|y_i - y_j\|^2)^{-1}}{\sum_{k \neq l} (1 + \|y_i - y_j\|^2)^{-1}}$
$C = \sum_i D_{KL}(P_i\|Q_i) = \sum_i \sum_j p_{j\|i} \log \frac{p_{j\|i}}{q_{j\|i}}$	$C = D_{KL}(P\|Q) = \sum_i \sum_j p_{ij} \log \frac{p_{ij}}{q_{ij}}$
$\frac{\partial C}{\partial y_i} = 2 \sum_j (p_{j\|i} - q_{j\|i} + p_{i\|j} - q_{i\|j})(y_i - y_j)$	$\frac{\partial C}{\partial y_i} = 4 \sum_j (p_{ij} - q_{ij})(y_i - y_j)(1 + \|y_i - y_j\|^2)^{-1}$

15.8 Application Fields

Dimensionality reduction is often used as a preprocessing step in machine learning. It can effectively eliminate irrelevant and redundant features in high-dimensional data, improve the efficiency of data analysis, enhance the performance of prediction, make the analysis results easy to visualize, and so on.

The application fields of dimensionality reduction are also very wide and can be used in image processing, face recognition, handwriting recognition, text classification, gene expression profiling, genetic chromosome engineering, etc. In addition, dimensionality reduction is also used in computational biology and proteomics.

Further Reading

1. Ian T. Jolliffe. *Principal Component Analysis*, Second Edition, Springer New York, 2002.
   [Notes] The first edition of the book was published in 1986 and is the first comprehensive book on principal component analysis. The second edition in 2002 was updated and significantly expanded, almost twice the length of the first edition, including core content, research progress, and its applications. The book is considered a classical one of principal component analysis.
2. Joshua Tenenbaum, Vin Silva, and John Langford. "A Global Geometric Framework for Nonlinear Dimensionality Reduction". *Science*, 290(5500), 2000.
   [Notes] This paper uses manifold learning to learn and obtain a global geometric framework from local metric information to solve the problem of nonlinear dimensionality reduction. Its algorithm is called Isomap, which is an abbreviation for Isometric Feature Mapping.
3. Sam Roweis, and Lawrence Saul. "Nonlinear Dimensionality Reduction by Locally Linear Embedding". *Science*, 290(5500), 2000.
   [Notes] The locally linear embedding proposed in the paper is an unsupervised learning algorithm that calculates the low-dimensional, neighborhood-preserving embedding contained in high-dimensional input data.
4. Geoffrey E. Hinton, and Sam Roweis. "Stochastic neighbor embedding." *Conference on Neural Information Processing Systems* (NIPS), 2002.
   [Notes] This is the first paper on stochastic neighbor embedding (SNE) that proposed a probabilistic approach to the task of dimensionality reduction through high-dimensional data points or by pairwise dissimilarities, in a low-dimensional space in a way that preserves neighbor identities.
5. Laurens Van der Maaten, and Geoffrey Hinton. "Visualizing data using t-SNE." *Journal of Machine Learning Research*, 9(11), 2008.

[Notes] This paper proposed an improved version of SNE using Student's t-distribution and is a more effective dimensionality reduction and visualization method.

6. Martin Wattenberg, Fernanda Viégas, and Ian Johnson. "How to Use t-SNE Effectively." *https://distill.pub/2016/misread-tsne/*.

[Notes] This article, written by three researchers from Google Brain and Google Cloud and published on *Distill.pub*, vividly introduces how to effectively use t-SNE, and provides some examples.

7. Geoffrey Hinton, and Ruslan Salakhutdinov. "Reducing the Dimensionality of Data with Neural Networks". *Science*, 313(5786), 2006.

[Notes] This paper proposed a dimensionality reduction method based on deep autoencoders. It can effectively extract low-dimensional features of data and is far superior to classical dimensionality reduction methods such as principal component analysis (PCA).

References

Belkin, M., and P. Niyogi. (2003). Laplacian eigenmaps for dimensionality reduction and data representation. *Neural Computation* 15(6): 1373–1396.

Harandi, M., M. Salzmann, and R. Hartley. (2017). Dimensionality reduction on SPD manifolds: The emergence of geometry-aware methods. *IEEE Transactions on Pattern Analysis and Machine Intelligence* 40(1): 48–62.

Hinton, G. E., and S. Roweis. (2002). Stochastic neighbor embedding. In *Conference on Neural Information Processing Systems (NIPS)*.

Hinton, G. E., and R. R. Salakhutdinov. (2006). Reducing the dimensionality of data with neural networks. *Science* 313(5786): 504–507.

Jolliffe, I. T. (2002). *Principal component analysis* (2nd edn.). Berlin: Springer.

Kim, K., and D. Lee. (2014). Inductive manifold learning using structured support vector machine. *Pattern Recognition* 47(1): 470–479.

Kingma, D. P., and M. Welling. (2013). Auto-Encoding Variational Bayes. arXiv preprint arXiv:1312.6114.

McInnes, L., J. Healy, and J. Melville. (2018). UMAP: Uniform Manifold Approximation and Projection for Dimension Reduction. arXiv preprint arXiv:1802.03426.

Roweis, S. T., and L. K. Saul. (2000). Nonlinear dimensionality reduction by locally linear embedding. *Science* 290(5500): 2323–2326.

Tenenbaum, J. B., S. V. De, and J. C. Langford. (2000). A global geometric framework for nonlinear dimensionality reduction. *Science* 290(5500): 2319–2323.

Van der Maaten, L., and G. Hinton. (2008). Visualizing data using t-SNE. *Journal of Machine Learning Research* 9: 2579–2605.

Wattenberg, M., F. Viégas, and I. Johnson. (2016). How to use t-SNE effectively. *Distill* 1(10): e2.

Wickelmaier, F. (2003). An introduction to MDS. *Sound Quality Research Unit, Aalborg University, Denmark* 46(5): 1–26.

Zhang, Z., and H. Zha. (2004). Principal manifolds and nonlinear dimensionality reduction via local tangent space alignment. *SIAM Journal on Scientific Computing* 26(1): 313–338.

Appendix A
Common Terms

Appendix A will introduce some common terms involved in the field of machine learning.

A.1 Algorithm, Model, and Hypothesis

The terms algorithm, model, and hypothesis in machine learning are easily confused. The model and hypothesis are related to the algorithm but not equivalent to it.

A.1.1 Algorithm

The concept of an algorithm originates from mathematics and computer science. In these fields, an algorithm is a sequence of processing steps used to solve a specific type of computational problem.

Similarly, an algorithm in machine learning is also a sequence of processing steps used to solve a specific type of machine learning problem, for example, the support vector machine (SVM) algorithm for classification, the k-means algorithm for clustering, and the convolutional neural network (CNN) algorithm for image processing.

A.1.2 Model

A model in machine learning is a program for machine learning that has been trained using some specific datasets. It can be used to predict unseen data or make decisions.

It is worth pointing out that some mathematical models are employed in machine learning, such as Gaussian mixture model. However, it is essential to recognize that while these mathematical models serve as building algorithms for implementing machine learning models, they are not necessarily equivalent to each other.

A.1.3 Hypothesis

In machine learning, the terms hypothesis and model are often used interchangeably. And in statistics, a hypothesis is defined as a function or statement, which gives the relationship between two or more variables of the specified population.

A.2 Data

A.2.1 Datasets

A dataset is a set of data selected for a specific machine learning task.

For example, in the case of handwritten digit recognition, which is a classification task, i.e., assigning a handwritten Arabic numeral to one of the ten categories from 0 to 9, the dataset consists of images scanned from handwritten Arabic numerals.

To more effectively solve the task of handwritten digit recognition, it is often necessary to collect Arabic numerals written by different people. The quality and quantity of the collected data will directly determine the effectiveness of machine learning. Therefore, the size, resolution, and clarity of each handwritten Arabic numeral image should be as consistent as possible, and there should also be a certain scale of data volume.

A.2.2 Training, Validation, and Test Dataset

For supervised learning, the dataset will be further divided into training dataset, validation dataset, and test dataset, as depicted in Fig. A.1.

Fig. A.1 Datasets and their training, validation, and testing datasets

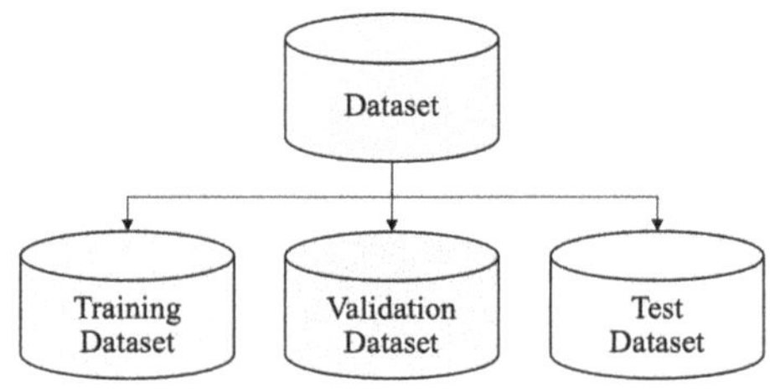

The roles of these three types of datasets are as follows:

- Training dataset:
  It is a set of samples used for the training of machine learning algorithms, adjusting the corresponding parameters, such as the weights of the connections between neurons in artificial neural networks.
- Validation dataset:
  It is used to evaluate the results of training. By calculating the loss (the difference between predicted and target outputs), the accuracy of the model on any given validation data can be known. The corresponding parameters are then further adjusted based on the evaluation results of the validation data.
- Test dataset:
  It is used to evaluate the final model, i.e., the model adjusted based on the training data and validation data. This step is crucial for the generalizability of the model. Using this set of data, the working accuracy of the model can be obtained.

It generally suggests that the ratios of training dataset, validation dataset, and test dataset will be 60%, 20%, and 20% respectively, which can be adjusted as needed.

Using open datasets or public datasets is also a way to collect datasets.[1]

A.2.3   Labeling

Labeling of data is the result of annotating the input observational data, forming a data pair, which serves as an element in the training sample set.

For example:

- In handwritten Arabic numeral recognition, it is to label each handwritten numeral image with its actual number.
- In computer vision, data labeling involves adding tags to raw data such as images and videos. Each tag represents an object class associated with the data.

The work of labeling is mainly done manually, and for some machine learning tasks, such as medical image segmentation, domain experts are needed to complete the labeling work.

A.2.4   Ground-Truth

In the field of machine learning, the ground-truth refers to correct answers or true labels made for a given dataset and the specific task, used to verify the correctness of machine learning results.

---

https://paperswithcode.com/datasets

For example, to classify images of animals, the ground-truth would be the correct labels for each image or each object in an image, such as "cat" "dog," or "tiger." Your model would be trained on a dataset that includes both the images and their corresponding ground-truth, and its performance would be evaluated based on how accurately it is able to predict the correct ground-truth for new, unseen images.

The ground-truth can be true labels, but not just labels, which also are correct answers. Since the term "ground-truth" originates from meteorology and remote sensing fields, it refers to information obtained from actual scenes.

A.2.5   Dataset vs. Database

It is worth noting that a dataset is not equivalent to a database:

- Dataset:
  It is a concept of a collection, referring to a group of similar data, and is usually closely related to a specific machine learning task.
- Database:
  It is a type of computer software system, the purpose of which is to allow the computer to store large amounts of data in an organized manner and to query, add, delete, and update these data, such as a student information management system.

A.3   Features

Features refer to unique, easily measurable information in the data, sometimes also referred to as attributes.

There are mainly two ways to extract features: hand-crafted features and learned features. The former is also known as manually extracted features, and the latter is also known as automatically extracted features through deep learning.

A.3.1   Hand-Crafted Features

Hand-crafted features are used in conjunction with traditional machine learning, as shown in Fig. A.2a.

Different types of data have different representations of hand-crafted features, and the same type of data has several methods of feature representation.

For data in the field of image processing and computer vision, it is often necessary to use some feature descriptor and corresponding algorithm to extract features, for example, local binary pattern (LBP), scale-invariant feature transform

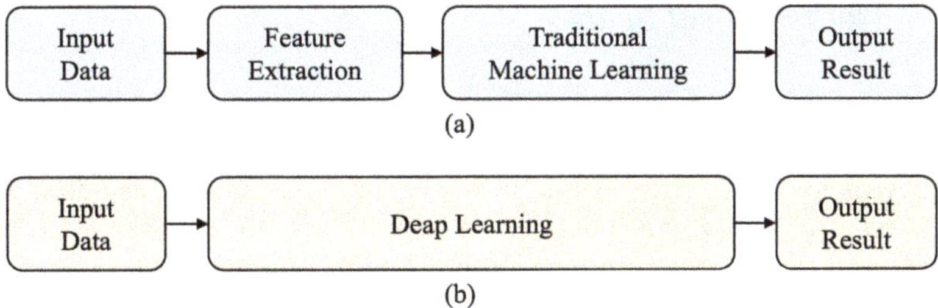

Fig. A.2 Difference between hand-crafted features and learned features

(SIFT), and histogram of oriented gradient (HOG). These feature descriptors and their feature extraction algorithms often have poor real-time performance, large limitations, narrow application range, and complex calculations.

For data in the field of natural language processing, some representation methods have been designed to represent corresponding features, for example: BoW (bag-of-words), Word2vec (word to vector), and GloVe (global vectors).

A.3.2   Learned Features

Learned features are features obtained by machine learning automatically. Deep artificial neural networks, especially convolutional neural networks, are very suitable for automatically obtaining image features through learning. Because learned features are automatically extracted during the deep learning process and used to solve specific tasks, there is no need for a manual feature extraction stage. In fact, deep learning models that perform feature extraction and classification have greater advantages than traditional classification models that manually extract features, as shown in Fig. A.2b. This is also one of the main reasons why deep learning has been popular since the deep convolutional neuron network AlexNet won the ImageNet challenge in 2012 (Krizhevsky et al. 2012).

However, learned features are only good at tasks close to the trained dataset, and they cannot control what features the model extracts from the data, and they are not highly interpretable.

A.4   Dimensionality

In machine learning, the dimensionality is often involved. Sometimes it refers to the dimension of mathematical (physical) space, but most of the time it refers to the dimension of data space.

A.4.1 Mathematical Space

The dimension of a mathematical space refers to the minimum number of coordinates needed to represent any point in that space.

For example, in Euclidean space in mathematics: a point in one-dimensional space is represented by one coordinate, a point in two-dimensional space is represented by two coordinates, and a point in three-dimensional space is represented by three coordinates.

A.4.2 Data Space

The dimension of a data space refers to the number of features or attributes in the data.

High-dimensional data refers to data with more than three dimensions.

Taking a digital image as an example, the width of the image is represented as w, and the height is represented as h.

- From the viewpoint of Euclidean space, each point in the image can be represented by two coordinates, so it is two-dimension.
- From the viewpoint of data space, this image has $m \times n$ pixels; if each pixel is represented as an RGB feature vector, then the dimension of this image is $m \times n$ dimension.

A.5   Normalization and Regularization

Normalization is one of the methods to adjust data, whereas regularization is a method used when adjusting the model.

A.5.1 Normalization

Many types of data need to be preprocessed, the main one is data scaling, which is called normalization.

For raw data with a large range of values, if it is not normalized, the model will not work properly. There are several ways to normalize. The simplest is to adjust the values measured on different scales to a small range of common scales, and the complex one is to adjust the values to maintain consistency on a certain probability distribution.

Feature scaling is a typical normalization method used to adjust the range of data values. The main methods are as follows.

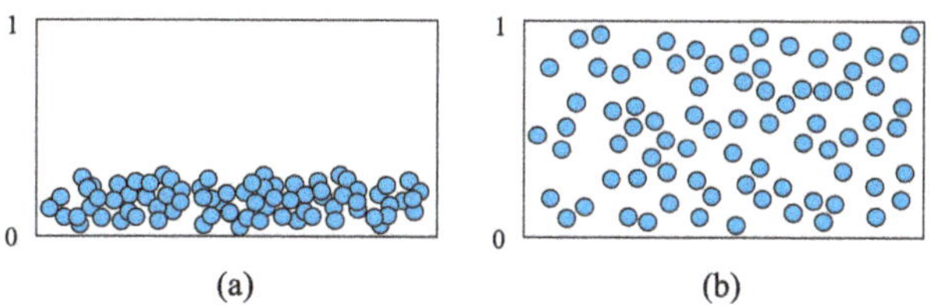

Fig. A.3 Schematic diagrams of a raw data and its min-max normalization

The most common is min-max normalization: that is, adjusting the data to the interval [0, 1], the expression is as follows:

$$x' = \frac{x - \min(x)}{\max(x) - \min(x)}, \tag{A.1}$$

where x is the value of the original data and x' is the adjusted value. In the case where the outliers in the input data are not obvious, the min-max normalization method is very useful; otherwise, the problem of outlier loss may occur. Figure A.3a, b shows the schematic diagrams before and after normalization.

In some cases, it is necessary to adjust the data to the $[-1, 1]$ interval, in which case the following mean normalization method is used, that is:

$$x' = \frac{x - \mean(x)}{\max(x) - \min(x)}. \tag{A.2}$$

In this way, it is more convenient to use other techniques such as matrix decomposition.

A.5.2   Regularization

Regularization is a method used when adjusting the model, specifically to solve the ill-posed problem of machine learning models or to prevent overfitting and underfitting problems.

Before introducing the ill-posed problem, let's first understand what a well-posed problem is. French mathematician Jacques Hadamard proposed the well-posed problem for mathematical models in 1922 and defined that a well-posed problem must satisfy the following three properties: (i) A solution exists; (ii) The solution is unique; and (iii) The solution's behaviour changes continuously with the initial conditions.

After knowing the definition of a well-posed problem, it can be inferred that if one of the three properties of a well-posed problem is not satisfied, it is an ill-posed problem.

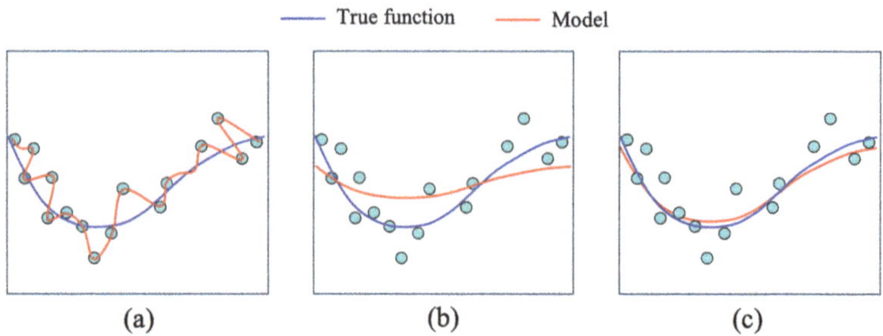

Fig. A.4 Schematic diagrams of (**a**) overfitting, (**b**) underfitting, and (**c**) best-fitting

For example, the solution to the equation $y = x + 50$ is not unique, so it is an ill-posed problem. The inverse problem is the reverse solving process of the forward problem, which usually belongs to the ill-posed problem. In the tasks of machine learning, different results will be obtained due to different algorithms, different parameters in the same algorithm, or different training times, all of which are ill-posed problems. Many scholars have proposed regularization methods to solve ill-posed problems (Cheng and Hofmann 2011; Clason et al. 2019).

Overfitting is caused by overtraining the model, that is, when the function curve obtained from training is too close to the training data points, this problem will occur, as shown by the red curve in Fig. A.4a. Overfitting makes the model too dependent on the data used and lacks generalization ability. In addition, data often have a certain degree of error and random noise, so making it too consistent with these data will also bring a lot of errors to the model.

It should be pointed out that overtraining will lead to overfitting, and insufficient training of the model will lead to underfitting. The issue of underfitting, where the function curve does not sufficiently fit the set of data points, is depicted in Fig. A.4b where the red curve is almost a straight line. This is an extreme case of underfitting. The predictive performance of underfitting models is often poor.

A.6   Baseline

A baseline usually refers to a line used as a base or benchmark for measurement or construction.

In the field of machine learning, a baseline refers to a reference model and can also serve as a base model.

A prerequisite for a machine learning baseline is firstly to have its papers (documents), source code, datasets, etc. Secondly, the source code is needed to build and debug in a specified hardware environment to form a runnable program. Thirdly, several tests need to be conducted using the dataset. If the performance indicators described in the paper (document) are basically met, then it can be used as a baseline.

Once the baseline is established, your research, improvements, and enhancements can be made on its basis to improve the performance of the baseline model. It can also be innovated upon to solve similar but different machine learning problems.

The usual way to obtain a machine learning baseline is through papers with open-source code. In addition, there is a "Papers With Code[2]" website, which currently has about 90,000 machine learning papers with source code, from which the required baseline can be retrieved.

References

Cheng, J., and B. Hofmann. (2011). Regularization Methods for Ill-Posed Problems. Handbook of Mathematical Methods in Imaging, Springer.

Clason, C., B. Kaltenbacher and E. Resmerita. (2019). Regularization of Ill-Posed Problems with Non-Negative Solutions. *Splitting Algorithms, Modern Operator Theory, and Applications,* Springer: 113–135.

Krizhevsky, A., I. Sutskever, and G. E. Hinton. (2012). Imagenet classification with deep convolutional neural networks. In *Conference on Neural Information Processing Systems (NIPS).*

---

https://paperswithcode.com/

Glossary

Autoencoder A neural network is known as an autoencoder (AE), if it satisfies $x' = g(z) = g(f(x))$ and $x' \not\equiv x$, where $x, x' \in X$ are the input and output data, respectively, and $z \in Z$ is the intermediate code in the latent space.

Undercomplete Autoencoder An undercomplete autoencoder (UAE) is an autoencoder, and the dimension of its latent space $Z \subseteq \mathbb{R}^v$ is less than the dimension of the data space $X \subseteq \mathbb{R}^u$, i.e., $v < u$.

Sparse Autoencoder A sparse autoencoder (SAE) is an autoencoder with a sparsity constraint, but the dimension of its latent space is not limited and can be equal to or even greater than the dimension of the input space.

Denoising Autoencoder A denoising autoencoder (DAE) learns useful features in the input data by changing the reconstruction error term in the loss function $\mathcal{L}(x, x') = \mathcal{L}(x, g(f(\tilde{x})))$, where $\tilde{x} = x \oplus \text{noise}$.

Variational Autoencoder A variational autoencoder (VAE) is an artificial neural network that uses probabilistic graphical models and variational Bayes method.

Autoregressive Model An autoregressive model (AR) is a type of generative model for dealing with time series problems.

Autoregressive Property Given a set of variables $\{x_1, x_2, \ldots, x_t\}$ over time t, it is called autoregressive property if the variable x_t depends linearly on some or all its previous variables $\{x_1, x_2, \ldots, x_{t-1}\}$.

k-Order Autoregressive Model A k-order autoregressive model AR (k) is a statistical model with the autoregressive property, which is defined as $x_t = \sum_{i=1}^{k} \varphi_i x_{t-i} + \varepsilon_t$, where $\varphi_1, \ldots, \varphi_k$ are the parameters of the model and ε_t is white noise.

Deep Autoregressive Model A deep autoregressive model is a deep network model with autoregressive property, also known as an autoregressive generative model.

Bayesian Decision Network A Bayesian decision network (BDN) is an ordered 4-tuple, $\mathrm{BDN} = \langle \mathrm{BDG}, P_{DG}, R, \mathbb{E} \rangle$, where (i) BDG is standing for directed acyclic Bayesian decision graph and $\mathrm{BDG} = \langle \widetilde{X}, \overrightarrow{E} \rangle$, where $\widetilde{X} = X \cup D \cup U$, X is the set of random variable nodes, D is the set of decision variable nodes, U is the set of utility variable nodes, and $X \cap D \cap U = \emptyset$; $\overrightarrow{E}$ is a directed edge, and it satisfies the condition of a directed acyclic graph, $\forall X_i \in \widetilde{X}, \nexists (X_i \to X_i)$. (ii) P_{DG} is the set of probability distributions of directed graphs, $P_{DG} = \{P_{X_i} | X_i \in X\}$. (iii) R is the set of reward values. And (iv) $\mathbb{E}$ is the expected utility.

Bayesian Model A Bayesian model is based on Bayes theorem, also known as Bayesian reasoning or Bayesian inference, which obtains a hypothesis from data and then uses the hypothesis for prediction.

> **Bayesian Learning** The learning to infer the cause based on the known effect, $P\,(\text{Cause}|\text{Effect}) = \frac{P\,(\text{Effect}|\text{Cause})\,P\,(\text{Cause})}{P\,(\text{Effect})}$, where $P\,(\text{Effect}|\text{Cause})$ is the likelihood of the causal relationship, $P\,(\text{Cause})$ is the prior probability of the cause, and $P\,(\text{Effect})$ is the evidence of the effect.
>
> **PAC-Bayesian Learning** A product of combining Bayesian learning with PAC learning theory, also known as PAC-Bayesian theory. It provides a convenient method of utilizing prior knowledge based on Bayesian learning, while also providing strict and explicit generalization guarantees based on PAC learning theory.

Behavioral Decision Process The behavior of intelligent agents is learned through interaction with the environment. Behavioral learning can be seen as a behavioral decision process adopted by intelligent agents.

> **Decision Process** The decision process (DP) of an intelligent agent can be formalized as a 3-tuple $\mathrm{DP} = \langle S, A, P \rangle$, where S and A are both random variables, representing the set of states and the set of actions of the decision process, respectively, and $P\,(s)$ denotes the probability of the state $s \in S$.
>
> **Utility** Given a decision process $\mathrm{DP} = \langle S, A, P \rangle$, the current state $s \in S$, the action to be taken $a \in A$, then its utility U is a function, denoted as $U\,(s, a)$, its result is a value in the set of real number $\mathbb{R}$, that is $U : S \times A \to \mathbb{R}$.
>
> **Expected Utility** Given a decision process $\mathrm{DP} = \langle S, A, P \rangle$ and utility $U\,(s, a)$, then its expected utility (EU) is $\mathrm{EU}\,(S, A) = \mathbb{E}[S, A] = \sum_{s \in S, a \in A} P\,(s)\,U\,(s, a)$, where P denotes the probability of state $s \in S$ occurring, $P\,(s) \geq 0$, and $\sum P\,(s) = 1$.
>
> **Maximum Expected Utility** Given the expected utility of a decision process $\mathrm{EU}\,(S, A)$, the maximum expected utility (MEU) of the decision process is $\mathrm{MEU}\,(S, A) = \arg\max_{a \in A} \mathrm{EU}\,(S, A)$.

Behavioral Decision Types The behavioral decisions of intelligent agents can be divided into three types, namely: single-stage decisions, multi-stage decisions, and sequential decisions.

Single-Stage Decision Given a decision process DP $= \langle S, A, P \rangle$, let the initial state be $s_0 \in S$, the target state be $s_g \in S$. Then the decision process is called single-stage decision, if taking an action $a \in A$ the initial state s_0 be transformed into the target state s_g with only one step, that is, $P\left(s_g | s_0, a\right)$.

Multi-Stage Decision Given a decision process DP $= \langle S, A, P \rangle$, initial state $s_0 \in S$, and the action of the ith step $a_i \in A$, the target state $S_g \subset S$. Then the decision process is called multistage decision, if its action space can be formed from the initial state to the target state, after several state transitions $P\left(s_{i+1} | s_i, a_i\right)$ can reach any target state from the initial state s_0 to $s_{i+1} = s_g \in S_g$.

Sequential Decision Given a Markov decision process, MDP $= \langle S, A, P, R \rangle$, initial state $s_0 \in S$, the action of ith step $a_i \in A$, and the reward of the environment $r_i \in R$, then according to each step's state and rewards, the current action is determined, that is, $P\left(s_{i+1} | s_i, a_i, r_i\right)$, finally reaching the target state $s_{i+1} = s_g$, known as sequential decisions or sequential decision-making.

Bias-Variance Problem In supervised learning, the hypothesis obtained after training will have errors when predicting unknown data, namely, the so-called bias and variance. These are a pair of parameters that can easily conflict, and these parameters have such a characteristic, that is: reducing bias will lead to an increase in variance and vice versa. This phenomenon is called the "bias-variance problem" or the "bias-variance dilemma."

Bias The difference between the expected value of prediction and the target value of the label, that is, $\text{Bias}\left[h\left(x\right)\right] = \mathbb{E}\left[h\left(x\right)\right] - y$.

Variance The expectation of the square of the difference between the predicted value and the expected value, that is, $\text{Var}\left[h\left(x\right)\right] = \mathbb{E}\left[\left(h\left(x\right) - \mathbb{E}\left[h\left(x\right)\right]\right)^2\right]$.

Classification A task in machine learning, which trains a classifier with the labeled samples whose categories are known, and then uses it to identify which categories the other data belongs to.

Linear Separable Let $\mathbb{R}^m$ be a set of m-dimensional real-valued vectors, $X \subseteq \mathbb{R}^m$ be an input space, $D \subset X$ and $D' \subset X$ be two sets of data points, $w \in \mathbb{R}^m$ be weight vector, and $b \in \mathbb{R}$ be bias. Then D and D' are called linear separable, if there exists a linear function $u(x) = w^\mathsf{T} x - b$ such that every data point $x \in D$ satisfies $u(x) > 0$ and every data point $x' \in D'$ satisfies $u(x') < 0$.

Clustering A task in machine learning, which is based on certain criteria to analyze input data and divide it into several groups called clusters. These clusters are unknown in advance, but there is some correlation between the data within the same cluster.

Computational Learning Theory A mathematical analysis theory about learnability, mainly used for the design and analysis of machine learning algorithms. Its mathematical analysis process adopts the basic ideas of probability theory.

Learnability Theory The theoretical basis of computational learning theory that concerns the questions such as: why is machine learnable, what information is required to support learning, and what computation is required for learning to be possible.

PAC Learning A concept class C is said to be PAC learnable, if there exists an algorithm $\mathcal{A}$ and a polynomial function poly $(\cdot, \cdot, \cdot, \cdot)$, such that for any $\epsilon > 0$ and $\delta > 0$, for all distributions D on X, and for any target concept $c \in C$, then for any sample size $m \geq$ poly $(1/\epsilon, 1/\delta, n, \text{size}(c))$, the following expression holds: $P_{S \sim D^n}[R(h) \leq \epsilon] \geq 1 - \delta$. If $\mathcal{A}$ further runs in poly $(1/\epsilon, 1/\delta, n, \text{size}(c))$, then C is said to be efficiently PAC learnable. When such an algorithm $\mathcal{A}$ exists, it is called a PAC learning algorithm for C.

Occam Learning Be named after Occam's razor. It is based on the law of parsimony of Occam's razor and applies it to the theory and mathematical proof of machine learning.

Let C be the concept class containing the target concept $c \in C$, and H be the hypothesis set. Then for constants $\alpha \geq 0$ and $0 \leq \beta \leq 1$, the learning algorithm $\mathcal{A}$ is an alpha-beta Occam algorithm that uses H to learn C if and only if: given a set of n samples $S = \{x_i\}_{i=1}^n$, using the concept $c(x)$ for labeling, we get n training samples $\{(x_i, c(x_i))\}_{i=1}^n$, which are used to train the learning algorithm $\mathcal{A}$ to get a hypothesis $h \in H$, so that (i) h on S is consistent with c, that is, $\forall x \in S$, $h(x)$ is consistent with $c(x)$, and (ii) $\text{size}(h) \leq (m \cdot \text{size}(c))^\alpha n^\beta$, where m is the maximum length for any sample $x \in S$.

Diffusion Model A generation model in machine learning which is inspired by nonequilibrium thermodynamics diffusion mechanisms.

Diffusion The net process of movement of molecules or particles under a concentration gradient, generally from a region of higher concentration to a region of lower concentration.

Diffusion Probabilistic Model A type of diffusion models, which consists of (i) the forward diffusion process to slowly add random noise to data and (ii) the reverse diffusion process to reconstruct desired data from the noise.

Dimensionality Reduction A task in machine learning which maps the data in high-dimensional space to a low-dimensional space, while retaining the basic features of the original high-dimensional data.

High-Dimensional Data Given a dataset, let n denote the number of its data points and p denote the number of its features (attributes); if n is very large and $p \gg 3$, then the dataset is referred to as high-dimensional data, where 3 is the dimension of Euclidean space.

Manifold An n-dimensional manifold is a topological space with the property that each point has a neighborhood that is homeomorphic to an open subset of n-dimensional Euclidean space.

Ensemble Learning A quasi-paradigm of machine learning that organically combines several base learners to form a strong learner, whose performance surpasses any of the base learners before the combination.

Generative Adversarial Model One of the generative models composed of generator and discriminator modules, based on the minimax theorem of two-player games.

Generative Adversarial Network (GAN) A class of neural network frameworks that belongs to the generative adversarial model.

Kernel Method A method to map the similarity between data pairs in nonlinear data in the original space to a linearly separable space through a specific kernel function.

Hilbert Space An inner product space with separability and completeness, denoted as $\mathcal{F}$.

Kernel Function Let $\mathcal{X}$ be a non-empty input space, the symmetric function $\kappa : \mathcal{X} \times \mathcal{X} \to \mathbb{R}$ is known as the kernel function, also simply called the kernel, where $\mathbb{R}$ is a set of real numbers.

Gram Matrix Given a kernel function $\kappa : \mathcal{X} \times \mathcal{X} \to \mathbb{R}$ and inputs $x_1, x_2, \ldots, x_n \in \mathcal{X}$, then $\mathbf{K} = \left[\kappa \left(x_i, x_j \right) \right]_{i,j} \in \mathbb{R}^{n \times n}$ is known as the Gram matrix (or kernel matrix) corresponding to $x_1, x_2, \ldots, x_n$.

Positive-Definite Matrix Let the column vector be $c = (c_1, \ldots, c_n)^\mathsf{T} \in \mathbb{R}^{n \times 1}$, $\mathbf{K}$ is a $n \times n$ Gram matrix, if the following $n \times n$ matrix, $c^\mathsf{T} \mathbf{K} c = \sum_{i=1}^{n} \sum_{j=1}^{n} c_i c_j \kappa \left(x_i, x_j \right) \geq 0$ is satisfied.

Positive-Definite Kernel For a kernel function $\kappa : \mathcal{X} \times \mathcal{X} \to \mathbb{R}$ and inputs $x_1, x_2, \ldots, x_n \in \mathcal{X}$, its kernel function is called a positive-definite kernel if it satisfies the condition of a positive-definite matrix.

Reproducing Kernel A kernel function is known as a reproducing kernel function if it satisfies $f(x) = \langle f, \kappa(x, \cdot) \rangle$.

Reproducing Kernel Hilbert Space The Hilbert space dealing with the reproducing kernel function $\mathcal{F}_\mathcal{X}$.

Logic Learning Modes The learning modes based on logic reasoning, which are abductive learning, deductive learning, and inductive learning.

Abductive Learning The learning that starts from observed facts (events, experiences, etc.), seeks the most likely or optimal premises (hypotheses, theories, etc.), and generates new knowledge from this learning.

Deductive Learning The learning that proposes hypotheses based on some existing facts or rules and then confirms them through observation.

Inductive Learning The learning of general facts or rules by observing certain patterns, forming hypotheses based on these patterns, and then deriving from these hypotheses.

Machine Learning A branch of artificial intelligence that studies and builds models that can learn from data or the environment to make predictions about unknown data or make decisions based on environmental information.

Markov Models A class of stochastic models used to model the systems with random changes. It is named after the Russian mathematician Andrey Markov.

Stochastic Process A stochastic process (SP) is usually defined as a set of random variables on a probability space, that is, $\text{SP} = \{S(t) \mid t \in T\}$, where T is an index set, also known as a parameter set; usually, T is interpreted as time, and each t is considered a point in time; $S(t)$ is a random variable at time t, also known as the state of the stochastic process at time t.

Markov Property Given the current state of stochastic process $S(t)$ and past states $\{S(0), \ldots, S(t-1)\}$, if the conditional probability of the next state $S(t+1)$ only depends on the current state $S(t)$ and is irrelevant to the past states $\{S(0), \ldots, S(t-1)\}$, then the stochastic process is said to have Markov properties. It is expressed as: $P(S(t+1) \mid S(0), \ldots, S(t-1), S(t)) = P(S(t+1) \mid S(t))$.

Markov Process A Markov process (MP) is a stochastic process with the Markov property, which can be represented as a 2-tuple: $\text{MP} = \langle S, P \rangle$, where S is a state set and P is the transition probability of the current state transitioning to the next state.

Hidden Markov Model Let $\{Y_i\}_{i=1}^n$ and $\{X_i\}_{i=1}^n$ be discrete-time stochastic processes, if it satisfies: i) Y_i is a Markov process with hidden states, and ii) $P(X_i \mid Y_1 = y_1, \ldots, Y_{i-1} = y_{i-1}, Y_i = y_i) = P(X_i \mid Y_i = y_i)$, then the set of pair $(\{Y_i\}, \{X_i\})_{i=1}^n$ is a hidden Markov model. $P(X_i \mid Y_i = y_i)$ is referred to as the output probability.

Markov Decision Process A Markov decision process (MDP) is a stochastic decision process that satisfies the Markov property and can be represented as a 4-tuple, $\text{MDP} = \langle S, A, P, R \rangle$, where S is a state set, A is an action set, P is a transition probability, and R is a reward function.

Model-Based A method is called model-based, if all the elements of MDP model are known and it uses a planning method such as dynamic programming (DP) to do the optimal control of the model.

Model-Free A method is called model-free, if some environment-dependent elements of MDP model are unknown and it uses a learning method such as reinforcement learning (RL) to do the optimal control of the model.

Partially Observable Markov Decision Process A partially observable Markov decision process (POMDP) can be represented as a 5-tuple, namely, $\text{POMDP} = \langle S, A, P, R, P_o \rangle$, where P_o is the conditional observation function and the other elements in the 5-tuple have the same definitions as in the Markov decision process.

Meta-Learning A quasi-paradigm of machine learning, which is dedicated to acquiring meta-knowledge, using it to train the meta-model and then applying the meta-model to solve new problems and tasks in machine learning. Meta-learning is also known as "learning to learn" or "learning how to learn."

Normalizing Flow Models A class of generative model frameworks for modeling probability distributions, which uses the change of variables in probability to transform simple distributions into complex distributions, and then performs inverse transformations to restore them to simple distributions.

Parametric and Nonparametric For a statistical model (SM), from the point of view of the parameters of the model, it can be divided into the parametric model and nonparametric model.

Parametric Models For a parameterized statistical model, $\text{PSM} = \langle \mathcal{X}, \mathcal{Y}, P, \Theta \rangle$, it is called a parametric model, if the Θ be a set of parameters with a finite number of parameters.

Nonparametric Models For a parameterized statistical model $\text{PSM} = \langle \mathcal{X}, \mathcal{Y}, P, \Theta \rangle$, it is called a nonparametric model if the Θ is a parameter set without fixed number of parameters.

Perspectives on Machine Learning For a comprehensive and multifaceted study of machine learning, this book proposes the three perspectives: learning frameworks, learning paradigms, and learning tasks.

Learning Frameworks A learning framework refers to a theoretical framework for the design of machine learning algorithms. It belongs to the theory level of machine learning rather than the data level or task level, and it is also different from the software framework used to build machine learning algorithms.

Learning Paradigms A learning paradigm to machine learning refers to the pattern or style taken by learning algorithms. It is a division of the paradigm of machine learning, independent of specific applications. The main basis to distinguish a learning paradigm is how to learn from data or how to interact with the environment.

Learning Tasks A learning task refers to a basic problem that can be solved by machine learning. It is a common problem abstracted from domain problems, and the algorithms of machine learning with the same mechanism can be used to complete this task.

Probabilistic Graphical Models Be also known as graphical models or structured probabilistic models. They are probability models that use graphs to represent the conditional dependency structure between random variables, mainly divided into directed graphical models, undirected graphical models, and factor graph models.

Directed Graphs A directed graph (DG) is an ordered 2-tuple, $\text{DG} = \left\langle X, \overrightarrow{E} \right\rangle$, where $X = \{X_i \mid i = 1, 2, \ldots, n\}$ is a set of random variables, X_i is a random variable, $\overrightarrow{E} = \{(X_i \rightarrow X_j) \mid X_i, X_j \in X\}$ is a set of directed edges, and $(X_i \rightarrow X_j)$ denotes a directed edge from X_i to X_j.

Directed Graphical Models A directed graphical model (DGM) is an ordered 2-tuple, $\text{DGM} = \langle \text{DG}, P_{\text{DG}} \rangle$, where DG is a directed graph and P_{DG} is a set of directed graph probability distributions, $P_{\text{DG}} = \{P_{X_i} \mid X_i \in X\}$.

Bayesian Networks A Bayesian network (BN) is an ordered 2-tuple, $\text{BN} = \langle \text{DAG}, P_{\text{BN}} \rangle$, where DAG is a directed acyclic graph, each random variable probability P_{X_i} in P_{BN} is a conditional probability distribution: $P_{X_i} = P(X_i | X_{i-1}, \ldots, X_1) = P(X_i | \text{Parents}(X_i))$.

Undirected Graphs An undirected graph (UG) is an ordered 2-tuple, $UG = \langle X, \overline{E} \rangle$, where $X = \{X_i \mid i = 1, 2, \ldots, n\}$ is a set of random variables, X_i is a random variable, $\overline{E} = \{(X_i - X_j) | X_i, X_j \in X\}$ is the set of undirected edges, and $(X_i - X_j)$ denotes the undirected edge between X_i and X_j; conversely $(X_j - X_i)$ also holds.

Undirected Graphical Models An undirected graphical model (UGM) is an ordered 2-tuple, $UGM = \langle UG, P_{MN} \rangle$, where UG is an undirected graph, P_{UG} is a collection of probability distributions in the directed graph, $P_{MN} = \{\phi_{c_k}(X_{c_k}) | X_{c_k} \subseteq X \text{ and } k = 1, \ldots, K\}$ where X_{c_k} is a clique composed of several random variables, K is the maximum number of cliques, $\phi_{c_k}(X_{c_k})$ is a non-negative potential function, and $\phi_{c_k} : \text{Val}(X_{c_k}) \rightarrow \mathbb{R}_+$ is a mapping from the clique X_{c_k} mapping to the non-negative real number space $\mathbb{R}_+$.

Markov Networks A Markov network (MN) is an ordered 2-tuple, $MN = \langle UG, P_{MN} \rangle$, where UG is an undirected graph, P_{MN} is a joint probability distribution, and the calculation equation is $P_{MN}(X_1, \ldots, X_n) = \frac{1}{Z_{MN}} \prod_{k \in K} \phi_{c_k}(X_{c_k})$.

Regression Regression in machine learning is the process of training a regression algorithm based on samples of corresponding values between labeled input and output data, obtaining an optimal regression hypothesis and then using it to predict unknown input data. The output is a continuous corresponding value.

Regression Model A regression model can be formalized as: $y = f(x, \theta) + \varepsilon$, where y is the *dependent variable*, x is the *independent variable*, θ is the parameter, ε is the error term, and $f(x, \theta)$ is referred to as the regression function.

Linear Combination of Parameters A linear combination of parameters is an expression composed of a set of terms, each of which is multiplied by parameter and variable and then added together. The value of the first variable is usually 1.

Linear Regression Model A regression model $y = f(x, \theta) + \varepsilon$ is referred to as a linear regression model, if the expression $f(x, \theta)$ is a linear combination of parameters.

Nonlinear Regression Model A regression model $y = f(x, \theta) + \varepsilon$ is referred to as a nonlinear regression, if the expression $f(x, \theta)$ is not a linear combination of parameters.

Parametric Regression Model A regression model $y = f(x, \theta) + \varepsilon$ is referred to as a parametric regression model, if the θ is a finite and fixed number of parameters.

Nonparametric Regression Model A regression model $y = f(x, \theta) + \varepsilon$ is referred to as a nonparametric regression model, if the number of parameters of θ is not predetermined but adjusted according to the size of the dataset.

Reinforcement Learning (RL) A learning paradigm in machine learning which is modeled as Markov decision process. The intelligent agent in RL interacts with the environment, taking corresponding actions based on the current state and feedback from the environment, and learns to maximize its cumulative rewards.

Policy Given a Markov decision process MDP $= \langle S, A, P, R \rangle$, its control policy π at time point t is the action a_t to be taken under the state s_t, represented as: $\pi\left(a_t | s_t\right) = \mathbb{P}\left[A\left(t\right) = a_t | S\left(t\right) = s_t\right]$, where $\mathbb{P}[\cdot]$ is a transition matrix.

Return It refers to the weighted cumulative rewards in reinforcement learning.

On-Policy In on-policy learning, its policy for evaluation and improvement is the same as the policy for decision-making.

Off-Policy In off-policy learning, its policy for evaluation and improvement is different from the policy for decision-making.

Risk Minimization Principles The important principles to consider in supervised learning that provide the theoretical performance boundaries for a supervised learning algorithm.

Expected Risk The expected risk of a hypothesis $h \in H$ is represented by risk function $R\left(h\right)$, which means the expected risk of loss function $\mathcal{L}\left(h\left(x\right), y\right)$, that is: $R\left(h\right) = \mathbb{E}_{\mathcal{X},\mathcal{Y}}\left[\mathcal{L}\left(h\left(x\right), y\right)\right] = \int_{\mathcal{X},\mathcal{Y}} \mathcal{L}\left(h\left(x\right), y\right) p\left(x, y\right) dx dy$.

Expected Risk Minimization It refers to finding an optimal hypothesis h^* in the hypothesis set H through training, so that its expected risk $R\left(h\right)$ is minimized: $h^* = \underset{h \in H}{\arg\min}\, R\left(h\right)$.

Empirical Risk For a given training sample set, $S = \{(x_i, y_i) \mid i = 1, \ldots, n\}$, its empirical risk is denoted as $R_{emp}\left(h\right)$, then its computing equation is as follows: $R_{emp}\left(h\right) = \frac{1}{n} \sum_{i=1}^{n} \mathcal{L}\left(h\left(x_i\right), y_i\right)$.

Empirical Risk Minimization (ERM) Be also known as the principle of empirical risk minimization, it means that the learning algorithm should choose a hypothesis that minimizes empirical risk h_{ERM}, that is: $h_{ERM} = \underset{h \in H}{\arg\min}\, R_{emp}\left(h\right)$.

Structural Risk In supervised learning, it is necessary to train a generalized model based on a finite training dataset. The problem that arises is overfitting, that is, the model is too suitable for the specificity of the training dataset, and the generalization effect on new unknown data is poor. This is the so-called structural risk.

Structural Risk Minimization (SRM) To solve the problem of structural risk by balancing the complexity of the model and the fit of the training data. This is an inductive principle in supervised learning.

Rule Learning Be also known as rule-based learning, or rule-based machine learning, including association rule learning, decision trees, and random forests.

Association Rule Learning A rule-based machine learning method used to discover interesting relations between variables in large transaction datasets is also known as association analysis.

Self-Supervised Learning (SSL) A quasi-paradigm of machine learning that automatically extracts *pseudo-labels* from raw data and then uses them for supervised learning in the next stage.

Statistical Learning Theory A theoretical framework for supervised learning and can be seen as a theory of inductive learning, originated from statistics and functional analysis.

Statistical Model A statistical model (SM) can be represented as a 3-tuple $SM = \langle \mathcal{X}, \mathcal{Y}, P \rangle$, where $\mathcal{X}$ is the input space of the data; $\mathcal{Y}$ is the output space, also known as the target space of the data; and P denotes the probability function.

Statistical Learning Model A statistical learning model (SLM) can be seen as a 4-tuple $SLM = \langle \mathcal{X}, \mathcal{Y}, P, H \rangle$, where $\langle \mathcal{X}, \mathcal{Y}, P \rangle$ is the same as the statistical model defined in the above and H is the hypothesis set, also known as the hypothesis function set.

Constraint Set Let $H : \mathcal{X} \rightarrow \{0, 1\}$ be the hypothesis set, $D \subseteq \mathcal{X}$ be a finite dataset, then the constraint set H_D is defined as the set of H acting on D, that is: $H_D = \{(h(x_1), \ldots, h(x_n)) \mid x_i \in D, h(x_i) \in H, n \in \mathbb{N}_+\}$, where $\mathbb{N}_+$ denotes the set of non-negative natural numbers.

Growth Function The growth function $G_H : \mathbb{N} \rightarrow \mathbb{N}$ is defined as $G_H(m) = \max_{n=|D|} |H_D|$, i.e., the growth function G_H is the maximum number of elements in the constraint set H_D.

Shattering The dataset D can be shattered by the hypothesis set H, if there exists a hypothesis set H and a dataset D, such that $|H_D| = 2^n$.

VC Dimension The VC dimension of a hypothesis set H, denoted as $VCdim(H)$, is the maximum value of n in the growth function that can be shattered by the hypothesis set H, that is: $VCdim(H) = \max\{n \mid G_H(n) = 2^n\}$.

Rademacher Average Let $\mathcal{Z}$ be input space, $S = \{z_1, z_2, \ldots, z_n\} \subseteq \mathcal{Z}$ be a sample data, F be a real-valued function class, and $F : \mathcal{Z} \rightarrow \sigma_{i=1}^n$. The Rademacher average of the function class F over the sample data S is as:

$$\text{Red}_S(F) = \mathbb{E}_\sigma \left[\sup_{f \in F} \left(\tfrac{1}{n} \sum_{i=1}^n \sigma_i f(z_i) \right) \right], \text{ where } \sigma_i \text{ denotes independent}$$

random noise, $\sigma_i \in \{-1, +1\}$, and $P(\sigma_i = +1) = P(\sigma_i = -1) = 0.5$.

Rademacher Complexity Let P be a Rademacher distribution over sample data S and the Rademacher complexity of the function class F over the distribution P be $\text{Red}_P(F) = \mathbb{E}_{S \sim P}[\text{Red}_S(F)]$.

Supervised Learning (SL) A learning paradigm with labeled data that is used to train the learning algorithm to obtain the best learner called the hypothesis. This hypothesis is then used to predict unknown input data to obtain corresponding output results.

Symbolic Logic A development of formal logic employing a special symbol capable of manipulation in accordance with precise rules, which consists of propositional logic and first-order logic.

> **Propositional Logic** A formal system for dealing with statements that have a truth value of true or false and for constructing rules for proving theorems.
>
> **First-Order Logic** A formal system built on top of propositional logic, which can use quantified variables of non-logical objects and statements containing variables. Be also known as predicate logic or first-order predicate calculus.

Transfer Learning A quasi-paradigm of machine learning that transfers the knowledge or functionality of a source model to a different but related target model.

Unsupervised Learning (UL) A learning paradigm to machine learning with unlabeled data, which analyzes or processes the data based on a certain established criterion to obtain the desired results.

Printed in the USA
CPSIA information can be obtained
at www.ICGtesting.com
CBHW071814021224
PP15928300001B/13